# LIST OF NUMBERED THEOREMS (*continued*)

# FIFTH EDITION

# *Mathematical Statistics*

## JOHN E. FREUND

Arizona State University

Prentice Hall, Englewood Cliffs, New Jersey 07632

*Library of Congress Cataloging-in-Publication Data*

FREUND, JOHN E.
   Mathematical statistics/John E. Freund.—5th ed.
      p.   cm.
   Includes index.
   ISBN 0-13-563834-8
   1. Mathematical statistics.  I. Title.
QA276.F692    1992          91-43357
519.5—dc20                CIP

Editorial/production supervision
   and interior design: *Kathleen M. Lafferty*
Cover design: *Joe DiDomenico*
Prepress buyer: *Paula Massenaro*
Manufacturing buyer: *Lori Bulwin*
Acquisitions editor: *Steven R. Comny*

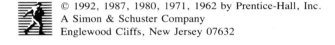 © 1992, 1987, 1980, 1971, 1962 by Prentice-Hall, Inc.
A Simon & Schuster Company
Englewood Cliffs, New Jersey 07632

Printed in the United States of America

10  9  8  7  6  5  4  3

ISBN   0-13-563834-8

Prentice-Hall International (UK) Limited, *London*
Prentice-Hall of Australia Pty. Limited, *Sydney*
Prentice-Hall Canada Inc., *Toronto*
Prentice-Hall Hispanoamericana, S.A., *Mexico*
Prentice-Hall of India Private Limited, *New Delhi*
Prentice-Hall of Japan, Inc., *Tokyo*
Simon & Schuster Asia Pte. Ltd., *Singapore*
Editora Prentice-Hall do Brasil, Ltda., *Rio de Janeiro*

To the memory of Dr. Ronald E. Walpole, who collaborated with me on the third edition and part of the fourth edition of this book.

*John E. Freund*

# Contents

# *Preface*

Like the first four editions, the fifth edition of *Mathematical Statistics* is designed for a two-semester or three-quarter calculus-based introduction to the mathematics of statistics. Most of the differences between this edition and the preceding one reflect the changes that have taken place in recent years in statistical thinking and in the teaching of statistics.

Most notable is the change in notation used in connection with random variables. A survey of instructors who have used the fourth edition showed a nearly unanimous preference for the use of capital letters instead of boldface type. So, we now write $P(X = x)$, for example, instead of $P(\mathbf{x} = x)$. This avoids the obvious problems one runs into with boldface type when working on a blackboard or when taking notes, but it can also lead to confusion. For instance, $X$ might be a random variable with values $x$ or a random variable with values $\chi$ (chi), and $A$ might be a random variable with values $a$ or a random variable with values $\alpha$ (alpha). Also, if one consistently uses capital letters for random variables, one runs into unfamiliar notation such as $\hat{\Sigma}$ for the random variable with values $\hat{\sigma}$ or $\hat{B}$ for the random variable with values $\hat{\beta}$. In some instances we deviated from this practice to retain notation traditionally used in statistics. For instance, in nonparametric statistics we continue using $u$ for the total number of runs and $U$ in connection with the Mann-Whitney test based on rank sums.

There have also been substantial changes in the basic material on statistical inference. The material on estimation has been expanded by giving more illustrations in the text and also more problems, and the material on tests of hypotheses has been expanded by stressing the dual approach of basing decisions on $P$-values or on the values of test statistics and critical regions.

There have also been major changes in the problem material. The problems are now labeled Exercises and Applications—the former being largely mathematical and the latter dealing with numbers. Also, there are more than 1,100 problems in this fifth edition, quite a few of them new, and they are about evenly divided between the two kinds. For convenience, the problems are numbered consecutively throughout each chapter.

Otherwise, there are minor changes and additions throughout the book. For instance, some new material on robustness may be found in Sections 10.6 and 16.1; some material on the use of computers may be found in Sections 11.8 and 13.9; and some suggestions about the analysis of $r \times c$ tables with ordered categories may be found in Applications 14.59 and 15.12. The appendix on Boolean Algebra has been deleted, but some of the exercises dealing with sets have been included in Chapter 2.

The author would like to express his appreciation for the many constructive comments which he received from students and colleagues, especially Arnold Adelberg, Grinnell College; Charles E. Antle, Pennsylvania State University; Daniel Martinez, California State University, Long Beach; Donald E. Meyers, University of Arizona; and Larry J. Ringer, Texas A&M University. Also, he would like to express his appreciation to the Robert E. Krieger Publishing Company for permission to base Table II on E. C. Molina's *Poisson's Exponential Binomial Limit;* to Prentice-Hall, Inc., for permission to reproduce part of Table 2 from R. A. Johnson and D. W. Wichern's *Applied Mulitvariate Statistical Analysis;* to Professor E. S. Pearson and the *Biometrika* trustees to reproduce the material in Tables V and VI; to the American Cynamid Company to reproduce the material in Table IX from F. Wilcoxon and R. A. Wilcox's *Some Rapid Approximate Statistical Procedures;* to D. Auble to base Table X on his "Extended Tables for the Mann-Whitney Statistics," *Bulletin of the Institute of Educational Research at Indiana University;* to the editor of the *Annals of Mathematical Statistics* to reproduce the material in Table XI; and to MINITAB® to reproduce the computer printouts shown in the text.

The author would also like to express his appreciation to Mrs. Rita Ewer for carefully reading the manuscript and helping with the proofreading and to Kathleen Lafferty, Prentice Hall production editor, for her courteous cooperation in the production of this book.

John E. Freund
*Scottsdale, Arizona*

# *About the Author*

**JOHN E. FREUND**
*Professor of Mathematics Emeritus*
*Arizona State University*

Educated at the University of London, U.C.L.A., Columbia University, and the University of Pittsburgh, Doctor Freund's interest in Mathematics, Logic, and the Philosophy of Science led him to a career in statistics. Keynoted by his approach to statistics as a way of thinking, and as such a refinement of everyday thinking, his textbooks in statistics at various levels and for various fields of application have been bestsellers for nearly forty years.

# 1

# *Introduction*

## 1.1 INTRODUCTION

In recent years, the growth of statistics has made itself felt in almost every phase of human activity. Statistics no longer consists merely of the collection of data and their presentation in charts and tables; it is now considered to encompass the science of basing inferences on observed data and the entire problem of making decisions in the face of uncertainty. This covers considerable ground since uncertainties are met when we flip a coin, when a dietician experiments with food additives, when an actuary determines life insurance premiums, when a quality control engineer accepts or rejects manufactured products, when a teacher compares the abilities of students, when an economist forecasts trends, when a newspaper predicts an election, and so forth.

It would be presumptuous to say that statistics, in its present state of development, can handle all situations involving uncertainties, but new techniques are constantly being developed and modern statistics can, at least, provide the framework for looking at these situations in a logical and systematic fashion. In other words, statistics provides the models that are needed to study situations involving uncertainties, in the same way as calculus provides the models that are needed to describe, say, the concepts of Newtonian physics.

The beginnings of the mathematics of statistics may be found in mid-eighteenth-century studies in probability motivated by interest in games of chance. The theory thus developed for "heads or tails" or "red or black" soon found applications in situations where the outcomes were "boy or girl," "life or death," or "pass or fail," and scholars began to apply probability theory to actuarial problems and some aspects of the social sciences. Later, probability and statistics were introduced into physics by L. Boltzmann, J. Gibbs, and J. Maxwell, and in this century they have found applications in all phases of human endeavor which in some way involve an element of uncertainty or risk. The names which are connected most prominently with the growth of mathematical statistics in the first half of this century are those of R. A. Fisher, J. Neyman, E. S. Pearson, and A. Wald. More recently, the work of R. Schlaifer, L. J. Savage, and others has given impetus to statistical theories based essentially on methods which date back to the eighteenth-century English clergyman Thomas Bayes.

The approach to statistics presented in this book is essentially the classical approach, with methods of inference based largely on the work of J. Neyman and E. S. Pearson. However, the more general decision-theory approach is introduced in Chapter 9 and some Bayesian methods are presented in Chapter 10. This material may be omitted without loss of continuity.

## 1.2  COMBINATORIAL METHODS

In many problems of statistics we must list all the alternatives that are possible in a given situation, or at least determine how many different possibilities there are. In connection with the latter, we often use the following theorem, sometimes called the **basic principle of counting**, the **counting rule for compound events**, or the **rule for the multiplication of choices**.

---

**THEOREM 1.1**  If an operation consists of two steps, of which the first can be done in $n_1$ ways and for each of these the second can be done in $n_2$ ways, then the whole operation can be done in $n_1 \cdot n_2$ ways.

---

Here, "operation" stands for any kind of procedure, process, or method of selection.

To justify this theorem, let us define the ordered pair $(x_i, y_j)$ to be the outcome which arises when the first step results in possibility $x_i$ and the second step results in possibility $y_j$. Then, the set of all possible outcomes is composed of the following $n_1 \cdot n_2$ pairs:

$$(x_1, y_1), (x_1, y_2), \ldots , (x_1, y_{n_2})$$

$$(x_2, y_1), (x_2, y_2), \ldots , (x_2, y_{n_2})$$

$$\ldots$$

$$\ldots$$

$$\ldots$$

$$(x_{n_1}, y_1), (x_{n_1}, y_2), \ldots , (x_{n_1}, y_{n_2})$$

## EXAMPLE 1.1

Suppose that someone wants to go by bus, by train, or by plane on a week's vacation to one of the five East North Central States. Find the number of different ways in which this can be done.

### Solution

The particular state can be chosen in $n_1 = 5$ ways and the means of transportation can be chosen in $n_2 = 3$ ways. Therefore, the trip can be carried out in $5 \cdot 3 = 15$ possible ways. If an actual listing of all the possibilities is desirable, a **tree diagram** like that in Figure 1.1 provides a systematic approach. This diagram shows that there are $n_1 = 5$ branches (possibilities) for the number of states and for each of these branches there are $n_2 = 3$ branches (possibilities) for the different means of transportation. It is apparent that the 15 possible ways of taking the vacation are represented by the 15 distinct paths along the branches of the tree.    ▲

## EXAMPLE 1.2

How many possible outcomes are there when we roll a pair of dice, one red and one green?

### Solution

The red die can land in any one of six ways, and for each of these six ways the green die can also land in six ways. Therefore, the pair of dice can land in $6 \cdot 6 = 36$ ways.    ▲

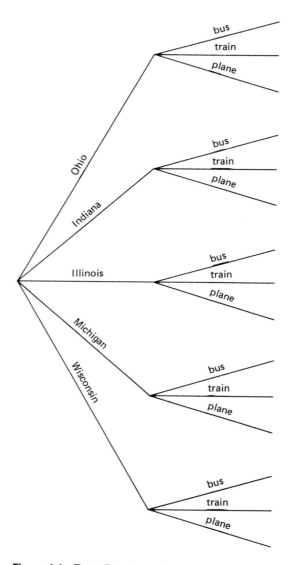

**Figure 1.1**   Tree diagram.

Theorem 1.1 may be extended to cover situations where an operation consists of two or more steps. In that case,

---

**THEOREM 1.2** If an operation consists of $k$ steps, of which the first can be done in $n_1$ ways, for each of these the second step can be done in $n_2$ ways, for each of the first two the third step can be done in $n_3$ ways, and so forth, then the whole operation can be done in $n_1 \cdot n_2 \cdot \ldots \cdot n_k$ ways.

---

## EXAMPLE 1.3

In how many different ways can one order soup, a sandwich, a dessert, and a drink for lunch, if there is a choice of four different soups, three kinds of sandwiches, five desserts, and four drinks?

### Solution

The total number of ways is $4 \cdot 3 \cdot 5 \cdot 4 = 240$.     ▲

## EXAMPLE 1.4

In how many different ways can one answer all the questions of a true-false test consisting of 20 questions?

### Solution

Altogether there are

$$\underbrace{2 \cdot 2 \cdot 2 \cdot 2 \cdot \ldots \cdot 2 \cdot 2}_{20 \text{ factors}} = 1,048,576$$

different ways in which one can answer all the questions; only one of these corresponds to the case where all the answers are correct and only one corresponds to the case where all the answers are wrong.     ▲

Frequently, we are interested in situations where the outcomes are the different ways in which a group of objects can be ordered or arranged. For instance, we might want to know in how many different ways the 24 members of a club can elect a president, a vice president, a treasurer, and a secretary, or we might want to know in how many different ways six persons can be seated around a table. Different arrangements like these are called **permutations**.

## EXAMPLE 1.5

How many permutations are there of the letters $a$, $b$, and $c$?

### Solution

The possible arrangements are *abc*, *acb*, *bac*, *bca*, *cab*, and *cba*, so the number of distinct permutations is six. Using Theorem 1.2, we could have arrived at this answer without actually listing the different permutations. Since there are three choices to select a letter for the first position, then two for the second position, leaving only one letter for the third position, the total number of permutations is $3 \cdot 2 \cdot 1 = 6$.    ▲

Generalizing the argument used in the preceding example, we find that $n$ distinct objects can be arranged in $n(n - 1)(n - 2) \cdot \ldots \cdot 3 \cdot 2 \cdot 1$ different ways. To simplify our notation, we represent this product by the symbol $n!$, which is read "$n$ factorial." Thus, $1! = 1$, $2! = 2 \cdot 1 = 2$, $3! = 3 \cdot 2 \cdot 1 = 6$, $4! = 4 \cdot 3 \cdot 2 \cdot 1 = 24$, $5! = 5 \cdot 4 \cdot 3 \cdot 2 \cdot 1 = 120$, and so on. Also, by definition we let $0! = 1$.

---

**THEOREM 1.3**    The number of permutations of $n$ distinct objects is $n!$.

---

## EXAMPLE 1.6

In how many different ways can the five starting players of a basketball team be introduced to the public?

### Solution

There are $5! = 5 \cdot 4 \cdot 3 \cdot 2 \cdot 1 = 120$ ways in which they can be introduced.    ▲

## EXAMPLE 1.7

The number of permutations of the four letters $a$, $b$, $c$, and $d$ is 24, but what is the number of permutations if we take only two of the four letters, or as it is usually put, if we take the four letters two at a time?

### Solution

We have two positions to fill, with four choices for the first and then three choices for the second. Therefore, by Theorem 1.1, the number of permutations is $4 \cdot 3 = 12$.    ▲

Generalizing the argument that we used in the preceding example, we find that $n$ distinct objects taken $r$ at a time, for $r > 0$, can be arranged in $n(n - 1) \cdot \ldots \cdot (n - r + 1)$ ways. We denote this product by $_nP_r$, and we let $_nP_0 = 1$ by definition. Therefore, we can write

---

**THEOREM 1.4** The number of permutations of $n$ distinct objects taken $r$ at a time is

$$_nP_r = \frac{n!}{(n - r)!}$$

for $r = 0, 1, 2, \ldots, n$.

---

**Proof.**    The formula $_nP_r = n(n - 1) \cdot \ldots \cdot (n - r + 1)$ cannot be used for $r = 0$, but we do have

$$_nP_0 = \frac{n!}{(n - 0)!} = 1$$

For $r = 1, 2, \ldots, n$, we have

$$_nP_r = n(n - 1)(n - 2) \cdot \ldots \cdot (n - r + 1)$$

$$= \frac{n(n - 1)(n - 2) \cdot \ldots \cdot (n - r + 1)(n - r)!}{(n - r)!}$$

$$= \frac{n!}{(n - r)!} \qquad \blacktriangledown$$

In problems concerning permutations, it is usually easier to proceed by using Theorem 1.2 as in Example 1.7, but the factorial formula of Theorem 1.4 is somewhat easier to remember. There are many statistical software packages that provide values of $_nP_r$ and other combinatorial quantities upon simple commands. Indeed, these quantities are also preprogrammed in many hand-held statistical (or scientific) calculators.

# EXAMPLE 1.8

Four names are drawn from among the 24 members of a club for the offices of president, vice president, treasurer, and secretary. In how many different ways can this be done?

*Solution*

The number of permutations of 24 distinct objects taken four at a time is

$$_{24}P_4 = \frac{24!}{20!} = 24 \cdot 23 \cdot 22 \cdot 21 = 255,024 \qquad \blacktriangle$$

## EXAMPLE 1.9

In how many ways can a local chapter of the American Chemical Society schedule three speakers for three different meetings, if they are all available on any of five possible dates?

*Solution*

Since we must choose three of the five dates and the order in which they are chosen (assigned to the three speakers) matters, we get

$$_5P_3 = \frac{5!}{2!} = \frac{120}{2} = 60$$

We might also argue that the first speaker can be scheduled in five ways, the second speaker in four ways, and the third speaker in three ways, so that the answer is $5 \cdot 4 \cdot 3 = 60$. $\qquad \blacktriangle$

Permutations that occur when objects are arranged in a circle are called **circular permutations**. Two circular permutations are not considered different (and are counted only once) if corresponding objects in the two arrangements have the same objects to their left and to their right. For example, if four persons are playing bridge, we do not get a different permutation if everyone moves to the chair at his or her right.

## EXAMPLE 1.10

How many circular permutations are there of four persons playing bridge?

*Solution*

If we arbitrarily consider the position of one of the four players as fixed, we can seat (arrange) the other three players in $3! = 6$ different ways. In other words, there are six different circular permutations. $\qquad \blacktriangle$

Generalizing the argument used in the preceding example, we get the following theorem.

---

**THEOREM 1.5** The number of permutations of $n$ distinct objects arranged in a circle is $(n - 1)!$.

---

We have been assuming until now that the $n$ objects from which we select $r$ objects and form permutations are all distinct. Thus, the various formulas cannot be used, for example, to determine the number of ways in which we can arrange the letters in the word "book," or the number of ways in which three copies of one novel and one copy each of four other novels can be arranged on a shelf.

# EXAMPLE 1.11

How many different permutations are there of the letters in the word "book"?

## Solution

If we distinguish for the moment between the two $o$'s by labeling them $o_1$ and $o_2$, there are $4! = 24$ different permutations of the symbols $b$, $o_1$, $o_2$, and $k$. However, if we drop the subscripts, then $bo_1ko_2$ and $bo_2ko_1$, for instance, both yield *boko*, and since each pair of permutations with subscripts yields but one arrangement without subscripts, the total number of arrangements of the letters in the word "book" is $\dfrac{24}{2} = 12$.    ▲

# EXAMPLE 1.12

In how many different ways can three copies of one novel and one copy each of four other novels be arranged on a shelf?

## Solution

If we denote the three copies of the first novel by $a_1$, $a_2$, and $a_3$ and the other four novels by $b$, $c$, $d$, and $e$, we find that *with subscripts* there are $7!$ different permutations of $a_1$, $a_2$, $a_3$, $b$, $c$, $d$, and $e$. However, since there are $3!$ permutations of $a_1$, $a_2$, and $a_3$ which lead to the same permutation of $a$, $a$, $a$, $b$, $c$, $d$, and $e$, we find that there are only $\dfrac{7!}{3!} = 7 \cdot 6 \cdot 5 \cdot 4 = 840$ ways in which the seven books can be arranged on a shelf.    ▲

Generalizing the argument that we used in the two preceding examples, we get the following theorem.

---

**THEOREM 1.6**  The number of permutations of $n$ objects of which $n_1$ are of one kind, $n_2$ are of a second kind, ... , $n_k$ are of a $k$th kind, and $n_1 + n_2 + \cdots + n_k = n$, is

$$\frac{n!}{n_1! \cdot n_2! \cdot \ldots \cdot n_k!}$$

---

## EXAMPLE 1.13

In how many ways can two paintings by Monet, three paintings by Renoir, and two paintings by Degas be hung side by side on a museum wall, if we do not distinguish between the paintings by the same artists?

### Solution

Substituting $n = 7$, $n_1 = 2$, $n_2 = 3$, and $n_3 = 2$ into the formula of Theorem 1.6, we get

$$\frac{7!}{2! \cdot 3! \cdot 2!} = 210 \qquad \blacktriangle$$

There are many problems in which we are interested in determining the number of ways in which $r$ objects can be selected from among $n$ distinct objects *without regard to the order in which they are selected*. Such selections (arrangements) are called **combinations**.

## EXAMPLE 1.14

In how many different ways can a person gathering data for a market research organization select three of the 20 households living in a certain apartment complex?

### Solution

If we care about the order in which the households are selected, the answer is

$$_{20}P_3 = 20 \cdot 19 \cdot 18 = 6,840$$

but each set of three households would then be counted 3! = 6 times. If we do not care about the order in which the households are selected, there are only $\dfrac{6,840}{6} = 1,140$ ways in which the person gathering the data can do his or her job. ▲

Actually, "combination" means the same as "subset," and when we ask for the number of combinations of r objects selected from a set of n distinct objects, we are simply asking for the total number of subsets of r objects that can be selected from a set of n distinct objects. In general, there are r! permutations of the objects in a subset of r objects, so that the $_nP_r$ permutations of r objects selected from a set of n distinct objects contain each subset r! times. Dividing $_nP_r$ by r! and denoting the result by the symbol $\dbinom{n}{r}$, we thus have

---

**THEOREM 1.7**  The number of combinations of n distinct objects taken r at a time is

$$\binom{n}{r} = \frac{n!}{r!(n-r)!}$$

for r = 0, 1, 2, ... , n.

---

# EXAMPLE 1.15

In how many different ways can six tosses of a coin yield two heads and four tails?

## Solution

This question is the same as asking for the number of ways in which we can select the two tosses on which heads is to occur. Therefore, applying Theorem 1.7, we find that the answer is

$$\binom{6}{2} = \frac{6!}{2! \cdot 4!} = 15$$

This result could also have been obtained by the rather tedious process of enumerating the various possibilities, HHTTTT, TTHTHT, HTHTTT, ... , where H stands for head and T for tail. ▲

## EXAMPLE 1.16

How many different committees of two chemists and one physicist can be formed from the four chemists and three physicists on the faculty of a small college?

### Solution

Since two of four chemists can be selected in $\binom{4}{2} = \dfrac{4!}{2! \cdot 2!} = 6$ ways and one of three physicists can be selected in $\binom{3}{1} = \dfrac{3!}{1! \cdot 2!} = 3$ ways, Theorem 1.1 shows that the number of committees is $6 \cdot 3 = 18$.   ▲

A combination of $r$ objects selected from a set of $n$ distinct objects may be considered a **partition** of the $n$ objects into two subsets containing, respectively, the $r$ objects that are selected and the $n - r$ objects that are left. Often, we are concerned with the more general problem of partitioning a set of $n$ distinct objects into $k$ subsets, which requires that each of the $n$ objects must belong to one and only one of the subsets.[†] The order of the objects within a subset is of no importance.

## EXAMPLE 1.17

In how many ways can a set of four objects be partitioned into three subsets containing, respectively, two, one, and one of the objects?

### Solution

Denoting the four objects by $a$, $b$, $c$, and $d$, we find by enumeration that there are the twelve possibilities:

| | | | |
|---|---|---|---|
| $ab\|c\|d$ | $ab\|d\|c$ | $ac\|b\|d$ | $ac\|d\|b$ |
| $ad\|b\|c$ | $ad\|c\|b$ | $bc\|a\|d$ | $bc\|d\|a$ |
| $bd\|a\|c$ | $bd\|c\|a$ | $cd\|a\|b$ | $cd\|b\|a$ |

---

[†]Symbolically, the subsets $A_1, A_2, \ldots, A_k$ constitute a partition of set $A$ if $A_1 \cup A_2 \cup \cdots \cup A_k = A$ and $A_i \cap A_j = \varnothing$ for all $i \neq j$.

The number of partitions for this example is denoted by the symbol

$$\binom{4}{2,\ 1,\ 1} = 12$$

where the number at the top represents the total number of objects and the numbers at the bottom represent the number of objects going into each subset.  ▲

Had we not wanted to enumerate all the possibilities in the preceding example, we could have argued that the two objects going into the first subset can be chosen in $\binom{4}{2} = 6$ ways, the object going into the second subset can then be chosen in $\binom{2}{1} = 2$ ways, and the object going into the third subset can then be chosen in $\binom{1}{1} = 1$ way. Thus, by Theorem 1.2 there are $6 \cdot 2 \cdot 1 = 12$ partitions.

Generalizing this argument, we have the following theorem.

---

**THEOREM 1.8** The number of ways in which a set of $n$ distinct objects can be partitioned into $k$ subsets with $n_1$ objects in the first subset, $n_2$ objects in the second subset, ..., and $n_k$ objects in the $k$th subset is

$$\binom{n}{n_1,\ n_2,\ \ldots,\ n_k} = \frac{n!}{n_1! \cdot n_2! \cdot \ldots \cdot n_k!}$$

---

*Proof.*   Since the $n_1$ objects going into the first subset can be chosen in $\binom{n}{n_1}$ ways, the $n_2$ objects going into the second subset can then be chosen in $\binom{n - n_1}{n_2}$ ways, the $n_3$ objects going into the third subset can then be chosen in $\binom{n - n_1 - n_2}{n_3}$ ways, and so forth, it follows by Theorem 1.2 that the total number of partitions is

$$\binom{n}{n_1, n_2, \ldots, n_k} = \binom{n}{n_1} \cdot \binom{n-n_1}{n_2} \cdot \ldots \cdot \binom{n-n_1-n_2-\cdots-n_{k-1}}{n_k}$$

$$= \frac{n!}{n_1! \cdot (n-n_1)!} \cdot \frac{(n-n_1)!}{n_2! \cdot (n-n_1-n_2)!} \cdot \ldots \cdot$$

$$\frac{(n-n_1-n_2-\cdots-n_{k-1})!}{n_k! \cdot 0!}$$

$$= \frac{n!}{n_1! \cdot n_2! \cdot \ldots \cdot n_k!} \qquad \blacktriangledown$$

## EXAMPLE 1.18

In how many ways can seven businessmen attending a convention be assigned to one triple and two double hotel rooms?

### Solution

Substituting $n = 7$, $n_1 = 3$, $n_2 = 2$, and $n_3 = 2$ into the formula of Theorem 1.8, we get

$$\binom{7}{3, 2, 2} = \frac{7!}{3! \cdot 2! \cdot 2!} = 210 \qquad \blacktriangle$$

## 1.3  BINOMIAL COEFFICIENTS

If $n$ is a positive integer and we multiply out $(x + y)^n$ term by term, each term will be the product of $x$'s and $y$'s, with an $x$ or a $y$ coming from each of the $n$ factors $x + y$. For instance, the expansion

$$(x + y)^3 = (x + y)(x + y)(x + y)$$

$$= x \cdot x \cdot x + x \cdot x \cdot y + x \cdot y \cdot x + x \cdot y \cdot y$$

$$+ y \cdot x \cdot x + y \cdot x \cdot y + y \cdot y \cdot x + y \cdot y \cdot y$$

$$= x^3 + 3x^2y + 3xy^2 + y^3$$

yields terms of the form $x^3$, $x^2y$, $xy^2$, and $y^3$. Their coefficients are 1, 3, 3, and 1, and the coefficient of $xy^2$, for example, is $\binom{3}{2} = 3$, the number of ways in which we can choose the two factors providing the $y$'s. Similarly, the coefficient of $x^2y$ is $\begin{pmatrix} 3 \\ 1 \end{pmatrix} = 3$, the number of ways in which we can choose the one factor

providing the $y$, and the coefficients of $x^3$ and $y^3$ are $\begin{pmatrix} 3 \\ 0 \end{pmatrix} = 1$ and $\begin{pmatrix} 3 \\ 3 \end{pmatrix} = 1$.

More generally, if $n$ is a positive integer and we multiply out $(x + y)^n$ term by term, the coefficient of $x^{n-r}y^r$ is $\begin{pmatrix} n \\ r \end{pmatrix}$, the number of ways in which we can choose the $r$ factors providing the $y$'s. Accordingly, we refer to $\begin{pmatrix} n \\ r \end{pmatrix}$ as a **binomial coefficient**. We can now state the following theorem.

---

**THEOREM 1.9**

$$(x + y)^n = \sum_{r=0}^{n} \begin{pmatrix} n \\ r \end{pmatrix} x^{n-r}y^r \qquad \text{for any positive integer } n$$

---

(For readers who are not familiar with the $\Sigma$ notation, a brief explanation is given in the appendix at the end of the book.)

The calculation of binomial coefficients can often be simplified by making use of the three theorems that follow.

---

**THEOREM 1.10**  For any positive integers $n$ and $r = 0, 1, 2, \ldots, n$,

$$\begin{pmatrix} n \\ r \end{pmatrix} = \begin{pmatrix} n \\ n - r \end{pmatrix}$$

---

***Proof.***  We might argue that when we select a subset of $r$ objects from a set of $n$ distinct objects we leave a subset of $n - r$ objects, and, hence, there are as many ways of selecting $r$ objects as there are ways of leaving (or selecting) $n - r$ objects. To prove the theorem algebraically, we write

$$\begin{pmatrix} n \\ n - r \end{pmatrix} = \frac{n!}{(n - r)![n - (n - r)]!} = \frac{n!}{(n - r)!r!}$$

$$= \frac{n!}{r!(n - r)!} = \begin{pmatrix} n \\ r \end{pmatrix} \qquad \blacktriangledown$$

Theorem 1.10 implies that if we calculate the binomial coefficients for $r = 0, 1, \ldots , \dfrac{n}{2}$ when $n$ is even and for $r = 0, 1, \ldots , \dfrac{n-1}{2}$ where $n$ is odd, the remaining binomial coefficients can be obtained by making use of the theorem.

## EXAMPLE 1.19

Given $\dbinom{4}{0} = 1$, $\dbinom{4}{1} = 4$, and $\dbinom{4}{2} = 6$, find $\dbinom{4}{3}$ and $\dbinom{4}{4}$.

*Solution*

$$\dbinom{4}{3} = \dbinom{4}{4-3} = \dbinom{4}{1} = 4 \quad \text{and} \quad \dbinom{4}{4} = \dbinom{4}{4-4} = \dbinom{4}{0} = 1 \qquad \blacktriangle$$

## EXAMPLE 1.20

Given $\dbinom{5}{0} = 1$, $\dbinom{5}{1} = 5$, and $\dbinom{5}{2} = 10$, find $\dbinom{5}{3}$, $\dbinom{5}{4}$, and $\dbinom{5}{5}$.

*Solution*

$$\dbinom{5}{3} = \dbinom{5}{5-3} = \dbinom{5}{2} = 10, \quad \dbinom{5}{4} = \dbinom{5}{5-4} = \dbinom{5}{1} = 5, \text{ and}$$

$$\dbinom{5}{5} = \dbinom{5}{5-5} = \dbinom{5}{0} = 1. \qquad \blacktriangle$$

It is precisely in this fashion that Theorem 1.10 may have to be used in connection with Table VII at the end of the book.

## EXAMPLE 1.21

Find $\dbinom{20}{12}$ and $\dbinom{17}{10}$.

*Solution*

Since $\dbinom{20}{12}$ is not given in Table VII, we make use of the fact that $\dbinom{20}{12} = \dbinom{20}{8}$, look up $\dbinom{20}{8}$, and get $\dbinom{20}{12} = 125{,}970$. Similarly, to find

$\binom{17}{10}$, we make use of the fact that $\binom{17}{10} = \binom{17}{7}$, look up $\binom{17}{7}$, and get $\binom{17}{10} = 19{,}448.$    ▲

---

**THEOREM 1.11**   For any positive integer $n$ and $r = 1, 2, \ldots, n - 1,$

$$\binom{n}{r} = \binom{n-1}{r} + \binom{n-1}{r-1}$$

---

***Proof.***   Substituting $x = 1$ into $(x + y)^n$, let us write

$$(1 + y)^n = (1 + y)(1 + y)^{n-1} = (1 + y)^{n-1} + y(1 + y)^{n-1}$$

and equate the coefficient of $y^r$ in $(1 + y)^n$ with that in $(1 + y)^{n-1} + y(1 + y)^{n-1}$. Since the coefficient of $y^r$ in $(1 + y)^n$ is $\binom{n}{r}$ and the coefficient of $y^r$ in $(1 + y)^{n-1} + y(1 + y)^{n-1}$ is the sum of the coefficient of $y^r$ in $(1 + y)^{n-1}$, namely, $\binom{n-1}{r}$, and the coefficient of $y^{r-1}$ in $(1 + y)^{n-1}$, namely, $\binom{n-1}{r-1}$, we obtain

$$\binom{n}{r} = \binom{n-1}{r} + \binom{n-1}{r-1}$$

which completes the proof.    ▼

Alternatively, take any one of the $n$ objects. If it is not to be included among the $r$ objects, there are $\binom{n-1}{r}$ ways of selecting the $r$ objects; if it is to be included, there are $\binom{n-1}{r-1}$ ways of selecting the other $r - 1$ objects. Therefore, there are $\binom{n-1}{r} + \binom{n-1}{r-1}$ ways of selecting the $r$ objects; that is,

$$\binom{n}{r} = \binom{n-1}{r} + \binom{n-1}{r-1}$$

Theorem 1.11 can also be proved by expressing the binomial coefficients on both sides of the equation in terms of factorials and then proceeding algebraically, but we shall leave this to the reader in Exercise 1.12. An important application of Theorem 1.11 is given in Exercise 1.11, where it provides the key for the construction of what is known as **Pascal's triangle**.

To state the third theorem about binomial coefficients, let us make the following definition: $\binom{n}{r} = 0$ whenever $n$ is a positive integer and $r$ is a positive integer greater than $n$. (Clearly, there is no way in which we can select a subset which contains more elements than the whole set itself.)

---

**THEOREM 1.12**

$$\sum_{r=0}^{k} \binom{m}{r}\binom{n}{k-r} = \binom{m+n}{k}$$

---

***Proof.***   Using the same technique as in the proof of Theorem 1.11, let us prove this theorem by equating the coefficients of $y^k$ in the expressions on both sides of the equation

$$(1 + y)^{m+n} = (1 + y)^m (1 + y)^n$$

The coefficient of $y^k$ in $(1 + y)^{m+n}$ is $\binom{m+n}{k}$, and the coefficient of $y^k$

in

$$(1 + y)^m (1 + y)^n = \left[ \binom{m}{0} + \binom{m}{1}y + \cdots + \binom{m}{m}y^m \right]$$
$$\times \left[ \binom{n}{0} + \binom{n}{1}y + \cdots + \binom{n}{n}y^n \right]$$

is the sum of the products which we obtain by multiplying the constant term of the first factor by the coefficient of $y^k$ in the second factor, the coefficient of $y$ in the first factor by the coefficient of $y^{k-1}$ in the second factor, ... , and the coefficient of $y^k$ in the first factor by the constant term of the second factor. Thus, the coefficient of $y^k$ in $(1 + y)^m (1 + y)^n$ is

$$\binom{m}{0}\binom{n}{k} + \binom{m}{1}\binom{n}{k-1} + \binom{m}{2}\binom{n}{k-2} + \cdots + \binom{m}{k}\binom{n}{0}$$

$$= \sum_{r=0}^{k} \binom{m}{r}\binom{n}{k-r}$$

and this completes the proof.    ▼

## EXAMPLE 1.22

Verify Theorem 1.12 numerically for $m = 2$, $n = 3$, and $k = 4$.

*Solution*

Substituting these values we get

$$\binom{2}{0}\binom{3}{4} + \binom{2}{1}\binom{3}{3} + \binom{2}{2}\binom{3}{2} + \binom{2}{3}\binom{3}{1} + \binom{2}{4}\binom{3}{0} = \binom{5}{4}$$

and since $\binom{3}{4}$, $\binom{2}{3}$, and $\binom{2}{4}$ equal 0 according to the definition on page

18, the equation reduces to

$$\binom{2}{1}\binom{3}{3} + \binom{2}{2}\binom{3}{2} = \binom{5}{4}$$

which checks, since $2 \cdot 1 + 1 \cdot 3 = 5$.    ▲

Using Theorem 1.8, we can extend our discussion to **multinomial coefficients**, namely, to the coefficients that arise in the expansion of $(x_1 + x_2 + \cdots + x_k)^n$. The multinomial coefficient of the term $x_1^{r_1} \cdot x_2^{r_2} \cdot \ldots \cdot x_k^{r_k}$ in the expansion of $(x_1 + x_2 + \cdots + x_k)^n$ is

$$\binom{n}{r_1, r_2, \ldots, r_k} = \frac{n!}{r_1! \cdot r_2! \cdot \ldots \cdot r_k!}$$

## EXAMPLE 1.23

What is the coefficient of $x_1^3 x_2 x_3^2$ in the expansion of $(x_1 + x_2 + x_3)^6$?

*Solution*

Substituting $n = 6$, $r_1 = 3$, $r_2 = 1$, and $r_3 = 2$ into the formula above, we get

$$\frac{6!}{3! \cdot 1! \cdot 2!} = 60 \quad \blacktriangle$$

## *EXERCISES*

**1.1**  An operation consists of two steps, of which the first can be made in $n_1$ ways. If the first step is made in the $i$th way, the second step can be made in $n_{2i}$ ways.[†]

    (a)  Find a formula for the total number of ways in which the whole operation can be made.

    (b)  A student can study 0, 1, 2, or 3 hours for a history test on any given day. Use the formula obtained in part (a) to verify that there are 13 ways in which the student can study at most 4 hours for the test on two consecutive days.

**1.2**  With reference to the preceding exercise, verify that if $n_{2i}$ equals the constant $n_2$, the formula obtained in part (a) reduces to that of Theorem 1.1.

**1.3**  With reference to Exercise 1.1, suppose that there is a third step, and if the first step is made in the $i$th way and the second step in the $j$th way, the third step can be made in $n_{3ij}$ ways.

    (a)  Verify that the whole operation can be made in

$$\sum_{i=1}^{n_1} \sum_{j=1}^{n_{2i}} n_{3ij}$$

different ways.

    (b)  With reference to part (b) of Exercise 1.1, use the formula of part (a) to verify that there are 32 ways in which the student can study at most 4 hours for the test on three consecutive days.

**1.4**  With reference to the preceding exercise, verify that if $n_{2i}$ equals the constant $n_2$ and $n_{3ij}$ equals the constant $n_3$, the formula obtained in part (a) reduces to that of Theorem 1.2.

**1.5**  In a two-team basketball play-off, the winner is the first team to win $m$ games.

    (a)  Counting separately the number of play-offs requiring $m$, $m + 1$, ... ,

---

[†]The use of double subscripts is explained in the appendix at the end of the book.

and $2m - 1$ games, show that the total number of different outcomes (sequences of wins and losses by one of the teams) is

$$2\left[\binom{m-1}{m-1} + \binom{m}{m-1} + \cdots + \binom{2m-2}{m-1}\right]$$

(b) How many different outcomes are there in a "2 out of 3" play-off, a "3 out of 5" play-off, and a "4 out of 7" play-off?

**1.6** When $n$ is large, $n!$ can be approximated by means of the expression

$$\sqrt{2\pi n}\left(\frac{n}{e}\right)^n$$

called **Stirling's formula**, where $e$ is the base of natural logarithms. (A derivation of this formula may be found in the book by W. Feller cited among the references at the end of this chapter.)

(a) Use Stirling's formula to obtain approximations for 10! and 12!, and find the percentage errors of these approximations by comparing them with the exact values given in Table VII.

(b) Use Stirling's formula to obtain an approximation for the number of 13-card bridge hands that can be dealt with an ordinary deck of 52 playing cards.

**1.7** Use Stirling's formula (see preceding exercise) to show that

$$\lim_{n \to \infty} \frac{\binom{2n}{n}\sqrt{\pi n}}{2^{2n}} = 1$$

**1.8** In some problems of **occupancy theory** we are concerned with the number of ways in which certain *distinguishable* objects can be distributed among individuals, urns, boxes, or cells. Find an expression for the number of ways in which $r$ *distinguishable* objects can be distributed among $n$ cells, and use it to find the number of ways in which three different books can be distributed among the 12 students in an English literature class.

**1.9** In some problems of occupancy theory we are concerned with the number of ways in which certain *indistinguishable* objects can be distributed among individuals, urns, boxes, or cells. Find an expression for the number of ways in which $r$ *indistinguishable* objects can be distributed among $n$ cells, and use it to find the number of ways in which a baker can sell five (indistinguishable) loaves of bread to three customers. (*Hint:* We might argue that L|LLL|L represents the case where the three customers buy one loaf, three loaves, and one loaf, respectively, and that LLLL| |L represents the case where the three customers buy four loaves, none of the loaves, and one loaf. Thus, we must look for the number of ways in which we can arrange the five L's and the two vertical bars.)

**1.10** In some problems of occupancy theory we are concerned with the number of ways in which certain *indistinguishable* objects can be distributed among individuals, urns, boxes, or cells with at least one in each cell. Find an expression for the number of ways in which *r indistinguishable* objects can be distributed among *n* cells with at least one in each cell, and rework the numerical part of the preceding exercise with each of the three customers getting at least one loaf of bread.

**1.11** When no table is available, it is sometimes convenient to determine binomial coefficients by means of the following arrangement, called **Pascal's triangle**,

$$
\begin{array}{ccccccccccc}
 & & & & & 1 & & & & & \\
 & & & & 1 & & 1 & & & & \\
 & & & 1 & & 2 & & 1 & & & \\
 & & 1 & & 3 & & 3 & & 1 & & \\
 & 1 & & 4 & & 6 & & 4 & & 1 & \\
1 & & 5 & & 10 & & 10 & & 5 & & 1
\end{array}
$$

$$\cdots\cdots\cdots\cdots\cdots$$

where each row begins with a 1, ends with a 1, and each other entry is the sum of the nearest two entries in the row immediately above. In this triangle, the *r*th entry of the *n*th row is the binomial coefficient $\binom{n-1}{r-1}$. Construct the next two (seventh and eighth) rows of the triangle and write the binomial expansions of $(x + y)^6$ and $(x + y)^7$.

**1.12** Prove Theorem 1.11 by expressing all the binomial coefficients in terms of factorials and then simplifying algebraically.

**1.13** Expressing the binomial coefficients in terms of factorials and simplifying algebraically, show that

(a) $\binom{n}{r} = \dfrac{n-r+1}{r} \cdot \binom{n}{r-1}$;

(b) $\binom{n}{r} = \dfrac{n}{n-r} \cdot \binom{n-1}{r}$;

(c) $n\binom{n-1}{r} = (r+1)\binom{n}{r+1}$.

**1.14** Substituting appropriate values for *x* and *y* into the formula of Theorem 1.9, show that

(a) $\displaystyle\sum_{r=0}^{n} \binom{n}{r} = 2^n;$

(b) $\displaystyle\sum_{r=0}^{n} (-1)^r \binom{n}{r} = 0;$

(c) $\displaystyle\sum_{r=0}^{n} \binom{n}{r}(a-1)^r = a^n.$

**1.15** Repeatedly applying Theorem 1.11, show that

$$\binom{n}{r} = \sum_{i=1}^{r+1} \binom{n-i}{r-i+1}$$

**1.16** Use Theorem 1.12 to show that

$$\sum_{r=0}^{n} \binom{n}{r}^2 = \binom{2n}{n}$$

**1.17** Show that $\displaystyle\sum_{r=0}^{n} r\binom{n}{r} = n2^{n-1}$ by setting $x = 1$ in Theorem 1.9, then differentiating the expressions on both sides with respect to $y$, and finally substituting $y = 1$.

**1.18** Rework the preceding exercise by making use of part (a) of Exercise 1.14 and part (c) of Exercise 1.13.

**1.19** If $n$ is not a positive integer or zero, the binomial expansion of $(1 + y)^n$ yields, for $-1 < y < 1$, the infinite series

$$1 + \binom{n}{1}y + \binom{n}{2}y^2 + \binom{n}{3}y^3 + \cdots + \binom{n}{r}y^r + \cdots$$

where $\displaystyle\binom{n}{r} = \frac{n(n-1)\cdot\ldots\cdot(n-r+1)}{r!}$ for $r = 1, 2, 3, \ldots$ . Use this

**generalized definition of binomial coefficients** (which agrees with the one on page 15 for positive integral values of $n$) to evaluate

(a) $\displaystyle\binom{\frac{1}{2}}{4}$ and $\displaystyle\binom{-3}{3}$;

(b) $\sqrt{5}$ by writing $\sqrt{5} = 2(1 + \frac{1}{4})^{\frac{1}{2}}$ and using the first four terms of the binomial expansion of $(1 + \frac{1}{4})^{\frac{1}{2}}$.

**1.20** With reference to the generalized definition of binomial coefficients in the preceding exercise, show that

(a) $\dbinom{-1}{r} = (-1)^r$;

(b) $\dbinom{-n}{r} = (-1)^r \dbinom{n+r-1}{r}$ for $n > 0$.

**1.21** Find the coefficient of $x^2y^3z^3$ in the expansion of $(x + y + z)^8$.

**1.22** Find the coefficient of $x^3y^2z^3w$ in the expansion of $(2x + 3y - 4z + w)^9$.

**1.23** Show that

$$\binom{n}{n_1, n_2, \ldots, n_k} = \binom{n-1}{n_1-1, n_2, \ldots, n_k} + \binom{n-1}{n_1, n_2-1, \ldots, n_k}$$

$$+ \cdots + \binom{n-1}{n_1, n_2, \ldots, n_k-1}$$

by expressing all these multinomial coefficients in terms of factorials and simplifying algebraically.

## APPLICATIONS

**1.24** There are four routes, $A$, $B$, $C$, and $D$, between a person's home and the place where he works, but route $B$ is one-way, so he cannot take it on the way to work, and route $C$ is one-way, so he cannot take it on the way home.

(a) Draw a tree diagram showing the various ways the person can go to and from work.

(b) Draw a tree diagram showing the various ways he can go to and from work without taking the same route both ways.

**1.25** A person with $2 in her pocket bets $1, even money, on the flip of a coin, and she continues to bet $1 so long as she has any money. Draw a tree diagram to show the various things that can happen during the first four flips of the coin. After the fourth flip of the coin, in how many of the cases will she be

(a) exactly even;

(b) exactly $2 ahead?

**1.26** Suppose that in a baseball World Series (in which the winner is the first team to win four games) the National League champion leads the American League champion three games to two. Construct a tree diagram to show the number of ways in which these teams may win or lose the remaining game or games.

**1.27** The pro at a golf course stocks two identical sets of women's clubs, reordering at the end of each day (for delivery early the next morning) if and only if he has sold them both. Construct a tree diagram to show that if he starts on a Monday with two sets of the clubs, there are altogether eight different ways in which he can make sales on the first two days of that week.

**1.28** Counting the number of outcomes in games of chance has been a popular pastime for many centuries. This was of interest not only because of the gambling that was involved, but also because the outcomes of games of chance were often interpreted as divine intent. Thus, it was just about a thousand years ago that a bishop in what is now Belgium determined that there are 56 different ways in which three dice can fall *provided one is interested only in the overall result and not in which die does what*. He assigned a virtue to each of these possibilities and each sinner had to concentrate for some time on the virtue which corresponded to his cast of the dice.

    (a)   Find the number of ways in which three dice can all come up with the same number of points.

    (b)   Find the number of ways in which two of the three dice can come up with the same number of points, while the third comes up with a different number of points.

    (c)   Find the number of ways in which all three of the dice can come up with a different number of points.

    (d)   Use the results of parts (a), (b), and (c) to verify the bishop's calculations that there are altogether 56 possibilities.

**1.29** If the NCAA has applications from six universities for hosting its intercollegiate tennis championships in 1994 and 1995, in how many ways can they select the hosts for these championships

    (a)   if they are not both to be held at the same university;

    (b)   if they may both be held at the same university?

**1.30** The five finalists in the Miss Universe contest are Miss Argentina, Miss Belgium, Miss U.S.A., Miss Japan, and Miss Norway. In how many ways can the judges choose

    (a)   the winner and the first runner-up;

    (b)   the winner, the first runner-up, and the second runner-up?

**1.31** In a primary election, there are four candidates for mayor, five candidates for city treasurer, and two candidates for county attorney.

    (a)   In how many ways can a voter mark his ballot for all three of these offices?

    (b)   In how many ways can a person vote if he exercises his option of not voting for a candidate for any or all of these offices?

**1.32** A multiple-choice test consists of 15 questions, each permitting a choice of three alternatives. In how many different ways can a student check off her answers to these questions?

**1.33** The price of a European tour includes four stopovers to be selected from among ten cities. In how many different ways can one plan such a tour

    (a)  if the order of the stopovers matters;

    (b)  if the order of the stopovers does not matter?

**1.34** In how many ways can a television director schedule a sponsor's six different commercials during the six time slots allocated to commercials during an hour "special"?

**1.35** In how many ways can the television director of the preceding exercise fill the six time slots for commercials if the sponsor has three different commercials, each of which is to be shown twice?

**1.36** In how many ways can the television director of Exercise 1.34 fill the six time slots for commercials if the sponsor has two different commercials, each of which is to be shown three times?

**1.37** In how many ways can five persons line up to get on a bus? In how many ways can they line up if two of the persons refuse to follow each other?

**1.38** In how many ways can eight persons form a circle for a folk dance?

**1.39** How many permutations are there of the letters in the word

    (a)  "great";

    (b)  "greet"?

**1.40** How many distinct permutations are there of the letters in the word "statistics"? How many of these begin and end with the letter $s$?

**1.41** A college team plays 10 football games during a season. In how many ways can it end the season with five wins, four losses, and one tie?

**1.42** If eight persons are having dinner together, in how many different ways can three order chicken, four order steak, and one order lobster?

**1.43** In Example 1.4 we showed that a true-false test consisting of 20 questions can be marked in 1,048,576 different ways. In how many ways can each question be marked true or false so that

    (a)  7 are right and 13 are wrong;

    (b)  10 are right and 10 are wrong;

    (c)  at least 17 are right?

**1.44** Among the seven nominees for two vacancies on a city council are three men and four women. In how many ways can these vacancies be filled

    (a)  with any two of the seven nominees;

    (b)  with any two of the four women;

    (c)  with one of the men and one of the women?

**1.45** A shipment of 10 television sets includes three that are defective. In how many ways can a hotel purchase four of these sets and receive at least two of the defective sets?

**1.46** Ms. Jones has four skirts, seven blouses, and three sweaters. In how many ways can she choose two of the skirts, three of the blouses, and one of the sweaters to take along on a trip?

**1.47** How many different bridge hands are possible containing five spades, three diamonds, three clubs, and two hearts?

**1.48** Find the number of ways in which one A, three B's, two C's, and one F can be distributed among seven students taking a course in statistics.

**1.49** An art collector, who owns 10 paintings by famous artists, is preparing her will. In how many different ways can she leave these paintings to her three heirs?

**1.50** A baseball fan has a pair of tickets for six different home games of the Chicago Cubs. If he has five friends who like baseball, in how many different ways can he take one of them along to each of the six games?

**1.51** At the end of the day, a bakery gives everything that is unsold to food banks for the needy. If it has 12 apple pies left at the end of a given day, in how many different ways can it distribute these pies among six food banks for the needy?

**1.52** With reference to the preceding exercise, in how many different ways can the bakery distribute the 12 apple pies, if each of the six food banks is to receive at least one of the pies?

**1.53** On a Friday morning, the pro shop of a tennis club has 14 identical cans of tennis balls. If they are all sold by Sunday night and we are interested only in how many were sold on each day, in how many different ways could the tennis balls have been sold on Friday, Saturday, and Sunday?

**1.54** Rework the preceding exercise given that at least two of the cans of tennis balls were sold on each of the three days.

# REFERENCES

Among the few books on the history of statistics there are

WALKER, H. M., *Studies in the History of Statistical Method*. Baltimore: The Williams & Wilkins Company, 1929,

WESTERGAARD, H., *Contributions to the History of Statistics*. London: P. S. King & Son, 1932,

and the more recent publications

KENDALL, M. G., and PLACKETT, R. L., eds., *Studies in the History of Statistics and Probability*, Vol. II. New York: Macmillan Publishing Co., Inc., 1977,

PEARSON, E. S., and KENDALL, M. G., eds., *Studies in the History of Statistics and Probability*. Darien, Conn.: Hafner Publishing Co., Inc., 1970,

PORTER, T. M., *The Rise of Statistical Thinking, 1820–1900*. Princeton: Princeton University Press, 1986,

STIGLER, S. M., *The History of Statistics*. Cambridge: Harvard University Press, 1986.

A wealth of material on combinatorial methods can be found in

COHEN, D. A., *Basic Techniques of Combinatorial Theory*. New York: John Wiley & Sons, Inc., 1978,

EISEN, M., *Elementary Combinatorial Analysis*. New York: Gordon and Breach, Science Publishers, Inc., 1970,

FELLER, W., *An Introduction to Probability Theory and Its Applications*, Vol. I, 3rd ed. New York: John Wiley & Sons, Inc., 1968,

NIVEN, J., *Mathematics of Choice*. New York: Random House, Inc., 1965,

ROBERTS, F. S., *Applied Combinatorics*. Englewood Cliffs, N.J.: Prentice Hall, Inc., 1984,

and in

WHITWORTH, W. A., *Choice and Chance*, 5th ed. New York: Hafner Publishing Co., Inc., 1959,

which has become a classic in this field. More advanced treatments may be found in

BECKENBACH, E. F., ed., *Applied Combinatorial Mathematics*. New York: John Wiley & Sons, Inc., 1964,

DAVID, F. N., and BARTON, D. E., *Combinatorial Chance*. New York: Hafner Publishing Co., Inc., 1962,

and

RIORDAN, J., *An Introduction to Combinatorial Analysis*. New York: John Wiley & Sons, Inc., 1958.

# 2

# *Probability*

## 2.1 INTRODUCTION

Historically, the oldest way of defining probabilities, the **classical probability concept**, applies when all possible outcomes are equally likely, as is presumably the case in most games of chance. We can then say that *if there are N equally likely possibilities, of which one must occur and n are regarded as favorable, or as a "success," then the probability of a "success" is given by the ratio $\frac{n}{N}$.*

## EXAMPLE 2.1

What is the probability of drawing an ace from an ordinary deck of 52 playing cards?

*Solution*

Since there are $n = 4$ aces among the $N = 52$ cards, the probability of drawing an ace is $\frac{4}{52} = \frac{1}{13}$. (It is assumed, of course, that each card has the same chance of being drawn.)    ▲

Although equally likely possibilities are found mostly in games of chance, the classical probability concept applies also in a great variety of situations where gambling devices are used to make random selections—when office space is assigned to teaching assistants by lot, when some of the families in a township are chosen in such a way that each one has the same chance of being included in a sample study, when machine parts are chosen for inspection so that each part produced has the same chance of being selected, and so forth.

A major shortcoming of the classical probability concept is its limited applicability, for there are many situations in which the possibilities that arise cannot all be regarded as equally likely. This would be the case, for instance, if we are concerned with the question whether it will rain on a given day, if we are concerned with the outcome of an election, or if we are concerned with a person's recovery from a disease.

Among the various probability concepts, most widely held is the **frequency interpretation**, according to which *the probability of an event (outcome or happening) is the proportion of the time that events of the same kind will occur in the long run.* If we say that the probability is 0.84 that a jet from Los Angeles to San Francisco will arrive on time, we mean (in accordance with the frequency interpretation) that such flights arrive on time 84 percent of the time. Similarly, if the weather bureau predicts that there is a 30 percent chance for rain (namely, a probability of 0.30), this means that under the same weather conditions it will rain 30 percent of the time. More generally, we say that an event has a probability of, say, 0.90, in the same sense in which we might say that our car will start in cold weather 90 percent of the time. We cannot guarantee what will happen on any particular occasion—the car may start and then it may not—but if we kept records over a long period of time, we should find that the proportion of "successes" is very close to 0.90.

An alternative point of view, which is currently gaining in favor, is to interpret probabilities as **personal** or **subjective evaluations**. Such probabilities express the strength of one's belief with regard to the uncertainties that are involved, and they apply especially when there is little or no direct evidence, so that there is no choice but to consider collateral (indirect) evidence, "educated guesses," and perhaps intuition and other subjective factors.

The approach to probability we shall use in this chapter is the **axiomatic approach**, in which probabilities are defined as "mathematical objects" which behave according to certain well-defined rules. Then, any one of the above probability concepts, or interpretations, can be used in applications, so long as it is consistent with these rules.

## 2.2  SAMPLE SPACES

Since all probabilities pertain to the occurrence or nonoccurrence of events, let us explain first what we mean here by *event* and by the related terms *experiment*, *outcome*, and *sample space*.

It is customary in statistics to refer to any process of observation or measurement as an **experiment**. In this sense, an experiment may consist of the simple process of checking whether a switch is turned on or off; it may consist of counting the imperfections in a piece of cloth; or it may consist of the very complicated process of determining the mass of an electron. The results one obtains from an experiment, whether they are instrument readings, counts, "yes" or "no" answers, or values obtained through extensive calculations, are called the **outcomes** of the experiment.

The set of all possible outcomes of an experiment is called the **sample space** and it is usually denoted by the letter $S$. Each outcome in a sample space is called an **element** of the sample space or simply a **sample point**. If a sample space has a finite number of elements, we may list the elements in the usual set notation; for instance, the sample space for the possible outcomes of one flip of a coin may be written

$$S = \{H, T\}$$

where H and T stand for head and tail. Sample spaces with a large or infinite number of elements are best described by a statement or rule; for example, if the possible outcomes of an experiment are the set of automobiles equipped with citizen band radios, the sample space may be written

$$S = \{x \mid x \text{ is an automobile with a CB radio}\}$$

This is read "$S$ is the set of all $x$ such that $x$ is an automobile with a CB radio." Similarly, if $S$ is the set of odd positive integers, we write

$$S = \{2k + 1 \mid k = 0, 1, 2, \ldots \}$$

How we formulate the sample space for a given situation will depend on the problem at hand. If an experiment consists of one roll of a die and we are interested in which face is turned up, we would use the sample space

$$S_1 = \{1, 2, 3, 4, 5, 6\}$$

However, if we are interested only in whether the face turned up is even or odd, we would use the sample space

$$S_2 = \{\text{even, odd}\}$$

This demonstrates that different sample spaces may well be used to describe an experiment. In general, *it is desirable to use sample spaces whose elements cannot be divided (partitioned or separated) into more primitive or more elementary kinds of outcomes.* In other words, *it is preferable that an element of a sample space does not represent two or more outcomes which are distinguishable in some way.* Thus, in the preceding illustration $S_1$ would be preferable to $S_2$.

## EXAMPLE 2.2

Describe a sample space that might be appropriate for an experiment in which we roll a pair of dice, one red and one green.

### Solution

The sample space that provides the most information consists of the 36 points given by

$$S_1 = \{(x, y) | x = 1, 2, \ldots, 6; y = 1, 2, \ldots, 6\}$$

where $x$ represents the number turned up by the red die and $y$ represents the number turned up by the green die. A second sample space, adequate for most purposes (though less desirable in general as it provides less information) is given by

$$S_2 = \{2, 3, 4, \ldots, 12\}$$

where the elements are the totals of the numbers turned up by the two dice.    ▲

Sample spaces are usually classified according to the number of elements they contain. In the preceding example the sample spaces $S_1$ and $S_2$ contained a **finite** number of elements, but if a coin is flipped until a head appears for the first time, this could happen on the first flip, the second flip, the third flip, the fourth flip, ... , and there are infinitely many possibilities. For this experiment we obtain the sample space

$$S = \{H, TH, TTH, TTTH, TTTTH, \dots \}$$

with an unending sequence of elements. But even here the number of elements can be matched one-to-one with the whole numbers, and in this sense the sample space is said to be **countable**. If a sample space contains a finite number of elements, or an infinite though countable number of elements, it is said to be **discrete**.

The outcomes of some experiments are neither finite nor countably infinite. Such is the case, for example, when one conducts an investigation to determine the distance that a certain make of car will travel over a prescribed test course on 5 liters of gasoline. If we assume that distance is a variable that can be measured to any desired degree of accuracy, there is an infinity of possibilities (distances) that cannot be matched one-to-one with the whole numbers. Also, if we want to measure the amount of time it takes for two chemicals to react, the amounts making up the sample space are infinite in number and not countable. Thus, sample spaces need not be discrete. If a sample space consists of a continuum, such as all the points of a line segment or all the points in a plane, it is said to be **continuous**. Continuous sample spaces arise in practice whenever the outcomes of experiments are measurements of physical properties such as temperature, speed, pressure, length, ... , that are measured on continuous scales.

## 2.3  EVENTS

In many problems we are interested in results that are not given directly by a specific element of a sample space.

### EXAMPLE 2.3

With reference to the first sample space $S_1$ on page 32, describe the event $A$ that the number of points rolled with the die is divisible by 3.

#### Solution

Among 1, 2, 3, 4, 5, and 6, only 3 and 6 are divisible by 3. Therefore, $A$ is represented by the subset $\{3, 6\}$ of the sample space $S_1$.    ▲

### EXAMPLE 2.4

With reference to the sample space $S_1$ of Example 2.2, describe the event $B$ that the total number of points rolled with the pair of dice is 7.

*Solution*

Among the 36 possibilities, only (1, 6), (2, 5), (3, 4), (4, 3), (5, 2), and (6, 1) yield a total of 7. So, we write

$$B = \{(1, 6), (2, 5), (3, 4), (4, 3), (5, 2), (6, 1)\}$$

Note that in Figure 2.1 the event of rolling a total of 7 with the two dice is represented by the set of points inside the region bounded by the dotted line.   ▲

In the same way, any event (outcome or result) can be identified with a collection of points, which constitute a subset of an appropriate sample space. Such a subset consists of all the elements of the sample space for which the event occurs, and in probability and statistics we identify the subset with the event. Thus, by definition, an **event** is a subset of a sample space.

## EXAMPLE 2.5

If someone takes three shots at a target and we care only whether each shot is a hit or a miss, describe a suitable sample space, the elements of the sample space that constitute event $M$ that the person will miss the target three times in a row, and the elements of event $N$ that the person will hit the target once and miss it twice.

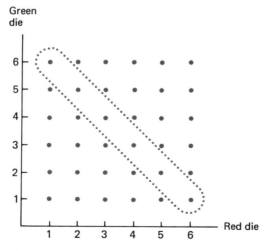

**Figure 2.1**  Rolling a total of 7 with a pair of dice.

*Solution*

If we let 0 and 1 represent a miss and a hit, respectively, the eight possibilities (0, 0, 0), (1, 0, 0), (0, 1, 0), (0, 0, 1), (1, 1, 0), (1, 0, 1), (0, 1, 1), and (1, 1, 1) may be displayed as in Figure 2.2. Thus, it can be seen that

$$M = \{(0, 0, 0)\}$$

and

$$N = \{(1, 0, 0), (0, 1, 0), (0, 0, 1)\} \quad \blacktriangle$$

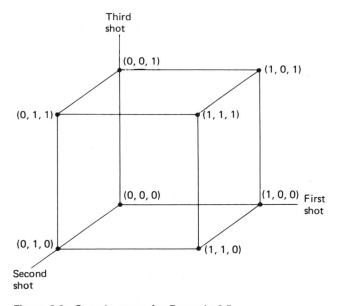

**Figure 2.2**  Sample space for Example 2.5.

## EXAMPLE 2.6

Construct a sample space for the length of the useful life of a certain electronic component and indicate the subset which represents the event $F$ that the component fails before the end of the sixth year.

*Solution*

If $t$ is the length of the component's useful life in years, the sample space may be written $S = \{t \mid t \geq 0\}$, and the subset $F = \{t \mid 0 \leq t < 6\}$ is the event that the component fails before the end of the sixth year.    $\blacktriangle$

According to our definition, any event is a subset of an appropriate sample space, but it should be observed that the converse is not necessarily true. For discrete sample spaces all subsets are events, but in the continuous case some rather abstruse point sets must be excluded for mathematical reasons. This is discussed further in some of the more advanced texts listed among the references at the end of this chapter, but it is of no consequence so far as the work of this book is concerned.

In many problems of probability we are interested in events that are actually combinations of two or more events, formed by taking **unions**, **intersections**, and **complements**. Although the reader must surely be familiar with these terms, let us review briefly that if $A$ and $B$ are any two subsets of a sample space $S$, their union $A \cup B$ is the subset of $S$ that contains all the elements that are either in $A$, in $B$, or in both; their intersection $A \cap B$ is the subset of $S$ that contains all the elements that are in both $A$ and $B$; and the complement $A'$ of $A$ is the subset of $S$ that contains all the elements of $S$ that are not in $A$. Some of the rules that control the formation of unions, intersections, and complements may be found in Exercises 2.1 through 2.4.

Sample spaces and events, particularly relationships among events, are often depicted by means of **Venn diagrams**, in which the sample space is represented by a rectangle, while events are represented by regions within the rectangle, usually by circles or parts of circles. For instance, the shaded regions of the four Venn diagrams of Figure 2.3 represent, respectively, event $A$, the complement of event $A$, the union of events $A$ and $B$, and the intersection of events $A$ and $B$.

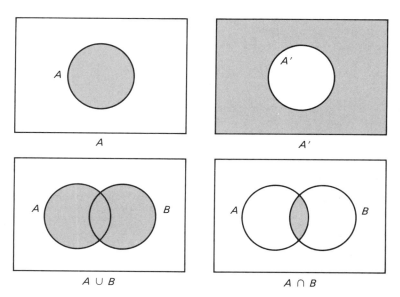

**Figure 2.3** Venn diagrams.

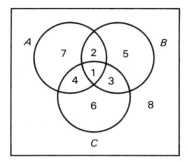

**Figure 2.4**  Venn diagram.

When we are dealing with three events, we usually draw the circles as in Figure 2.4. Here, the regions are numbered 1 through 8 for easy reference.

To indicate special relationships among events, we sometimes draw diagrams like those of Figure 2.5. Here, the one on the left serves to indicate that events $A$ and $B$ are **mutually exclusive**, namely, that the two sets have no elements in common (or that the two events cannot both occur.) When $A$ and $B$ are mutually exclusive, we write $A \cap B = \varnothing$ where $\varnothing$ denotes the **empty set**, which has no elements at all. The diagram on the right serves to indicate that $A$ is contained in $B$, and symbolically we express this by writing $A \subset B$.

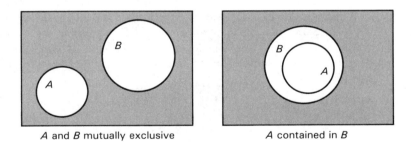

A and B mutually exclusive             A contained in B

**Figure 2.5**  Diagrams showing special relationships among events.

*EXERCISES*

**2.1**  Use Venn diagrams to verify that

(a)  $(A \cup B) \cup C$ is the same event as $A \cup (B \cup C)$;

(b)  $A \cap (B \cup C)$ is the same event as $(A \cap B) \cup (A \cap C)$;

(c)  $A \cup (B \cap C)$ is the same event as $(A \cup B) \cap (A \cup C)$.

**2.2** Use Venn diagrams to verify the two **de Morgan laws**:
(a) $(A \cap B)' = A' \cup B'$;
(b) $(A \cup B)' = A' \cap B'$.

**2.3** Use Venn diagrams to verify that if $A$ is contained in $B$, then $A \cap B = A$ and $A \cap B' = \emptyset$.

**2.4** Use Venn diagrams to verify that
(a) $(A \cap B) \cup (A \cap B') = A$;
(b) $(A \cap B) \cup (A \cap B') \cup (A' \cap B) = A \cup B$;
(c) $A \cup (A' \cap B) = A \cup B$.

## APPLICATIONS

**2.5** If $S = \{1, 2, 3, 4, 5, 6, 7, 8, 9\}$, $A = \{1, 3, 5, 7\}$, $B = \{6, 7, 8, 9\}$, $C = \{2, 4, 8\}$, and $D = \{1, 5, 9\}$, list the elements of the subsets of $S$ corresponding to the following events:

(a) $A' \cap B$;
(b) $(A' \cap B) \cap C$;
(c) $B' \cup C$;
(d) $(B' \cup C) \cap D$;
(e) $A' \cap C$;
(f) $(A' \cap C) \cap D$.

**2.6** An electronics firm plans to build a research laboratory in Southern California, and its management has to decide between sites in Los Angeles, San Diego, Long Beach, Pasadena, Santa Barbara, Anaheim, Santa Monica, and Westwood. If $A$ represents the event that they will choose a site in San Diego or Santa Barbara, $B$ represents the event that they will choose a site in San Diego or Long Beach, $C$ represents the event that they will choose a site in Santa Barbara or Anaheim, and $D$ represents the event that they will choose a site in Los Angeles or Santa Barbara, list the elements of each of the following subsets of the sample space, which consists of the eight site selections:

(a) $A'$;
(b) $D'$;
(c) $C \cap D$;
(d) $B \cap C$;
(e) $B \cup C$;
(f) $A \cup B$;
(g) $C \cup D$;
(h) $(B \cup C)'$;
(i) $B' \cap C'$.

**2.7** Among the eight cars which a dealer has in his showroom, Car 1 is new, has air-conditioning, power steering, and bucket seats, Car 2 is one year old, has air-conditioning, but neither power steering nor bucket seats, Car 3 is two years old, has air-conditioning and power steering, but no bucket seats, Car 4 is three years old, has air-conditioning, but neither power steering nor bucket seats, Car 5 is new, has no air-conditioning, no power steering, and no bucket seats, Car 6 is one year old, has power steering, but neither air-conditioning nor bucket seats, Car 7 is two years old, has no air-conditioning, no power steering, and no bucket seats, and Car 8 is three years old, has no air-conditioning, but has power steering as well as bucket seats. If a customer buys one of these cars and the event that he chooses a new car, for example, is represented by the set {Car 1, Car 5}, indicate similarly the sets which represent the events that

(a)  he chooses a car without air-conditioning;
(b)  he chooses a car without power steering;
(c)  he chooses a car with bucket seats;
(d)  he chooses a car that is either two or three years old.

**2.8**  With reference to the preceding exercise, state in words what kind of car the customer will choose, if his choice is given by

(a)  the complement of the set of part (a);
(b)  the union of the sets of parts (b) and (c);
(c)  the intersection of the sets of parts (c) and (d);
(d)  the intersection of parts (b) and (c) of this exercise.

**2.9**  If Ms. Brown buys one of the houses advertised for sale in a Seattle newspaper (on a given Sunday), $T$ is the event that the house has three or more baths, $U$ is the event that it has a fireplace, $V$ is the event that it costs more than \$100,000, and $W$ is the event that it is new, describe (in words) each of the following events:

(a)  $T'$;            (b)  $U'$;            (c)  $V'$;
(d)  $W'$;            (e)  $T \cap U$;      (f)  $T \cap V$;
(g)  $U' \cap V$;     (h)  $V \cup W$;      (i)  $V' \cup W$;
(j)  $T \cup U$;      (k)  $T \cup V$;      (l)  $V \cap W$.

**2.10**  A resort hotel has two station wagons, which it uses to shuttle its guests to and from the airport. If the larger of the two station wagons can carry five passengers and the smaller can carry four passengers, the point (0, 3) represents the event that at a given moment the larger station wagon is empty while the smaller one has three passengers, the point (4, 2) represents the event that at the given moment the larger station wagon has four passengers while the smaller one has two passengers, ... , draw a figure showing the 30 points of the corresponding sample space. Also, if $E$ stands for the event that at least one of the station wagons is empty, $F$ stands for the event that together they carry two, four, or six passengers, and $G$ stands for the event that each carries the same number of passengers, list the points of the sample space that correspond to each of the following events:

(a)  $E$;             (b)  $F$;             (c)  $G$;
(d)  $E \cup F$;      (e)  $E \cap F$;      (f)  $F \cup G$;
(g)  $E \cup F'$;     (h)  $E \cap G'$;     (i)  $F' \cap E'$.

**2.11**  A coin is tossed once. Then, if it comes up heads, a die is thrown once; if it comes up tails, it is tossed twice more. Using the notation in which (H, 2), for example, denotes the event that the coin comes up heads and then the die comes up 2, and (T, T, T) denotes the event that the coin comes up tails three times in a row, list

(a)  the ten elements of the sample space $S$;

(b)    the elements of $S$ corresponding to event $A$ that exactly one head occurs;

(c)    the elements of $S$ corresponding to event $B$ that at least two tails occur or a number greater than 4 occurs.

**2.12** An electronic game contains three components arranged in the series-parallel circuit shown in Figure 2.6. At any given time, each component may or may not be operative, and the game will operate only if there is a continuous circuit from $P$ to $Q$. Let $A$ be the event that the game will operate; let $B$ be the event that the game will operate though component $x$ is not operative; and let $C$ be the event that the game will operate though component $y$ is not operative. Using the notation in which (0, 0, 1), for example, denotes that component $z$ is operative but components $x$ and $y$ are not,

(a)    list the elements of the sample space $S$ and also the elements of $S$ corresponding to events $A$, $B$, and $C$;

(b)    determine which pairs of events, $A$ and $B$, $A$ and $C$, or $B$ and $C$, are mutually exclusive.

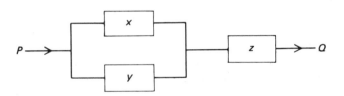

**Figure 2.6**    Diagram for Exercise 2.12.

**2.13** An experiment consists of rolling a die until a 3 appears. Describe the sample space and determine

(a)    how many elements of the sample space correspond to the event that the 3 appears on the $k$th roll of the die;

(b)    how many elements of the sample space correspond to the event that the 3 appears not later than the $k$th roll of the die.

**2.14** If $S = \{x | 0 < x < 10\}$, $M = \{x | 3 < x \leqslant 8\}$, and $N = \{x | 5 < x < 10\}$, find

(a)    $M \cup N$;

(b)    $M \cap N$;

(c)    $M \cap N'$;

(d)    $M' \cup N$.

**2.15** Express symbolically the sample space $S$ that consists of all the points $(x, y)$ on or in the circle of radius 3 centered at the point $(2, -3)$.

**2.16** In Figure 2.7, $L$ is the event that a driver has liability insurance and $C$ is the event that she has collision insurance. Express in words what events are represented by regions 1, 2, 3, and 4.

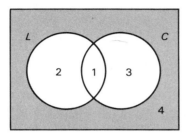

**Figure 2.7**  Venn diagram for Exercise 2.16.

**2.17** With reference to Exercise 2.16 and Figure 2.7, what events are represented by
    (a)   regions 1 and 2 together;
    (b)   regions 2 and 4 together;
    (c)   regions 1, 2, and 3 together;
    (d)   regions 2, 3, and 4 together?

**2.18** In Figure 2.8, $E$, $T$, and $N$ are the events that a car brought to a garage needs an engine overhaul, transmission repairs, or new tires. Express in words the events represented by
    (a)   region 1;
    (b)   region 3;
    (c)   region 7;
    (d)   regions 1 and 4 together;
    (e)   regions 2 and 5 together;
    (f)   regions 3, 5, 6, and 8 together.

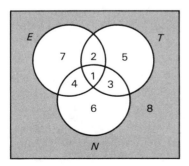

**Figure 2.8**  Venn diagram for Exercise 2.18.

**2.19** With reference to the preceding exercise and Figure 2.8, list the region or combinations of regions representing the events that a car brought to the garage needs

(a) transmission repairs, but neither an engine overhaul nor new tires;

(b) an engine overhaul and transmission repairs;

(c) transmission repairs or new tires, but not an engine overhaul;

(d) new tires.

**2.20** In a group of 200 college students 138 are enrolled in a course in psychology, 115 are enrolled in a course in sociology, and 91 are enrolled in both. How many of these students are not enrolled in either course? (*Hint*: Draw a suitable Venn diagram and fill in the numbers associated with the various regions.)

**2.21** A market research organization claims that among 500 shoppers interviewed, 308 regularly buy Product $X$, 266 regularly buy Product $Y$, 103 regularly buy both, and 59 buy neither on a regular basis. Using a Venn diagram and filling in the number of shoppers associated with the various regions, check whether the results of this study should be questioned.

**2.22** Among 120 visitors to Disneyland, 74 stayed for at least 3 hours, 86 spent at least $20, 64 went on the Matterhorn ride, 60 stayed for at least 3 hours and spent at least $20, 52 stayed for at least 3 hours and went on the Matterhorn ride, 54 spent at least $20 and went on the Matterhorn ride, and 48 stayed for at least 3 hours, spent at least $20, and went on the Matterhorn ride. Drawing a Venn diagram with three circles (like that of Figure 2.4) and filling in the numbers associated with the various regions, find how many of the 120 visitors to Disneyland

(a) stayed for at least 3 hours, spent at least $20, but did not go on the Matterhorn ride;

(b) went on the Matterhorn ride, but stayed less than 3 hours and spent less than $20;

(c) stayed less than 3 hours, spent at least $20, but did not go on the Matterhorn ride.

## 2.4 THE PROBABILITY OF AN EVENT

To formulate the postulates of probability we shall follow the practice of denoting events by means of capital letters, and we shall write the probability of event $A$ as $P(A)$, the probability of event $B$ as $P(B)$, and so forth. As before, we shall denote the set of all possible outcomes, the sample space, by the letter $S$.

Probabilities are values of a set function, also called a **probability measure**, for as we shall see, this function assigns real numbers to the various subsets of a sample space $S$. As we shall formulate them here, the postulates of probability apply only when the sample space $S$ is discrete.

---

**POSTULATE 1**    The probability of an event is a nonnegative real number; that is, $P(A) \geq 0$ for any subset $A$ of $S$.

**POSTULATE 2**    $P(S) = 1$.

**POSTULATE 3**    If $A_1, A_2, A_3, \ldots$ , is a finite or infinite sequence of mutually exclusive events of $S$, then

$$P(A_1 \cup A_2 \cup A_3 \cup \cdots) = P(A_1) + P(A_2) + P(A_3) + \cdots$$

---

Postulates *per se* require no proof, but if the resulting theory is to be applied, we must show that the postulates are satisfied when we give probabilities a "real" meaning. Let us illustrate this here in connection with the frequency interpretation; the relationship between the postulates and the classical probability concept will be discussed on page 47, while the relationship between the postulates and subjective probabilities is left for the reader to examine in Exercises 2.34 and 2.56.

Since proportions are always positive or zero, the first postulate is in complete agreement with the frequency interpretation. The second postulate states indirectly that certainty is identified with a probability of 1—after all, it is always assumed that one of the possibilities in $S$ must occur, and it is to this certain event that we assign a probability of 1. So far as the frequency interpretation is concerned, a probability of 1 implies that the event in question will occur 100 percent of the time, or in other words, that it is certain to occur.

Taking the third postulate in the simplest case, namely, for two mutually exclusive events $A_1$ and $A_2$, it can easily be seen that it is satisfied by the frequency interpretation. If one event occurs, say, 28 percent of the time, another event occurs 39 percent of the time, and the two events cannot both occur at the same time (that is, they are mutually exclusive), then one or the other will occur $28 + 39 = 67$ percent of the time. Thus, the third postulate is satisfied, and the same kind of argument applies when there are more than two mutually exclusive events.

Before we study some of the immediate consequences of the postulates of probability, let us emphasize the point that the three postulates do not tell us how to assign probabilities to events, they merely restrict the ways in which it can be done.

## EXAMPLE 2.7

An experiment has four possible outcomes, $A$, $B$, $C$, and $D$, which are mutually exclusive. Explain why the following assignments of probabilities are not permissible:

(a)  $P(A) = 0.12$, $P(B) = 0.63$, $P(C) = 0.45$, $P(D) = -0.20$;

(b)  $P(A) = \dfrac{9}{120}$, $P(B) = \dfrac{45}{120}$, $P(C) = \dfrac{27}{120}$, $P(D) = \dfrac{46}{120}$.

*Solution*

(a)  $P(D) = -0.20$ violates Postulate 1;

(b)  $P(S) = P(A \cup B \cup C \cup D) = \dfrac{9}{120} + \dfrac{45}{120} + \dfrac{27}{120} + \dfrac{46}{120} = \dfrac{127}{120} \neq 1$,

and this violates Postulate 2.  ▲

Of course, in actual practice probabilities are assigned on the basis of past experience, on the basis of a careful analysis of all underlying conditions, on the basis of subjective judgments, or on the basis of assumptions—sometimes the assumption that all possible outcomes are equiprobable.

To assign a probability measure to a sample space, it is not necessary to specify the probability of each possible subset. This is fortunate, for a sample space with as few as 20 possible outcomes has already $2^{20} = 1,048,576$ subsets [the general formula follows directly from part (a) of Exercise 1.14], and the number of subsets grows very rapidly when there are 50 possible outcomes, 100 possible outcomes, or more. Instead of listing the probabilities of all possible subsets, we often list the probabilities of the individual outcomes, or sample points of $S$, and then make use of the following theorem.

---

**THEOREM 2.1**  If $A$ is an event in a discrete sample space $S$, then $P(A)$ equals the sum of the probabilities of the individual outcomes comprising $A$.

---

*Proof.*  Let $O_1$, $O_2$, $O_3$, ... , be the finite or infinite sequence of outcomes that comprise the event $A$. Thus,

$$A = O_1 \cup O_2 \cup O_3 \dots$$

and since the individual outcomes, the $O$'s, are mutually exclusive, the third postulate of probability yields

$$P(A) = P(O_1) + P(O_2) + P(O_3) + \cdots$$

This completes the proof.    ▼

To use this theorem, we must be able to assign probabilities to the individual outcomes of experiments. How this is done in some special situations is illustrated by the following examples.

## EXAMPLE 2.8

If we twice flip a balanced coin, what is the probability of getting at least one head?

### Solution

The sample space is $S = \{HH, HT, TH, TT\}$, where H and T denote head and tail. Since we assume that the coin is balanced, these outcomes are equally likely and we assign to each sample point the probability $\frac{1}{4}$. Letting $A$ denote the event that we will get at least one head, we get $A = \{HH, HT, TH\}$ and

$$P(A) = P(HH) + P(HT) + P(TH)$$
$$= \tfrac{1}{4} + \tfrac{1}{4} + \tfrac{1}{4}$$
$$= \tfrac{3}{4}    ▲$$

## EXAMPLE 2.9

A die is loaded in such a way that each odd number is twice as likely to occur as each even number. Find $P(G)$, where $G$ is the event that a number greater than 3 occurs on a single roll of the die.

### Solution

The sample space is $S = \{1, 2, 3, 4, 5, 6\}$. Hence, if we assign probability $w$ to each even number and probability $2w$ to each odd number, we find that $2w + w + 2w + w + 2w + w = 9w = 1$ in accordance with Postulate 2. It follows that $w = \frac{1}{9}$ and

$$P(G) = \tfrac{1}{9} + \tfrac{2}{9} + \tfrac{1}{9} = \tfrac{4}{9}    ▲$$

If a sample space is countably infinite, probabilities will have to be assigned to the individual outcomes by means of a mathematical rule, preferably by means of a formula or equation.

## EXAMPLE 2.10

If, for a given experiment, $O_1, O_2, O_3, \ldots$ , is an infinite sequence of outcomes, verify that

$$P(O_i) = \left(\frac{1}{2}\right)^i \qquad \text{for } i = 1, 2, 3, \ldots$$

is, indeed, a probability measure.

### Solution

Since the probabilities are all positive, it remains to be shown that $P(S) = 1$. Getting

$$P(S) = \tfrac{1}{2} + \tfrac{1}{4} + \tfrac{1}{8} + \tfrac{1}{16} + \cdots$$

and making use of the formula for the sum of the terms of an infinite geometric progression, we find that

$$P(S) = \frac{\frac{1}{2}}{1 - \frac{1}{2}} = 1 \qquad \blacktriangle$$

In connection with the preceding example, the word "sum" in Theorem 2.1 will have to be interpreted so that it includes the value of an infinite series.

As we shall see in Chapter 5, the probability measure of Example 2.10 would be appropriate, for example, if $O_i$ is the event that a person flipping a balanced coin will get a tail for the first time on the $i$th flip of the coin. Thus, the probability that the first tail will come on the third, fourth, or fifth flip of the coin is

$$\left(\frac{1}{2}\right)^3 + \left(\frac{1}{2}\right)^4 + \left(\frac{1}{2}\right)^5 = \frac{7}{32}$$

and the probability that the first tail will come on an odd-numbered flip of the coin is

$$\left(\frac{1}{2}\right)^1 + \left(\frac{1}{2}\right)^3 + \left(\frac{1}{2}\right)^5 + \cdots = \frac{\frac{1}{2}}{1 - \frac{1}{4}} = \frac{2}{3}$$

Here again we made use of the formula for the sum of the terms of an infinite geometric progression.

If an experiment is such that we can assume equal probabilities for all the sample points, as was the case in Example 2.8, we can take advantage of the following special case of Theorem 2.1.

---

**THEOREM 2.2** If an experiment can result in any one of $N$ different equally likely outcomes, and if $n$ of these outcomes together constitute event $A$, then the probability of event $A$ is

$$P(A) = \frac{n}{N}$$

---

***Proof.*** Let $O_1, O_2, \ldots, O_N$ represent the individual outcomes in $S$, each with probability $\frac{1}{N}$. If $A$ is the union of $n$ of these mutually exclusive outcomes, and it does not matter which ones, then

$$P(A) = P(O_1 \cup O_2 \cup \cdots \cup O_n)$$

$$= P(O_1) + P(O_2) + \cdots + P(O_n)$$

$$= \underbrace{\frac{1}{N} + \frac{1}{N} + \cdots + \frac{1}{N}}_{n \text{ terms}}$$

$$= \frac{n}{N} \qquad \blacktriangledown$$

Observe that the formula $P(A) = \frac{n}{N}$ of Theorem 2.2 is identical with the one for the classical probability concept (see page 29). Indeed, what we have shown here is that the classical probability concept is consistent with the postulates of probability—it follows from the postulates in the special case where the individual outcomes are all equiprobable.

## EXAMPLE 2.11

A five-card poker hand dealt from a deck of 52 playing cards is said to be a full house if it consists of three of a kind and a pair. If all the five-card hands are equally likely, what is the probability of being dealt a full house?

### Solution

The number of ways in which we can be dealt a particular full house, say three kings and two aces, is $\binom{4}{3}\binom{4}{2}$. Since there are 13 ways of selecting the face value for the three of a kind and for each of these there are 12 ways of selecting the face value for the pair, there are altogether

$$n = 13 \cdot 12 \cdot \binom{4}{3}\binom{4}{2}$$

different full houses. Also, the total number of equally likely five-card poker hands is

$$N = \binom{52}{5}$$

and it follows by Theorem 2.2 that the probability of getting a full house is

$$P(A) = \frac{n}{N} = \frac{13 \cdot 12 \binom{4}{3}\binom{4}{2}}{\binom{52}{5}} = 0.0014 \quad \blacktriangle$$

## 2.5 SOME RULES OF PROBABILITY

Based on the three postulates of probability, we can derive many other rules which have important applications. Among them, the next four theorems are immediate consequences of the postulates.

---

**THEOREM 2.3** If $A$ and $A'$ are complementary events in a sample space $S$, then

$$P(A') = 1 - P(A)$$

---

*Proof.* In the second and third steps of the proof that follows, we make use of the definition of a complement, according to which $A$ and $A'$ are mutually exclusive and $A \cup A' = S$. Thus, we write

$$1 = P(S) \qquad \text{(by Postulate 2)}$$
$$= P(A \cup A')$$
$$= P(A) + P(A') \qquad \text{(by Postulate 3)}$$

and it follows that $P(A') = 1 - P(A)$. ▼

In connection with the frequency interpretation, this result implies that if an event occurs, say, 37 percent of the time, then it does not occur 63 percent of the time.

---

**THEOREM 2.4** $P(\varnothing) = 0$ for any sample space $S$.

---

*Proof.* Since $S$ and $\varnothing$ are mutually exclusive and $S \cup \varnothing = S$ in accordance with the definition of the empty set $\varnothing$, it follows that

$$P(S) = P(S \cup \varnothing)$$
$$= P(S) + P(\varnothing) \qquad \text{(by Postulate 3)}$$

and, hence, that $P(\varnothing) = 0$. ▼

It is important to note that it does not necessarily follow from $P(A) = 0$ that $A = \varnothing$. In practice, we often assign 0 probability to events which, in colloquial terms, would not happen in a million years. For instance, there is the classical example that we assign a probability of 0 to the event that a monkey set loose on a typewriter will type Plato's *Republic* word for word without a mistake. As we shall see in Chapters 3 and 6, the fact that $P(A) = 0$ does not imply $A = \varnothing$ is of relevance, especially, in the continuous case.

---

**THEOREM 2.5** If $A$ and $B$ are events in a sample space $S$ and $A \subset B$, then $P(A) \le P(B)$.

---

*Proof.* Since $A \subset B$, we can write

$$B = A \cup (A' \cap B)$$

as can easily be verified by means of a Venn diagram. Then, since $A$ and $A' \cap B$ are mutually exclusive, we get

$$P(B) = P(A) + P(A' \cap B) \qquad \text{(by Postulate 3)}$$
$$\geqslant P(A) \qquad\qquad\qquad \text{(by Postulate 1)} \qquad \blacktriangledown$$

In words, this theorem states that if $A$ is a subset of $B$, then $P(A)$ cannot be greater than $P(B)$. For instance, the probability of drawing a heart from an ordinary deck of 52 playing cards cannot be greater than the probability of drawing a red card. Indeed, the probability is $\frac{1}{4}$, compared with $\frac{1}{2}$.

---

**THEOREM 2.6**   $0 \leqslant P(A) \leqslant 1$ for any event $A$.

---

**Proof.**   Using Theorem 2.5 and the fact that $\varnothing \subset A \subset S$ for any event $A$ in $S$, we have

$$P(\varnothing) \leqslant P(A) \leqslant P(S)$$

Then, $P(\varnothing) = 0$ and $P(S) = 1$ leads to the result that

$$0 \leqslant P(A) \leqslant 1 \qquad \blacktriangledown$$

The third postulate of probability is sometimes referred to as the **special addition rule**; it is special in the sense that events $A_1$, $A_2$, $A_3$, ... , must all be mutually exclusive. For any two events $A$ and $B$ there exists the **general addition rule**:

---

**THEOREM 2.7**   If $A$ and $B$ are any two events in a sample space $S$, then
$$P(A \cup B) = P(A) + P(B) - P(A \cap B)$$

---

**Proof.**   Assigning the probabilities $a$, $b$, and $c$ to the mutually exclusive events $A \cap B$, $A \cap B'$, and $A' \cap B$ as in the Venn diagram of Figure 2.9, we find that

$$P(A \cup B) = a + b + c$$
$$= (a + b) + (c + a) - a$$
$$= P(A) + P(B) - P(A \cap B) \qquad \blacktriangledown$$

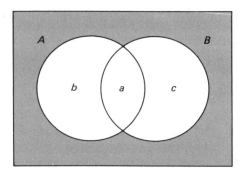

**Figure 2.9**  Venn diagram for proof of Theorem 2.7.

## EXAMPLE 2.12

In a large metropolitan area, the probabilities are 0.86, 0.35, and 0.29 that a family (randomly chosen for a sample survey) owns a color television set, a black-and-white set, or both kinds of sets. What is the probability that a family owns either or both kinds of sets?

### Solution

If $A$ is the event that a family in this metropolitan area owns a color television set and $B$ is the event that it owns a black-and-white set, we have $P(A) = 0.86$, $P(B) = 0.35$, and $P(A \cap B) = 0.29$; substitution into the formula of Theorem 2.7 yields

$$P(A \cup B) = 0.86 + 0.35 - 0.29$$
$$= 0.92 \quad \blacktriangle$$

## EXAMPLE 2.13

Near a certain exit of I-17, the probabilities are 0.23 and 0.24 that a truck stopped at a roadblock will have faulty brakes or badly worn tires. Also, the probability is 0.38 that a truck stopped at the roadblock will have faulty brakes and/or badly worn tires. What is the probability that a truck stopped at this roadblock will have faulty brakes as well as badly worn tires?

### Solution

If $B$ is the event that a truck stopped at the roadblock will have faulty brakes and $T$ is the event that it will have badly worn tires, we have $P(B) = 0.23$,

$P(T) = 0.24$, and $P(B \cup T) = 0.38$; substitution into the formula of Theorem 2.7 yields

$$0.38 = 0.23 + 0.24 - P(B \cap T)$$

Solving for $P(B \cap T)$, we thus get

$$P(B \cap T) = 0.23 + 0.24 - 0.38 = 0.09 \qquad \blacktriangle$$

Repeatedly using the formula of Theorem 2.7, we can generalize this addition rule so that it will apply to any number of events. For instance, for three events we get

---

**THEOREM 2.8**  If $A$, $B$, and $C$ are any three events in a sample space $S$, then

$$P(A \cup B \cup C) = P(A) + P(B) + P(C) - P(A \cap B) - P(A \cap C)$$
$$- P(B \cap C) + P(A \cap B \cap C)$$

---

*Proof.*   Writing $A \cup B \cup C$ as $A \cup (B \cup C)$ and using the formula of Theorem 2.7 twice, once for $P[A \cup (B \cup C)]$ and once for $P(B \cup C)$, we get

$$P(A \cup B \cup C) = P[A \cup (B \cup C)]$$
$$= P(A) + P(B \cup C) - P[A \cap (B \cup C)]$$
$$= P(A) + P(B) + P(C) - P(B \cap C)$$
$$- P[A \cap (B \cup C)]$$

Then, using the distributive law which the reader was asked to verify in part (b) of Exercise 2.1, we find that

$$P[A \cap (B \cup C)] = P[(A \cap B) \cup (A \cap C)]$$
$$= P(A \cap B) + P(A \cap C) - P[(A \cap B) \cap (A \cap C)]$$
$$= P(A \cap B) + P(A \cap C) - P(A \cap B \cap C)$$

and hence that

$$P(A \cup B \cup C) = P(A) + P(B) + P(C) - P(A \cap B) - P(A \cap C)$$
$$- P(B \cap C) + P(A \cap B \cap C) \qquad \blacktriangledown$$

(In Exercise 2.30 the reader will be asked to give an alternative proof of this theorem, based on the method used in the text to prove Theorem 2.7.)

## EXAMPLE 2.14

Suppose that if a person visits his dentist, the probability that he will have his teeth cleaned is 0.44, the probability that he will have a cavity filled is 0.24, the probability that he will have a tooth extracted is 0.21, the probability that he will have his teeth cleaned and a cavity filled is 0.08, the probability that he will have his teeth cleaned and a tooth extracted is 0.11, the probability that he will have a cavity filled and a tooth extracted is 0.07, and the probability that he will have his teeth cleaned, a cavity filled, and a tooth extracted is 0.03. What is the probability that a person visiting his dentist will have at least one of these things done to him?

### Solution

If $C$ is the event that the person will have his teeth cleaned, $F$ is the event that he will have a cavity filled, and $E$ is the event that he will have a tooth extracted, we are given $P(C) = 0.44$, $P(F) = 0.24$, $P(E) = 0.21$, $P(C \cap F) = 0.08$, $P(C \cap E) = 0.11$, $P(F \cap E) = 0.07$, and $P(C \cap F \cap E) = 0.03$, and substitution into the formula of Theorem 2.8 yields

$$P(C \cup F \cup E) = 0.44 + 0.24 + 0.21 - 0.08 - 0.11 - 0.07 + 0.03$$

$$= 0.66 \quad \blacktriangle$$

### EXERCISES

**2.23** Use parts (a) and (b) of Exercise 2.4 to show that
  (a)  $P(A) \geqslant P(A \cap B)$;
  (b)  $P(A) \leqslant P(A \cup B)$.

**2.24** Referring to Figure 2.9, verify that

$$P(A \cap B') = P(A) - P(A \cap B)$$

**2.25** Referring to Figure 2.9 and letting $P(A' \cap B') = d$, verify that

$$P(A' \cap B') = 1 - P(A) - P(B) + P(A \cap B)$$

**2.26** The event that "$A$ or $B$ but not both" will occur can be written as

$$(A \cap B') \cup (A' \cap B)$$

Express the probability of this event in terms of $P(A)$, $P(B)$, and $P(A \cap B)$.

$A \cap B \subset A$    $\cdot \ P(A \cap B) \leq P(A)$
$\leq P(A) + P(B)$

**2.27** Use the formula of Theorem 2.7 to show that

   (a)  $P(A \cap B) \leq P(A) + P(B)$;

   (b)  $P(A \cap B) \geq P(A) + P(B) - 1$.

**2.28** Use the Venn diagram of Figure 2.10 with the probabilities $a, b, c, d, e,$ $f$, and $g$ assigned to $A \cap B \cap C, A \cap B \cap C', \ldots ,$ and $A \cap B' \cap C'$ to show that if $P(A) = P(B) = P(C) = 1$, then $P(A \cap B \cap C) = 1$. (*Hint:* Start with the argument that since $P(A) = 1$, it follows that $e = c = f = 0$.)

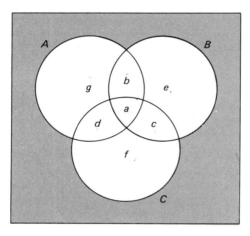

**Figure 2.10**  Diagram for Exercises 2.28, 2.30, and 2.31.

**2.29** Give an alternative proof of Theorem 2.7 by making use of the relationships $A \cup B = A \cup (A' \cap B)$ and $B = (A \cap B) \cup (A' \cap B)$.

**2.30** Use the Venn diagram of Figure 2.10 and the method by which we proved Theorem 2.7 to prove Theorem 2.8.

**2.31** Duplicate the method of proof used in the preceding exercise to show that

$$P(A \cup B \cup C \cup D) = P(A) + P(B) + P(C) + P(D) - P(A \cap B)$$

$$- P(A \cap C) - P(A \cap D) - P(B \cap C) - P(B \cap D)$$

$$- P(C \cap D) + P(A \cap B \cap C) + P(A \cap B \cap D)$$

$$+ P(A \cap C \cap D) + P(B \cap C \cap D)$$

$$- P(A \cap B \cap C \cap D)$$

(*Hint:* With reference to the Venn diagram of Figure 2.10, divide each of the eight regions into two parts, designating one to be inside $D$ and the other outside $D$ and letting $a$, $b$, $c$, $d$, $e$, $f$, $g$, $h$, $i$, $j$, $k$, $l$, $m$, $n$, $o$, and $p$ be the probabilities associated with the resulting sixteen regions.)

**2.32** Prove by induction that

$$P(E_1 \cup E_2 \cup \cdots \cup E_n) \leqslant \sum_{i=1}^{n} P(E_i)$$

for any finite sequence of events $E_1$, $E_2$, ... , and $E_n$.

**2.33** The **odds** that an event will occur are given by the ratio of the probability that the event will occur to the probability that it will not occur, provided neither probability is zero. Odds are usually quoted in terms of positive integers having no common factor. Show that if the odds are $a$ to $b$ that an event will occur, its probability is

$$p = \frac{a}{a + b}$$

**2.34** Subjective probabilities may be determined by exposing persons to risk-taking situations and finding the odds at which they would consider it fair to bet on the outcome. The odds are then converted into probabilities by means of the formula of the preceding exercise. For instance, if a person feels that 3 to 2 are fair odds that a business venture will succeed (or that it would be fair to bet $30 against $20 that it will succeed), the probability is $\frac{3}{3 + 2} = 0.6$ that the business venture will succeed. Show that if subjective probabilities are determined in this way, they satisfy
   (a)   Postulate 1 on page 43;
   (b)   Postulate 2.
   See also Exercise 2.56.


## APPLICATIONS

**2.35** An experiment has five possible outcomes, $A$, $B$, $C$, $D$, and $E$, which are mutually exclusive. Check whether the following assignments of probabilities are permissible and explain your answers:
   (a)   $P(A) = 0.20$, $P(B) = 0.20$, $P(C) = 0.20$, $P(D) = 0.20$, and $P(E) = 0.20$;
   (b)   $P(A) = 0.21$, $P(B) = 0.26$, $P(C) = 0.58$, $P(D) = 0.01$, and $P(E) = 0.06$;
   (c)   $P(A) = 0.18$, $P(B) = 0.19$, $P(C) = 0.20$, $P(D) = 0.21$, and $P(E) = 0.22$;

(d) $P(A) = 0.10$, $P(B) = 0.30$, $P(C) = 0.10$, $P(D) = 0.60$, and $P(E) = -0.10$;

(e) $P(A) = 0.23$, $P(B) = 0.12$, $P(C) = 0.05$, $P(D) = 0.50$, and $P(E) = 0.08$.

**2.36** If $A$ and $B$ are mutually exclusive, $P(A) = 0.37$, and $P(B) = 0.44$, find

|         |                    |
|---------|--------------------|
| (a) $P(A')$;        | (b) $P(B')$;         |
| (c) $P(A \cup B)$;  | (d) $P(A \cap B)$;   |
| (e) $P(A \cap B')$; | (f) $P(A' \cap B')$. |

**2.37** Explain why there must be a mistake in each of the following statements:

(a) The probability that Jean will pass the bar examination is 0.66 and the probability that she will not pass is $-0.34$.

(b) The probability that the home team will win an upcoming football game is 0.77, the probability that it will tie the game is 0.08, and the probability that it will win or tie the game is 0.95.

(c) The probabilities that a secretary will make 0, 1, 2, 3, 4, or 5 *or more* mistakes in typing a report are, respectively, 0.12, 0.25, 0.36, 0.14, 0.09, and 0.07.

(d) The probabilities that a bank will get 0, 1, 2, or 3 *or more* bad checks on any given day are, respectively, 0.08, 0.21, 0.29, and 0.40.

**2.38** Suppose that each of the 30 points of the sample space of Exercise 2.10 is assigned the probability $\frac{1}{30}$. Find the probabilities that at a given moment

(a) at least one of the station wagons is empty;

(b) each of the two station wagons carries the same number of passengers;

(c) the larger station wagon carries more passengers than the smaller station wagon;

(d) together they carry at least six passengers.

**2.39** The probabilities that the serviceability of a new X-ray machine will be rated very difficult, difficult, average, easy, or very easy are, respectively, 0.12, 0.17, 0.34, 0.29, and 0.08. Find the probabilities that the serviceability of the machine will be rated

(a) difficult or very difficult;

(b) neither very difficult nor very easy;

(c) average or worse;

(d) average or better.

**2.40** A police department needs new tires for its patrol cars and the probabilities are 0.15, 0.24, 0.03, 0.28, 0.22, and 0.08 that it will buy Uniroyal tires, Goodyear tires, Michelin tires, General tires, Goodrich tires, or Armstrong tires. Find the probabilities that it will buy

(a) Goodyear or Goodrich tires;

(b) Uniroyal, Michelin, or Goodrich tires;

(c) Michelin or Armstrong tires;

(d) Uniroyal, Michelin, General, or Goodrich tires.

**2.41** A hat contains twenty white slips of paper numbered from 1 through 20, ten red slips of paper numbered from 1 through 10, forty yellow slips of paper numbered from 1 through 40, and ten blue slips of paper numbered from 1 through 10. If these 80 slips of paper are thoroughly shuffled so that each slip has the same probability of being drawn, find the probabilities of drawing a slip of paper which is

(a)   blue or white;

(b)   numbered 1, 2, 3, 4, or 5;

(c)   red or yellow and numbered 1, 2, 3, or 4;

(d)   numbered 5, 15, 25, or 35;

(e)   white and numbered higher than 12 or yellow and numbered higher than 26.

**2.42** Four candidates are seeking a vacancy on a school board. If $A$ is twice as likely to be elected as $B$, and $B$ and $C$ are given about the same chance of being elected, while $C$ is twice as likely to be elected as $D$, what are the probabilities that

(a)   $C$ will win;

(b)   $A$ will not win?

**2.43** Two cards are randomly drawn from a deck of 52 playing cards. Find the probability that both cards will be greater than 3 and less than 8.

**2.44** In a poker game five cards are dealt at random from an ordinary deck of 52 playing cards. Find the probabilities of getting

(a)   two pairs (any two distinct face values occurring exactly twice);

(b)   four of a kind (four cards of equal face value).

**2.45** In a game of Yahtzee five balanced dice are rolled simultaneously. Find the probabilities of getting

(a)   two pairs;

(b)   three of a kind;

(c)   a full house (three of a kind and a pair);

(d)   four of a kind.

**2.46** Among the 78 doctors on the staff of a hospital, 64 carry malpractice insurance, 36 are surgeons, and 34 of the surgeons carry malpractice insurance. If one of these doctors is chosen by lot to represent the hospital staff at an A.M.A. convention (that is, each doctor has a probability of $\frac{1}{78}$ of being selected), what is the probability that the one chosen is not a surgeon and does not carry malpractice insurance?

**2.47** Explain on the basis of the various rules of Exercises 2.23 through 2.27 why there is a mistake in each of the following statements:

(a)   The probability that it will rain is 0.67 and the probability that it will rain or snow is 0.55.

(b)   The probability that a student will get a passing grade in English is 0.82 and the probability that she will get a passing grade in English and French is 0.86.

(c)   The probability that a person visiting the San Diego Zoo will see the giraffes is 0.72, the probability that he will see the bears is 0.84, and the probability that he will see both is 0.52.

**2.48** Given $P(A) = 0.59$, $P(B) = 0.30$, and $P(A \cap B) = 0.21$, find

(a)   $P(A \cup B)$;  

(b)   $P(A \cap B')$;

(c)   $P(A' \cup B')$;  

(d)   $P(A' \cap B')$.

**2.49** For married couples living in a certain suburb, the probability that the husband will vote in a school board election is 0.21, the probability that the wife will vote in the election is 0.28, and the probability that they will both vote is 0.15. What is the probability that at least one of them will vote?

**2.50** A biology professor has two graduate assistants helping her with her research. The probability that the older of the two assistants will be absent on any given day is 0.08, the probability that the younger of the two will be absent on any given day is 0.05, and the probability that they will both be absent on any given day is 0.02. Find the probabilities that

(a)   either or both of the graduate assistants will be absent on any given day;

(b)   at least one of the two graduate assistants will not be absent on any given day;

(c)   only one of the two graduate assistants will be absent on any given day.

**2.51** At Roanoke College it is known that $\frac{1}{3}$ of the students live off campus. It is also known that $\frac{5}{9}$ of the students are from within the state of Virginia and that $\frac{3}{4}$ of the students are from out-of-state or live on campus. What is the probability that a student selected at random from Roanoke College is from out-of-state and lives on campus?

**2.52** Suppose that if a person visits Disneyland, the probability that he will go on the Jungle Cruise is 0.74, the probability that he will ride the Monorail is 0.70, the probability that he will go on the Matterhorn ride is 0.62, the probability that he will go on the Jungle Cruise and ride the Monorail is 0.52, the probability that he will go on the Jungle Cruise as well as the Matterhorn ride is 0.46, the probability that he will ride the Monorail and go on the Matterhorn ride is 0.44, and the probability that he will go on all three of these rides is 0.34. What is the probability that a person visiting Disneyland will go on at least one of these three rides?

**2.53** Suppose that if a person travels to Europe for the first time, the probability that he will see London is 0.70, the probability that he will see Paris is 0.64, the probability that he will see Rome is 0.58, the probability that he will see Amsterdam is 0.58, the probability that he will see London and Paris is 0.45, the probability that he will see London and Rome is 0.42, the probability that he will see London and Amsterdam is 0.41, the probability that he will see Paris and Rome is 0.35, the probability that he will see Paris and Amsterdam is 0.39, the probability that he will see Rome and Amsterdam is 0.32, the probability that he will see London, Paris, and Rome is 0.23, the probability that he will see London, Paris, and Amsterdam is 0.26, the probability that he will see London, Rome, and Amsterdam is 0.21, the probability that he will see Paris, Rome, and Amsterdam is 0.20, and the probability that he will see all four of these cities is 0.12. What is the probability that a person traveling to Europe for the first time will see at least one of these four cities? (*Hint*: Use the formula of Exercise 2.31.)

**2.54** Use the formula of Exercise 2.33 to convert each of the following odds to probabilities:

(a) If three eggs are randomly chosen from a carton of twelve eggs of which three are cracked, the odds are 34 to 21 that at least one of them will be cracked.

(b) If a person has eight $1 bills, five $5 bills, and one $20 bill, and randomly selects three of them, the odds are 11 to 2 that they will not all be $1 bills.

(c) If we arbitrarily arrange the letters in the word "nest," the odds are 5 to 1 that we will not get a meaningful word in the English language.

**2.55** Use the definition of "odds" given in Exercise 2.33 to convert each of the following probabilities to odds:

(a) The probability that the last digit of a car's license plate is a 2, 3, 4, 5, 6, or 7 is $\frac{6}{10}$.

(b) The probability of getting at least two heads in four flips of a balanced coin is $\frac{11}{16}$.

(c) The probability of rolling "7 or 11" with a pair of balanced dice is $\frac{2}{9}$.

**2.56** If subjective probabilities are determined by the method suggested in Exercise 2.34, the third postulate of probability may not be satisfied. However, proponents of the subjective probability concept usually impose this postulate as a **consistency criterion**; in other words, they regard subjective probabilities which do not satisfy the postulate as inconsistent.

(a) A high school principal feels that the odds are 7 to 5 against her getting a $1,000 raise and 11 to 1 against her getting a $2,000 raise. Furthermore, she feels that it is an even-money bet that she will get one of these raises or the other. Discuss the consistency of the corresponding subjective probabilities.

(b) Asked about his political future, a party official replies that the odds are 2 to 1 that he will not run for the House of Representatives and 4 to 1 that he will not run for the Senate. Furthermore, he feels that the odds are 7 to 5 that he will run for one or the other. Are the corresponding probabilities consistent?

**2.57** There are two Porsches in a road race in Italy, and a reporter feels that the odds against their winning are 3 to 1 and 5 to 3. To be consistent (see preceding exercise), what odds should the reporter assign to the event that either car will win?

## 2.6   CONDITIONAL PROBABILITY

Difficulties can easily arise when probabilities are quoted without specification of the sample space. For instance, if we ask for the probability that a lawyer makes more than $50,000 per year, we may well get several different answers, and they may all be correct. One of them might apply to all law school graduates, another might apply to all persons licensed to practice law, a third might apply to all those who are actively engaged in the practice of law, and so forth. Since the choice of the sample space (namely, the set of all possibilities under consideration) is by no means always self-evident, it often helps to use the symbol $P(A|S)$ to denote the **conditional probability** of event $A$ relative to the sample space $S$, or as we also call it "the probability of $A$ given $S$." The symbol $P(A|S)$ makes it explicit that we are referring to a particular sample space $S$, and it is preferable to the abbreviated notation $P(A)$ unless the tacit choice of $S$ is clearly understood. It is also preferable when we want to refer to several sample spaces in the same example. If $A$ is the event that a person makes more than $50,000 per year, $G$ is the event that a person is a law school graduate, $L$ is the event that a person is licensed to practice law, and $E$ is the event that a person is actively engaged in the practice of law, then $P(A|G)$ is the probability that a law school graduate makes more than $50,000 per year, $P(A|L)$ is the probability that a person licensed to practice law makes more than $50,000 per year, and $P(A|E)$ is the probability that a person actively engaged in the practice of law makes more than $50,000 per year.

Some ideas connected with conditional probabilities are illustrated in the following example.

## EXAMPLE 2.15

A consumer research organization has studied the services under warranty provided by the 50 new-car dealers in a certain city, and its findings are summarized in the following table.

|  | Good service under warranty | Poor service under warranty |
|---|---|---|
| *In business 10 years or more* | 16 | 4 |
| *In business less than 10 years* | 10 | 20 |

If a person randomly selects one of these new-car dealers, what is the probability that he gets one who provides good service under warranty? Also, if a person randomly selects one of the dealers who has been in business for 10 years or more, what is the probability that he gets one who provides good service under warranty?

*Solution*

By "randomly" we mean that, in each case, all possible selections are equally likely, and we can therefore use the formula of Theorem 2.2. If we let $G$ denote the selection of a dealer who provides good service under warranty, and if we let $n(G)$ denote the number of elements in $G$, and $n(S)$ the number of elements in the whole sample space, we get

$$P(G) = \frac{n(G)}{n(S)} = \frac{16 + 10}{50} = 0.52$$

This answers the first question.

For the second question, we limit ourselves to the reduced sample space which consists of the first line of the table, namely, the $16 + 4 = 20$ dealers who have been in business 10 years or more. Of these, 16 provide good service under warranty, and we get

$$P(G|T) = \frac{16}{20} = 0.80$$

where $T$ denotes the selection of a dealer who has been in business 10 years or more. This answers the second question, and as should have been expected, $P(G|T)$ is considerably higher than $P(G)$.    ▲

Since the numerator of $P(G|T)$ is $n(T \cap G) = 16$ in the preceding example, the number of dealers who have been in business for 10 years or more and provide good service under warranty, and the denominator is $n(T)$, the number of dealers who have been in business 10 years or more, we can write symbolically

$$P(G|T) = \frac{n(T \cap G)}{n(T)}$$

Then, if we divide the numerator and the denominator by $n(S)$, the total number of new-car dealers in the given city, we get

$$P(G|T) = \frac{\dfrac{n(T \cap G)}{n(S)}}{\dfrac{n(T)}{n(S)}} = \frac{P(T \cap G)}{P(T)}$$

and we have, thus, expressed the conditional probability $P(G|T)$ in terms of two probabilities defined for the whole sample space $S$.

Generalizing from the above, let us now make the following definition of conditional probability.

---

**DEFINITION 2.1**  If $A$ and $B$ are any two events in a sample space $S$ and $P(A) \neq 0$, the **conditional probability** of $B$ given $A$ is

$$P(B|A) = \frac{P(A \cap B)}{P(A)}$$

---

## EXAMPLE 2.16

With reference to Example 2.15, what is the probability that one of the dealers who has been in business less than 10 years will provide good service under warranty?

*Solution*

Since $P(T' \cap G) = \dfrac{10}{50} = 0.20$ and $P(T') = \dfrac{10 + 20}{50} = 0.60$, substitution into the formula yields

$$P(G|T') = \frac{P(T' \cap G)}{P(T')} = \frac{0.20}{0.60} = \frac{1}{3} \qquad \blacktriangle$$

Although we introduced the formula for $P(B|A)$ by means of an example in which the possibilities were all equally likely, this is not a requirement for its use.

## EXAMPLE 2.17

With reference to the loaded die of Example 2.9, what is the probability that the number of points rolled is a perfect square? Also, what is the probability that it is a perfect square given that it is greater than 3?

*Solution*

If $A$ is the event that the number of points rolled is greater than 3 and $B$ is the event that it is a perfect square, we have $A = \{4, 5, 6\}$, $B = \{1, 4\}$, and $A \cap B = \{4\}$. Since the probabilities of rolling a 1, 2, 3, 4, 5, or 6 with the die are $\frac{2}{9}, \frac{1}{9}, \frac{2}{9}, \frac{1}{9}, \frac{2}{9}$, and $\frac{1}{9}$ (see page 45), we find that the answer to the first question is

$$P(B) = \tfrac{2}{9} + \tfrac{1}{9} = \tfrac{1}{3}$$

To determine $P(B|A)$, we first calculate

$$P(A \cap B) = \tfrac{1}{9} \quad \text{and} \quad P(A) = \tfrac{1}{9} + \tfrac{2}{9} + \tfrac{1}{9} = \tfrac{4}{9}$$

Then, substituting into the formula of Definition 2.1, we get

$$P(B|A) = \frac{P(A \cap B)}{P(A)} = \frac{\tfrac{1}{9}}{\tfrac{4}{9}} = \frac{1}{4} \quad \blacktriangle$$

To verify that the formula of Definition 2.1 has yielded the "right" answer in the preceding example, we have only to assign probability $v$ to the two even numbers in the reduced sample space $A$ and probability $2v$ to the odd number, such that the sum of the three probabilities is equal to 1. We thus have $v + 2v + v = 1$, $v = \tfrac{1}{4}$, and, hence, $P(B|A) = \tfrac{1}{4}$ as before.

## EXAMPLE 2.18

A manufacturer of airplane parts knows from past experience that the probability is 0.80 that an order will be ready for shipment on time, and it is 0.72 that an order will be ready for shipment on time and will also be delivered on time. What is the probability that such an order will be delivered on time given that it was ready for shipment on time?

*Solution*

If we let $R$ stand for the event that an order is ready for shipment on time and $D$ for the event that it is delivered on time, we have $P(R) = 0.80$ and $P(R \cap D) = 0.72$, and it follows that

$$P(D|R) = \frac{P(R \cap D)}{P(R)} = \frac{0.72}{0.80} = 0.90$$

Thus, 90 percent of the shipments will be delivered on time provided they are shipped on time. Note that $P(R|D)$, the probability that a shipment which

is delivered on time was also ready for shipment on time, cannot be determined without further information; for this purpose we would also have to
know $P(D)$.    ▲

If we multiply the expressions on both sides of the formula of Definition
2.1 by $P(A)$, we obtain the following **multiplication rule**.

---

**THEOREM 2.9**   If $A$ and $B$ are any two events in a sample space $S$ and $P(A) \neq 0$,
then

$$P(A \cap B) = P(A) \cdot P(B|A)$$

---

In words, the probability that $A$ and $B$ will both occur is the product of the probability of $A$ and the conditional probability of $B$ given $A$. Alternatively, if $P(B) \neq 0$,
the probability that $A$ and $B$ will both occur is the product of the probability of
$B$ and the conditional probability of $A$ given $B$; symbolically,

$$P(A \cap B) = P(B) \cdot P(A|B)$$

To derive this alternative multiplication rule, we interchange $A$ and $B$ in the formula of Theorem 2.9 and make use of the fact that $A \cap B = B \cap A$.

## EXAMPLE 2.19

If we randomly pick two television tubes in succession from a shipment of 240
television tubes of which 15 are defective, what is the probability that they will
both be defective?

### Solution

If we assume equal probabilities for each selection (which is what we mean
by "randomly" picking the tubes), the probability that the first tube will be
defective is $\frac{15}{240}$, and the probability that the second tube will be defective
given that the first tube is defective is $\frac{14}{239}$. Thus, the probability that both
tubes will be defective is $\frac{15}{240} \cdot \frac{14}{239} = \frac{7}{1,912}$. This assumes that we are **sampling
without replacement**, namely, that the first tube is not replaced before the
second tube is selected.    ▲

## EXAMPLE 2.20

Find the probabilities of randomly drawing two aces in succession from an ordinary deck of 52 playing cards, if we sample

(a)  without replacement;

(b)  with replacement.

*Solution*

(a)  If the first card is not replaced before the second card is drawn, the probability of getting two aces in succession is

$$\frac{4}{52} \cdot \frac{3}{51} = \frac{1}{221}$$

(b)  If the first card is replaced before the second card is drawn, the corresponding probability is

$$\frac{4}{52} \cdot \frac{4}{52} = \frac{1}{169} \quad \blacktriangle$$

In the situations described in the two preceding examples there is a definite temporal order between the two events $A$ and $B$. In general, this need not be the case when we write $P(A|B)$ or $P(B|A)$. For instance, we could ask for the probability that the first card drawn was an ace given that the second card drawn (without replacement) is an ace—the answer would also be $\frac{3}{51}$.

Theorem 2.9 can easily be generalized so that it applies to more than two events; for instance, for three events we have

---

**THEOREM 2.10**  If $A$, $B$, and $C$ are any three events in a sample space $S$ such that $P(A \cap B) \neq 0$, then

$$P(A \cap B \cap C) = P(A) \cdot P(B|A) \cdot P(C|A \cap B)$$

---

***Proof.***  Writing $A \cap B \cap C$ as $(A \cap B) \cap C$ and using the formula of Theorem 2.9 twice, we get

$$P(A \cap B \cap C) = P[(A \cap B) \cap C]$$

$$= P(A \cap B) \cdot P(C|A \cap B)$$

$$= P(A) \cdot P(B|A) \cdot P(C|A \cap B) \quad \blacktriangledown$$

## EXAMPLE 2.21

A box of fuses contains 20 fuses, of which 5 are defective. If 3 of the fuses are selected at random and removed from the box in succession without replacement, what is the probability that all three fuses are defective?

### Solution

If $A$ is the event that the first fuse is defective, $B$ is the event that the second fuse is defective, and $C$ is the event that the third fuse is defective, then $P(A) = \frac{5}{20}$, $P(B|A) = \frac{4}{19}$, $P(C|A \cap B) = \frac{3}{18}$, and substitution into the formula yields

$$P(A \cap B \cap C) = \frac{5}{20} \cdot \frac{4}{19} \cdot \frac{3}{18}$$
$$= \frac{1}{114} \quad \blacktriangle$$

Further generalization of Theorems 2.9 and 2.10 to $k$ events is straightforward, and the resulting formula can be proved by mathematical induction.

## 2.7 INDEPENDENT EVENTS

Informally speaking, two events $A$ and $B$ are **independent** if the occurrence or nonoccurrence of either one does not affect the probability of the occurrence of the other. For instance, in the preceding example the selections would all have been independent had each fuse been replaced before the next one was selected; the probability of getting a defective fuse would have remained $\frac{5}{20}$.

Symbolically, two events $A$ and $B$ are independent if $P(B|A) = P(B)$ and $P(A|B) = P(A)$, and it can be shown that either of these equalities implies the other when both of the conditional probabilities exist, namely, when neither $P(A)$ nor $P(B)$ equals zero (see Exercise 2.62).

Now, if we substitute $P(B)$ for $P(B|A)$ into the formula of Theorem 2.9, we get

$$P(A \cap B) = P(A) \cdot P(B|A)$$
$$= P(A) \cdot P(B)$$

and we shall use this as our formal definition of independence.

---

**DEFINITION 2.2** Two events $A$ and $B$ are **independent** if and only if

$$P(A \cap B) = P(A) \cdot P(B)$$

---

Reversing the steps, we can also show that Definition 2.2 implies the definition of independence that we earlier gave on page 66.

If two events are not independent, they are said to be **dependent**. In the derivation of the formula of Definition 2.2 we assume that $P(B|A)$ exists and, hence, that $P(A) \neq 0$. For mathematical convenience, we shall let the definition apply also when $P(A) = 0$ and/or $P(B) = 0$.

## EXAMPLE 2.22

A coin is tossed three times and the eight possible outcomes, HHH, HHT, HTH, THH, HTT, THT, TTH, and TTT, are assumed to be equally likely. If $A$ is the event that a head occurs on each of the first two tosses, $B$ is the event that a tail occurs on the third toss, and $C$ is the event that exactly two tails occur in the three tosses, show that

    (a)   events $A$ and $B$ are independent;

    (b)   events $B$ and $C$ are dependent.

### Solution

Since

$$A = \{HHH, HHT\}$$

$$B = \{HHT, HTT, THT, TTT\}$$

$$C = \{HTT, THT, TTH\}$$

$$A \cap B = \{HHT\}$$

$$B \cap C = \{HTT, THT\}$$

the assumption that the eight possible outcomes are all equiprobable yields $P(A) = \frac{1}{4}$, $P(B) = \frac{1}{2}$, $P(C) = \frac{3}{8}$, $P(A \cap B) = \frac{1}{8}$, and $P(B \cap C) = \frac{1}{4}$.

    (a)   Since $P(A) \cdot P(B) = \frac{1}{4} \cdot \frac{1}{2} = \frac{1}{8} = P(A \cap B)$, events $A$ and $B$ are independent.

    (b)   Since $P(B) \cdot P(C) = \frac{1}{2} \cdot \frac{3}{8} = \frac{3}{16} \neq P(B \cap C)$, events $B$ and $C$ are not independent.   ▲

In connection with Definition 2.2, it can be shown that if $A$ and $B$ are independent, then so are $A$ and $B'$, $A'$ and $B$, and $A'$ and $B'$. For instance,

---

**THEOREM 2.11** If $A$ and $B$ are independent, then $A$ and $B'$ are also independent.

**Proof.** Since $A = (A \cap B) \cup (A \cap B')$, as the reader was asked to show in part (a) of Exercise 2.4, $A \cap B$ and $A \cap B'$ are mutually exclusive, and $A$ and $B$ are independent by assumption, we have

$$P(A) = P[(A \cap B) \cup (A \cap B')]$$
$$= P(A \cap B) + P(A \cap B')$$
$$= P(A) \cdot P(B) + P(A \cap B')$$

It follows that

$$P(A \cap B') = P(A) - P(A) \cdot P(B)$$
$$= P(A) \cdot [1 - P(B)]$$
$$= P(A) \cdot P(B')$$

and hence that $A$ and $B'$ are independent.    ▼

In Exercises 2.63 and 2.64 the reader will be asked to show that if $A$ and $B$ are independent, then $A'$ and $B$ are independent and so are $A'$ and $B'$, and if $A$ and $B$ are dependent, then $A$ and $B'$ are dependent.

To extend the concept of independence to more than two events, let us make the following definition.

---

**DEFINITION 2.3**  Events $A_1, A_2, \ldots$ , and $A_k$ are **independent** if and only if the probability of the intersection of any 2, 3, $\ldots$ , or $k$ of these events equals the product of their respective probabilities.

---

For three events $A$, $B$, and $C$, for example, independence requires that

$$P(A \cap B) = P(A) \cdot P(B)$$
$$P(A \cap C) = P(A) \cdot P(C)$$
$$P(B \cap C) = P(B) \cdot P(C)$$

and

$$P(A \cap B \cap C) = P(A) \cdot P(B) \cdot P(C)$$

It is of interest to note that three or more events can be **pairwise independent** without being independent.

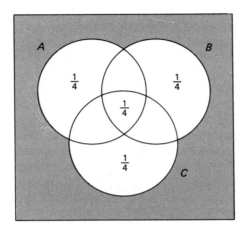

**Figure 2.11**   Venn diagram for Example 2.23.

## EXAMPLE 2.23

Figure 2.11 shows a Venn diagram with probabilities assigned to its various regions. Verify that $A$ and $B$ are independent, $A$ and $C$ are independent, and $B$ and $C$ are independent, but $A$, $B$, and $C$ are not independent.

### Solution

As can be seen from the diagram, $P(A) = P(B) = P(C) = \frac{1}{2}$, $P(A \cap B) = P(A \cap C) = P(B \cap C) = \frac{1}{4}$, and $P(A \cap B \cap C) = \frac{1}{4}$. Thus,

$$P(A) \cdot P(B) = \tfrac{1}{4} = P(A \cap B)$$

$$P(A) \cdot P(C) = \tfrac{1}{4} = P(A \cap C)$$

$$P(B) \cdot P(C) = \tfrac{1}{4} = P(B \cap C)$$

but

$$P(A) \cdot P(B) \cdot P(C) = \tfrac{1}{8} \neq P(A \cap B \cap C) \quad \blacktriangle$$

Incidentally, the preceding example can be given a "real" interpretation by considering a large room that has three separate switches controlling the ceiling lights. These lights will be on when all three switches are "up" and hence also when one of the switches is "up" and the other two are "down." If $A$ is the event that the first switch is "up," $B$ is the event that the second switch is "up," and $C$ is the event that the third switch is "up," the Venn diagram of Figure 2.11 shows a possible set of probabilities associated with the switches being "up" or "down" when the ceiling lights are on.

It can also happen that $P(A \cap B \cap C) = P(A) \cdot P(B) \cdot P(C)$ without $A$, $B$, and $C$ being pairwise independent—this the reader will be asked to verify in Exercise 2.65.

Of course, if we are given that certain events are independent, the probability that they will all occur is simply the product of their respective probabilities.

## EXAMPLE 2.24

Find the probabilities of getting

(a) three heads in three random tosses of a balanced coin;

(b) four sixes and then another number in five random rolls of a balanced die.

### Solution

(a) Multiplying the respective probabilities, we get

$$\frac{1}{2} \cdot \frac{1}{2} \cdot \frac{1}{2} = \frac{1}{8}$$

(b) Multiplying the respective probabilities, we get

$$\frac{1}{6} \cdot \frac{1}{6} \cdot \frac{1}{6} \cdot \frac{1}{6} \cdot \frac{5}{6} = \frac{5}{7,776} \quad \blacktriangle$$

## 2.8 BAYES' THEOREM

There are many situations where the outcome of an experiment depends on what happens in various intermediate stages. The following is a simple example in which there is one intermediate stage consisting of two alternatives:

## EXAMPLE 2.25

The completion of a construction job may be delayed because of a strike. The probabilities are 0.60 that there will be a strike, 0.85 that the construction job will be completed on time if there is no strike, and 0.35 that the construction job will be completed on time if there is a strike. What is the probability that the construction job will be completed on time?

### Solution

If $A$ is the event that the construction job will be completed on time and $B$ is the event that there will be a strike, we are given $P(B) = 0.60$,

$P(A|B') = 0.85$, and $P(A|B) = 0.35$. Making use of the formula of part (a) of Exercise 2.4, the fact that $A \cap B$ and $A \cap B'$ are mutually exclusive, and the alternative form of the multiplication rule, we can write

$$P(A) = P[(A \cap B) \cup (A \cap B')]$$
$$= P(A \cap B) + P(A \cap B')$$
$$= P(B) \cdot P(A|B) + P(B') \cdot P(A|B')$$

Then, substituting the given numerical values, we get

$$P(A) = (0.60)(0.35) + (1 - 0.60)(0.85)$$
$$= 0.55 \quad \blacktriangle$$

An immediate generalization of this kind of situation is the case where the intermediate stage permits $k$ different alternatives (whose occurrence is denoted by $B_1, B_2, \ldots, B_k$). It requires the following theorem, sometimes called the **rule of total probability** or the **rule of elimination**.

---

**THEOREM 2.12**  If the events $B_1, B_2, \ldots$, and $B_k$ constitute a partition of the sample space $S$ and $P(B_i) \neq 0$ for $i = 1, 2, \ldots, k$, then for any event $A$ in $S$

$$P(A) = \sum_{i=1}^{k} P(B_i) \cdot P(A|B_i)$$

---

As was defined in the footnote to page 12, the $B$'s constitute a partition of the sample space if they are pairwise mutually exclusive and if their union equals $S$. A formal proof of Theorem 2.12 consists, essentially, of the same steps we used in Example 2.25, and it is left to the reader in Exercise 2.71.

## EXAMPLE 2.26

The members of a consulting firm rent cars from three rental agencies: 60 percent from agency 1, 30 percent from agency 2, and 10 percent from agency 3. If 9 percent of the cars from agency 1 need a tune-up, 20 percent of the cars from agency 2 need a tune-up, and 6 percent of the cars from agency 3 need a tune-up, what is the probability that a rental car delivered to the firm will need a tune-up?

*Solution*

If $A$ is the event that the car needs a tune-up, and $B_1$, $B_2$, and $B_3$ are the events that the car comes from rental agencies 1, 2, or 3, we have $P(B_1) = 0.60$, $P(B_2) = 0.30$, $P(B_3) = 0.10$, $P(A|B_1) = 0.09$, $P(A|B_2) = 0.20$, and $P(A|B_3) = 0.06$. Substituting these values into the formula of Theorem 2.12, we get

$$P(A) = (0.60)(0.09) + (0.30)(0.20) + (0.10)(0.06)$$

$$= 0.12$$

Thus, 12 percent of all the rental cars delivered to this firm will need a tune-up.   ▲

With reference to the preceding example, suppose that we are interested in the following question: If a rental car delivered to the consulting firm needs a tune-up, what is the probability that it came from rental agency 2? To answer questions of this kind, we need the following theorem, called **Bayes' theorem**:

THEOREM 2.13   If $B_1$, $B_2$, ... , $B_k$ constitute a partition of the sample space $S$ and $P(B_i) \neq 0$ for $i = 1, 2, \ldots , k$, then for any event $A$ in $S$ such that $P(A) \neq 0$

$$P(B_r|A) = \frac{P(B_r) \cdot P(A|B_r)}{\sum_{i=1}^{k} P(B_i) \cdot P(A|B_i)}$$

for $r = 1, 2, \ldots , k$.

In words, the probability that event $A$ was reached via the $r$th branch of the tree diagram of Figure 2.12, given that it was reached via one of its $k$ branches, is the *ratio* of the probability associated with the $r$th branch to the sum of the probabilities associated with all $k$ branches of the tree.

***Proof.***   Writing $P(B_r|A) = \dfrac{P(A \cap B_r)}{P(A)}$ in accordance with the definition of conditional probability, we have only to substitute $P(B_r) \cdot P(A|B_r)$ for $P(A \cap B_r)$ and the formula of Theorem 2.12 for $P(A)$.   ▼

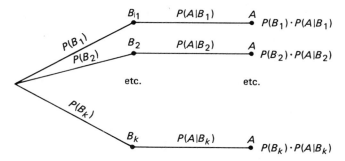

**Figure 2.12**  Tree diagram for Bayes' theorem.

## EXAMPLE 2.27

With reference to Example 2.26, if a rental car delivered to the consulting firm needs a tune-up, what is the probability that it came from rental agency 2?

### Solution

Substituting the probabilities on page 72 into the formula of Theorem 2.13, we get

$$P(B_2|A) = \frac{(0.30)(0.20)}{(0.60)(0.09) + (0.30)(0.20) + (0.10)(0.06)}$$

$$= \frac{0.060}{0.120}$$

$$= 0.5$$

Observe that although only 30 percent of the cars delivered to the firm come from agency 2, 50 percent of those requiring a tune-up come from that agency.    ▲

## EXAMPLE 2.28

In a certain state, 25 percent of all cars emit excessive amounts of pollutants. If the probability is 0.99 that a car emitting excessive amounts of pollutants will fail the state's vehicular emission test, and the probability is 0.17 that a car not emitting excessive amounts of pollutants will nevertheless fail the test, what is the probability that a car which fails the test actually emits excessive amounts of pollutants?

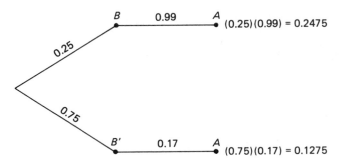

**Figure 2.13** Tree diagram for Example 2.28.

## Solution

Picturing this situation as in Figure 2.13, we find that the probabilities associated with the two branches of the tree diagram are $(0.25)(0.99) = 0.2475$ and $(1 - 0.25)(0.17) = 0.1275$. Thus, the probability that a car which fails the test actually emits excessive amounts of pollutants is

$$\frac{0.2475}{0.2475 + 0.1275} = 0.66$$

Of course, this result could also have been obtained without the diagram, by substituting directly into the formula of Bayes' theorem. ▲

Although Bayes' theorem follows from the postulates of probability and the definition of conditional probability, it has been the subject of extensive controversy. There can be no question about the validity of Bayes' theorem, but considerable arguments have been raised about the assignment of the **prior probabilities** $P(B_i)$. Also, a good deal of mysticism surrounds Bayes' theorem because it entails a "backward," or "inverse," sort of reasoning, namely, reasoning "from effect to cause." For instance, in Example 2.28 failing the test is the effect and emitting excessive amounts of pollutants is a possible cause.

## EXERCISES

**2.58** Show that the postulates of probability are satisfied by conditional probabilities. In other words, show that if $P(B) \neq 0$, then

    (a)  $P(A|B) \geqslant 0$;

    (b)  $P(B|B) = 1$;

    (c)  $P(A_1 \cup A_2 \cup \cdots |B) = P(A_1|B) + P(A_2|B) + \cdots$ for any sequence of mutually exclusive events $A_1, A_2, \ldots$ .

**2.59** Show by means of numerical examples that $P(B|A) + P(B|A')$
  (a)  may be equal to 1;
  (b)  need not be equal to 1.

**2.60** Duplicating the method of proof of Theorem 2.10, show that

$$P(A \cap B \cap C \cap D) = P(A) \cdot P(B|A) \cdot P(C|A \cap B) \cdot P(D|A \cap B \cap C)$$

  provided that $P(A \cap B \cap C) \neq 0$.

**2.61** Given three events $A$, $B$, and $C$ such that $P(A \cap B \cap C) \neq 0$ and $P(C|A \cap B) = P(C|B)$, show that $P(A|B \cap C) = P(A|B)$.

**2.62** Show that if $P(B|A) = P(B)$ and $P(B) \neq 0$, then $P(A|B) = P(A)$.

**2.63** Show that if events $A$ and $B$ are independent, then
  (a)  events $A'$ and $B$ are independent;
  (b)  events $A'$ and $B'$ are independent.

**2.64** Show that if events $A$ and $B$ are dependent, then events $A$ and $B'$ are dependent.

**2.65** Refer to Figure 2.14 to show that $P(A \cap B \cap C) = P(A) \cdot P(B) \cdot P(C)$ does not necessarily imply that $A$, $B$, and $C$ are all pairwise independent.

**2.66** Refer to Figure 2.14 to show that if $A$ is independent of $B$ and $A$ is independent of $C$, then $B$ is not necessarily independent of $C$.

**2.67** Refer to Figure 2.14 to show that if $A$ is independent of $B$ and $A$ is independent of $C$, then $A$ is not necessarily independent of $B \cup C$.

**2.68** If events $A$, $B$, and $C$ are independent, show that
  (a)  $A$ and $B \cap C$ are independent;
  (b)  $A$ and $B \cup C$ are independent.

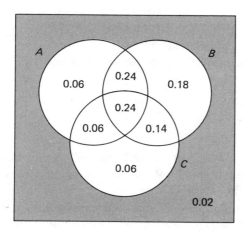

**Figure 2.14**  Diagram for Exercises 2.65, 2.66, and 2.67.

**2.69**  Show that $2^k - k - 1$ conditions must be satisfied for $k$ events to be independent.

**2.70**  For any event $A$, show that $A$ and $\varnothing$ are independent.

**2.71**  Prove Theorem 2.12 by making use of the following generalization of the distributive law given in part (b) of Exercise 2.1:

$$A \cap (B_1 \cup B_2 \cup \cdots \cup B_k) = (A \cap B_1) \cup (A \cap B_2) \cup \cdots \cup (A \cap B_k)$$

## APPLICATIONS

**2.72**  There are 90 applicants for a job with the news department of a television station. Some of them are college graduates and some are not, some of them have at least three years' experience and some have not, with the exact breakdown being

|  | *College graduates* | *Not college graduates* |
|---|---|---|
| *At least three years' experience* | 18 | 9 |
| *Less than three years' experience* | 36 | 27 |

If the order in which the applicants are interviewed by the station manager is random, $G$ is the event that the first applicant interviewed is a college graduate, and $T$ is the event that the first applicant interviewed has at least three years' experience, determine each of the following probabilities directly from the entries and the row and column totals of the table:

(a)  $P(G)$;          (b)  $P(T')$;

(c)  $P(G \cap T)$;          (d)  $P(G' \cap T')$;

(e)  $P(T|G)$;          (f)  $P(G'|T')$.

**2.73**  Use the results of the preceding exercise to verify that

(a)  $P(T|G) = \dfrac{P(G \cap T)}{P(G)}$;

(b)  $P(G'|T') = \dfrac{P(G' \cap T')}{P(T')}$.

**2.74**  With reference to Exercise 2.46, what is the probability that the doctor chosen to represent the hospital staff at the convention carries malpractice insurance given that he or she is a surgeon?

**2.75**  With reference to Exercise 2.49, what is the probability that a husband will vote in the election given that his wife is going to vote?

**2.76**  With reference to Exercise 2.51, what is the probability that one of the students will be living on campus given that he or she is from out-of-state?

**2.77**  It is felt that the probabilities are 0.20, 0.40, 0.30, and 0.10 that the basketball teams of four universities, $T$, $U$, $V$, and $W$, will win their conference championship. If university $U$ is placed on probation and declared ineligible for the championship, what is the probability that university $T$ will win the conference championship?

**2.78**  With reference to Exercise 2.52, find the probabilities that a person who visits Disneyland will

(a)  ride the Monorail given that he will go on the Jungle Cruise;
(b)  go on the Matterhorn ride given that he will go on the Jungle Cruise and ride the Monorail;
(c)  not go on the Jungle Cruise given that he will ride the Monorail and/or go on the Matterhorn ride;
(d)  go on the Matterhorn ride and the Jungle Cruise given that he will not ride the Monorail.

(*Hint*: Draw a Venn diagram and fill in the probabilities associated with the various regions.)

**2.79**  The probability of surviving a certain transplant operation is 0.55. If a patient survives the operation, the probability that his or her body will reject the transplant within a month is 0.20. What is the probability of surviving both of these critical stages?

**2.80**  Crates of eggs are inspected for blood clots by randomly removing three eggs in succession and examining their contents. If all three eggs are good, the crate is shipped; otherwise it is rejected. What is the probability that a crate will be shipped if it contains 120 eggs, of which 10 have blood clots?

**2.81**  Suppose that in Vancouver, B.C., the probability that a rainy fall day is followed by a rainy day is 0.80 and the probability that a sunny fall day is followed by a rainy day is 0.60. Find the probabilities that a rainy fall day is followed by

(a)  a rainy day, a sunny day, and another rainy day;
(b)  two sunny days and then a rainy day;
(c)  two rainy days and then two sunny days;
(d)  rain two days later.

[*Hint*: In part (c) use the formula of Exercise 2.60.]

**2.82**  Use the formula of Exercise 2.60 to find the probability of randomly choosing (without replacement) four healthy guinea pigs from a cage containing 20 guinea pigs, of which 15 are healthy and 5 are diseased.

**2.83**  A balanced die is tossed twice. If $A$ is the event that an even number comes up on the first toss, $B$ is the event that an even number comes up on the second toss, and $C$ is the event that both tosses result in the same number, are the events $A$, $B$, and $C$

(a)  pairwise independent;
(b)  independent?

**2.84** A sharpshooter hits a target with probability 0.75. Assuming independence, find the probabilities of getting

(a)  a hit followed by two misses;

(b)  two hits and a miss in any order.

**2.85** A coin is loaded so that the probabilities of heads and tails are 0.52 and 0.48, respectively. If the coin is tossed three times, what are the probabilities of getting

(a)  all heads;

(b)  two tails and a head in that order?

**2.86** Medical records show that one out of 10 persons in a certain town has a thyroid deficiency. If 12 persons in this town are randomly chosen and tested, what is the probability that at least one of them will have a thyroid deficiency?

**2.87** If five of a company's 10 delivery trucks do not meet emission standards and three of them are chosen for inspection, what is the probability that none of the trucks chosen will meet emission standards?

**2.88** If a person randomly picks four of the 15 gold coins a dealer has in stock, and six of the coins are counterfeits, what is the probability that the coins picked will all be counterfeits?

**2.89** A department store which bills its charge-account customers once a month has found that if a customer pays promptly one month, the probability is 0.90 that he or she will also pay promptly the next month; however, if a customer does not pay promptly one month, the probability that he or she will pay promptly the next month is only 0.40.

(a)  What is the probability that a customer who pays promptly one month will also pay promptly the next three months?

(b)  What is the probability that a customer who does not pay promptly one month will also not pay promptly the next two months and then make a prompt payment the month after that?

**2.90** With reference to Figure 2.15, verify that events $A$, $B$, $C$, and $D$ are independent. Note that the region representing $A$ consists of two circles, and so do the regions representing $B$ and $C$.

**2.91** At an electronics plant, it is known from past experience that the probability is 0.84 that a new worker who has attended the company's training program will meet the production quota, and that the corresponding probability is 0.49 for a new worker who has not attended the company's training program. If 70 percent of all new workers attend the training program, what is the probability that a new worker will meet the production quota?

**2.92** In a T-maze, a rat is given food if it turns left and an electric shock if it turns right. On the first trial there is a 50-50 chance that a rat will turn either way; then, if it receives food on the first trial the probability is 0.68 that it will turn left on the next trial, and if it receives a shock on the first

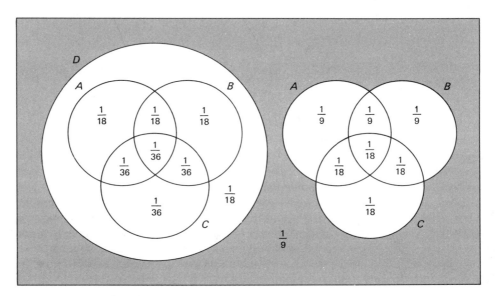

**Figure 2.15** Diagram for Exercise 2.90.

trial the probability is 0.84 that it will turn left on the next trial. What is the probability that a rat will turn left on the second trial?

**2.93** It is known from experience that in a certain industry 60 percent of all labor-management disputes are over wages, 15 percent are over working conditions, and 25 percent are over fringe issues. Also, 45 percent of the disputes over wages are resolved without strikes, 70 percent of the disputes over working conditions are resolved without strikes, and 40 percent of the disputes over fringe issues are resolved without strikes. What is the probability that a labor-management dispute in this industry will be resolved without a strike?

**2.94** With reference to the preceding exercise, what is the probability that if a labor-management dispute in this industry is resolved without a strike, it was over wages?

**2.95** The probability that a one-car accident is due to faulty brakes is 0.04, the probability that a one-car accident is correctly attributed to faulty brakes is 0.82, and the probability that a one-car accident is incorrectly attributed to faulty brakes is 0.03. What is the probability that

(a)   a one-car accident will be attributed to faulty brakes;

(b)   a one-car accident attributed to faulty brakes was actually due to faulty brakes?

**2.96** In a certain community, 8 percent of all adults over 50 have diabetes. If a health service in this community correctly diagnoses 95 percent of all persons with diabetes as having the disease and incorrectly diagnoses 2 percent of all persons without diabetes as having the disease, find the probabilities that

(a)   the community health service will diagnose an adult over 50 as having diabetes;

(b)   a person over 50 diagnosed by the health service as having diabetes actually has the disease.

**2.97**   With reference to Example 2.25, suppose that we discover later that the job was completed on time. What is the probability that there had been a strike?

**2.98**   A mail-order house employs three stock clerks, $U$, $V$, and $W$, who pull items from shelves and assemble them for subsequent verification and packaging. $U$ makes a mistake in an order (gets a wrong item or the wrong quantity) one time in a hundred, $V$ makes a mistake in an order five times in a hundred, and $W$ makes a mistake in an order three times in a hundred. If $U$, $V$, and $W$ fill, respectively, 30, 40, and 30 percent of all orders, what are the probabilities that

(a)   a mistake will be made in an order;

(b)   if a mistake is made in an order, the order was filled by $U$;

(c)   if a mistake is made in an order, the order was filled by $V$?

**2.99**   An explosion at a construction site could have occurred as the result of static electricity, malfunctioning of equipment, carelessness, or sabotage. Interviews with construction engineers analyzing the risks involved led to the estimates that such an explosion would occur with probability 0.25 as a result of static electricity, 0.20 as a result of malfunctioning of equipment, 0.40 as a result of carelessness, and 0.75 as a result of sabotage. It is also felt that the prior probabilities of the four causes of the explosion are 0.20, 0.40, 0.25, and 0.15. Based on all this information, what is

(a)   the most likely cause of the explosion;

(b)   the least likely cause of the explosion?

**2.100**   An art dealer receives a shipment of five old paintings from abroad, and, on the basis of past experience, she feels that the probabilities are, respectively, 0.76, 0.09, 0.02, 0.01, 0.02, and 0.10 that 0, 1, 2, 3, 4, or all 5 of them are forgeries. Since the cost of authentication is fairly high, she decides to select one of the five paintings at random and send it away for authentication. If it turns out that this painting is a forgery, what probability should she now assign to the possibility that all the other paintings are also forgeries?

**2.101**   To get answers to sensitive questions, we sometimes use a method called the **randomized response technique**. Suppose, for instance, that we want to determine what percentage of the students at a large university smoke marijuana. We construct 20 flash cards, write "I smoke marijuana at least once a week" on 12 of the cards, where 12 is an arbitrary choice, and "I do not smoke marijuana at least once a week" on the others. Then, we let each student (in the sample interviewed) select one of the 20 cards at random, and respond "yes" or "no" without divulging the question.

(a)  Establish a relationship between $P(Y)$, the probability that a student will give a "yes" response, and $P(M)$, the probability that a student randomly selected at that university smokes marijuana at least once a week.

(b)  If 106 of 250 students answered "yes" under these conditions, use the result of part (a) and $\frac{106}{250}$ as an estimate of $P(Y)$ to estimate $P(M)$.

## REFERENCES

Among the numerous textbooks on probability theory published in recent years, one of the most popular is

FELLER, W., *An Introduction to Probability Theory and Its Applications,* Vol. I, 3rd ed. New York: John Wiley & Sons, Inc., 1968.

More elementary treatments may be found in

BARR, D. R., and ZEHNA, P. W., *Probability: Modeling Uncertainty.* Reading, Mass.: Addison-Wesley Publishing Company, Inc., 1983,

DRAPER, N. R., and LAWRENCE, W. E., *Probability: An Introductory Course.* Chicago: Markam Publishing Company, 1970,

FREUND, J. E., *Introduction to Probability.* Encino, Calif.: Dickenson Publishing Co., Inc., 1973,

GOLDBERG, S., *Probability—An Introduction.* Mineola, N.Y.: Dover Publications, Inc. (republication of 1960 edition),

HODGES, J. L., and LEHMANN, E. L., *Elements of Finite Probability.* San Francisco: Holden-Day, Inc., 1965,

NOSAL, M., *Basic Probability and Applications.* Philadelphia: W. B. Saunders Company, 1977.

More advanced treatments are given in many texts—for instance, in

HOEL, P., PORT, S. C., and STONE, C. J., *Introduction to Probability Theory.* Boston: Houghton Mifflin Company, 1971,

KHAZANIE, R., *Basic Probability Theory and Applications.* Pacific Palisades, Calif.: Goodyear Publishing Company, Inc., 1976,

PARZEN, E., *Modern Probability Theory and Its Applications.* New York: John Wiley & Sons, Inc., 1960,

ROSS, S., *A First Course in Probability*, 3rd ed. New York: Macmillan Publishing Company, 1988.

SOLOMON, F., *Probability and Stochastic Processes.* Englewood Cliffs, N.J.: Prentice Hall, Inc., 1987.

# 3

# *Probability Distributions and Probability Densities*

## 3.1 INTRODUCTION

In most applied problems involving probabilities we are interested only in a particular aspect (or in two or a few particular aspects) of the outcomes of experiments. For instance, when we roll a pair of dice we are usually interested only in the total, and not in the outcome for each die; when we interview a randomly chosen married couple we may be interested in the size of their family and in their joint income, but not in the number of years they have been married or their total assets; and when we sample mass-produced light bulbs we may be interested in their durability or their brightness, but not in their price.

In each of these examples we are interested in numbers that are associated with the outcomes of chance experiments, namely, in the values taken on by so-called **random variables**. In the language of probability and statistics, the total we roll with a pair of dice is a random variable, the size of the family of a randomly chosen married couple and their joint income are random variables, and so are the durability and the brightness of a light bulb randomly picked for inspection.

To be more explicit, consider Figure 3.1, which (like Figure 2.1 on page 34) pictures the sample space for an experiment in which we roll a pair of dice, and let us assume that each of the 36 possible outcomes has the probability $\frac{1}{36}$. Note, however, that in Figure 3.1 we have attached a number to each point: for instance, we attached the number 2 to the point (1, 1), the number 6 to the point (1, 5), the number 8 to the point (6, 2), the number 11 to the point (5, 6), and so forth. Evidently, we associated with each point the value of a random variable, namely, the corresponding total rolled with the pair of dice.

Since "associating a number with each point (element) of a sample space" is merely another way of saying that we are "defining a function over the points of a sample space," let us now make the following definition.

> **DEFINITION 3.1** If $S$ is a sample space with a probability measure and $X$ is a real-valued function defined over the elements of $S$, then $X$ is called a **random variable**.[†]

In this book we shall always denote random variables by capital letters and their values by the corresponding lowercase letters; for instance, we shall write $x$ to denote a value of the random variable $X$. This departs from the notation used in previous editions of this book, where we wrote **x** and $x$ instead of $X$ and $x$; nowadays, the $X$ and $x$ notation is most widely used in statistics.

With reference to the preceding example and Figure 3.1, observe that the random variable $X$ takes on the value 9, and we write $X = 9$, for the subset

$$\{(6, 3), (5, 4), (4, 5), (3, 6)\}$$

of the sample space $S$. Thus, $X = 9$ is to be interpreted as the set of elements of $S$ for which the total is 9, and more generally, $X = x$ is to be interpreted as the set of elements of the sample space for which the random variable $X$ takes on the value $x$. This may seem confusing, but it reminds one of mathematicians who say "$f(x)$ is a function of $x$" instead of "$f(x)$ is the value of a function at $x$."

---

[†]Instead of "random variable," the terms "chance variable," "stochastic variable," and "variate" are also used in some books.

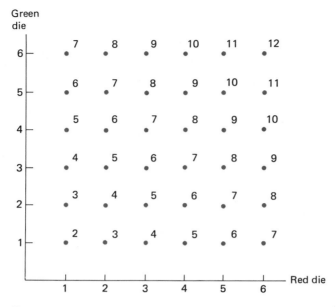

**Figure 3.1**   The total number of points rolled with a pair of dice.

## EXAMPLE 3.1

Two socks are selected at random and removed in succession from a drawer containing five brown socks and three green socks. List the elements of the sample space, the corresponding probabilities, and the corresponding values $w$ of the random variable $W$, where $W$ is the number of brown socks selected.

### Solution

If $B$ and $G$ stand for brown and green, the probabilities for $BB$, $BG$, $GB$, and $GG$ are, respectively, $\frac{5}{8} \cdot \frac{4}{7} = \frac{5}{14}$, $\frac{5}{8} \cdot \frac{3}{7} = \frac{15}{56}$, $\frac{3}{8} \cdot \frac{5}{7} = \frac{15}{56}$, and $\frac{3}{8} \cdot \frac{2}{7} = \frac{3}{28}$, and the results are shown in the following table:

| Element of sample space | Probability | $w$ |
|:---:|:---:|:---:|
| $BB$ | $\frac{5}{14}$ | 2 |
| $BG$ | $\frac{15}{56}$ | 1 |
| $GB$ | $\frac{15}{56}$ | 1 |
| $GG$ | $\frac{3}{28}$ | 0 |

Also, we can write $P(W = 2) = \frac{5}{14}$, for example, for the probability of the event that the random variable $W$ will take on the value 2.  ▲

## EXAMPLE 3.2

A balanced coin is tossed four times. List the elements of the sample space that are presumed to be equally likely, as this is what we mean by a coin being balanced, and the corresponding values $x$ of the random variable $X$, the total number of heads.

### Solution

If H and T stand for heads and tails, the results are as shown in the following table:

| Element of sample space | Probability | $x$ |
|:---:|:---:|:---:|
| HHHH | $\frac{1}{16}$ | 4 |
| HHHT | $\frac{1}{16}$ | 3 |
| HHTH | $\frac{1}{16}$ | 3 |
| HTHH | $\frac{1}{16}$ | 3 |
| THHH | $\frac{1}{16}$ | 3 |
| HHTT | $\frac{1}{16}$ | 2 |
| HTHT | $\frac{1}{16}$ | 2 |
| HTTH | $\frac{1}{16}$ | 2 |
| THHT | $\frac{1}{16}$ | 2 |
| THTH | $\frac{1}{16}$ | 2 |
| TTHH | $\frac{1}{16}$ | 2 |
| HTTT | $\frac{1}{16}$ | 1 |
| THTT | $\frac{1}{16}$ | 1 |
| TTHT | $\frac{1}{16}$ | 1 |
| TTTH | $\frac{1}{16}$ | 1 |
| TTTT | $\frac{1}{16}$ | 0 |

Thus, we can write $P(X = 3) = \frac{4}{16}$, for example, for the probability of the event that the random variable $X$ will take on the value 3.  ▲

The fact that Definition 3.1 is limited to real-valued functions does not impose any restrictions. If the numbers we want to assign to the outcomes of an experiment are complex numbers, we can always look upon the real and the imaginary parts separately as values taken on by two random variables. Also, if we want to describe the outcomes of an experiment quantitatively, say, by giving the color of a person's hair, we can arbitrarily make the descriptions real-valued by coding the various colors; perhaps, by representing them with the numbers 1, 2, 3, etc.

In all of the examples of this section we have limited our discussion to discrete sample spaces, and hence to **discrete random variables**, namely, random variables whose range is finite or countably infinite. Continuous random variables defined over continuous sample spaces will be taken up in Section 3.3.

## 3.2 PROBABILITY DISTRIBUTIONS

As we already saw in Examples 3.1 and 3.2, the probability measure defined over a discrete sample space automatically provides the probabilities that a random variable will take on any given value within its range.

For instance, having assigned the probability $\frac{1}{36}$ to each element of the sample space of Figure 3.1, we immediately find that the random variable $X$, the total rolled with the pair of dice, takes on the value 9 with probability $\frac{4}{36}$; as described on page 83, $X = 9$ contains four of the equally likely elements of the sample space. The probabilities associated with all possible values of $X$ are shown in the following table:

| $x$ | $P(X = x)$ |
|:---:|:---:|
| 2 | $\frac{1}{36}$ |
| 3 | $\frac{2}{36}$ |
| 4 | $\frac{3}{36}$ |
| 5 | $\frac{4}{36}$ |
| 6 | $\frac{5}{36}$ |
| 7 | $\frac{6}{36}$ |
| 8 | $\frac{5}{36}$ |
| 9 | $\frac{4}{36}$ |
| 10 | $\frac{3}{36}$ |
| 11 | $\frac{2}{36}$ |
| 12 | $\frac{1}{36}$ |

Instead of displaying the probabilities associated with the values of a random variable in a table, as we did in the preceding illustration, it is usually preferable to give a formula, namely, to express the probabilities by means of a function such that its values, $f(x)$, equal $P(X = x)$ for each $x$ within the range of the random variable $X$. For instance, for the total rolled with a pair of dice we could write

$$f(x) = \frac{6 - |x - 7|}{36} \qquad \text{for } x = 2, 3, \ldots, 12$$

as can easily be verified by substitution. Clearly,

$$f(2) = \frac{6 - |2 - 7|}{36} = \frac{6 - 5}{36} = \frac{1}{36}$$

$$f(3) = \frac{6 - |3 - 7|}{36} = \frac{6 - 4}{36} = \frac{2}{36}$$

$$\cdots \cdots \cdots \cdots \cdots \cdots \cdots \cdots \cdots$$

$$f(12) = \frac{6 - |12 - 7|}{36} = \frac{6 - 5}{36} = \frac{1}{36}$$

and all these values agree with the ones shown in the preceding table.

---

**DEFINITION 3.2** If $X$ is a discrete random variable, the function given by $f(x) = P(X = x)$ for each $x$ within the range of $X$ is called the **probability distribution** of $X$.

---

Based on the postulates of probability, it immediately follows that

---

**THEOREM 3.1** A function can serve as the probability distribution of a discrete random variable $X$ if and only if its values, $f(x)$, satisfy the conditions

1. $f(x) \geq 0$ for each value within its domain;

2. $\sum_x f(x) = 1$, where the summation extends over all the values within its domain.

---

## EXAMPLE 3.3

Find a formula for the probability distribution of the total number of heads obtained in four tosses of a balanced coin.

### Solution

Based on the probabilities in the table on page 85, we find that $P(X = 0)$ $= \frac{1}{16}$, $P(X = 1) = \frac{4}{16}$, $P(X = 2) = \frac{6}{16}$, $P(X = 3) = \frac{4}{16}$, and $P(X = 4) = \frac{1}{16}$. Observing that the numerators of these five fractions, 1, 4, 6, 4, and 1, are the binomial coefficients $\binom{4}{0}$, $\binom{4}{1}$, $\binom{4}{2}$, $\binom{4}{3}$, and $\binom{4}{4}$, we find that the formula for the probability distribution can be written as

$$f(x) = \frac{\binom{4}{x}}{16} \qquad \text{for } x = 0, 1, 2, 3, 4$$

A theoretical justification for this formula and a more general treatment for $n$ tosses of a balanced coin are given in Section 5.4.   ▲

## EXAMPLE 3.4

Check whether the function given by

$$f(x) = \frac{x + 2}{25} \qquad \text{for } x = 1, 2, 3, 4, 5$$

can serve as the probability distribution of a discrete random variable.

### Solution

Substituting the different values of $x$, we get $f(1) = \frac{3}{25}$, $f(2) = \frac{4}{25}$, $f(3) = \frac{5}{25}$, $f(4) = \frac{6}{25}$, and $f(5) = \frac{7}{25}$. Since these values are all nonnegative, the first condition of Theorem 3.1 is satisfied, and since

$$f(1) + f(2) + f(3) + f(4) + f(5) = \frac{3}{25} + \frac{4}{25} + \frac{5}{25} + \frac{6}{25} + \frac{7}{25}$$

$$= 1$$

the second condition of Theorem 3.1 is satisfied. Thus, the given function can serve as the probability distribution of a random variable having the range {1, 2, 3, 4, 5}. Of course, whether any given random variable actually has this probability distribution is an entirely different matter. ▲

In some problems it is desirable to present probability distributions graphically, and two kinds of graphical presentations used for this purpose are shown in Figures 3.2 and 3.3. The one shown in Figure 3.2, called a **probability histogram**, represents the probability distribution of Example 3.3. The height of each rectangle equals the probability that $X$ takes on the value which corresponds to the midpoint of its base. By representing 0 with the interval from $-0.5$ to $0.5$, 1 with the interval from $0.5$ to $1.5$, ... , and 4 with the interval from $3.5$ to $4.5$, we are, so to speak, "spreading" the values of the given discrete random variable over a continuous scale.

Since each rectangle of the histogram of Figure 3.2 has unit width, we could have said that the *areas* of the rectangles, rather than their heights, equal the corresponding probabilities. There are certain advantages to identifying the areas of the rectangles with the probabilities; for instance, when we wish to approximate the graph of a discrete probability distribution with a continuous curve. This can be done even when the rectangles of a histogram do not all have unit width, by adjusting the heights of the rectangles or by modifying the vertical scale.

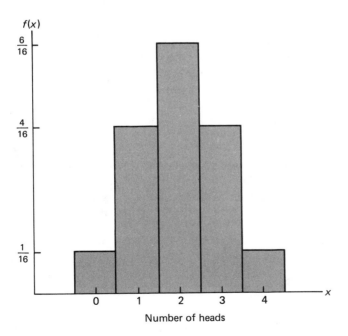

**Figure 3.2** Probability histogram.

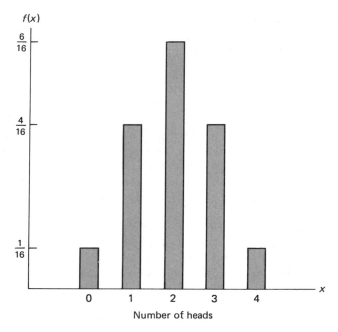

**Figure 3.3**  Bar chart.

The graph of Figure 3.3 is called a **bar chart**. As in Figure 3.2, the height of each rectangle, or bar, equals the probability of the corresponding value of the random variable, but there is no pretense of having a continuous horizontal scale. Although there are several occasions where we shall use such charts in this text, histograms and bar charts are used mainly in descriptive statistics to convey visually the information provided by a probability distribution or a distribution of actual data.

There are many problems in which it is of interest to know the probability that the value of a random variable is less than or equal to some real number $x$. Thus, let us write the probability that $X$ takes on a value less than or equal to $x$ as $F(x) = P(X \leq x)$ and refer to this function defined for all real numbers $x$ as the **distribution function**, or the **cumulative distribution**, of $X$.

---

**DEFINITION 3.3**  If $X$ is a discrete random variable, the function given by

$$F(x) = P(X \leq x) = \sum_{t \leq x} f(t) \qquad \text{for } -\infty < x < \infty$$

where $f(t)$ is the value of the probability distribution of $X$ at $t$, is called the **distribution function**, or the **cumulative distribution**, of $X$.

---

Based on the postulates of probability and some of their immediate consequences, it follows that

---

**THEOREM 3.2**  The values $F(x)$ of the distribution function of a discrete random variable $X$ satisfy the conditions

1.  $F(-\infty) = 0$ and $F(\infty) = 1$;
2.  if $a < b$, then $F(a) \leq F(b)$ for any real numbers $a$ and $b$.

---

If we are given the probability distribution of a discrete random variable, the corresponding distribution function is generally easy to find.

## EXAMPLE 3.5

Find the distribution function of the total number of heads obtained in four tosses of a balanced coin.

### Solution

Given $f(0) = \frac{1}{16}$, $f(1) = \frac{4}{16}$, $f(2) = \frac{6}{16}$, $f(3) = \frac{4}{16}$, and $f(4) = \frac{1}{16}$ from Example 3.3, it follows that

$$F(0) = f(0) = \frac{1}{16}$$

$$F(1) = f(0) + f(1) = \frac{5}{16}$$

$$F(2) = f(0) + f(1) + f(2) = \frac{11}{16}$$

$$F(3) = f(0) + f(1) + f(2) + f(3) = \frac{15}{16}$$

$$F(4) = f(0) + f(1) + f(2) + f(3) + f(4) = 1$$

Hence, the distribution function is given by

$$F(x) = \begin{cases} 0 & \text{for } x < 0 \\ \frac{1}{16} & \text{for } 0 \leq x < 1 \\ \frac{5}{16} & \text{for } 1 \leq x < 2 \\ \frac{11}{16} & \text{for } 2 \leq x < 3 \\ \frac{15}{16} & \text{for } 3 \leq x < 4 \\ 1 & \text{for } x \geq 4 \end{cases}$$

Observe that this distribution function is defined not only for the values taken on by the given random variable, but for all real numbers. For instance, we can write $F(1.7) = \frac{5}{16}$ and $F(100) = 1$, although the probabilities of getting "at most 1.7 heads" or "at most 100 heads" in four tosses of a balanced coin may not be of any real significance.    ▲

## EXAMPLE 3.6

Find the distribution function of the random variable $W$ of Example 3.1 and plot its graph.

### Solution

Based on the probabilities given in the table on page 84, we can write $f(0) = \frac{3}{28}$, $f(1) = \frac{15}{56} + \frac{15}{56} = \frac{15}{28}$, and $f(2) = \frac{5}{14}$, so that

$$F(0) = f(0) = \frac{3}{28}$$
$$F(1) = f(0) + f(1) = \frac{9}{14}$$
$$F(2) = f(0) + f(1) + f(2) = 1$$

Hence, the distribution function of $W$ is given by

$$F(w) = \begin{cases} 0 & \text{for } w < 0 \\ \frac{3}{28} & \text{for } 0 \leq w < 1 \\ \frac{9}{14} & \text{for } 1 \leq w < 2 \\ 1 & \text{for } w \geq 2 \end{cases}$$

The graph of this distribution function, shown in Figure 3.4, was obtained by first plotting the points $(w, F(w))$ for $w = 0, 1,$ and 2 and then completing the step function as indicated. Note that at all points of discontinuity the distribution function takes on the greater of the two values.    ▲

We can also reverse the process illustrated in the two preceding examples, namely, obtain values of the probability distribution of a random variable from its distribution function. To this end, we use the following result:

---

**THEOREM 3.3** If the range of a random variable $X$ consists of the values $x_1 < x_2 < x_3 < \cdots < x_n$, then $f(x_1) = F(x_1)$ and

$$f(x_i) = F(x_i) - F(x_{i-1}) \qquad \text{for } i = 2, 3, \ldots, n$$

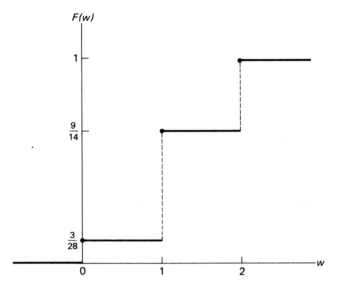

**Figure 3.4**  Graph of the distribution function of Example 3.6.

## EXAMPLE 3.7

If the distribution function of $X$ is given by

$$F(x) = \begin{cases} 0 & \text{for } x < 2 \\ \frac{1}{36} & \text{for } 2 \leq x < 3 \\ \frac{3}{36} & \text{for } 3 \leq x < 4 \\ \frac{6}{36} & \text{for } 4 \leq x < 5 \\ \frac{10}{36} & \text{for } 5 \leq x < 6 \\ \frac{15}{36} & \text{for } 6 \leq x < 7 \\ \frac{21}{36} & \text{for } 7 \leq x < 8 \\ \frac{26}{36} & \text{for } 8 \leq x < 9 \\ \frac{30}{36} & \text{for } 9 \leq x < 10 \\ \frac{33}{36} & \text{for } 10 \leq x < 11 \\ \frac{35}{36} & \text{for } 11 \leq x < 12 \\ 1 & \text{for } x \geq 12 \end{cases}$$

find the probability distribution of this random variable.

## Solution

Making use of Theorem 3.3, we get $f(2) = \frac{1}{36}$, $f(3) = \frac{3}{36} - \frac{1}{36} = \frac{2}{36}$, $f(4) = \frac{6}{36} - \frac{3}{36} = \frac{3}{36}$, $f(5) = \frac{10}{36} - \frac{6}{36} = \frac{4}{36}$, ..., $f(12) = 1 - \frac{35}{36} = \frac{1}{36}$, and comparison with the probabilities in the table on page 86 reveals that the random variable with which we are concerned here is the total number of points rolled with a pair of dice.    ▲

In the remainder of this chapter we will be concerned with continuous random variables and their distributions, and with problems relating to the simultaneous occurrence of the values of two or more random variables. In Chapter 5 we shall return to probability distributions of discrete random variables; in fact, all of that chapter will be devoted to probability distributions that provide especially important models for applications.

## EXERCISES

**3.1** For each of the following, determine whether the given values can serve as the values of a probability distribution of a random variable with the range $x = 1$, 2, 3, and 4:

(a) $f(1) = 0.25$, $f(2) = 0.75$, $f(3) = 0.25$, and $f(4) = -0.25$;

(b) $f(1) = 0.15$, $f(2) = 0.27$, $f(3) = 0.29$, and $f(4) = 0.29$;

(c) $f(1) = \frac{1}{19}$, $f(2) = \frac{10}{19}$, $f(3) = \frac{2}{19}$, and $f(4) = \frac{5}{19}$.

**3.2** For each of the following, determine whether the given function can serve as the probability distribution of a random variable with the given range:

(a) $f(x) = \dfrac{x-2}{5}$    for $x = 1$, 2, 3, 4, 5;

(b) $f(x) = \dfrac{x^2}{30}$    for $x = 0$, 1, 2, 3, 4;

(c) $f(x) = \frac{1}{5}$    for $x = 0$, 1, 2, 3, 4, 5.

**3.3** Verify that $f(x) = \dfrac{2x}{k(k+1)}$ for $x = 1$, 2, 3, ..., $k$ can serve as the probability distribution of a random variable with the given range.

**3.4** For each of the following, determine $c$ so that the function can serve as the probability distribution of a random variable with the given range:

(a) $f(x) = cx$    for $x = 1$, 2, 3, 4, 5;

(b) $f(x) = c\dbinom{5}{x}$    for $x = 0$, 1, 2, 3, 4, 5;

(c) $f(x) = cx^2$    for $x = 1$, 2, 3, ..., $k$;

(d)   $f(x) = c\left(\dfrac{1}{4}\right)^{x}$    for $x = 1, 2, 3, \ldots$ .

[*Hint*: For part (c) refer to the appendix at the end of the book.]

**3.5** For what values of $k$ can

$$f(x) = (1 - k)k^{x}$$

serve as the values of the probability distribution of a random variable with the countably infinite range $x = 0, 1, 2, \ldots$ ?

**3.6** Show that there are no values of $c$ such that

$$f(x) = \frac{c}{x}$$

can serve as the values of the probability distribution of a random variable with the countably infinite range $x = 1, 2, 3, \ldots$ .

**3.7** Construct a probability histogram for each of the following probability distributions:

(a)   $f(x) = \dfrac{\dbinom{2}{x}\dbinom{4}{3-x}}{\dbinom{6}{3}}$    for $x = 0, 1, 2$;

(b)   $f(x) = \dbinom{5}{x}\left(\dfrac{1}{5}\right)^{x}\left(\dfrac{4}{5}\right)^{5-x}$    for $x = 0, 1, 2, 3, 4, 5$.

**3.8** Prove Theorem 3.2.

**3.9** For each of the following, determine whether the given values can serve as the values of a distribution function of a random variable with the range $x = 1, 2, 3$, and 4:

(a)   $F(1) = 0.3$, $F(2) = 0.5$, $F(3) = 0.8$, and $F(4) = 1.2$;

(b)   $F(1) = 0.5$, $F(2) = 0.4$, $F(3) = 0.7$, and $F(4) = 1.0$;

(c)   $F(1) = 0.25$, $F(2) = 0.61$, $F(3) = 0.83$, and $F(4) = 1.0$.

**3.10** Find the distribution function of the random variable of part (a) of Exercise 3.7 and plot its graph.

**3.11** Find the distribution function of the random variable which has the probability distribution

$$f(x) = \frac{x}{15}    \text{for } x = 1, 2, 3, 4, 5$$

**3.12** If $X$ has the distribution function

$$F(x) = \begin{cases} 0 & \text{for } x < 1 \\ \frac{1}{3} & \text{for } 1 \leqslant x < 4 \\ \frac{1}{2} & \text{for } 4 \leqslant x < 6 \\ \frac{5}{6} & \text{for } 6 \leqslant x < 10 \\ 1 & \text{for } x \geqslant 10 \end{cases}$$

find

(a)  $P(2 < X \leqslant 6)$;
(b)  $P(X = 4)$;
(c)  the probability distribution of $X$.

**3.13** If $X$ has the distribution function

$$F(x) = \begin{cases} 0 & \text{for } x < -1 \\ \frac{1}{4} & \text{for } -1 \leqslant x < 1 \\ \frac{1}{2} & \text{for } 1 \leqslant x < 3 \\ \frac{3}{4} & \text{for } 3 \leqslant x < 5 \\ 1 & \text{for } x \geqslant 5 \end{cases}$$

find

(a)  $P(X \leqslant 3)$;    (b)  $P(X = 3)$;    (c)  $P(X < 3)$;
(d)  $P(X \geqslant 1)$;    (e)  $P(-0.4 < X < 4)$;    (f)  $P(X = 5)$.

**3.14** With reference to Example 3.4, verify that the values of the distribution function are given by

$$F(x) = \frac{x^2 + 5x}{50}$$

for $x = 1, 2, 3, 4,$ and 5.

**3.15** With reference to Theorem 3.3, verify that

(a)  $P(X > x_i) = 1 - F(x_i)$    for $i = 1, 2, 3, \ldots, n$;
(b)  $P(X \geqslant x_i) = 1 - F(x_{i-1})$    for $i = 2, 3, \ldots, n$, and $P(X \geqslant x_1) = 1$.

## APPLICATIONS

**3.16** With reference to Example 3.3, find the probability distribution of $Y$, the difference between the number of heads and the number of tails obtained in four tosses of a balanced coin.

**3.17** An urn contains four balls numbered 1, 2, 3, and 4. If two balls are drawn from the urn at random (that is, each pair has the same chance of being selected) and $Z$ is the sum of the numbers on the two balls drawn, find

(a)  the probability distribution of $Z$ and draw a histogram;

(b)  the distribution function of $Z$ and draw its graph.

**3.18** A coin is biased so that heads is twice as likely as tails. For three independent tosses of the coin, find

(a)  the probability distribution of $X$, the total number of heads;

(b)  the probability of getting at most two heads.

**3.19** With reference to Exercise 3.18, find the distribution function of the random variable $X$ and plot its graph. Use the distribution function of $X$ to find

(a)  $P(1 < X \leq 3)$;

(b)  $P(X > 2)$.

**3.20** The probability distribution of $V$, the weekly number of accidents at a certain intersection, is given by $g(0) = 0.40$, $g(1) = 0.30$, $g(2) = 0.20$, and $g(3) = 0.10$. Construct the distribution function of $V$ and draw its graph.

**3.21** With reference to Exercise 3.20, find the probability that there will be at least two accidents in any one week, using

(a)  the original probabilities;

(b)  the values of the distribution function.

## 3.3  CONTINUOUS RANDOM VARIABLES

In Section 3.1 we introduced the concept of a random variable as a real-valued function defined over the points of a sample space with a probability measure, and in Figure 3.1 we illustrated this by assigning the total rolled with a pair of dice to each of the 36 equally likely points of the sample space. In the continuous case, where random variables can take on values on a continuous scale, the procedure is very much the same. The outcomes of experiments are represented by the points on line segments or lines, and the values of random variables are numbers appropriately assigned to the points by means of rules or equations. When the value of a random variable is given directly by a measurement or observation, we generally do not bother to distinguish between the value of the random variable (the measurement which we obtain) and the outcome of the experiment (the corresponding point on the real axis). Thus, if an experiment consists of determining the actual content of a 230-gram jar of instant coffee, the result itself, say, 225.3 grams, is the value of the random variable with which we are concerned, and there is no real need to add that the sample space consists of a certain continuous interval of points on the positive real axis.

The problem of defining probabilities in connection with continuous sample spaces and continuous random variables involves some complications. To illustrate, let us consider the following situation.

## EXAMPLE 3.8

Suppose that we are concerned with the possibility that an accident will occur on a freeway which is 200 kilometers long and that we are interested in the probability that it will occur at a given location, or perhaps on a given stretch of the road. The sample space of this "experiment" consists of a continuum of points, those on the interval from 0 to 200, and we shall assume, for the sake of argument, that the probability that an accident will occur on any interval of length $d$ is $\dfrac{d}{200}$, with $d$ measured in kilometers. Note that this assignment of probabilities is consistent with Postulates 1 and 2 on page 43—the probabilities $\dfrac{d}{200}$ are all non-negative and $P(S) = \frac{200}{200} = 1$. So far this assignment of probabilities applies only to intervals on the line segment from 0 to 200, but if we use Postulate 3, we can also obtain probabilities for the union of any finite or countably infinite sequence of nonoverlapping intervals. For instance, the probability that an accident will occur on either of two nonoverlapping intervals of length $d_1$ and $d_2$ is

$$\frac{d_1 + d_2}{200}$$

and the probability that it will occur on any one of a countably infinite sequence of nonoverlapping intervals of length $d_1, d_2, d_3, \ldots$ is

$$\frac{d_1 + d_2 + d_3 + \cdots}{200}$$

Then, if we apply Theorem 2.7, we can extend the probability assignment to the union of intervals that overlap, and since the intersection of two intervals is an interval and the complement of an interval is either an interval or the union of two intervals, we can extend the probability assignment to any subset of the sample space which can be obtained by forming unions or intersections of finitely many or countably many intervals, or by forming complements.    ▲

Thus, in extending the concept of probability to the continuous case, we have again used Postulates 1, 2, and 3, but to do this in general we must exclude from our definition of "event" all subsets of the sample space which cannot be obtained by forming unions or intersections of finitely many or countably many intervals, or by forming complements. Practically speaking, this is of no consequence, for we simply do not assign probabilities to such abstruse kinds of sets.

With reference to Example 3.8, observe also that the probability of the accident occurring on a very short interval, say, an interval of 1 centimeter, is only 0.00000005, which is very small. As the length of the interval approaches zero, the probability that an accident will occur on it also approaches zero; indeed, in the continuous case we always assign zero probability to individual points. This does not mean that the corresponding events cannot occur—after all, when an accident occurs on the 200-kilometer stretch of road, it has to occur at some point even though each point has zero probability.

## 3.4 PROBABILITY DENSITY FUNCTIONS

The way in which we assigned probabilities in Example 3.8 is very special, and it is similar in nature to the way in which we assign equal probabilities to the six faces of a die, heads and tails, the 52 playing cards in a standard deck, and so forth. To treat the problem of associating probabilities with values of continuous random variables more generally, suppose that a bottler of soft drinks is concerned about the actual amount of a soft drink that his bottling machine puts into 16-ounce bottles. Evidently, the amount will vary somewhat from bottle to bottle; it is, in fact, a continuous random variable. However, if he rounds the amounts to the nearest tenth of an ounce, he will be dealing with a discrete random variable which has a probability distribution, and this probability distribution may be pictured as a histogram in which the probabilities are given by the areas of rectangles, say, as in the diagram at the top of Figure 3.5. If he rounds the amounts to the nearest hundredth of an ounce, he will again be dealing with a discrete random variable (a different one) which has a probability distribution, and this probability distribution may be pictured as a histogram in which the probabilities are given by the areas of rectangles, say, as in the diagram in the middle of Figure 3.5.

It should be apparent that if he rounded the amounts to the nearest thousandth of an ounce or to the nearest ten-thousandth of an ounce, the histograms of the probability distributions of the corresponding discrete random variables will approach the continuous curve shown in the diagram at the bottom of Figure 3.5, and the sum of the areas of the rectangles which represent the probability that the amount falls within any specified interval approaches the corresponding area under the curve.

Indeed, the definition of probability in the continuous case presumes for each random variable the existence of a function, called a **probability density function**, so that areas under the curve give the probabilities associated with the corresponding intervals along the horizontal axis. In other words, a probability density function, integrated from $a$ to $b$ (with $a \leqslant b$), gives the probability that the corresponding random variable will take on a value on the interval from $a$ to $b$.

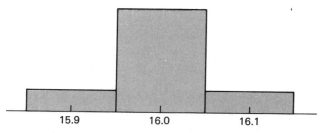

Amounts rounded to nearest tenth of an ounce

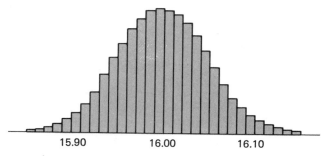

Amounts rounded to nearest hundredth of an ounce

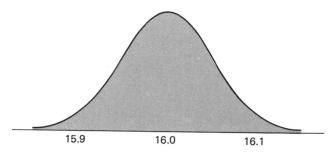

**Figure 3.5**  Definition of probability in the continuous case.

**DEFINITION 3.4**  A function with values $f(x)$, defined over the set of all real numbers, is called a **probability density function** of the continuous random variable $X$ if and only if

$$P(a \leq X \leq b) = \int_a^b f(x)\, dx$$

for any real constants $a$ and $b$ with $a \leq b$.

Probability density functions are also referred to, more briefly, as **probability densities, density functions, densities,** or **p.d.f.'s.**

Note that $f(c)$, the value of the probability density of $X$ at $c$, does not give $P(X = c)$ as in the discrete case. In connection with continuous random variables, probabilities are always associated with intervals and $P(X = c) = 0$ for any real constant $c$. This agrees with what we said on page 99 and it also follows directly from Definition 3.4 with $a = b = c$.

Because of this property, the value of a probability density function can be changed for some of the values of a random variable without changing the probabilities, and this is why we said in Definition 3.4 that $f(x)$ is the value of *a* probability density, not *the* probability density, of the random variable $X$ at $x$. Also, in view of this property, it does not matter whether we include the endpoints of the interval from $a$ to $b$; symbolically,

---

**THEOREM 3.4** If $X$ is a continuous random variable and $a$ and $b$ are real constants with $a \leq b$, then

$$P(a \leq X \leq b) = P(a \leq X < b) = P(a < X \leq b) = P(a < X < b)$$

---

Analogous to Theorem 3.1, let us now state the following properties of probability densities, which again follow directly from the postulates of probability.

---

**THEOREM 3.5** A function can serve as a probability density of a continuous random variable $X$ if its values, $f(x)$, satisfy the conditions[†]

1. $f(x) \geq 0$      for $-\infty < x < \infty$;

2. $\int_{-\infty}^{\infty} f(x) \, dx = 1$.

---

[†]The conditions are not "if and only if" as in Theorem 3.1 because $f(x)$ could be negative for some values of the random variable without affecting any of the probabilities. However, both conditions of Theorem 3.5 will be satisfied by nearly all the probability densities used in practice and studied in this text.

## EXAMPLE 3.9

If $X$ has the probability density

$$f(x) = \begin{cases} k \cdot e^{-3x} & \text{for } x > 0 \\ 0 & \text{elsewhere} \end{cases}$$

find $k$ and $P(0.5 \leq X \leq 1)$.

*Solution*

To satisfy the second condition of Theorem 3.5, we must have

$$\int_{-\infty}^{\infty} f(x)\, dx = \int_{0}^{\infty} k \cdot e^{-3x}\, dx = k \cdot \lim_{t \to \infty} \left. \frac{e^{-3x}}{-3} \right|_{0}^{t} = \frac{k}{3} = 1$$

and it follows that $k = 3$. For the probability we get

$$P(0.5 \leq X \leq 1) = \int_{0.5}^{1} 3e^{-3x}\, dx = \left. -e^{-3x} \right|_{0.5}^{1} = -e^{-3} + e^{-1.5} = 0.173$$

▲

Although the random variable of the preceding example cannot take on negative values, we artificially extended the domain of its probability density to include all the real numbers. This is a practice we shall follow throughout this text.

As in the discrete case, there are many problems in which it is of interest to know the probability that the value of a continuous random variable $X$ is less than or equal to some real number $x$. Thus, let us make the following definition analogous to Definition 3.3.

---

**DEFINITION 3.5** If $X$ is a continuous random variable and the value of its probability density at $t$ is $f(t)$, then the function given by

$$F(x) = P(X \leq x) = \int_{-\infty}^{x} f(t)\, dt \qquad \text{for } -\infty < x < \infty$$

is called the **distribution function**, or the **cumulative distribution**, of $X$.

---

The properties of distribution functions given in Theorem 3.2 hold also for the continuous case; that is, $F(-\infty) = 0$, $F(\infty) = 1$, and $F(a) \leq F(b)$ when $a < b$. Furthermore, it follows directly from Definition 3.5 that

> **THEOREM 3.6** If $f(x)$ and $F(x)$ are the values of the probability density and the distribution function of $X$ at $x$, then
>
> $$P(a \leqslant X \leqslant b) = F(b) - F(a)$$
>
> for any real constants $a$ and $b$ with $a \leqslant b$, and
>
> $$f(x) = \frac{dF(x)}{dx}$$
>
> where the derivative exists.

## EXAMPLE 3.10

Find the distribution function of the random variable $X$ of Example 3.9, and use it to reevaluate $P(0.5 \leqslant X \leqslant 1)$.

### Solution

For $x > 0$,

$$F(x) = \int_{-\infty}^{x} f(t)\, dt = \int_{0}^{x} 3e^{-3t}\, dt = -e^{-3t}\Big|_{0}^{x} = 1 - e^{-3x}$$

and since $F(x) = 0$ for $x \leqslant 0$, we can write

$$F(x) = \begin{cases} 0 & \text{for } x \leqslant 0 \\ 1 - e^{-3x} & \text{for } x > 0 \end{cases}$$

To determine the probability $P(0.5 \leqslant X \leqslant 1)$, we use the first part of Theorem 3.6, getting

$$P(0.5 \leqslant X \leqslant 1) = F(1) - F(0.5)$$
$$= (1 - e^{-3}) - (1 - e^{-1.5})$$
$$= 0.173$$

This agrees with the result obtained by using the probability density directly in Example 3.9.    ▲

## EXAMPLE 3.11

Find a probability density function for the random variable whose distribution function is given by

$$F(x) = \begin{cases} 0 & \text{for } x \leqslant 0 \\ x & \text{for } 0 < x < 1 \\ 1 & \text{for } x \geqslant 1 \end{cases}$$

and plot its graph.

### Solution

Since the given density function is differentiable everywhere except at $x = 0$ and $x = 1$, we differentiate for $x < 0$, $0 < x < 1$, and $x > 1$, getting 0, 1, and 0. Thus, according to the second part of Theorem 3.6, we can write

$$f(x) = \begin{cases} 0 & \text{for } x < 0 \\ 1 & \text{for } 0 < x < 1 \\ 0 & \text{for } x > 1 \end{cases}$$

To fill the gaps at $x = 0$ and $x = 1$, we let $f(0)$ and $f(1)$ both equal zero. Actually, it does not matter how the probability density is defined at these two points, but there are certain advantages (which are explained on page 265) for choosing the values in such a way that the probability density is nonzero over an open interval. Thus, we can write the probability density of the original random variable as

$$f(x) = \begin{cases} 1 & \text{for } 0 < x < 1 \\ 0 & \text{elsewhere} \end{cases}$$

Its graph is shown in Figure 3.6.    ▲

In most practical applications we encounter random variables that are either discrete or continuous, so that the corresponding distribution functions have a steplike appearance as in Figure 3.4 or they are continuous curves as in Figure 3.7, which shows the graph of the distribution function of Example 3.11. Discontinuous distribution functions like the one shown in Figure 3.8 arise when random variables are **mixed**. Such a distribution function will be discontinuous at each point having a nonzero probability and continuous elsewhere. As in the discrete case, the height of the step at a point of discontinuity gives the probability that the random variable will take on that particular value. With reference to Figure 3.8, $P(X = 0.5) = \frac{3}{4} - \frac{1}{4} = \frac{1}{2}$, but otherwise the random variable is like a continuous random variable.

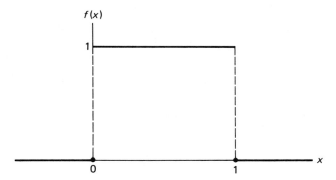

**Figure 3.6** Probability density of Example 3.11.

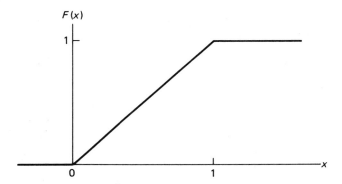

**Figure 3.7** Distribution function of Example 3.11.

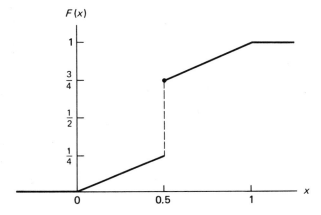

**Figure 3.8** Distribution function of a mixed random variable.

In this text we shall limit ourselves to random variables that are discrete or continuous, with the latter having distribution functions that are differentiable for all but a finite set of values of the random variables.

## EXERCISES

**3.22** The probability density of the continuous random variable $X$ is given by

$$f(x) = \begin{cases} \frac{1}{5} & \text{for } 2 < x < 7 \\ 0 & \text{elsewhere} \end{cases}$$

(a) Draw its graph and verify that the total area under the curve (above the $x$-axis) is equal to 1.

(b) Find $P(3 < X < 5)$.

**3.23** Find the distribution function of the random variable $X$ of Exercise 3.22 and use it to reevaluate part (b).

**3.24** The probability density of the random variable $Y$ is given by

$$f(y) = \begin{cases} \frac{1}{8}(y + 1) & \text{for } 2 < y < 4 \\ 0 & \text{elsewhere} \end{cases}$$

Find $P(Y < 3.2)$ and $P(2.9 < Y < 3.2)$.

**3.25** Find the distribution function of the random variable $Y$ of Exercise 3.24 and use it to determine the two probabilities asked for in that exercise.

**3.26** The p.d.f. of the random variable $X$ is given by

$$f(x) = \begin{cases} \dfrac{c}{\sqrt{x}} & \text{for } 0 < x < 4 \\ 0 & \text{elsewhere} \end{cases}$$

Find

(a) the value of $c$;

(b) $P(X < \frac{1}{4})$ and $P(X > 1)$.

**3.27** Find the distribution function of the random variable $X$ of Exercise 3.26 and use it to determine the two probabilities asked for in part (b) of that exercise.

**3.28** The probability density of the random variable $Z$ is given by

$$f(z) = \begin{cases} kze^{-z^2} & \text{for } z > 0 \\ 0 & \text{for } z \leqslant 0 \end{cases}$$

Find $k$ and draw the graph of this probability density.

**3.29** With reference to Exercise 3.28, find the distribution function of $Z$ and draw its graph.

**3.30** The density function of the random variable $X$ is given by

$$g(x) = \begin{cases} 6x(1-x) & \text{for } 0 < x < 1 \\ 0 & \text{elsewhere} \end{cases}$$

Find $P(X < \frac{1}{4})$ and $P(X > \frac{1}{2})$.

**3.31** With reference to Exercise 3.30, find the distribution function of $X$ and use it to reevaluate the two probabilities asked for in that exercise.

**3.32** Find the distribution function of the random variable $X$ whose probability density is given by

$$f(x) = \begin{cases} \frac{1}{3} & \text{for } 0 < x < 1 \\ \frac{1}{3} & \text{for } 2 < x < 4 \\ 0 & \text{elsewhere} \end{cases}$$

Also sketch the graphs of the probability density and distribution functions.

**3.33** Find the distribution function of the random variable $X$ whose probability density is given by

$$f(x) = \begin{cases} x & \text{for } 0 < x < 1 \\ 2 - x & \text{for } 1 \leq x < 2 \\ 0 & \text{elsewhere} \end{cases}$$

Also sketch the graphs of the probability density and distribution functions.

**3.34** With reference to Exercise 3.33, find $P(0.8 < X < 1.2)$ using
   (a) the probability density;
   (b) the distribution function.

**3.35** Find the distribution function of the random variable $X$ whose probability density is given by

$$f(x) = \begin{cases} \dfrac{x}{2} & \text{for } 0 < x \leq 1 \\[2mm] \dfrac{1}{2} & \text{for } 1 < x \leq 2 \\[2mm] \dfrac{3-x}{2} & \text{for } 2 < x < 3 \\[2mm] 0 & \text{elsewhere} \end{cases}$$

Also sketch the graphs of these probability density and distribution functions.

**3.36** The distribution function of the random variable $X$ is given by

$$F(x) = \begin{cases} 0 & \text{for } x < -1 \\ \dfrac{x+1}{2} & \text{for } -1 \leqslant x < 1 \\ 1 & \text{for } x \geqslant 1 \end{cases}$$

Find $P(-\frac{1}{2} < X < \frac{1}{2})$ and $P(2 < X < 3)$.

**3.37** With reference to Exercise 3.36, find the probability density of $X$ and use it to recalculate the two probabilities.

**3.38** The distribution function of the random variable $Y$ is given by

$$F(y) = \begin{cases} 1 - \dfrac{9}{y^2} & \text{for } y > 3 \\ 0 & \text{elsewhere} \end{cases}$$

Find $P(Y \leqslant 5)$ and $P(Y > 8)$.

**3.39** With reference to Exercise 3.38, find the probability density of $Y$ and use it to recalculate the two probabilities.

**3.40** With reference to Exercise 3.38 and the result of Exercise 3.39, sketch the graphs of the distribution function and the probability density of $Y$, letting $f(3) = 0$.

**3.41** The distribution function of the random variable $X$ is given by

$$F(x) = \begin{cases} 1 - (1 + x)e^{-x} & \text{for } x > 0 \\ 0 & \text{for } x \leqslant 0 \end{cases}$$

Find $P(X \leqslant 2)$, $P(1 < X < 3)$, and $P(X > 4)$.

**3.42** With reference to Exercise 3.41, find the probability density of $X$.

**3.43** With reference to Figure 3.8, find expressions for the values of the distribution function of the mixed random variable $X$ for

(a) $x \leqslant 0$;

(b) $0 < x < 0.5$;

(c) $0.5 \leqslant x < 1$;

(d) $x \geqslant 1$.

**3.44** Use the results of Exercise 3.43 to find expressions for the values of the probability density of the mixed random variable $X$ for

(a) $x < 0$;

(b) $0 < x < 0.5$;

(c)  $0.5 < x < 1$;

(d)  $x > 1$.

$P(X = 0.5) = \frac{1}{2}$, as we already indicated on page 104, and $f(0)$ and $f(1)$ are undefined.

**3.45** The distribution function of the mixed random variable $Z$ is given by

$$F(z) = \begin{cases} 0 & \text{for } z < -2 \\ \dfrac{z + 4}{8} & \text{for } -2 \leqslant z < 2 \\ 1 & \text{for } z \geqslant 2 \end{cases}$$

Find $P(Z = -2)$, $P(Z = 2)$, $P(-2 < Z < 1)$, and $P(0 \leqslant Z \leqslant 2)$.

## APPLICATIONS

**3.46** The actual amount of coffee (in grams) in a 230-gram jar filled by a certain machine is a random variable whose probability density is given by

$$f(x) = \begin{cases} 0 & \text{for } x \leqslant 227.5 \\ \frac{1}{5} & \text{for } 227.5 < x < 232.5 \\ 0 & \text{for } x \geqslant 232.5 \end{cases}$$

Find the probabilities that a 230-gram jar filled by this machine will contain

(a)  at most 228.65 grams of coffee;

(b)  anywhere from 229.34 to 231.66 grams of coffee;

(c)  at least 229.85 grams of coffee.

**3.47** The number of minutes that a flight from Phoenix to Tucson is early or late is a random variable whose probability density is given by

$$f(x) = \begin{cases} \frac{1}{288}(36 - x^2) & \text{for } -6 < x < 6 \\ 0 & \text{elsewhere} \end{cases}$$

where negative values are indicative of the flight's being early and positive values are indicative of its being late. Find the probabilities that one of these flights will be

(a)  at least 2 minutes early;

(b)  at least 1 minute late;

(c)  anywhere from 1 to 3 minutes early;

(d)  exactly 5 minutes late.

**3.48** The shelf life (in hours) of a certain perishable packaged food is a random variable whose probability density function is given by

$$f(x) = \begin{cases} \dfrac{20,000}{(x + 100)^3} & \text{for } x > 0 \\ 0 & \text{elsewhere} \end{cases}$$

Find the probabilities that one of these packages will have a shelf life of

(a)   at least 200 hours;
(b)   at most 100 hours;
(c)   anywhere from 80 to 120 hours.

**3.49** The tread wear (in thousands of kilometers) which car owners get with a certain kind of tire is a random variable whose probability density is given by

$$f(x) = \begin{cases} \frac{1}{30} e^{-\frac{x}{30}} & \text{for } x > 0 \\ 0 & \text{for } x \leqslant 0 \end{cases}$$

Find the probabilities that one of these tires will last

(a)   at most 18,000 kilometers;
(b)   anywhere from 27,000 to 36,000 kilometers;
(c)   at least 48,000 kilometers.

**3.50** In a certain city the daily consumption of water (in millions of liters) is a random variable whose probability density is given by

$$f(x) = \begin{cases} \frac{1}{9} xe^{-\frac{x}{3}} & \text{for } x > 0 \\ 0 & \text{elsewhere} \end{cases}$$

What are the probabilities that on a given day

(a)   the water consumption in this city is no more than 6 million liters;
(b)   the water supply is inadequate if the daily capacity of this city is 9 million liters?

**3.51** The total lifetime (in years) of five-year-old dogs of a certain breed is a random variable whose distribution function is given by

$$F(x) = \begin{cases} 0 & \text{for } x \leqslant 5 \\ 1 - \dfrac{25}{x^2} & \text{for } x > 5 \end{cases}$$

Find the probabilities that such a five-year-old dog will live

(a)   beyond 10 years;
(b)   less than eight years;
(c)   anywhere from 12 to 15 years.

## 3.5  MULTIVARIATE DISTRIBUTIONS

In the beginning of this chapter we defined a random variable as a real-valued function defined over a sample space with a probability measure, and it stands to reason that many different random variables can be defined over one and the same sample space. With reference to the sample space of Figure 3.1, for example, we considered only the random variable whose values were the totals rolled with a pair of dice, but we could also have considered the random variable whose values are the products of the numbers rolled with the two dice, the random variable whose values are the differences between the numbers rolled with the red die and the green die, the random variable whose values are 0, 1, or 2 depending on the number of dice which come up 2, and so forth. Closer to life, an experiment may consist of randomly choosing some of the 345 students attending an elementary school, and the principal may be interested in their I.Q.'s, the school nurse in their weights, their teachers in the number of days they have been absent, and so forth.

In this section we shall be concerned first with the **bivariate case**, that is, with situations where we are interested at the same time in a pair of random variables defined over a joint sample space. Later, we shall extend this discussion to the **multivariate case**, covering any finite number of random variables.

If $X$ and $Y$ are discrete random variables, we write the probability that $X$ will take on the value $x$ and $Y$ will take on the value $y$ as $P(X = x, Y = y)$. Thus, $P(X = x, Y = y)$ is the probability of the intersection of the events $X = x$ and $Y = y$. As in the **univariate case**, where we dealt with one random variable and could display the probabilities associated with all values of $X$ by means of a table, we can now, in the bivariate case, display the probabilities associated with all pairs of values of $X$ and $Y$ by means of a table.

## EXAMPLE 3.12

Two caplets are selected at random from a bottle containing three aspirin, two sedative, and four laxative caplets. If $X$ and $Y$ are, respectively, the numbers of aspirin and sedative caplets included among the two caplets drawn from the bottle, find the probabilities associated with all possible pairs of values of $X$ and $Y$.

### Solution

The possible pairs are (0, 0), (0, 1), (1, 0), (1, 1), (0, 2), and (2, 0). To find the probability associated with (1, 0), for example, observe that we are concerned with the event of getting one of the three aspirin caplets, none of the two sedative caplets, and, hence, one of the four laxative caplets. The number of ways in which this can be done is $\binom{3}{1}\binom{2}{0}\binom{4}{1} = 12$ and the total number of ways in which two of the nine caplets can be selected

is $\binom{9}{2}$ = 36. Since these possibilities are all equally likely by virtue of the

assumption that the selection is random, it follows from Theorem 2.2 that the probability associated with (1, 0) is $\frac{12}{36} = \frac{1}{3}$. Similarly, the probability associated with (1, 1) is

$$\frac{\binom{3}{1}\binom{2}{1}\binom{4}{0}}{36} = \frac{6}{36} = \frac{1}{6}$$

and, continuing this way, we obtain the values shown in the following table:

|       |   | $x$ | | |
|-------|---|-----|-----|-----|
|       |   | 0   | 1   | 2   |
|       | 0 | $\frac{1}{6}$  | $\frac{1}{3}$ | $\frac{1}{12}$ |
| $y$   | 1 | $\frac{2}{9}$  | $\frac{1}{6}$ |     |
|       | 2 | $\frac{1}{36}$ |     |     |

▲

Actually, as in the univariate case, it is generally preferable to represent probabilities such as these by means of a formula. In other words, it is preferable to express the probabilities by means of a function with the values $f(x, y) = P(X = x, Y = y)$ for any pair of values $(x, y)$ within the range of the random variables $X$ and $Y$. For instance, we shall see in Chapter 5 that for the two random variables of Example 3.12 we can write

$$f(x, y) = \frac{\binom{3}{x}\binom{2}{y}\binom{4}{2 - x - y}}{\binom{9}{2}} \qquad \text{for } x = 0, 1, 2; \quad y = 0, 1, 2; \\ 0 \leqslant x + y \leqslant 2$$

---

**DEFINITION 3.6** If $X$ and $Y$ are discrete random variables, the function given by $f(x, y) = P(X = x, Y = y)$ for each pair of values $(x, y)$ within the range of $X$ and $Y$ is called the **joint probability distribution** of $X$ and $Y$.

---

Analogous to Theorem 3.1, it follows from the postulates of probability that

---

**THEOREM 3.7**  A bivariate function can serve as the joint probability distribution of a pair of discrete random variables $X$ and $Y$ if and only if its values, $f(x, y)$, satisfy the conditions

1. $f(x, y) \geq 0$ for each pair of values $(x, y)$ within its domain;

2. $\sum_x \sum_y f(x, y) = 1$, where the double summation extends over all possible pairs $(x, y)$ within its domain.

---

## EXAMPLE 3.13

Determine the value of $k$ for which the function given by

$$f(x, y) = kxy \qquad \text{for } x = 1, 2, 3; \quad y = 1, 2, 3$$

can serve as a joint probability distribution.

### Solution

Substituting the various values of $x$ and $y$, we get $f(1, 1) = k$, $f(1, 2) = 2k$, $f(1, 3) = 3k$, $f(2, 1) = 2k$, $f(2, 2) = 4k$, $f(2, 3) = 6k$, $f(3, 1) = 3k$, $f(3, 2) = 6k$, and $f(3, 3) = 9k$. To satisfy the first condition of Theorem 3.7, the constant $k$ must be nonnegative, and to satisfy the second condition,

$$k + 2k + 3k + 2k + 4k + 6k + 3k + 6k + 9k = 1$$

so that $36k = 1$ and $k = \frac{1}{36}$.    ▲

As in the univariate case, there are many problems in which it is of interest to know the probability that the values of two random variables are less than or equal to some real numbers $x$ and $y$.

---

**DEFINITION 3.7**  If $X$ and $Y$ are discrete random variables, the function given by

$$F(x, y) = P(X \leq x, Y \leq y) = \sum_{s \leq x} \sum_{t \leq y} f(s, t) \qquad \begin{array}{l} \text{for } -\infty < x < \infty, \\ -\infty < y < \infty \end{array}$$

where $f(s, t)$ is the value of the joint probability distribution of $X$ and $Y$ at $(s, t)$, is called the **joint distribution function**, or the **joint cumulative distribution**, of $X$ and $Y$.

---

In Exercise 3.58 the reader will be asked to prove properties of joint distribution functions which are analogous to those of Theorem 3.2.

## EXAMPLE 3.14

With reference to Example 3.12, find $F(1, 1)$.

*Solution*

$$F(1, 1) = P(X \leq 1, Y \leq 1)$$
$$= f(0, 0) + f(0, 1) + f(1, 0) + f(1, 1)$$
$$= \tfrac{1}{6} + \tfrac{2}{9} + \tfrac{1}{3} + \tfrac{1}{6}$$
$$= \tfrac{8}{9} \quad \blacktriangle$$

As in the univariate case, the joint distribution function of two random variables is defined for all real numbers. For instance, for Example 3.12 we also get $F(-2, 1) = P(X \leq -2, Y \leq 1) = 0$ and $F(3.7, 4.5) = P(X \leq 3.7, Y \leq 4.5) = 1$.

Let us now extend the various concepts introduced in this section to the continuous case.

---

**DEFINITION 3.8**  A bivariate function with values $f(x, y)$, defined over the $xy$-plane, is called a **joint probability density function** of the continuous random variables $X$ and $Y$ if and only if

$$P[(X, Y) \in A] = \int_A \int f(x, y) \, dx \, dy$$

for any region $A$ in the $xy$-plane.

---

Analogous to Theorem 3.5, it follows from the postulates of probability that

---

**THEOREM 3.8**  A bivariate function can serve as a joint probability density function of a pair of continuous random variables $X$ and $Y$ if its values, $f(x, y)$, satisfy the conditions

1. $f(x, y) \geq 0$    for $-\infty < x < \infty,\quad -\infty < y < \infty$;

2. $\displaystyle\int_{-\infty}^{\infty} \int_{-\infty}^{\infty} f(x, y) \, dx \, dy = 1.$

---

## EXAMPLE 3.15

Given the joint probability density function

$$f(x, y) = \begin{cases} \frac{3}{5} x(y + x) & \text{for } 0 < x < 1, 0 < y < 2 \\ 0 & \text{elsewhere} \end{cases}$$

of two random variables $X$ and $Y$, find $P[(X, Y) \in A]$, where $A$ is the region $\{(x, y)|0 < x < \frac{1}{2}, 1 < y < 2\}$.

*Solution*

$$P[(X, Y) \in A] = P(0 < X < \tfrac{1}{2}, 1 < Y < 2)$$

$$= \int_1^2 \int_0^{\frac{1}{2}} \frac{3}{5} x\,(y + x)\, dx\, dy$$

$$= \int_1^2 \frac{3x^2 y}{10} + \frac{3x^3}{15} \Big|^{x=\frac{1}{2}}\, dy$$

$$= \int_1^2 \left( \frac{3y}{40} + \frac{1}{40} \right) dy = \frac{3y^2}{80} + \frac{y}{40} \Big|_1^2$$

$$= \frac{11}{80} \qquad \blacktriangle$$

Analogous to Definition 3.7, we have the following definition of the joint distribution function of two continuous random variables.

---

**DEFINITION 3.9** If $X$ and $Y$ are continuous random variables, the function given by

$$F(x, y) = P(X \leqslant x, Y \leqslant y) = \int_{-\infty}^y \int_{-\infty}^x f(s, t)\, ds\, dt \qquad \text{for } -\infty < x < \infty, \\ -\infty < y < \infty$$

where $f(s, t)$ is the value of the joint probability density of $X$ and $Y$ at $(s, t)$, is called the **joint distribution function** of $X$ and $Y$.

---

Note that the properties of joint distribution functions, which the reader will be asked to prove in Exercise 3.58 for the discrete case, hold also for the continuous case.

As in Section 3.4, we shall limit our discussion here to random variables whose joint distribution function is continuous everywhere and partially differentiable with respect to each variable for all but a finite set of values of the two random variables.

Analogous to the relationship $f(x) = \dfrac{dF(x)}{dx}$ of Theorem 3.6, partial differentiation in Definition 3.9 leads to

$$f(x, y) = \frac{\partial^2}{\partial x \, \partial y} F(x, y)$$

wherever these partial derivatives exist. As in Section 3.4, the joint distribution function of two continuous random variables determines their **joint density** (short for joint probability density function) at all points $(x, y)$ where the joint density is continuous. Also as in Section 3.4, we generally let the values of joint probability densities equal zero wherever they are not defined by the above relationship.

## EXAMPLE 3.16

If the joint probability density of $X$ and $Y$ is given by

$$f(x, y) = \begin{cases} x + y & \text{for } 0 < x < 1, 0 < y < 1 \\ 0 & \text{elsewhere} \end{cases}$$

find the joint distribution function of these two random variables.

### Solution

If either $x < 0$ or $y < 0$, it follows immediately that $F(x, y) = 0$. For $0 < x < 1$ and $0 < y < 1$ (Region I of Figure 3.9) we get

$$F(x, y) = \int_0^y \int_0^x (s + t) \, ds \, dt = \tfrac{1}{2} xy(x + y)$$

for $x > 1$ and $0 < y < 1$ (Region II of Figure 3.9) we get

$$F(x, y) = \int_0^y \int_0^1 (s + t) \, ds \, dt = \tfrac{1}{2} y(y + 1)$$

for $0 < x < 1$ and $y > 1$ (Region III of Figure 3.9) we get

$$F(x, y) = \int_0^1 \int_0^x (s + t) \, ds \, dt = \tfrac{1}{2} x(x + 1)$$

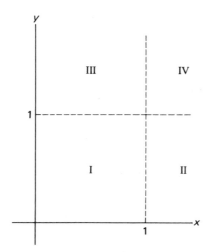

**Figure 3.9** Diagram for Example 3.16.

and for $x > 1$ and $y > 1$ (Region IV of Figure 3.9) we get

$$F(x, y) = \int_0^1 \int_0^1 (s + t) \, ds \, dt = 1$$

Since the joint distribution function is everywhere continuous, the boundaries between any two of these regions can be included in either one, and we can write

$$F(x, y) = \begin{cases} 0 & \text{for } x \leq 0 \text{ or } y \leq 0 \\ \frac{1}{2} xy(x + y) & \text{for } 0 < x < 1, \, 0 < y < 1 \\ \frac{1}{2} y(y + 1) & \text{for } x \geq 1, \, 0 < y < 1 \\ \frac{1}{2} x(x + 1) & \text{for } 0 < x < 1, \, y \geq 1 \\ 1 & \text{for } x \geq 1, \, y \geq 1 \quad \blacktriangle \end{cases}$$

## EXAMPLE 3.17

Find the joint probability density of the two random variables $X$ and $Y$ whose joint distribution function is given by

$$F(x, y) = \begin{cases} (1 - e^{-x})(1 - e^{-y}) & \text{for } x > 0 \text{ and } y > 0 \\ 0 & \text{elsewhere} \end{cases}$$

Also use the joint probability density to determine $P(1 < X < 3, 1 < Y < 2)$.

### Solution

Since partial differentiation yields

$$\frac{\partial^2}{\partial x\, \partial y} F(x, y) = e^{-(x+y)}$$

for $x > 0$ and $y > 0$ and 0 elsewhere, we find that the joint probability density of $X$ and $Y$ is given by

$$f(x, y) = \begin{cases} e^{-(x+y)} & \text{for } x > 0 \text{ and } y > 0 \\ 0 & \text{elsewhere} \end{cases}$$

Thus, integration yields

$$\int_1^2 \int_1^3 e^{-(x+y)}\, dx\, dy = (e^{-1} - e^{-3})(e^{-1} - e^{-2})$$

$$= e^{-2} - e^{-3} - e^{-4} + e^{-5}$$

$$= 0.074$$

for $P(1 < X < 3, 1 < Y < 1)$.    ▲

For two random variables, the joint probability is, geometrically speaking, a surface, and the probability that we calculated in the preceding example is given by the volume under this surface, as shown in Figure 3.10.

All the definitions of this section can be generalized to the **multivariate** case, where there are $n$ random variables. Corresponding to Definition 3.6, the values of the joint probability distribution of $n$ discrete random variables $X_1$, $X_2$, ..., and $X_n$ are given by

$$f(x_1, x_2, \ldots, x_n) = P(X_1 = x_1, X_2 = x_2, \ldots, X_n = x_n)$$

for each $n$-tuple $(x_1, x_2, \ldots, x_n)$ within the range of the random variables; and corresponding to Definition 3.7, the values of their joint distribution function are given by

$$F(x_1, x_2, \ldots, x_n) = P(X_1 \leq x_1, X_2 \leq x_2, \ldots, X_n \leq x_n)$$

for $-\infty < x_1 < \infty$, $-\infty < x_2 < \infty$, ..., $-\infty < x_n < \infty$.

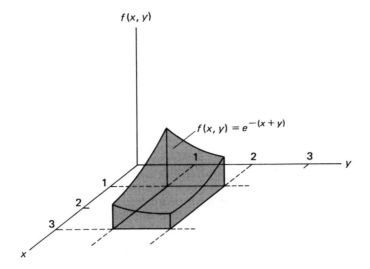

**Figure 3.10**  Diagram for Example 3.17.

## EXAMPLE 3.18

If the joint probability distribution of three discrete random variables $X$, $Y$, and $Z$ is given by

$$f(x, y, z) = \frac{(x + y)z}{63} \qquad \text{for } x = 1, 2; \quad y = 1, 2, 3; \quad z = 1, 2$$

find $P(X = 2, Y + Z \leq 3)$.

*Solution*

$$P(X = 2, Y + Z \leq 3) = f(2, 1, 1) + f(2, 1, 2) + f(2, 2, 1)$$

$$= \tfrac{3}{63} + \tfrac{6}{63} + \tfrac{4}{63}$$

$$= \tfrac{13}{63} \qquad \blacktriangle$$

In the continuous case, probabilities are again obtained by integrating the joint probability density, and the joint distribution function is given by

$$F(x_1, x_2, \ldots, x_n) = \int_{-\infty}^{x_n} \cdots \int_{-\infty}^{x_2} \int_{-\infty}^{x_1} f(t_1, t_2, \ldots, t_n) \, dt_1 \, dt_2 \cdots dt_n$$

for $-\infty < x_1 < \infty$, $-\infty < x_2 < \infty$, ... , $-\infty < x_n < \infty$ analogous to Definition 3.9. Also, partial differentiation yields

$$f(x_1, x_2, \ldots , x_n) = \frac{\partial^n}{\partial x_1 \, \partial x_2 \cdots \partial x_n} F(x_1, x_2, \ldots , x_n)$$

wherever these partial derivatives exist.

## EXAMPLE 3.19

If the **trivariate** probability density of $X_1$, $X_2$, and $X_3$ is given by

$$f(x_1, x_2, x_3) = \begin{cases} (x_1 + x_2)e^{-x_3} & \text{for } 0 < x_1 < 1, 0 < x_2 < 1, x_3 > 0 \\ 0 & \text{elsewhere} \end{cases}$$

find $P[(X_1, X_2, X_3) \in A]$, where $A$ is the region

$$\{(x_1, x_2, x_3)|0 < x_1 < \tfrac{1}{2}, \tfrac{1}{2} < x_2 < 1, x_3 < 1\}$$

*Solution*

$$P[(X_1, X_2, X_3) \in A] = P(0 < X_1 < \tfrac{1}{2}, \tfrac{1}{2} < X_2 < 1, X_3 < 1)$$

$$= \int_0^1 \int_{\frac{1}{2}}^1 \int_0^{\frac{1}{2}} (x_1 + x_2)e^{-x_3} \, dx_1 \, dx_2 \, dx_3$$

$$= \int_0^1 \int_{\frac{1}{2}}^1 \left( \frac{1}{8} + \frac{x_2}{2} \right) e^{-x_3} \, dx_2 \, dx_3$$

$$= \int_0^1 \tfrac{1}{4} e^{-x_3} \, dx_3$$

$$= \tfrac{1}{4} (1 - e^{-1}) = 0.158 \qquad \blacktriangle$$

## EXERCISES

**3.52** If the values of the joint probability distribution of $X$ and $Y$ are as shown in the table

|   | $x$ | | |
|---|---|---|---|
| $y$ | 0 | 1 | 2 |
| 0 | $\frac{1}{12}$ | $\frac{1}{6}$ | $\frac{1}{24}$ |
| 1 | $\frac{1}{4}$ | $\frac{1}{4}$ | $\frac{1}{40}$ |
| 2 | $\frac{1}{8}$ | $\frac{1}{20}$ | |
| 3 | $\frac{1}{120}$ | | |

find

(a)  $P(X = 1, Y = 2)$;
(b)  $P(X = 0, 1 \leq Y < 3)$;
(c)  $P(X + Y \leq 1)$;
(d)  $P(X > Y)$.

**3.53** With reference to Exercise 3.52, find the following values of the joint distribution function of the two random variables:

(a)  $F(1.2, 0.9)$;
(b)  $F(-3, 1.5)$;
(c)  $F(2, 0)$;
(d)  $F(4, 2.7)$.

**3.54** If the joint probability distribution of $X$ and $Y$ is given by

$$f(x, y) = c(x^2 + y^2) \qquad \text{for } x = -1, 0, 1, 3; \quad y = -1, 2, 3$$

find the value of $c$.

**3.55** With reference to Exercise 3.54 and the value obtained for $c$, find

(a)  $P(X \leq 1, Y > 2)$;
(b)  $P(X = 0, Y \leq 2)$;
(c)  $P(X + Y > 2)$.

**3.56** Show that there is no value of $k$ for which

$$f(x, y) = ky(2y - x) \qquad \text{for } x = 0, 3; \quad y = 0, 1, 2$$

can serve as the joint probability distribution of two random variables.

**3.57** If the joint probability distribution of $X$ and $Y$ is given by

$$f(x, y) = \tfrac{1}{30}(x + y) \qquad \text{for } x = 0, 1, 2, 3; \quad y = 0, 1, 2$$

construct a table showing the values of the joint distribution function of the two random variables at the twelve points $(0, 0)$, $(0, 1)$, ... , $(3, 2)$.

**3.58** If $F(x, y)$ is the value of the joint distribution function of two discrete random variables $X$ and $Y$ at $(x, y)$, show that

(a)  $F(-\infty, -\infty) = 0$;
(b)  $F(\infty, \infty) = 1$;
(c)  if $a < b$ and $c < d$, then $F(a, c) \leq F(b, d)$.

**3.59** Determine $k$ so that

$$f(x, y) = \begin{cases} kx(x - y) & \text{for } 0 < x < 1, -x < y < x \\ 0 & \text{elsewhere} \end{cases}$$

can serve as a joint probability density.

**3.60** If the joint probability density of $X$ and $Y$ is given by

$$f(x, y) = \begin{cases} 24xy & \text{for } 0 < x < 1, 0 < y < 1, \quad x + y < 1 \\ 0 & \text{elsewhere} \end{cases}$$

find $P(X + Y < \frac{1}{2})$.

**3.61** If the joint probability density of $X$ and $Y$ is given by

$$f(x, y) = \begin{cases} 2 & \text{for } x > 0, y > 0, x + y < 1 \\ 0 & \text{elsewhere} \end{cases}$$

find
(a)  $P(X \leq \frac{1}{2}, Y \leq \frac{1}{2})$;
(b)  $P(X + Y > \frac{2}{3})$;
(c)  $P(X > 2Y)$.

**3.62** With reference to Exercise 3.61, find an expression for the values of the joint distribution function of $X$ and $Y$ when $x > 0$, $y > 0$, and $x + y < 1$, and use it to verify the result of part (a).

**3.63** If the joint probability density of $X$ and $Y$ is given by

$$f(x, y) = \begin{cases} \dfrac{1}{y} & \text{for } 0 < x < y, 0 < y < 1 \\ 0 & \text{elsewhere} \end{cases}$$

find the probability that the sum of the values of $X$ and $Y$ will exceed $\frac{1}{2}$.

**3.64** Find the joint probability density of the two random variables $X$ and $Y$ whose joint distribution function is given by

$$F(x, y) = \begin{cases} (1 - e^{-x^2})(1 - e^{-y^2}) & \text{for } x > 0, y > 0 \\ 0 & \text{elsewhere} \end{cases}$$

**3.65** Use the joint probability density obtained in Exercise 3.64 to find $P(1 < X \leq 2, 1 < Y \leq 2)$.

**3.66** Find the joint probability density of the two random variables $X$ and $Y$ whose joint distribution function is given by

$$F(x, y) = \begin{cases} 1 - e^{-x} - e^{-y} + e^{-x-y} & \text{for } x > 0, y > 0 \\ 0 & \text{elsewhere} \end{cases}$$

**3.67** Use the joint probability density obtained in Exercise 3.66 to find $P(X + Y > 3)$.

**3.68** If $F(x, y)$ is the value of the joint distribution function of the two continuous random variables $X$ and $Y$ at $(x, y)$, express $P(a < X \leq b, c < Y \leq d)$ in terms of $F(a, c)$, $F(a, d)$, $F(b, c)$, and $F(b, d)$. Observe that the result holds also for discrete random variables.

**3.69** Use the formula obtained in Exercise 3.68 to verify the result, 0.074, of Example 3.17.

**3.70** Use the formula obtained in Exercise 3.68 to verify the result of Exercise 3.65.

**3.71** Use the formula obtained in Exercise 3.68 to verify the result of Exercise 3.67.

**3.72** Find $k$, if the joint probability distribution of $X$, $Y$, and $Z$ is given by

$$f(x, y, z) = kxyz$$

for $x = 1, 2; y = 1, 2, 3; z = 1, 2$.

**3.73** With reference to Exercise 3.72, find

(a) $P(X = 1, Y \leq 2, Z = 1)$;
(b) $P(X = 2, Y + Z = 4)$.

**3.74** With reference to Exercise 3.72, find the following values of the joint distribution function of the three random variables:

(a) $F(2, 1, 2)$;
(b) $F(1, 0, 1)$;
(c) $F(4, 4, 4)$.

**3.75** Find $k$ if the joint probability density of $X$, $Y$, and $Z$ is given by

$$f(x, y, z) = \begin{cases} kxy(1 - z) & \text{for } 0 < x < 1, 0 < y < 1, 0 < z < 1, x + y + z < 1 \\ 0 & \text{elsewhere} \end{cases}$$

**3.76** With reference to Exercise 3.75, find $P(X + Y < \frac{1}{2})$.

**3.77** Use the result of Example 3.16 to verify that the joint distribution function of the random variables $X_1$, $X_2$, and $X_3$ of Example 3.19 is given by

$$F(x_1, x_2, x_3) = \begin{cases} 0 & \text{for } x_1 \leq 0, x_2 \leq 0, \text{ or } x_3 \leq 0 \\ \frac{1}{2} x_1 x_2 (x_1 + x_2)(1 - e^{-x_3}) & \text{for } 0 < x_1 < 1, 0 < x_2 < 1, x_3 > 0 \\ \frac{1}{2} x_2 (x_2 + 1)(1 - e^{-x_3}) & \text{for } x_1 \geq 1, 0 < x_2 < 1, x_3 > 0 \\ \frac{1}{2} x_1 (x_1 + 1)(1 - e^{-x_3}) & \text{for } 0 < x_1 < 1, x_2 \geq 1, x_3 > 0 \\ 1 - e^{-x_3} & \text{for } x_1 \geq 1, x_2 \geq 1, x_3 > 0 \end{cases}$$

**3.78** If the joint probability density of $X$, $Y$, and $Z$ is given by

$$f(x, y, z) = \begin{cases} \frac{1}{3}(2x + 3y + z) & \text{for } 0 < x < 1, 0 < y < 1, 0 < z < 1 \\ 0 & \text{elsewhere} \end{cases}$$

find
(a)  $P(X = \frac{1}{2}, Y = \frac{1}{2}, Z = \frac{1}{2})$;
(b)  $P(X < \frac{1}{2}, Y < \frac{1}{2}, Z < \frac{1}{2})$.

## APPLICATIONS

**3.79** Suppose we roll a pair of balanced dice, $X$ is the number of dice that come up 1, and $Y$ is the number of dice that come up 4, 5, or 6.
(a)  Draw a diagram like that of Figure 3.1, showing the values of $X$ and $Y$ associated with each of the 36 equally likely points of the sample space.
(b)  Construct a table showing the values of the joint probability distribution of $X$ and $Y$.

**3.80** Two textbooks are selected at random from a shelf that contains three statistics texts, two mathematics texts, and three physics texts. If $X$ is the number of statistics text and $Y$ the number of mathematics texts actually chosen, construct a table showing the values of the joint probability distribution of $X$ and $Y$.

**3.81** If $X$ is the number of heads and $Y$ the number of heads minus the number of tails obtained in three flips of a balanced coin, construct a table showing the values of the joint probability distribution of $X$ and $Y$.

**3.82** A sharpshooter is aiming at a circular target with radius 1. If we draw a rectangular system of coordinates with its origin at the center of the target, the coordinates of the point of impact, $(X, Y)$, are random variables having the joint probability density

$$f(x, y) = \begin{cases} \dfrac{1}{\pi} & \text{for } 0 < x^2 + y^2 < 1 \\ 0 & \text{elsewhere} \end{cases}$$

Find

(a)  $P[(X, Y) \in A]$, where $A$ is the sector of the circle in the first quadrant bounded by the lines $y = 0$ and $y = x$;

(b)  $P[(X, Y) \in B]$, where $B = \{(x, y)|0 < x^2 + y^2 < \frac{1}{2}\}$.

**3.83** A certain college gives aptitude tests in the sciences and the humanities to all entering freshmen. If $X$ and $Y$ are, respectively, the proportions of correct answers a student gets on the tests in the two subjects, the joint probability distribution of these random variables can be approximated with the joint probability density

$$f(x, y) = \begin{cases} \frac{2}{5}(2x + 3y) & \text{for } 0 < x < 1, 0 < y < 1 \\ 0 & \text{elsewhere} \end{cases}$$

What are the probabilities that a student will get

(a)  less than 0.40 on both tests;

(b)  more than 0.80 on the science test and less than 0.50 on the humanities test?

**3.84** Suppose that $P$, the price of a certain commodity (in dollars), and $S$, its total sales (in 10,000 units), are random variables whose joint probability distribution can be approximated closely with the joint probability density

$$f(p, s) = \begin{cases} 5pe^{-ps} & \text{for } 0.20 < p < 0.40, s > 0 \\ 0 & \text{elsewhere} \end{cases}$$

Find the probabilities that

(a)  the price will be less than 30 cents and sales will exceed 20,000 units;

(b)  the price will be between 25 cents and 30 cents and sales will be less than 10,000 units.

## 3.6  MARGINAL DISTRIBUTIONS

To introduce the concept of a **marginal distribution**, let us consider the following example.

## EXAMPLE 3.20

In Example 3.12 we derived the joint probability distribution of two random variables $X$ and $Y$, the number of aspirin caplets and the number of sedative caplets included among two caplets drawn at random from a bottle containing three aspirin, two sedative, and four laxative caplets. Find the probability distribution of $X$ alone and that of $Y$ alone.

*Solution*

The results of Example 3.12 are shown in the following table, together with the **marginal totals**, that is, the totals of the respective rows and columns:

| | | $x$ | | | |
|---|---|---|---|---|---|
| | | 0 | 1 | 2 | |
| $y$ | 0 | $\frac{1}{6}$ | $\frac{1}{3}$ | $\frac{1}{12}$ | $\frac{7}{12}$ |
| | 1 | $\frac{2}{9}$ | $\frac{1}{6}$ | | $\frac{7}{18}$ |
| | 2 | $\frac{1}{36}$ | | | $\frac{1}{36}$ |
| | | $\frac{5}{12}$ | $\frac{1}{2}$ | $\frac{1}{12}$ | |

The column totals are the probabilities that $X$ will take on the values 0, 1, and 2. In other words, they are the values

$$g(x) = \sum_{y=0}^{2} f(x, y) \qquad \text{for } x = 0, 1, 2$$

of the probability distribution of $X$. By the same token, the row totals are the values

$$h(y) = \sum_{x=0}^{2} f(x, y) \qquad \text{for } y = 0, 1, 2$$

of the probability distribution of $Y$.    ▲

We are thus led to the following definition.

---

**DEFINITION 3.10** If $X$ and $Y$ are discrete random variables and $f(x, y)$ is the value of their joint probability distribution at $(x, y)$, the function given by

$$g(x) = \sum_{y} f(x, y)$$

for each $x$ within the range of $X$ is called the **marginal distribution** of $X$. Correspondingly, the function given by

$$h(y) = \sum_{x} f(x, y)$$

for each $y$ within the range of $Y$ is called the **marginal distribution** of $Y$.

---

When $X$ and $Y$ are continuous random variables, the probability distributions are replaced by probability densities, the summations are replaced by integrals, and we get

---

**DEFINITION 3.11** If $X$ and $Y$ are continuous random variables and $f(x, y)$ is the value of their joint probability density at $(x, y)$, the function given by

$$g(x) = \int_{-\infty}^{\infty} f(x, y) \, dy \qquad \text{for } -\infty < x < \infty$$

is called the **marginal density** of $X$. Correspondingly, the function given by

$$h(y) = \int_{-\infty}^{\infty} f(x, y) \, dx \qquad \text{for } -\infty < y < \infty$$

is called the **marginal density** of $Y$.

---

## EXAMPLE 3.21

Given the joint probability density

$$f(x, y) = \begin{cases} \frac{2}{3}(x + 2y) & \text{for } 0 < x < 1, \, 0 < y < 1 \\ 0 & \text{elsewhere} \end{cases}$$

find the marginal densities of $X$ and $Y$.

*Solution*

Performing the necessary integrations, we get

$$g(x) = \int_{-\infty}^{\infty} f(x, y) \, dy = \int_{0}^{1} \frac{2}{3}(x + 2y) \, dy = \frac{2}{3}(x + 1)$$

for $0 < x < 1$ and $g(x) = 0$ elsewhere. Likewise,

$$h(y) = \int_{-\infty}^{\infty} f(x, y) \, dx = \int_{0}^{1} \frac{2}{3}(x + 2y) \, dx = \frac{1}{3}(1 + 4y)$$

for $0 < y < 1$ and $h(y) = 0$ elsewhere.    ▲

When we are dealing with more than two random variables, we can speak not only of the marginal distributions of the individual random variables, but also of the **joint marginal distributions** of several of the random variables. If the joint probability distribution of the discrete random variables $X_1, X_2, \ldots,$ and $X_n$ has the values $f(x_1, x_2, \ldots, x_n)$, the marginal distribution of $X_1$ alone is given by

$$g(x_1) = \sum_{x_2} \cdots \sum_{x_n} f(x_1, x_2, \ldots, x_n)$$

for all values within the range of $X_1$, the joint marginal distribution of $X_1, X_2,$ and $X_3$ is given by

$$m(x_1, x_2, x_3) = \sum_{x_4} \cdots \sum_{x_n} f(x_1, x_2, \ldots, x_n)$$

for all values within the range of $X_1, X_2,$ and $X_3$, and other marginal distributions can be defined in the same way. For the continuous case, probability distributions are replaced by probability densities, summations are replaced by integrals, and if the joint probability density of the continuous random variables $X_1, X_2, \ldots,$ and $X_n$ has the values $f(x_1, x_2, \ldots, x_n)$, the marginal density of $X_2$ alone is given by

$$h(x_2) = \int_{-\infty}^{\infty} \cdots \int_{-\infty}^{\infty} f(x_1, x_2, \ldots, x_n) \, dx_1 \, dx_3 \cdots dx_n$$

for $-\infty < x_2 < \infty$, the joint marginal density of $X_1$ and $X_n$ is given by

$$\varphi(x_1, x_n) = \int_{-\infty}^{\infty} \cdots \int_{-\infty}^{\infty} f(x_1, x_2, \ldots, x_n) \, dx_2 \, dx_3 \cdots dx_{n-1}$$

for $-\infty < x_1 < \infty$ and $-\infty < x_n < \infty$, and so forth.

## EXAMPLE 3.22

Considering again the trivariate probability density of Example 3.19, namely,

$$f(x_1, x_2, x_3) = \begin{cases} (x_1 + x_2)e^{-x_3} & \text{for } 0 < x_1 < 1, 0 < x_2 < 1, x_3 > 0 \\ 0 & \text{elsewhere} \end{cases}$$

find the joint marginal density of $X_1$ and $X_3$ and the marginal density of $X_1$ alone.

*Solution*

Performing the necessary integration, we find that the joint marginal density of $X_1$ and $X_3$ is given by

$$m(x_1, x_3) = \int_0^1 (x_1 + x_2)e^{-x_3}\, dx_2 = (x_1 + \tfrac{1}{2})e^{-x_3}$$

for $0 < x_1 < 1$ and $x_3 > 0$ and $m(x_1, x_3) = 0$ elsewhere. Using this result, we find that the marginal density of $X_1$ alone is given by

$$g(x_1) = \int_0^\infty \int_0^1 f(x_1, x_2, x_3)\, dx_2\, dx_3 = \int_0^\infty m(x_1, x_3)\, dx_3$$

$$= \int_0^\infty (x_1 + \tfrac{1}{2})e^{-x_3}\, dx_3 = x_1 + \tfrac{1}{2}$$

for $0 < x_1 < 1$ and $g(x_1) = 0$ elsewhere.  ▲

Corresponding to the various marginal and joint marginal distributions and densities we have introduced in this section, we can also define **marginal** and **joint marginal distribution functions**. Some problems relating to such distribution functions will be left to the reader in Exercises 3.88, 3.95, and 3.96.

## 3.7 CONDITIONAL DISTRIBUTIONS

In Chapter 2 we defined the conditional probability of event $A$ given event $B$ as

$$P(A|B) = \frac{P(A \cap B)}{P(B)}$$

provided $P(B) \neq 0$. Suppose now that $A$ and $B$ are the events $X = x$ and $Y = y$, so that we can write

$$P(X = x|Y = y) = \frac{P(X = x, Y = y)}{P(Y = y)}$$

$$= \frac{f(x, y)}{h(y)}$$

provided $P(Y = y) = h(y) \neq 0$, where $f(x, y)$ is the value of the joint probability distribution of $X$ and $Y$ at $(x, y)$ and $h(y)$ is the value of the marginal distribution of $Y$ at $y$. Denoting the conditional probability by $f(x|y)$ to indicate that $x$ is a variable and $y$ is fixed, let us now make the following definition.

---

**DEFINITION 3.12** If $f(x, y)$ is the value of the joint probability distribution of the discrete random variables $X$ and $Y$ at $(x, y)$ and $h(y)$ is the value of the marginal distribution of $Y$ at $y$, the function given by

$$f(x|y) = \frac{f(x, y)}{h(y)} \qquad h(y) \neq 0$$

for each $x$ within the range of $X$, is called the **conditional distribution** of $X$ given $Y = y$. Correspondingly, if $g(x)$ is the value of the marginal distribution of $X$ at $x$, the function given by

$$w(y|x) = \frac{f(x, y)}{g(x)} \qquad g(x) \neq 0$$

for each $y$ within the range of $Y$, is called the **conditional distribution** of $Y$ given $X = x$.

---

### EXAMPLE 3.23

With reference to Examples 3.12 and 3.20, find the conditional distribution of $X$ given $Y = 1$.

*Solution*

Substituting the appropriate values from the table on page 126, we get

$$f(0|1) = \frac{\frac{2}{9}}{\frac{7}{18}} = \frac{4}{7}$$

$$f(1|1) = \frac{\frac{1}{6}}{\frac{7}{18}} = \frac{3}{7}$$

$$f(2|1) = \frac{0}{\frac{7}{18}} = 0 \qquad \blacktriangle$$

When $X$ and $Y$ are continuous random variables, the probability distributions are replaced by probability densities, and we get

> **DEFINITION 3.13** If $f(x, y)$ is the value of the joint density of the continuous random variables $X$ and $Y$ at $(x, y)$ and $h(y)$ is the value of the marginal density of $Y$ at $y$, the function given by
>
> $$f(x|y) = \frac{f(x, y)}{h(y)} \qquad h(y) \neq 0$$
>
> for $-\infty < x < \infty$, is called the **conditional density** of $X$ given $Y = y$. Correspondingly, if $g(x)$ is the value of the marginal density of $X$ at $x$, the function given by
>
> $$w(y|x) = \frac{f(x, y)}{g(x)} \qquad g(x) \neq 0$$
>
> for $-\infty < y < \infty$, is called the **conditional density** of $Y$ given $X = x$.

## EXAMPLE 3.24

With reference to Example 3.21, find the conditional density of $X$ given $Y = y$, and use it to evaluate $P(X \leq \frac{1}{2}| Y = \frac{1}{2})$.

### Solution

Using the results obtained on page 127, we have

$$f(x|y) = \frac{f(x, y)}{h(y)} = \frac{\frac{2}{3}(x + 2y)}{\frac{1}{3}(1 + 4y)}$$

$$= \frac{2x + 4y}{1 + 4y}$$

for $0 < x < 1$ and $f(x|y) = 0$ elsewhere. Now, $f(x|\frac{1}{2}) = \dfrac{2x + 4 \cdot \frac{1}{2}}{1 + 4 \cdot \frac{1}{2}} = \dfrac{2x + 2}{3}$ and we can write

$$P(X \leq \tfrac{1}{2}| Y = \tfrac{1}{2}) = \int_0^{\frac{1}{2}} \frac{2x + 2}{3} \, dx = \tfrac{5}{12}$$

It is of interest to note that in Figure 3.11 this probability is given by the ratio of the area of trapezoid $ABCD$ to the area of trapezoid $AEFD$.  ▲

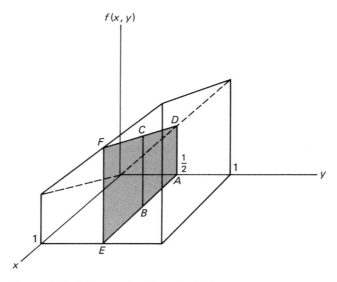

**Figure 3.11**  Diagram for Example 3.24.

## EXAMPLE 3.25

Given the joint probability density

$$f(x, y) = \begin{cases} 4xy & \text{for } 0 < x < 1, 0 < y < 1 \\ 0 & \text{elsewhere} \end{cases}$$

find the marginal densities of $X$ and $Y$ and the conditional density of $X$ given $Y = y$.

*Solution*

Performing the necessary integrations, we get

$$g(x) = \int_{-\infty}^{\infty} f(x, y)\, dy = \int_{0}^{1} 4xy\, dy$$

$$= 2xy^2 \Big|_{y=0}^{y=1} = 2x$$

for $0 < x < 1$, and $g(x) = 0$ elsewhere; also

$$h(y) = \int_{-\infty}^{\infty} f(x, y) \, dx = \int_{0}^{1} 4xy \, dx$$

$$= 2x^2 y \Big|_{x=0}^{x=1} = 2y$$

for $0 < y < 1$, and $h(y) = 0$ elsewhere. Then, substituting into the formula for a conditional density, we get

$$f(x|y) = \frac{f(x, y)}{h(y)} = \frac{4xy}{2y} = 2x$$

for $0 < x < 1$, and $f(x|y) = 0$ elsewhere.    ▲

When we are dealing with more than two random variables, whether continuous or discrete, we can consider various different kinds of conditional distributions or densities. For instance, if $f(x_1, x_2, x_3, x_4)$ is the value of the joint distribution of the discrete random variables $X_1$, $X_2$, $X_3$, and $X_4$ at $(x_1, x_2, x_3, x_4)$, we can write

$$p(x_3|x_1, x_2, x_4) = \frac{f(x_1, x_2, x_3, x_4)}{g(x_1, x_2, x_4)} \qquad g(x_1, x_2, x_4) \neq 0$$

for the value of the conditional distribution of $X_3$ at $x_3$ given $X_1 = x_1$, $X_2 = x_2$, and $X_4 = x_4$, where $g(x_1, x_2, x_4)$ is the value of the joint marginal distribution of $X_1$, $X_2$, and $X_4$ at $(x_1, x_2, x_4)$. We can also write

$$q(x_2, x_4|x_1, x_3) = \frac{f(x_1, x_2, x_3, x_4)}{m(x_1, x_3)} \qquad m(x_1, x_3) \neq 0$$

for the value of the **joint conditional distribution** of $X_2$ and $X_4$ at $(x_2, x_4)$ given $X_1 = x_1$ and $X_3 = x_3$, or

$$r(x_2, x_3, x_4|x_1) = \frac{f(x_1, x_2, x_3, x_4)}{b(x_1)} \qquad b(x_1) \neq 0$$

for the value of the joint conditional distribution of $X_2$, $X_3$, and $X_4$ at $(x_2, x_3, x_4)$ given $X_1 = x_1$.

When we are dealing with two or more random variables, questions of **independence** are usually of great importance. In Example 3.25 we see that $f(x|y) = 2x$ does not depend on the given value $Y = y$, but this is clearly not the case in Example 3.24, where $f(x|y) = \dfrac{2x + 4y}{1 + 4y}$. Whenever the values of the conditional distribution of $X$ given $Y = y$ do not depend on $y$, it follows that $f(x|y) = g(x)$, and hence the formulas of Definitions 3.12 and 3.13 yield

$$f(x, y) = f(x|y) \cdot h(y) = g(x) \cdot h(y)$$

That is, the values of the joint distribution are given by the products of the corresponding values of the two marginal distributions. Generalizing from this observation, let us now make the following definition.

---

**DEFINITION 3.14** If $f(x_1, x_2, \ldots, x_n)$ is the value of the joint probability distribution of the $n$ discrete random variables $X_1, X_2, \ldots, X_n$ at $(x_1, x_2, \ldots, x_n)$, and $f_i(x_i)$ is the value of the marginal distribution of $X_i$ at $x_i$ for $i = 1, 2, \ldots, n$, then the $n$ random variables are **independent** if and only if

$$f(x_1, x_2, \ldots, x_n) = f_1(x_1) \cdot f_2(x_2) \cdot \ldots \cdot f_n(x_n)$$

for all $(x_1, x_2, \ldots, x_n)$ within their range.

---

To give a corresponding definition for continuous random variables, we simply substitute the word *density* for the word *distribution*.

With this definition of independence, it can easily be verified that the three random variables of Example 3.22 are not independent, but that the two random variables $X_1$ and $X_3$ and also the two random variables $X_2$ and $X_3$ are **pairwise independent** (see Exercise 3.97).

The following examples serve to illustrate the use of Definition 3.14 in finding probabilities relating to several independent random variables.

## EXAMPLE 3.26

Considering $n$ independent flips of a balanced coin, let $X_i$ be the number of heads (0 or 1) obtained in the $i$th flip for $i = 1, 2, \ldots, n$. Find the joint probability distribution of these $n$ random variables.

### Solution

Since each of the random variables $X_i$, for $i = 1, 2, \ldots, n$, has the probability distribution

$$f_i(x_i) = \tfrac{1}{2} \qquad \text{for } x_i = 0, 1$$

and the $n$ random variables are independent, their joint probability distribution is given by

$$f(x_1, x_2, \ldots, x_n) = f_1(x_1) \cdot f_2(x_2) \cdot \ldots \cdot f_n(x_n)$$

$$= \frac{1}{2} \cdot \frac{1}{2} \cdot \ldots \cdot \frac{1}{2} = \left(\frac{1}{2}\right)^n$$

where $x_i = 0$ or $1$ for $i = 1, 2, \ldots, n$.    ▲

## EXAMPLE 3.27

Given the independent random variables $X_1$, $X_2$, and $X_3$ with the probability densities

$$f_1(x_1) = \begin{cases} e^{-x_1} & \text{for } x_1 > 0 \\ 0 & \text{elsewhere} \end{cases}$$

$$f_2(x_2) = \begin{cases} 2e^{-2x_2} & \text{for } x_2 > 0 \\ 0 & \text{elsewhere} \end{cases}$$

$$f_3(x_3) = \begin{cases} 3e^{-3x_3} & \text{for } x_3 > 0 \\ 0 & \text{elsewhere} \end{cases}$$

find their joint probability density, and use it to evaluate the probability $P(X_1 + X_2 \leq 1, X_3 > 1)$.

*Solution*

According to Definition 3.14, the values of the joint probability density are given by

$$f(x_1, x_2, x_3) = f_1(x_1) \cdot f_2(x_2) \cdot f_3(x_3)$$

$$= e^{-x_1} \cdot 2e^{-2x_2} \cdot 3e^{-3x_3}$$

$$= 6e^{-x_1 - 2x_2 - 3x_3}$$

for $x_1 > 0$, $x_2 > 0$, $x_3 > 0$, and $f(x_1, x_2, x_3) = 0$ elsewhere. Thus,

$$P(X_1 + X_2 \leq 1, X_3 > 1) = \int_1^\infty \int_0^1 \int_0^{1-x_2} 6e^{-x_1 - 2x_2 - 3x_3} \, dx_1 \, dx_2 \, dx_3$$

$$= (1 - 2e^{-1} + e^{-2})e^{-3}$$

$$= 0.020 \quad ▲$$

## EXERCISES

**3.85**  Given the values of the joint probability distribution of $X$ and $Y$ shown in the table

$$
\begin{array}{cc|cc}
 & & \multicolumn{2}{c}{x} \\
 & & -1 & 1 \\
\hline
 & -1 & \frac{1}{8} & \frac{1}{2} \\
y & 0 & 0 & \frac{1}{4} \\
 & 1 & \frac{1}{8} & 0 \\
\end{array}
$$

find

(a)  the marginal distribution of $X$;
(b)  the marginal distribution of $Y$;
(c)  the conditional distribution of $X$ given $Y = -1$.

**3.86**  With reference to Exercise 3.52, find

(a)  the marginal distribution of $X$;
(b)  the marginal distribution of $Y$;
(c)  the conditional distribution of $X$ given $Y = 1$;
(d)  the conditional distribution of $Y$ given $X = 0$.

**3.87**  Given the joint probability distribution

$$f(x, y, z) = \frac{xyz}{108} \qquad \text{for } x = 1, 2, 3; \quad y = 1, 2, 3; \quad z = 1, 2$$

find

(a)  the joint marginal distribution of $X$ and $Y$;
(b)  the joint marginal distribution of $X$ and $Z$;
(c)  the marginal distribution of $X$;
(d)  the conditional distribution of $Z$ given $X = 1$ and $Y = 2$;
(e)  the joint conditional distribution of $Y$ and $Z$ given $X = 3$.

**3.88**  With reference to Example 3.20, find

(a)  the **marginal distribution function** of $X$, namely, the function given by $G(x) = P(X \leqslant x)$ for $-\infty < x < \infty$;
(b)  the **conditional distribution function** of $X$ given $Y = 1$, namely, the function given by $F(x|1) = P(X \leqslant x|Y = 1)$ for $-\infty < x < \infty$.

**3.89**  Check whether $X$ and $Y$ are independent, if their joint probability distribution is given by

(a)  $f(x, y) = \frac{1}{4}$ for $x = -1$ and $y = -1$, $x = -1$ and $y = 1$, $x = 1$ and $y = -1$, and $x = 1$ and $y = 1$;

(b)  $f(x, y) = \frac{1}{3}$ for $x = 0$ and $y = 0$, $x = 0$ and $y = 1$, and $x = 1$ and $y = 1$.

**3.90**  If the joint probability density of $X$ and $Y$ is given by

$$f(x, y) = \begin{cases} \frac{1}{4}(2x + y) & \text{for } 0 < x < 1, 0 < y < 2 \\ 0 & \text{elsewhere} \end{cases}$$

find

(a)  the marginal density of $X$;
(b)  the conditional density of $Y$ given $X = \frac{1}{4}$.

**3.91**  With reference to Exercise 3.90, find

(a)  the marginal density of $Y$;
(b)  the conditional density of $X$ given $Y = 1$.

**3.92**  If the joint probability density of $X$ and $Y$ is given by

$$f(x, y) = \begin{cases} 24y(1 - x - y) & \text{for } x > 0, \quad y > 0, \quad x + y < 1 \\ 0 & \text{elsewhere} \end{cases}$$

find

(a)  the marginal density of $X$;
(b)  the marginal density of $Y$.

Also determine whether the two random variables are independent.

**3.93**  With reference to Exercise 3.63, find

(a)  the marginal density of $X$;
(b)  the marginal density of $Y$.

Also determine whether the two random variables are independent.

**3.94**  With reference to Example 3.22, find

(a)  the conditional density of $X_2$ given $X_1 = \frac{1}{3}$ and $X_3 = 2$;
(b)  the joint conditional density of $X_2$ and $X_3$ given $X_1 = \frac{1}{2}$.

**3.95**  If $F(x, y)$ is the value of the joint distribution function of $X$ and $Y$ at $(x, y)$, show that the **marginal distribution function** of $X$ is given by

$$G(x) = F(x, \infty) \qquad \text{for } -\infty < x < \infty$$

Use this result to find the marginal distribution function of $X$ for the random variables of Exercise 3.64.

**3.96**  If $F(x_1, x_2, x_3)$ is the value of the joint distribution function of $X_1$, $X_2$, and $X_3$ at $(x_1, x_2, x_3)$, show that the **joint marginal distribution function** of $X_1$ and $X_3$ is given by

$$M(x_1, x_3) = F(x_1, \infty, x_3) \qquad \text{for } -\infty < x_1 < \infty, -\infty < x_3 < \infty$$

and that the **marginal distribution function** of $X_1$ is given by

$$G(x_1) = F(x_1, \infty, \infty) \qquad \text{for } -\infty < x_1 < \infty$$

With reference to Example 3.19, use these results to find

(a) the joint marginal distribution function of $X_1$ and $X_3$;
(b) the marginal distribution function of $X_1$.

**3.97** With reference to Example 3.22, verify that the three random variables $X_1$, $X_2$, and $X_3$ are not independent, but that the two random variables $X_1$ and $X_3$, and also the two random variables $X_2$ and $X_3$, are **pairwise independent**.

**3.98** If the independent random variables $X$ and $Y$ have the marginal densities

$$f(x) = \begin{cases} \frac{1}{2} & \text{for } 0 < x < 2 \\ 0 & \text{elsewhere} \end{cases}$$

$$\pi(y) = \begin{cases} \frac{1}{3} & \text{for } 0 < y < 3 \\ 0 & \text{elsewhere} \end{cases}$$

find

(a) the joint probability density of $X$ and $Y$;
(b) the value of $P(X^2 + Y^2 > 1)$.

## APPLICATIONS

**3.99** With reference to Exercise 3.80, find

(a) the marginal distribution of $X$;
(b) the conditional distribution of $Y$ given $X = 0$.

**3.100** If two cards are randomly drawn (without replacement) from an ordinary deck of 52 playing cards, $Z$ is the number of aces obtained in the first draw and $W$ is the total number of aces obtained in both draws, find

(a) the joint probability distribution of $Z$ and $W$;
(b) the marginal distribution of $Z$;
(c) the conditional distribution of $W$ given $Z = 1$.

**3.101** If $X$ is the proportion of persons who will respond to one kind of mail-order solicitation, $Y$ is the proportion of persons who will respond to another kind of mail-order solicitation, and the joint probability density of $X$ and $Y$ is given by

$$f(x, y) = \begin{cases} \frac{2}{5}(x + 4y) & \text{for } 0 < x < 1, 0 < y < 1 \\ 0 & \text{elsewhere} \end{cases}$$

find the probabilities that

(a)   at least 30 percent will respond to the first kind of mail-order solicitation;

(b)   at most 50 percent will respond to the second kind of mail-order solicitation given that there has been a 20 percent response to the first kind of mail-order solicitation.

**3.102**  With reference to Exercise 3.84, find

(a)   the marginal density of $P$;
(b)   the conditional density of $S$ given $P = p$;
(c)   the probability that sales will be less than 30,000 units when $p = 25$ cents.

**3.103**  If $X$ is the amount of money (in dollars) that a salesperson spends on gasoline during a day and $Y$ is the corresponding amount of money (in dollars) for which he or she is reimbursed, the joint density of these two random variables is given by

$$f(x, y) = \begin{cases} \dfrac{1}{25}\left(\dfrac{20 - x}{x}\right) & \text{for } 10 < x < 20, \ \dfrac{x}{2} < y < x \\ 0 & \text{elsewhere} \end{cases}$$

find

(a)   the marginal density of $X$;
(b)   the conditional density of $Y$ given $X = 12$;
(c)   the probability that the salesperson will be reimbursed at least \$8 when spending \$12.

**3.104**  Show that the two random variables of Exercise 3.83 are not independent.

**3.105**  The useful life (in hours) of a certain kind of vacuum tube is a random variable having the probability density

$$f(x) = \begin{cases} \dfrac{20{,}000}{(x + 100)^3} & \text{for } x > 0 \\ 0 & \text{elsewhere} \end{cases}$$

If three of these tubes operate independently, find

(a)   the joint probability density of $X_1$, $X_2$, and $X_3$, representing the lengths of their useful lives;
(b)   the value of $P(X_1 < 100, X_2 < 100, X_3 \geq 200)$.

## REFERENCES

More advanced, or more detailed treatments of the material in this chapter may be found in

BRUNK, H. D., *An Introduction to Mathematical Statistics*, 3rd ed. Lexington, Mass.: Xerox College Publishing, 1975,

DEGROOT, M. H., *Probability and Statistics*, 2nd ed. Reading, Mass.: Addison-Wesley Publishing Company, Inc., 1986,

FRASER, D. A. S., *Probability and Statistics: Theory and Applications*. North Scituate, Mass.: Duxbury Press, 1976,

HOGG, R. V., and CRAIG, A. T., *Introduction to Mathematical Statistics*, 4th ed. New York: Macmillan Publishing Co., Inc., 1978,

KENDALL, M. G., and STUART, A., *The Advanced Theory of Statistics*, Vol. 1, 4th ed. New York: Macmillan Publishing Co., Inc., 1977,

KHAZANIE, R., *Basic Probability Theory and Applications*. Pacific Palisades, Calif.: Goodyear Publishing Company, Inc., 1976.

# 4

# *Mathematical Expectation*

## 4.1 INTRODUCTION

Originally, the concept of a **mathematical expectation** arose in connection with games of chance, and in its simplest form it is the product of the amount a player stands to win and the probability that he or she will win. For instance, if we hold one of 10,000 tickets in a raffle for which the grand prize is a trip worth $4,800, our mathematical expectation is $4,800 \cdot \dfrac{1}{10,000} = \$0.48$. This figure will have to be interpreted in the sense of an average—altogether the 10,000 tickets pay $4,800, or on the average $\dfrac{\$4,800}{10,000} = \$0.48$ per ticket.

If there is also a second prize worth $1,200 and a third prize worth $400, we can argue that altogether the 10,000 tickets pay $4,800 + $1,200 + $400 = $6,400, or on the average $\dfrac{\$6,400}{10,000} = \$0.64$ per ticket. Looking at this in a different way, we could argue that if the raffle is repeated many times, we would lose 99.97 percent of the time (or with probability 0.9997) and win each of the prizes 0.01 percent of the time (or with probability 0.0001). On the average we would thus win

$$0(0.9997) + 4,800(0.0001) + 1,200(0.0001) + 400(0.0001) = \$0.64$$

which is the sum of the products obtained by multiplying each amount by the corresponding probability.

## 4.2  THE EXPECTED VALUE OF A RANDOM VARIABLE

In the illustration of the preceding section, the amount we won was a random variable, and the mathematical expectation of this random variable was the sum of the products obtained by multiplying each value of the random variable by the corresponding probability. Referring to the mathematical expectation of a random variable simply as its **expected value**, and extending the definition to the continuous case by replacing the operation of summation by integration, we thus have

---

**DEFINITION 4.1**  If $X$ is a discrete random variable and $f(x)$ is the value of its probability distribution at $x$, the **expected value** of $X$ is

$$E(X) = \sum_x x \cdot f(x)$$

Correspondingly, if $X$ is a continuous random variable and $f(x)$ is the value of its probability density at $x$, the **expected value** of $X$ is

$$E(X) = \int_{-\infty}^{\infty} x \cdot f(x)\, dx$$

---

In this definition it is assumed, of course, that the sum or the integral exists; otherwise, the mathematical expectation is undefined.

## EXAMPLE 4.1

A lot of 12 television sets includes 2 with white cords. If three of the sets are chosen at random for shipment to a hotel, how many sets with white cords can the shipper expect to send to the hotel?

### Solution

Since $x$ of the two sets with white cords and $3 - x$ of the 10 other sets can be chosen in $\binom{2}{x}\binom{10}{3-x}$ ways, three of the 12 sets can be chosen in $\binom{12}{3}$ ways, and these $\binom{12}{3}$ possibilities are presumably equiprobable, we find that the probability distribution of $X$, the number of sets with white cords shipped to the hotel, is given by

$$f(x) = \frac{\binom{2}{x}\binom{10}{3-x}}{\binom{12}{3}} \qquad \text{for } x = 0, 1, 2$$

or, in tabular form,

| $x$ | 0 | 1 | 2 |
|---|---|---|---|
| $f(x)$ | $\frac{6}{11}$ | $\frac{9}{22}$ | $\frac{1}{22}$ |

Now,

$$E(X) = 0 \cdot \tfrac{6}{11} + 1 \cdot \tfrac{9}{22} + 2 \cdot \tfrac{1}{22} = \tfrac{1}{2}$$

and since half a set cannot possibly be shipped, it should be clear that the term "expect" is not used in its colloquial sense. Indeed, it should be interpreted as an average pertaining to repeated shipments made under the given conditions.  ▲

## EXAMPLE 4.2

Certain coded measurements of the pitch diameter of threads of a fitting have the probability density

$$
f(x) = \begin{cases} \dfrac{4}{\pi(1 + x^2)} & \text{for } 0 < x < 1 \\ 0 & \text{elsewhere} \end{cases}
$$

Find the expected value of this random variable.

### Solution

Using Definition 4.1, we have

$$
E(X) = \int_0^1 x \cdot \frac{4}{\pi(1 + x^2)}\, dx
$$

$$
= \frac{4}{\pi} \int_0^1 \frac{x}{1 + x^2}\, dx
$$

$$
= \frac{\ln 4}{\pi} = 0.4413 \qquad \blacktriangle
$$

There are many problems in which we are interested not only in the expected value of a random variable $X$, but also in the expected values of random variables related to $X$. Thus, we might be interested in the random variable $Y$, whose values are related to those of $X$ by means of the equation $y = g(x)$, and to simplify our notation we denote this random variable by $g(X)$. For instance, $g(X)$ might be $X^3$, so that when $X$ takes on the value 2, $g(X)$ takes on the value $2^3 = 8$. If we want to find the expected value of such a random variable $g(X)$, we could first determine its probability distribution or density (by one of the methods which will be discussed in Chapter 7) and then use Definition 4.1, but generally it is easier and more straightforward to use the following theorem.

**THEOREM 4.1** If $X$ is a discrete random variable and $f(x)$ is the value of its probability distribution at $x$, the expected value of $g(X)$ is given by

$$
E[g(X)] = \sum_x g(x) \cdot f(x)
$$

Correspondingly, if $X$ is a continuous random variable and $f(x)$ is the value of its probability density at $x$, the expected value of $g(X)$ is given by

$$E[g(X)] = \int_{-\infty}^{\infty} g(x) \cdot f(x) \, dx$$

***Proof.***  Since a more general proof is beyond the scope of this text, we shall prove this theorem here only for the case where $X$ is discrete and has a finite range. Since $y = g(x)$ does not necessarily define a one-to-one correspondence, suppose that $g(x)$ takes on the value $g_i$ when $x$ takes on the values $x_{i1}, x_{i2}, \ldots, x_{in_i}$. Then, the probability that $g(X)$ will take on the value $g_i$ is

$$P[g(X) = g_i] = \sum_{j=1}^{n_i} f(x_{ij})$$

and if $g(x)$ takes on the values $g_1, g_2, \ldots, g_m$, it follows that

$$E[g(X)] = \sum_{i=1}^{m} g_i \cdot P[g(X) = g_i]$$

$$= \sum_{i=1}^{m} g_i \cdot \sum_{j=1}^{n_i} f(x_{ij})$$

$$= \sum_{i=1}^{m} \sum_{j=1}^{n_i} g_i \cdot f(x_{ij})$$

$$= \sum_{x} g(x) \cdot f(x)$$

where the summation extends over all values of $X$.    ▼

# EXAMPLE 4.3

If $X$ is the number of points rolled with a balanced die, find the expected value of $g(X) = 2X^2 + 1$.

### Solution

Since each possible outcome has the probability $\frac{1}{6}$, we get

$$E[g(X)] = \sum_{x=1}^{6} (2x^2 + 1) \cdot \frac{1}{6}$$

$$= (2 \cdot 1^2 + 1) \cdot \tfrac{1}{6} + \cdots + (2 \cdot 6^2 + 1) \cdot \tfrac{1}{6}$$

$$= \tfrac{94}{3} \quad \blacktriangle$$

## EXAMPLE 4.4

If $X$ has the probability density

$$f(x) = \begin{cases} e^{-x} & \text{for } x > 0 \\ 0 & \text{elsewhere} \end{cases}$$

find the expected value of $g(X) = e^{3X/4}$.

*Solution*

According to Theorem 4.1, we have

$$E[e^{3X/4}] = \int_0^\infty e^{3x/4} \cdot e^{-x} \, dx$$

$$= \int_0^\infty e^{-x/4} \, dx$$

$$= 4 \quad \blacktriangle$$

The determination of mathematical expectations can often be simplified by using the following theorems, which enable us to calculate expected values from other known or easily computed expectations. Since the steps are essentially the same, some proofs will be given either for the discrete case or the continuous case; others are left for the reader as exercises.

---

**THEOREM 4.2**  If $a$ and $b$ are constants, then

$$E(aX + b) = aE(X) + b$$

---

***Proof.***   Using Theorem 4.1 with $g(X) = aX + b$, we get

$$E(aX + b) = \int_{-\infty}^\infty (ax + b) \cdot f(x) \, dx$$

$$= a \int_{-\infty}^{\infty} x \cdot f(x) \, dx + b \int_{-\infty}^{\infty} f(x) \, dx$$

$$= aE(X) + b \quad \blacktriangledown$$

If we set $b = 0$ or $a = 0$, it follows from Theorem 4.2 that

---

**COROLLARY 1**   If $a$ is a constant, then

$$E(aX) = aE(X)$$

**COROLLARY 2**   If $b$ is a constant, then

$$E(b) = b$$

---

Observe that if we write $E(b)$, the constant $b$ may be looked upon as a random variable that always takes on the value $b$.

---

**THEOREM 4.3**   If $c_1, c_2, \ldots,$ and $c_n$ are constants, then

$$E\left[ \sum_{i=1}^{n} c_i g_i(X) \right] = \sum_{i=1}^{n} c_i E[g_i(X)]$$

---

***Proof.***   According to Theorem 4.1 with $g(X) = \sum_{i=1}^{n} c_i g_i(X)$, we get

$$E\left[ \sum_{i=1}^{n} c_i g_i(X) \right] = \sum_{x} \left[ \sum_{i=1}^{n} c_i g_i(x) \right] f(x)$$

$$= \sum_{i=1}^{n} \sum_{x} c_i g_i(x) f(x)$$

$$= \sum_{i=1}^{n} c_i \sum_{x} g_i(x) f(x)$$

$$= \sum_{i=1}^{n} c_i E[g_i(X)] \quad \blacktriangledown$$

## EXAMPLE 4.5

Making use of the fact that

$$E(X^2) = (1^2 + 2^2 + 3^2 + 4^2 + 5^2 + 6^2) \cdot \tfrac{1}{6} = \tfrac{91}{6}$$

for the random variable of Example 4.3, rework that example.

*Solution*

$$E(2X^2 + 1) = 2E(X^2) + 1 = 2 \cdot \tfrac{91}{6} + 1 = \tfrac{94}{3} \qquad \blacktriangle$$

## EXAMPLE 4.6

If the probability density of $X$ is given by

$$f(x) = \begin{cases} 2(1 - x) & \text{for } 0 < x < 1 \\ 0 & \text{elsewhere} \end{cases}$$

(a) show that

$$E(X^r) = \frac{2}{(r + 1)(r + 2)}$$

(b) and use this result to evaluate

$$E[(2X + 1)^2]$$

*Solution*

(a) $\quad E(X^r) = \displaystyle\int_0^1 x^r \cdot 2(1 - x)\, dx = 2 \int_0^1 (x^r - x^{r+1})\, dx$

$$= 2\left(\frac{1}{r + 1} - \frac{1}{r + 2}\right) = \frac{2}{(r + 1)(r + 2)}$$

(b) Since $E[(2X + 1)^2] = 4E(X^2) + 4E(X) + 1$, and substitution of $r = 1$ and $r = 2$ into the preceding formula yields $E(X) = \dfrac{2}{2 \cdot 3} = \dfrac{1}{3}$ and

$$E(X^2) = \frac{2}{3 \cdot 4} = \frac{1}{6}, \text{ we get}$$

$$E[(2X + 1)^2] = 4 \cdot \tfrac{1}{6} + 4 \cdot \tfrac{1}{3} + 1 = 3 \qquad \blacktriangle$$

## EXAMPLE 4.7

Show that

$$E[(aX + b)^n] = \sum_{i=0}^{n} \binom{n}{i} a^{n-i} b^i E(X^{n-i})$$

*Solution*

Since $(ax + b)^n = \sum_{i=0}^{n} \binom{n}{i} (ax)^{n-i} b^i$ according to Theorem 1.9, it follows that

$$E[(aX + b)^n] = E\left[ \sum_{i=0}^{n} \binom{n}{i} a^{n-i} b^i X^{n-i} \right]$$

$$= \sum_{i=0}^{n} \binom{n}{i} a^{n-i} b^i E(X^{n-i}) \qquad \blacktriangle$$

The concept of a mathematical expectation can easily be extended to situations involving more than one random variable. For instance, if $Z$ is the random variable whose values are related to those of the two random variables $X$ and $Y$ by means of the equation $z = g(x, y)$, it can be shown that

---

**THEOREM 4.4** If $X$ and $Y$ are discrete random variables and $f(x, y)$ is the value of their joint probability distribution at $(x, y)$, the expected value of $g(X, Y)$ is

$$E[g(X, Y)] = \sum_{x} \sum_{y} g(x, y) \cdot f(x, y)$$

Correspondingly, if $X$ and $Y$ are continuous random variables and $f(x, y)$ is the value of their joint probability density at $(x, y)$, the expected value of $g(X, Y)$ is

$$E[g(X, Y)] = \int_{-\infty}^{\infty} \int_{-\infty}^{\infty} g(x, y) f(x, y) \, dx \, dy$$

---

Generalization of this theorem to functions of any finite number of random variables is straightforward.

## EXAMPLE 4.8

With reference to Example 3.12, find the expected value of $g(X, Y) = X + Y$.

*Solution*

$$E(X + Y) = \sum_{x=0}^{2} \sum_{y=0}^{2} (x + y) \cdot f(x, y)$$

$$= (0 + 0) \cdot \tfrac{1}{6} + (0 + 1) \cdot \tfrac{2}{9} + (0 + 2) \cdot \tfrac{1}{36} + (1 + 0) \cdot \tfrac{1}{3}$$

$$+ (1 + 1) \cdot \tfrac{1}{6} + (2 + 0) \cdot \tfrac{1}{12}$$

$$= \tfrac{10}{9} \quad \blacktriangle$$

## EXAMPLE 4.9

If the joint probability density of $X$ and $Y$ is given by

$$f(x, y) = \begin{cases} \tfrac{2}{7} (x + 2y) & \text{for } 0 < x < 1, 1 < y < 2 \\ 0 & \text{elsewhere} \end{cases}$$

find the expected value of $g(X, Y) = X/Y^3$.

*Solution*

$$E(X/Y^3) = \int_{1}^{2} \int_{0}^{1} \frac{2x(x + 2y)}{7y^3} \, dx \, dy$$

$$= \tfrac{2}{7} \int_{1}^{2} \left( \frac{1}{3y^3} + \frac{1}{y^2} \right) dy$$

$$= \tfrac{15}{84} \quad \blacktriangle$$

The following is another theorem which finds useful applications in subsequent work. It is a generalization of Theorem 4.3, and its proof parallels the one of that theorem.

**THEOREM 4.5**  If $c_1$, $c_2$, ... , and $c_n$ are constants, then

$$E\left[\sum_{i=1}^{n} c_i g_i(X_1, X_2, \ldots, X_k)\right] = \sum_{i=1}^{n} c_i E[g_i(X_1, X_2, \ldots, X_k)]$$

## EXERCISES

**4.1**  To illustrate the proof of Theorem 4.1, consider the random variable $X$ which takes on the values $-2$, $-1$, 0, 1, 2, and 3 with probabilities $f(-2)$, $f(-1)$, $f(0)$, $f(1)$, $f(2)$, and $f(3)$. If $g(X) = X^2$, find

 (a)  $g_1$, $g_2$, $g_3$, and $g_4$, the four possible values of $g(x)$;

 (b)  the probabilities $P[g(X) = g_i]$ for $i = 1, 2, 3, 4$;

 (c)  $E[g(X)] = \sum_{i=1}^{4} g_i \cdot P[g(X) = g_i]$, and show that it equals $\sum_x g(x) \cdot f(x)$.

**4.2**  Prove Theorem 4.2 for discrete random variables.

**4.3**  Prove Theorem 4.3 for continuous random variables.

**4.4**  Prove Theorem 4.5 for discrete random variables.

**4.5**  Given two continuous random variables $X$ and $Y$, use Theorem 4.4 to express $E(X)$ in terms of

 (a)  the joint density of $X$ and $Y$;

 (b)  the marginal density of $X$.

**4.6**  Find the expected value of the discrete random variable $X$ having the probability distribution

$$f(x) = \frac{|x - 2|}{7} \qquad \text{for } x = -1, 0, 1, 3$$

**4.7**  Find the expected value of the random variable $Y$ whose probability density is given by

$$f(y) = \begin{cases} \frac{1}{8}(y + 1) & \text{for } 2 < y < 4 \\ 0 & \text{elsewhere} \end{cases}$$

**4.8**  Find the expected value of the random variable $X$ whose probability density is given by

$$f(x) = \begin{cases} x & \text{for } 0 < x < 1 \\ 2 - x & \text{for } 1 \leqslant x < 2 \\ 0 & \text{elsewhere} \end{cases}$$

**4.9**  (a)  If $X$ takes on the values 0, 1, 2, and 3 with probabilities $\frac{1}{125}$, $\frac{12}{125}$, $\frac{48}{125}$, and $\frac{64}{125}$, find $E(X)$ and $E(X^2)$.

 (b)  Use the results of part (a) to determine the value of $E[(3X + 2)^2]$.

**4.10**  (a)  If the probability density of $X$ is given by

$$f(x) = \begin{cases} \dfrac{1}{x(\ln 3)} & \text{for } 1 < x < 3 \\ 0 & \text{elsewhere} \end{cases}$$

find $E(X)$, $E(X^2)$, and $E(X^3)$.

 (b)  Use the results of part (a) to determine $E(X^3 + 2X^2 - 3X + 1)$.

**4.11**  If the probability density of $X$ is given by

$$f(x) = \begin{cases} \dfrac{x}{2} & \text{for } 0 < x \le 1 \\[2mm] \dfrac{1}{2} & \text{for } 1 < x \le 2 \\[2mm] \dfrac{3-x}{2} & \text{for } 2 < x < 3 \\[2mm] 0 & \text{elsewhere} \end{cases}$$

find the expected value of $g(X) = X^2 - 5X + 3$.

**4.12**  With reference to Exercise 3.57, find $E(2X - Y)$.

**4.13**  With reference to Exercise 3.63, find $E(X/Y)$.

**4.14**  With reference to Exercise 3.72, find the expected value of $U = X + Y + Z$.

**4.15**  With reference to Exercise 3.78, find the expected value of $W = X^2 - YZ$.

**4.16**  If the probability distribution of $X$ is given by $f(x) = (\frac{1}{2})^x$ for $x = 1, 2, 3,$ $\dots$ , show that $E(2^X)$ does not exist. This is the famous **Petersburg paradox**, according to which a player's expectation is infinite (does not exist) if he or she is to receive $2^x$ dollars when, in a series of flips of a balanced coin, the first head appears on the $x$th flip.

## APPLICATIONS

**4.17**  The probability that Ms. Brown will sell a piece of property at a profit of $3,000 is $\frac{3}{20}$, the probability that she will sell it at a profit of $1,500 is $\frac{7}{20}$, the probability that she will break even is $\frac{7}{20}$, and the probability that she will lose $1,500 is $\frac{3}{20}$. What is her expected profit?

**4.18**  A game of chance is considered **fair**, or **equitable**, if each player's expectation is equal to zero. If someone pays us $10 each time we roll a 3 or a

4 with a balanced die, how much should we pay that person when we roll a 1, 2, 5, or 6 to make the game equitable?

**4.19** The manager of a bakery knows that the number of chocolate cakes he can sell on any given day is a random variable having the probability distribution $f(x) = \frac{1}{6}$ for $x = 0, 1, 2, 3, 4,$ and 5. He also knows that there is a profit of $1.00 for each cake which he sells and a loss (due to spoilage) of $0.40 for each cake he does not sell. Assuming that each cake can be sold only on the day it is made, find the baker's expected profit for a day on which he bakes

(a) one of the cakes;
(b) two of the cakes;
(c) three of the cakes;
(d) four of the cakes;
(e) five of the cakes.

How many should he bake in order to maximize his expected profit?

**4.20** If a contractor's profit on a construction job can be looked upon as a continuous random variable having the probability density

$$f(x) = \begin{cases} \frac{1}{18}(x + 1) & \text{for } -1 < x < 5 \\ 0 & \text{elsewhere} \end{cases}$$

where the units are in $1,000, what is her expected profit?

**4.21** With reference to Exercise 3.49, what tread wear can a car owner expect to get with one of the tires?

**4.22** With reference to Exercise 3.50, what is the city's expected water consumption for any given day?

**4.23** With reference to Exercise 3.84, find $E(PS)$, the expected receipts for the commodity.

**4.24** Mr. Adams and Ms. Smith are betting on repeated flips of a coin. At the start of the game Mr. Adams has $a$ dollars and Ms. Smith has $b$ dollars, at each flip the loser pays the winner one dollar, and the game continues until either player is "ruined." Making use of the fact that in an equitable game each player's mathematical expectation is zero, find the probability that Mr. Adams will win Ms. Smith's $b$ dollars before he loses his $a$ dollars.

## 4.3 MOMENTS

In statistics, the mathematical expectations defined here and in Definition 4.4, called the **moments** of the distribution of a random variable or simply the **moments** of a random variable, are of special importance.

---

**DEFINITION 4.2** The $r$th **moment about the origin** of a random variable $X$, denoted by $\mu_r'$, is the expected value of $X^r$; symbolically,

$$\mu_r' = E(X^r) = \sum_x x^r \cdot f(x)$$

for $r = 0, 1, 2, \ldots$ when $X$ is discrete, and

$$\mu_r' = E(X^r) = \int_{-\infty}^{\infty} x^r \cdot f(x)\, dx$$

when $X$ is continuous.

---

It is of interest to note that the term "moment" comes from the field of physics—if the quantities $f(x)$ in the discrete case were point masses acting perpendicularly to the $x$-axis at distances $x$ from the origin, $\mu_1'$ would be the $x$-coordinate of the center of gravity, namely, the first moment divided by $\Sigma f(x) = 1$, and $\mu_2'$ would be the moment of inertia. This also explains why the moments $\mu_r'$ are called moments about the origin—in the analogy to physics, the length of the lever arm is in each case the distance from the origin. The analogy applies also in the continuous case, where $\mu_1'$ and $\mu_2'$ might be the $x$-coordinate of the center of gravity and the moment of inertia of a rod of variable density.

When $r = 0$, we have $\mu_0' = E(X^0) = E(1) = 1$ by Corollary 2 of Theorem 4.2, and this result is as it should be in accordance with Theorems 3.1 and 3.5. When $r = 1$, we have $\mu_1' = E(X)$, which is just the expected value of the random variable $X$, and in view of its importance in statistics we give it a special symbol and a special name.

---

**DEFINITION 4.3** $\mu_1'$ is called the **mean** of the distribution of $X$, or simply the **mean** of $X$, and it is denoted by $\mu$.

---

The special moments we shall define next are of importance in statistics because they serve to describe the shape of the distribution of a random variable, namely, the shape of the graph of its probability distribution or probability density.

---

**DEFINITION 4.4** The $r$th **moment about the mean** of a random variable $X$, denoted by $\mu_r$, is the expected value of $(X - \mu)^r$; symbolically,

$$\mu_r = E[(X - \mu)^r] = \sum_x (x - \mu)^r \cdot f(x)$$

for $r = 0, 1, 2, \ldots$ when $X$ is discrete, and

$$\mu_r = E[(X - \mu)^r] = \int_{-\infty}^{\infty} (x - \mu)^r \cdot f(x) \, dx$$

when $X$ is continuous.

Note that $\mu_0 = 1$ and $\mu_1 = 0$ for any random variable for which $\mu$ exists (see Exercise 4.25).

The second moment about the mean is of special importance in statistics because it is indicative of the spread or dispersion of the distribution of a random variable; thus, it is given a special symbol and a special name.

**DEFINITION 4.5**  $\mu_2$ is called the **variance** of the distribution of $X$, or simply the **variance** of $X$, and it is denoted by $\sigma^2$, var($X$), or $V(X)$; $\sigma$, the positive square root of the variance, is called the **standard deviation**.

Figure 4.1 shows how the variance reflects the spread or dispersion of the distribution of a random variable. Here we show the histograms of the probability distributions of four random variables with the same mean $\mu = 5$, but variances equaling 5.26, 3.18, 1.66, and 0.88. As can be seen, a small value of $\sigma^2$ suggests that we are likely to get a value close to the mean, and a large value of $\sigma^2$ suggests that there is a greater probability of getting a value that is not close to the mean. This will be discussed further in Section 4.4. A brief discussion of how $\mu_3$, the third moment about the mean, describes the **symmetry** or **skewness** (lack of symmetry) of a distribution is given in Exercise 4.34.

In many instances moments about the mean are obtained by first calculating moments about the origin and then expressing the $\mu_r$ in terms of the $\mu_r'$. To serve this purpose, the reader will be asked to verify a general formula in Exercise 4.33. Here, let us merely derive the following computing formula for $\sigma^2$.

**THEOREM 4.6**

$$\sigma^2 = \mu_2' - \mu^2$$

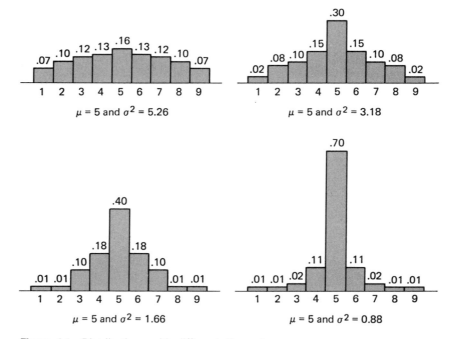

**Figure 4.1**   Distributions with different dispersions.

*Proof.*

$$\sigma^2 = E[(X - \mu)^2]$$
$$= E(X^2 - 2\mu X + \mu^2)$$
$$= E(X^2) - 2\mu E(X) + E(\mu^2)$$
$$= E(X^2) - 2\mu \cdot \mu + \mu^2$$
$$= \mu_2' - \mu^2 \quad \blacktriangledown$$

## EXAMPLE 4.10

Use Theorem 4.6 to calculate the variance of $X$, representing the number of points rolled with a balanced die.

### Solution

First we compute

$$\mu = E(X) = 1 \cdot \tfrac{1}{6} + 2 \cdot \tfrac{1}{6} + 3 \cdot \tfrac{1}{6} + 4 \cdot \tfrac{1}{6} + 5 \cdot \tfrac{1}{6} + 6 \cdot \tfrac{1}{6}$$

$$= \tfrac{7}{2}$$

Now,

$$\mu_2' = E(X^2) = 1^2 \cdot \tfrac{1}{6} + 2^2 \cdot \tfrac{1}{6} + 3^2 \cdot \tfrac{1}{6} + 4^2 \cdot \tfrac{1}{6} + 5^2 \cdot \tfrac{1}{6} + 6^2 \cdot \tfrac{1}{6}$$

$$= \tfrac{91}{6}$$

and it follows that $\sigma^2 = \tfrac{91}{6} - (\tfrac{7}{2})^2 = \tfrac{35}{12}$    ▲

## EXAMPLE 4.11

With reference to Example 4.2, find the standard deviation of the random variable $X$.

### Solution

In Example 4.2 we showed that $\mu = E(X) = 0.4413$. Now

$$\mu_2' = E(X^2) = \frac{4}{\pi} \int_0^1 \frac{x^2}{1 + x^2} \, dx$$

$$= \frac{4}{\pi} \int_0^1 \left( 1 - \frac{1}{1 + x^2} \right) dx$$

$$= \frac{4}{\pi} - 1$$

$$= 0.2732$$

and it follows that

$$\sigma^2 = 0.2732 - (0.4413)^2 = 0.0785$$

and $\sigma = \sqrt{0.0785} = 0.2802$.    ▲

The following is another theorem that is of importance in work connected with standard deviations or variances.

---

**THEOREM 4.7**  If $X$ has the variance $\sigma^2$, then

$$\text{var}(aX + b) = a^2\sigma^2$$

---

The proof of this theorem will be left to the reader, but let us point out the following corollaries: For $a = 1$ we find that the addition of a constant to the values of a random variable, resulting in a shift of all the values of $X$ to the left or to the right, in no way affects the spread of its distribution; for $b = 0$ we find that if the values of a random variable are multiplied by a constant, the variance is multiplied by the square of that constant, resulting in a corresponding change in the spread of the distribution.

## 4.4  CHEBYSHEV'S THEOREM

To demonstrate how $\sigma$ or $\sigma^2$ is indicative of the spread or dispersion of the distribution of a random variable, let us now prove the following theorem, called **Chebyshev's theorem** after the nineteenth-century Russian mathematician P. L. Chebyshev. We shall prove it here only for the continuous case, leaving the discrete case as an exercise.

---

**THEOREM 4.8**  (Chebyshev's Theorem) If $\mu$ and $\sigma$ are the mean and the standard deviation of a random variable $X$, then for any positive constant $k$ the probability is *at least* $1 - \dfrac{1}{k^2}$ that $X$ will take on a value within $k$ standard deviations of the mean; symbolically

$$P(|X - \mu| < k\sigma) \geqslant 1 - \frac{1}{k^2}$$

---

***Proof.***    According to Definitions 4.4 and 4.5, we write

$$\sigma^2 = E[(X - \mu)^2] = \int_{-\infty}^{\infty} (x - \mu)^2 \cdot f(x)\, dx$$

Then, dividing the integral into three parts as shown in Figure 4.2, we get

$$\sigma^2 = \int_{-\infty}^{\mu - k\sigma} (x - \mu)^2 \cdot f(x)\, dx + \int_{\mu - k\sigma}^{\mu + k\sigma} (x - \mu)^2 \cdot f(x)\, dx$$
$$+ \int_{\mu + k\sigma}^{\infty} (x - \mu)^2 \cdot f(x)\, dx$$

Since the integrand $(x - \mu)^2 \cdot f(x)$ is nonnegative, we can form the inequality

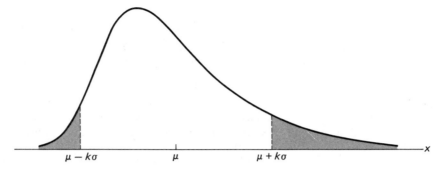

**Figure 4.2**  Diagram for proof of Chebyshev's theorem.

$$\sigma^2 \geq \int_{-\infty}^{\mu-k\sigma} (x - \mu)^2 \cdot f(x) \, dx + \int_{\mu+k\sigma}^{\infty} (x - \mu)^2 \cdot f(x) \, dx$$

by deleting the second integral. Now, since $(x - \mu)^2 \geq k^2\sigma^2$ for $x \leq \mu - k\sigma$ or $x \geq \mu + k\sigma$, it follows that

$$\sigma^2 \geq \int_{-\infty}^{\mu-k\sigma} k^2\sigma^2 \cdot f(x) \, dx + \int_{\mu+k\sigma}^{\infty} k^2\sigma^2 \cdot f(x) \, dx$$

and, hence, that

$$\frac{1}{k^2} \geq \int_{-\infty}^{\mu-k\sigma} f(x) \, dx + \int_{\mu+k\sigma}^{\infty} f(x) \, dx$$

provided $\sigma^2 \neq 0$. Since the sum of the two integrals on the right-hand side is the probability that $X$ will take on a value less than or equal to $\mu - k\sigma$ or greater than or equal to $\mu + k\sigma$, we have thus shown that

$$P(|X - \mu| \geq k\sigma) \leq \frac{1}{k^2}$$

and it follows that

$$P(|X - \mu| < k\sigma) \geq 1 - \frac{1}{k^2} \qquad \blacktriangledown$$

For instance, the probability is at least $1 - \frac{1}{2^2} = \frac{3}{4}$ that a random variable $X$ will take on a value within two standard deviations of the mean, the probability is at least $1 - \dfrac{1}{3^2} = \dfrac{8}{9}$ that it will take on a value within three standard deviations

of the mean, and the probability is at least $1 - \dfrac{1}{5^2} = \dfrac{24}{25}$ that it will take on a value

within five standard deviations of the mean. It is in this sense that $\sigma$ controls the spread or dispersion of the distribution of a random variable. Clearly, the probability given by Chebyshev's theorem is only a lower bound; whether the probability that a given random variable will take on a value within $k$ standard deviations of the mean is actually greater than $1 - \dfrac{1}{k^2}$, and if so by how much, we cannot say, but Chebyshev's theorem assures us that this probability cannot be less than $1 - \dfrac{1}{k^2}$. Only when the distribution of a random variable is known can we calculate the exact probability.

## EXAMPLE 4.12

If the probability density of $X$ is given by

$$f(x) = \begin{cases} 630x^4(1 - x)^4 & \text{for } 0 < x < 1 \\ 0 & \text{elsewhere} \end{cases}$$

find the probability that it will take on a value within two standard deviations of the mean and compare this probability with the lower bound provided by Chebyshev's theorem.

### Solution

Straightforward integration shows that $\mu = \frac{1}{2}$ and $\sigma^2 = \frac{1}{44}$, so that $\sigma = \sqrt{1/44}$, or approximately $0.15$. Thus, the probability that $X$ will take on a value within two standard deviations of the mean is the probability that it will take on a value between $0.20$ and $0.80$, namely,

$$P(0.20 < X < 0.80) = \int_{0.20}^{0.80} 630x^4(1 - x)^4 \, dx$$

$$= 0.96$$

Observe that the statement "the probability is $0.96$" is a much stronger statement than "the probability is at least $0.75$," which is provided by Chebyshev's theorem.    ▲

## 4.5  MOMENT-GENERATING FUNCTIONS

Although the moments of most distributions can be determined directly by evaluating the necessary integrals or sums, there is an alternative procedure which sometimes provides considerable simplifications. This technique utilizes **moment-generating functions**.

---

**DEFINITION 4.6**  The **moment-generating function** of a random variable $X$, where it exists, is given by

$$M_X(t) = E(e^{tX}) = \sum_x e^{tx} \cdot f(x)$$

when $X$ is discrete and

$$M_X(t) = E(e^{tX}) = \int_{-\infty}^{\infty} e^{tx} \cdot f(x) \, dx$$

when $X$ is continuous.

---

The independent variable is $t$, and we are usually interested in values of $t$ in the neighborhood of 0.

To explain why we refer to this function as a "moment-generating" function, let us substitute for $e^{tx}$ its Maclaurin's series expansion, namely,

$$e^{tx} = 1 + tx + \frac{t^2 x^2}{2!} + \frac{t^3 x^3}{3!} + \cdots + \frac{t^r x^r}{r!} + \cdots$$

For the discrete case, we thus get

$$M_X(t) = \sum_x \left[ 1 + tx + \frac{t^2 x^2}{2!} + \cdots + \frac{t^r x^r}{r!} + \cdots \right] f(x)$$

$$= \sum_x f(x) + t \cdot \sum_x x f(x) + \frac{t^2}{2!} \cdot \sum_x x^2 f(x) + \cdots + \frac{t^r}{r!} \cdot \sum_x x^r f(x) + \cdots$$

$$= 1 + \mu t + \mu_2' \cdot \frac{t^2}{2!} + \cdots + \mu_r' \cdot \frac{t^r}{r!} + \cdots$$

and it can be seen that in the Maclaurin's series of the moment-generating function of $X$, the coefficient of $\dfrac{t^r}{r!}$ is $\mu'_r$, the $r$th moment about the origin. In the continuous case, the argument is the same.

## EXAMPLE 4.13

Find the moment-generating function of the random variable whose probability density is given by

$$f(x) = \begin{cases} e^{-x} & \text{for } x > 0 \\ 0 & \text{elsewhere} \end{cases}$$

and use it to find an expression for $\mu'_r$.

*Solution*

By definition

$$M_X(t) = E(e^{tX}) = \int_0^\infty e^{tx} \cdot e^{-x}\, dx$$

$$= \int_0^\infty e^{-x(1-t)}\, dx$$

$$= \frac{1}{1-t} \qquad \text{for } t < 1$$

As is well known, when $|t| < 1$ the Maclaurin's series for this moment-generating function is

$$M_X(t) = 1 + t + t^2 + t^3 + \cdots + t^r + \cdots$$

$$= 1 + 1! \cdot \frac{t}{1!} + 2! \cdot \frac{t^2}{2!} + 3! \cdot \frac{t^3}{3!} + \cdots + r! \cdot \frac{t^r}{r!} + \cdots$$

and, hence, $\mu'_r = r!$ for $r = 0, 1, 2, \ldots$ .    ▲

The main difficulty in using the Maclaurin's series of a moment-generating function to determine the moments of a random variable is usually *not* that of finding the moment-generating function, but that of expanding it into a Maclaurin's series. If we are interested only in the first few moments of a random variable, say, $\mu'_1$ and $\mu'_2$, their determination can usually be simplified by using the following theorem.

> **THEOREM 4.9**
>
> $$\left. \frac{d^r M_X(t)}{dt^r} \right|_{t=0} = \mu'_r$$

This follows from the fact that if a function is expanded as a power series in $t$, the coefficient of $\dfrac{t^r}{r!}$ is the $r$th derivative of the function with respect to $t$ at $t = 0$.

# EXAMPLE 4.14

Given that $X$ has the probability distribution $f(x) = \frac{1}{8}\begin{pmatrix}3\\x\end{pmatrix}$ for $x = 0, 1, 2,$ and $3$, find the moment-generating function of this random variable and use it to determine $\mu'_1$ and $\mu'_2$.

## Solution

In accordance with Definition 4.6,

$$M_X(t) = E(e^{tX}) = \frac{1}{8} \cdot \sum_{x=0}^{3} e^{tx} \begin{pmatrix}3\\x\end{pmatrix}$$

$$= \frac{1}{8}(1 + 3e^t + 3e^{2t} + e^{3t})$$

$$= \frac{1}{8}(1 + e^t)^3$$

Then, by Theorem 4.9,

$$\mu'_1 = M'_X(0) = \left. \frac{3}{8}(1 + e^t)^2 e^t \right|_{t=0} = \frac{3}{2}$$

and

$$\mu'_2 = M''_X(0) = \left. \frac{3}{4}(1 + e^t)e^{2t} + \frac{3}{8}(1 + e^t)^2 e^t \right|_{t=0} = 3 \qquad \blacktriangle$$

Often the work involved in using moment-generating functions can be simplified by making use of the following theorem.

**THEOREM 4.10**  If $a$ and $b$ are constants, then

1. $M_{X+a}(t) = E[e^{(X+a)t}] = e^{at} \cdot M_X(t);$
2. $M_{bX}(t) = E(e^{bXt}) = M_X(bt);$
3. $M_{\frac{X+a}{b}}(t) = E[e^{\left(\frac{X+a}{b}\right)t}] = e^{\frac{a}{b}t} \cdot M_X\left(\frac{t}{b}\right).$

The proof of this theorem is left to the reader in Exercise 4.46. As we shall see later, the first part of the theorem is of special importance when $a = -\mu$, and the third part is of special importance when $a = -\mu$ and $b = \sigma$, in which case

$$M_{\frac{X-\mu}{\sigma}}(t) = e^{-\frac{\mu t}{\sigma}} \cdot M_X\left(\frac{t}{\sigma}\right)$$

## EXERCISES

**4.25** With reference to Definition 4.4, show that $\mu_0 = 1$ and that $\mu_1 = 0$ for any random variable for which $E(X)$ exists.

**4.26** Find $\mu$, $\mu_2'$, and $\sigma^2$ for the random variable $X$ which has the probability distribution $f(x) = \frac{1}{2}$ for $x = -2$ and $x = 2$.

**4.27** Find $\mu$, $\mu_2'$, and $\sigma^2$ for the random variable $X$ which has the probability density

$$f(x) = \begin{cases} \dfrac{x}{2} & \text{for } 0 < x < 2 \\ 0 & \text{elsewhere} \end{cases}$$

**4.28** Find $\mu_r'$ and $\sigma^2$ for the random variable $X$ which has the probability density

$$f(x) = \begin{cases} \dfrac{1}{\ln 3} \cdot \dfrac{1}{x} & \text{for } 1 < x < 3 \\ 0 & \text{elsewhere} \end{cases}$$

**4.29** Prove Theorem 4.7.

**4.30** With reference to Exercise 4.8, find the variance of $g(X) = 2X + 3$.

**4.31** If the random variable $X$ has the mean $\mu$ and the standard deviation $\sigma$, show that the random variable $Z$ whose values are related to those of $X$ by means of the equation $z = \dfrac{x - \mu}{\sigma}$ has

$$E(Z) = 0 \quad \text{and} \quad \text{var}(Z) = 1$$

A distribution which has the mean 0 and the variance 1 is said to be in **standard form,** and when we perform the above change of variable, we are said to be **standardizing** the distribution of $X$.

**4.32** If the probability density of $X$ is given by

$$f(x) = \begin{cases} 2x^{-3} & \text{for } x > 1 \\ 0 & \text{elsewhere} \end{cases}$$

check whether its mean and its variance exist.

**4.33** Show that

$$\mu_r = \mu_r' - \binom{r}{1}\mu_{r-1}' \cdot \mu + \cdots + (-1)^i \binom{r}{i} \mu_{r-i}' \cdot \mu^i + \cdots$$
$$+ (-1)^{r-1}(r - 1) \cdot \mu^r$$

for $r = 1, 2, 3, \ldots$, and use this formula to express $\mu_3$ and $\mu_4$ in terms of moments about the origin.

**4.34** The **symmetry** or **skewness** (lack of symmetry) of a distribution is often measured by means of the quantity

$$\alpha_3 = \frac{\mu_3}{\sigma^3}$$

Use the formula for $\mu_3$ obtained in Exercise 4.33 to determine $\alpha_3$ for each of the following distributions (which have equal means and standard deviations):

(a) $f(1) = 0.05, f(2) = 0.15, f(3) = 0.30, f(4) = 0.30, f(5) = 0.15$, and $f(6) = 0.05$;
(b) $f(1) = 0.05, f(2) = 0.20, f(3) = 0.15, f(4) = 0.45, f(5) = 0.10$, and $f(6) = 0.05$.

Also draw histograms of the two distributions and note that whereas the first is symmetrical, the second has a "tail" on the left-hand side and is said to be **negatively skewed.**

**4.35** The extent to which a distribution is peaked or flat, also called the **kurtosis** of the distribution, is often measured by means of the quantity

$$\alpha_4 = \frac{\mu_4}{\sigma^4}$$

Use the formula for $\mu_4$ obtained in Exercise 4.33 to find $\alpha_4$ for each of the following symmetrical distributions, of which the first is more peaked (narrow humped) than the second:

(a)  $f(-3) = 0.06, f(-2) = 0.09, f(-1) = 0.10, f(0) = 0.50, f(1) = 0.10,$
     $f(2) = 0.09,$ and $f(3) = 0.06;$
(b)  $f(-3) = 0.04, f(-2) = 0.11, f(-1) = 0.20, f(0) = 0.30, f(1) = 0.20,$
     $f(2) = 0.11,$ and $f(3) = 0.04.$

**4.36** Duplicate the steps used in the proof of Theorem 4.8 to prove Chebyshev's theorem for a discrete random variable $X$.

**4.37** Show that if $X$ is a random variable with the mean $\mu$ for which $f(x) = 0$ for $x < 0$, then for any positive constant $a$,

$$P(X \geq a) \leq \frac{\mu}{a}$$

This inequality is called **Markov's inequality**, and we have given it here mainly because it leads to a relatively simple alternative proof of Chebyshev's theorem.

**4.38** Use the inequality of Exercise 4.37 to prove Chebyshev's theorem. [*Hint*: Substitute $(X - \mu)^2$ for $X$.]

**4.39** What is the smallest value of $k$ in Chebyshev's theorem for which the probability that a random variable will take on a value between $\mu - k\sigma$ and $\mu + k\sigma$ is

(a)  at least 0.95;
(b)  at least 0.99?

**4.40** If we let $k\sigma = c$ in Chebyshev's theorem, what does this theorem assert about the probability that a random variable will take on a value between $\mu - c$ and $\mu + c$?

**4.41** Find the moment-generating function of the discrete random variable $X$ which has the probability distribution

$$f(x) = 2(\tfrac{1}{3})^x \qquad \text{for } x = 1, 2, 3, \ldots$$

and use it to determine the values of $\mu_1'$ and $\mu_2'$.

**4.42** Find the moment-generating function of the continuous random variable $X$ whose probability density is given by

$$f(x) = \begin{cases} 1 & \text{for } 0 < x < 1 \\ 0 & \text{elsewhere} \end{cases}$$

and use it to find $\mu_1'$, $\mu_2'$, and $\sigma^2$.

**4.43** If we let $R_X(t) = \ln M_X(t)$, show that $R_X'(0) = \mu$ and $R_X''(0) = \sigma^2$. Also, use these results to find the mean and the variance of a random variable $X$ having the moment-generating function

$$M_X(t) = e^{4(e^t - 1)}$$

**4.44** Explain why there can be no random variable for which $M_X(t) = \dfrac{t}{1-t}$.

**4.45** Show that if a random variable has the probability density

$$f(x) = \tfrac{1}{2} e^{-|x|} \qquad \text{for } -\infty < x < \infty$$

its moment-generating function is given by

$$M_X(t) = \frac{1}{1 - t^2}$$

**4.46** With reference to Exercise 4.45, find the variance of the random variable by

   (a)   expanding the moment-generating function as an infinite series and reading off the necessary coefficients;

   (b)   using Theorem 4.9.

**4.47** Prove the three parts of Theorem 4.10.

**4.48** Given the moment-generating function $M_X(t) = e^{3t + 8t^2}$, find the moment-generating function of the random variable $Z = \tfrac{1}{4}(X - 3)$, and use it to determine the mean and the variance of $Z$.

## APPLICATIONS

**4.49** With reference to Example 4.1, find the variance of the number of television sets with white cords.

**4.50** The amount of time it takes a person to be served at a given cafeteria is a random variable with the probability density

$$f(x) = \begin{cases} \tfrac{1}{4} e^{-\frac{x}{4}} & \text{for } x > 0 \\ 0 & \text{elsewhere} \end{cases}$$

Find the mean and the variance of this random variable.

**4.51** With reference to Exercise 3.47, find the mean and the variance of the random variable in question.

**4.52** With reference to Exercise 3.20, find the mean and the variance of the random variable $V$.

**4.53** The following are some applications of the Markov inequality of Exercise 4.37:

(a) The scores which high school juniors get on the verbal part of the PSAT/NMSQT test may be looked upon as values of a random variable with the mean $\mu = 41$. Find an upper bound to the probability that one of the students will get a score of 65 or more.

(b) The weight of certain animals may be looked upon as a random variable with a mean of 212 grams. If none of the animals weighs less than 165 grams, find an upper bound to the probability that such an animal will weigh at least 250 grams.

**4.54** The number of marriage licenses issued in a certain city during the month of June may be looked upon as a random variable with $\mu = 124$ and $\sigma = 7.5$. According to Chebyshev's theorem, with what probability can we assert that between 64 and 184 marriage licenses will be issued there during a month of June?

**4.55** A study of the nutritional value of a certain kind of bread shows that the amount of thiamine (vitamin B$_1$) in a slice may be looked upon as a random variable with $\mu = 0.260$ milligram and $\sigma = 0.005$ milligram. According to Chebyshev's theorem, between what values must be the thiamine content of

(a) at least $\frac{35}{36}$ of all slices of this bread;

(b) at least $\frac{143}{144}$ of all slices of this bread?

**4.56** With reference to Exercise 4.50, what can we assert about the amount of time it takes a person to be served at the given cafeteria if we use Chebyshev's theorem with $k = 1.5$? What is the corresponding probability rounded to four decimals?

## 4.6 PRODUCT MOMENTS

To continue the discussion of Section 4.3, let us now present the **product moments** of two random variables.

---

**DEFINITION 4.7** The $r$th and $s$th **product moment about the origin** of the random variables $X$ and $Y$, denoted by $\mu'_{r,s}$, is the expected value of $X^r Y^s$; symbolically,

$$\mu'_{r,s} = E(X^r Y^s) = \sum_x \sum_y x^r y^s \cdot f(x, y)$$

for $r = 0, 1, 2, \ldots$ and $s = 0, 1, 2, \ldots$ when $X$ and $Y$ are discrete, and

---

$$\mu'_{r,s} = E(X^r Y^s) = \int_{-\infty}^{\infty} \int_{-\infty}^{\infty} x^r y^s \cdot f(x, y) \, dx \, dy$$

when $X$ and $Y$ are continuous.

In the discrete case, the double summation extends over the entire joint range of the two random variables. Note that $\mu'_{1,0} = E(X)$, which we denote here by $\mu_X$, and that $\mu'_{0,1} = E(Y)$, which we denote here by $\mu_Y$.

Analogous to Definition 4.4, let us now make the following definition of product moments about the respective means.

**DEFINITION 4.8** The $r$th and $s$th **product moment about the means** of the random variables $X$ and $Y$, denoted by $\mu_{r,s}$, is the expected value of $(X - \mu_X)^r (Y - \mu_Y)^s$; symbolically,

$$\mu_{r,s} = E[(X - \mu_X)^r (Y - \mu_Y)^s]$$
$$= \sum_x \sum_y (x - \mu_X)^r (y - \mu_Y)^s \cdot f(x, y)$$

for $r = 0, 1, 2, \ldots$ and $s = 0, 1, 2, \ldots$ when $X$ and $Y$ are discrete, and

$$\mu_{r,s} = E[(X - \mu_X)^r (Y - \mu_Y)^s]$$
$$= \int_{-\infty}^{\infty} \int_{-\infty}^{\infty} (x - \mu_X)^r (y - \mu_Y)^s \cdot f(x, y) \, dx \, dy$$

when $X$ and $Y$ are continuous.

In statistics, $\mu_{1,1}$ is of special importance because it is indicative of the relationship, if any, between the values of $X$ and $Y$; thus, it is given a special symbol and a special name.

**DEFINITION 4.9** $\mu_{1,1}$ is called the **covariance** of $X$ and $Y$, and it is denoted by $\sigma_{XY}$, cov$(X, Y)$, or $C(X, Y)$.

Observe that if there is a high probability that large values of $X$ will go with large values of $Y$ and small values of $X$ with small values of $Y$, the covariance will be positive; if there is a high probability that large values of $X$ will go with small values of $Y$ and vice versa, the covariance will be negative. It is in this sense that the covariance measures the relationship, or association, between the values of $X$ and $Y$.

Let us now prove the following result, analogous to Theorem 4.6, which is useful in actually determining covariances.

---

**THEOREM 4.11**

$$\sigma_{XY} = \mu'_{1,1} - \mu_X \mu_Y$$

---

      ***Proof.*** Using the various theorems about expected values, we can write

$$\sigma_{XY} = E[(X - \mu_X)(Y - \mu_Y)]$$
$$= E(XY - X\mu_Y - Y\mu_X + \mu_X \mu_Y)$$
$$= E(XY) - \mu_Y E(X) - \mu_X E(Y) + \mu_X \mu_Y$$
$$= E(XY) - \mu_Y \mu_X - \mu_X \mu_Y + \mu_X \mu_Y$$
$$= \mu'_{1,1} - \mu_X \mu_Y \quad \blacktriangledown$$

## EXAMPLE 4.15

In Example 3.20, the joint and marginal probabilities of $X$ and $Y$, the numbers of aspirin and sedative caplets among two caplets drawn at random from a bottle containing three aspirin, two sedative, and four laxative caplets, were recorded as follows:

|   |   | $x$ | | | |
|---|---|---|---|---|---|
|   |   | 0 | 1 | 2 | |
|   | 0 | $\frac{1}{6}$ | $\frac{1}{3}$ | $\frac{1}{12}$ | $\frac{7}{12}$ |
| $y$ | 1 | $\frac{2}{9}$ | $\frac{1}{6}$ | | $\frac{7}{18}$ |
|   | 2 | $\frac{1}{36}$ | | | $\frac{1}{36}$ |
|   |   | $\frac{5}{12}$ | $\frac{1}{2}$ | $\frac{1}{12}$ | |

Find the covariance of $X$ and $Y$.

*Solution*

Referring to the joint probabilities given here, we get

$$\mu'_{1,1} = E(XY)$$
$$= 0\cdot0\cdot\tfrac{1}{6} + 0\cdot1\cdot\tfrac{2}{9} + 0\cdot2\cdot\tfrac{1}{36} + 1\cdot0\cdot\tfrac{1}{3} + 1\cdot1\cdot\tfrac{1}{6} + 2\cdot0\cdot\tfrac{1}{12}$$
$$= \tfrac{1}{6}$$

and using the marginal probabilities, we get

$$\mu_X = E(X) = 0\cdot\tfrac{5}{12} + 1\cdot\tfrac{1}{2} + 2\cdot\tfrac{1}{12} = \tfrac{2}{3}$$

and

$$\mu_Y = E(Y) = 0\cdot\tfrac{7}{12} + 1\cdot\tfrac{7}{18} + 2\cdot\tfrac{1}{36} = \tfrac{4}{9}$$

It follows that

$$\sigma_{XY} = \tfrac{1}{6} - \tfrac{2}{3}\cdot\tfrac{4}{9} = -\tfrac{7}{54}$$

The negative result suggests that the more aspirin tablets we get the fewer sedative tablets we will get and vice versa, and this, of course, makes sense. ▲

## EXAMPLE 4.16

Find the covariance of the random variables whose joint probability density is given by

$$f(x, y) = \begin{cases} 2 & \text{for } x > 0, y > 0, x + y < 1 \\ 0 & \text{elsewhere} \end{cases}$$

*Solution*

Evaluating the necessary integrals, we get

$$\mu_X = \int_0^1 \int_0^{1-x} 2x \, dy \, dx = \tfrac{1}{3}$$

$$\mu_Y = \int_0^1 \int_0^{1-x} 2y \, dy \, dx = \tfrac{1}{3}$$

and

$$\sigma'_{1,1} = \int_0^1 \int_0^{1-x} 2xy \, dy \, dx = \tfrac{1}{12}$$

It follows that

$$\sigma_{XY} = \tfrac{1}{12} - \tfrac{1}{3} \cdot \tfrac{1}{3} = -\tfrac{1}{36} \qquad \blacktriangle$$

So far as the relationship between $X$ and $Y$ is concerned, observe that if $X$ and $Y$ are independent, their covariance is zero; symbolically,

---

**THEOREM 4.12**  If $X$ and $Y$ are independent, then $E(XY) = E(X) \cdot E(Y)$ and $\sigma_{XY} = 0$.

---

*Proof.*    For the discrete case we have, by definition,

$$E(XY) = \sum_x \sum_y xy \cdot f(x,y)$$

Since $X$ and $Y$ are independent, we can write $f(x, y) = g(x) \cdot h(y)$, where $g(x)$ and $h(y)$ are the values of the marginal distributions of $X$ and $Y$, and we get

$$E(XY) = \sum_x \sum_y xy \cdot g(x)h(y)$$

$$= \left[ \sum_x x \cdot g(x) \right]\left[ \sum_y y \cdot h(y) \right]$$

$$= E(X) \cdot E(Y)$$

Hence,

$$\sigma_{XY} = \mu'_{1,1} - \mu_X \mu_Y$$

$$= E(X) \cdot E(Y) - E(X) \cdot E(Y)$$

$$= 0 \qquad \blacktriangledown$$

It is of interest to note that the independence of two random variables implies a zero covariance, but a zero covariance does not necessarily imply their independence. This is illustrated by the following example (see also Exercises 4.61 and 4.62).

## EXAMPLE 4.17

If the joint probability distribution of $X$ and $Y$ is given by

|  |  | $x$ | | | |
|---|---|:---:|:---:|:---:|:---:|
|  |  | $-1$ | $0$ | $1$ | |
|  | $-1$ | $\frac{1}{6}$ | $\frac{1}{3}$ | $\frac{1}{6}$ | $\frac{2}{3}$ |
| $y$ | $0$ | $0$ | $0$ | $0$ | $0$ |
|  | $1$ | $\frac{1}{6}$ | $0$ | $\frac{1}{6}$ | $\frac{1}{3}$ |
|  |  | $\frac{1}{3}$ | $\frac{1}{3}$ | $\frac{1}{3}$ | |

show that their covariance is zero even though the two random variables are not independent.

## Solution

Using the probabilities shown in the margins, we get

$$\mu_X = (-1) \cdot \tfrac{1}{3} + 0 \cdot \tfrac{1}{3} + 1 \cdot \tfrac{1}{3} = 0$$

$$\mu_Y = (-1) \cdot \tfrac{2}{3} + 0 \cdot 0 + 1 \cdot \tfrac{1}{3} = -\tfrac{1}{3}$$

and

$$\mu'_{1,1} = (-1)(-1) \cdot \tfrac{1}{6} + 0(-1) \cdot \tfrac{1}{3} + 1(-1) \cdot \tfrac{1}{6} + (-1)1 \cdot \tfrac{1}{6} + 1 \cdot 1 \cdot \tfrac{1}{6}$$

$$= 0$$

Thus, $\sigma_{XY} = 0 - 0(-\tfrac{1}{3}) = 0$, the covariance is zero, but the two random variables are not independent. For instance, $f(x, y) \neq g(x) \cdot h(y)$ for $x = -1$ and $y = -1$.    ▲

Product moments can also be defined for the case where there are more than two random variables. Here let us merely state the important result that

---

**THEOREM 4.13** If $X_1, X_2, \ldots, X_n$ are independent, then

$$E(X_1 X_2 \cdot \ldots \cdot X_n) = E(X_1) \cdot E(X_2) \cdot \ldots \cdot E(X_n)$$

---

This is a generalization of the first part of Theorem 4.12; in fact, the proof of this theorem, based on Definition 3.14, is essentially like that of the first part of Theorem 4.12.

## 4.7 MOMENTS OF LINEAR COMBINATIONS OF RANDOM VARIABLES

In this section we shall derive expressions for the mean and the variance of a linear combination of $n$ random variables and the covariance of two linear combinations of $n$ random variables. Applications of these results will be treated later in our discussion of sampling theory and problems of statistical inference.

---

**THEOREM 4.14** If $X_1, X_2, \ldots, X_n$ are random variables and

$$Y = \sum_{i=1}^{n} a_i X_i$$

where $a_1, a_2, \ldots a_n$ are constants, then

$$E(Y) = \sum_{i=1}^{n} a_i E(X_i)$$

and

$$\text{var}(Y) = \sum_{i=1}^{n} a_i^2 \cdot \text{var}(X_i) + 2 \sum\sum_{i<j} a_i a_j \cdot \text{cov}(X_i X_j)$$

where the double summation extends over all values of $i$ and $j$, from 1 to $n$, for which $i < j$.

---

***Proof.*** From Theorem 4.5 with $g_i(X_1, X_2, \ldots, X_k) = X_i$ for $i = 0$, $1, 2, \ldots, n$, it follows immediately that

$$E(Y) = E\left(\sum_{i=1}^{n} a_i X_i\right) = \sum_{i=1}^{n} a_i E(X_i)$$

and this proves the first part of the theorem. To obtain the expression for the variance of $Y$, let us write $\mu_i$ for $E(X_i)$, so that we get

$$\text{var}(Y) = E([Y - E(Y)]^2) = E\left\{\left[\sum_{i=1}^{n} a_i X_i - \sum_{i=1}^{n} a_i E(X_i)\right]^2\right\}$$

$$= E\left\{\left[\sum_{i=1}^{n} a_i(X_i - \mu_i)\right]^2\right\}$$

Then, expanding by means of the multinomial theorem according to which $(a + b + c + d)^2$, for example, equals $a^2 + b^2 + c^2 + d^2 + 2ab + 2ac + 2ad + 2bc + 2bd + 2cd$, and again referring to Theorem 4.5, we get

$$\text{var}(Y) = \sum_{i=1}^{n} a_i^2 E[(X_i - \mu_i)^2] + 2 \sum\sum_{i<j} a_i a_j E[(X_i - \mu_i)(X_j - \mu_j)]$$

$$= \sum_{i=1}^{n} a_i^2 \cdot \text{var}(X_i) + 2 \sum\sum_{i<j} a_i a_j \cdot \text{cov}(X_i, X_j)$$

Note that we have tacitly made use of the fact that $\text{cov}(X_i, X_j) = \text{cov}(X_j, X_i)$. ▼

Since $\text{cov}(X_i, X_j) = 0$ when $X_i$ and $X_j$ are independent, it follows immediately that

---

**COROLLARY**  If the random variables $X_1, X_2, \ldots, X_n$ are independent and $Y = \sum_{i=1}^{n} a_i X_i$, then

$$\text{var}(Y) = \sum_{i=1}^{n} a_i^2 \cdot \text{var}(X_i)$$

---

## EXAMPLE 4.18

If the random variables $X$, $Y$, and $Z$ have the means $\mu_X = 2$, $\mu_Y = -3$, and $\mu_Z = 4$, the variances $\sigma_X^2 = 1$, $\sigma_Y^2 = 5$, and $\sigma_Z^2 = 2$, and the covariances $\text{cov}(X, Y) = -2$, $\text{cov}(X, Z) = -1$, and $\text{cov}(Y, Z) = 1$, find the mean and the variance of $W = 3X - Y + 2Z$.

*Solution*

By Theorem 4.14, we get

$$E(W) = E(3X - Y + 2Z)$$
$$= 3E(X) - E(Y) + 2E(Z)$$
$$= 3 \cdot 2 - (-3) + 2 \cdot 4$$
$$= 17$$

and

$$\text{var}(W) = 9 \, \text{var}(X) + \text{var}(Y) + 4 \, \text{var}(Z) - 6 \, \text{cov}(X, Y)$$
$$+ 12 \, \text{cov}(X, Z) - 4 \, \text{cov}(Y, Z)$$
$$= 9 \cdot 1 + 5 + 4 \cdot 2 - 6(-2) + 12(-1) - 4 \cdot 1$$
$$= 18 \quad \blacktriangle$$

The following is another important theorem about linear combinations of random variables; it concerns the covariance of two linear combinations of $n$ random variables.

---

**THEOREM 4.15**  If $X_1, X_2, \ldots, X_n$ are random variables and

$$Y_1 = \sum_{i=1}^{n} a_i X_i \quad \text{and} \quad Y_2 = \sum_{i=1}^{n} b_i X_i$$

where $a_1, a_2, \ldots, a_n, b_1, b_2, \ldots, b_n$ are constants, then

$$\text{cov}(Y_1, Y_2) = \sum_{i=1}^{n} a_i b_i \cdot \text{var}(X_i) + \sum\sum_{i<j} (a_i b_j + a_j b_i) \cdot \text{cov}(X_i, X_j)$$

---

The proof of this theorem, which is very similar to that of Theorem 4.14, will be left to the reader in Exercise 4.67.

Since $\text{cov}(X_i, X_j) = 0$ when $X_i$ and $X_j$ are independent, it follows immediately that

---

**COROLLARY** If the random variables $X_1, X_2, \ldots, X_n$ are independent,

$$Y_1 = \sum_{i=1}^{n} a_i X_i \quad \text{and} \quad Y_2 = \sum_{i=1}^{n} b_i X_i, \text{ then}$$

$$\text{cov}(Y_1, Y_2) = \sum_{i=1}^{n} a_i b_i \cdot \text{var}(X_i)$$

---

## EXAMPLE 4.19

If the random variables $X$, $Y$, and $Z$ have the means $\mu_X = 3$, $\mu_Y = 5$, and $\mu_Z = 2$, the variances $\sigma_X^2 = 8$, $\sigma_Y^2 = 12$, and $\sigma_Z^2 = 18$, and $\text{cov}(X, Y) = 1$, $\text{cov}(X, Z) = -3$, and $\text{cov}(Y, Z) = 2$, find the covariance of

$$U = X + 4Y + 2Z \quad \text{and} \quad V = 3X - Y - Z$$

### Solution

By Theorem 4.15, we get

$$\text{cov}(U, V) = \text{cov}(X + 4Y + 2Z, 3X - Y - Z)$$
$$= 3 \text{ var}(X) - 4 \text{ var}(Y) - 2 \text{ var}(Z) + 11 \text{ cov}(X, Y)$$
$$+ 5 \text{ cov}(X, Z) - 6 \text{ cov}(Y, Z)$$
$$= 3 \cdot 8 - 4 \cdot 12 - 2 \cdot 18 + 11 \cdot 1 + 5(-3) - 6 \cdot 2$$
$$= -76 \quad \blacktriangle$$

## 4.8 CONDITIONAL EXPECTATIONS

In Section 3.7 we obtained conditional probabilities by adding the values of conditional probability distributions, or integrating the values of conditional probability densities. **Conditional expectations** of random variables are likewise defined in terms of their conditional distributions.

---

**DEFINITION 4.10** If $X$ is a discrete random variable and $f(x|y)$ is the value of the conditional probability distribution of $X$ given $Y = y$ at $x$, the **conditional expectation** of $u(X)$ given $Y = y$ is

$$E[u(X)|y] = \sum_x u(x) \cdot f(x|y)$$

Correspondingly, if $X$ is a continuous random variable and $f(x|y)$ is the value of the conditional probability density of $X$ given $Y = y$ at $x$, the **conditional expectation** of $u(X)$ given $Y = y$ is

$$E[u(X)|y] = \int_{-\infty}^{\infty} u(x) \cdot f(x|y)\, dx$$

---

Similar expressions based on the conditional probability distribution or density of $Y$ given $X = x$ define the conditional expectation of $v(Y)$ given $X = x$.

If we let $u(X) = X$ in Definition 4.10, we obtain the **conditional mean** of the random variable $X$ given $Y = y$, which we denote by

$$\mu_{X|y} = E(X|y)$$

Correspondingly, the **conditional variance** of $X$ given $Y = y$ is

$$\sigma_{X|y}^2 = E[(X - \mu_{X|y})^2 | y]$$
$$= E(X^2|y) - \mu_{X|y}^2$$

where $E(X^2|y)$ is given by Definition 4.10 with $u(X) = X^2$. The reader should not find it difficult to generalize Definition 4.10 for conditional expectations involving more than two random variables.

## EXAMPLE 4.20

With reference to Example 3.12 on page 111, find the conditional mean of $X$ given $Y = 1$.

### Solution

Making use of the results obtained in Example 3.23 on page 130, namely, $f(0|1) = \frac{4}{7}$, $f(1|1) = \frac{3}{7}$, and $f(2|1) = 0$, we get

$$E(X|1) = 0 \cdot \frac{4}{7} + 1 \cdot \frac{3}{7} + 2 \cdot 0 = \frac{3}{7} \qquad \blacktriangle$$

## EXAMPLE 4.21

If the joint probability density of $X$ and $Y$ is given by

$$f(x, y) = \begin{cases} \frac{2}{3}(x + 2y) & \text{for } 0 < x < 1, 0 < y < 1 \\ 0 & \text{elsewhere} \end{cases}$$

find the conditional mean and the conditional variance of $X$ given $Y = \frac{1}{2}$.

### Solution

In Example 3.24 we showed that for these random variables the conditional density of $X$ given $Y = y$ is

$$f(x|y) = \begin{cases} \dfrac{2x + 4y}{1 + 4y} & \text{for } 0 < x < 1 \\ 0 & \text{elsewhere} \end{cases}$$

so that

$$f(x|\tfrac{1}{2}) = \begin{cases} \frac{2}{3}(x + 1) & \text{for } 0 < x < 1 \\ 0 & \text{elsewhere} \end{cases}$$

Thus, $\mu_{X|\frac{1}{2}}$ is given by

$$E(X|\tfrac{1}{2}) = \int_0^1 \tfrac{2}{3} x(x + 1)\, dx$$

$$= \tfrac{5}{9}$$

Next we find

$$E(X^2|\tfrac{1}{2}) = \int_0^1 \tfrac{2}{3} x^2(x + 1)\, dx$$

$$= \tfrac{7}{18}$$

and it follows that

$$\sigma^2_{X|\frac{1}{2}} = \tfrac{7}{18} - (\tfrac{5}{9})^2 = \tfrac{13}{162} \qquad \blacktriangle$$

## EXERCISES

**4.57** If $X$ and $Y$ have the joint probability distribution $f(x, y) = \frac{1}{4}$ for $x = -3$ and $y = -5$, $x = -1$ and $y = -1$, $x = 1$ and $y = 1$, and $x = 3$ and $y = 5$, find $\text{cov}(X, Y)$.

**4.58** With reference to Exercise 3.52, find the covariance of $X$ and $Y$.

**4.59** With reference to Example 3.22, find the covariance of $X_1$ and $X_3$.

**4.60** With reference to Exercise 3.90, find $\text{cov}(X, Y)$.

**4.61** If $X$ and $Y$ have the joint probability distribution $f(-1, 0) = 0, f(-1, 1) = \frac{1}{4}, f(0, 0) = \frac{1}{6}, f(0, 1) = 0, f(1, 0) = \frac{1}{12}$, and $f(1, 1) = \frac{1}{2}$, show that
- (a)  $\text{cov}(X, Y) = 0$;
- (b)  the two random variables are not independent.

**4.62** If the probability density of $X$ is given by

$$f(x) = \begin{cases} 1 + x & \text{for } -1 < x \leqslant 0 \\ 1 - x & \text{for } 0 < x < 1 \\ 0 & \text{elsewhere} \end{cases}$$

and $U = X$ and $V = X^2$, show that
- (a)  $\text{cov}(U, V) = 0$;
- (b)  $U$ and $V$ are dependent.

**4.63** For $k$ random variables $X_1, X_2, \ldots X_k$, the values of their **joint moment-generating function** are given by

$$E(e^{t_1 X_1 + t_2 X_2 + \cdots + t_k X_k})$$

- (a)  Show for either the discrete case or the continuous case that the partial derivative of the joint moment-generating function with respect to $t_i$ at $t_1 = t_2 = \cdots = t_k = 0$ is $E(X_i)$.
- (b)  Show for either the discrete case or the continuous case that the second partial derivative of the joint moment-generating function with respect to $t_i$ and $t_j$, $i \neq j$, at $t_1 = t_2 = \cdots = t_k = 0$ is $E(X_i X_j)$.
- (c)  If two random variables have the joint density given by

$$f(x, y) = \begin{cases} e^{-x-y} & \text{for } x > 0, y > 0 \\ 0 & \text{elsewhere} \end{cases}$$

  find their joint moment-generating function and use it to determine the values of $E(XY)$, $E(X)$, $E(Y)$, and $\text{cov}(X, Y)$.

**4.64** If $X_1, X_2$, and $X_3$ are independent, have the means 4, 9, and 3, and the variances 3, 7, and 5, find the mean and the variance of
- (a)  $Y = 2X_1 - 3X_2 + 4X_3$;
- (b)  $Z = X_1 + 2X_2 - X_3$.

**4.65** Repeat both parts of Exercise 4.64, dropping the assumption of independence and using instead the information that $\text{cov}(X_1, X_2) = 1, \text{cov}(X_2, X_3) = -2$, and $\text{cov}(X_1, X_3) = -3$.

**4.66** If the joint probability density of $X$ and $Y$ is given by

$$f(x, y) = \begin{cases} \frac{1}{3}(x + y) & \text{for } 0 < x < 1, 0 < y < 2 \\ 0 & \text{elsewhere} \end{cases}$$

find the variance of $W = 3X + 4Y - 5$.

**4.67** Prove Theorem 4.15.

**4.68** Express $\text{var}(X + Y)$, $\text{var}(X - Y)$, and $\text{cov}(X + Y, X - Y)$ in terms of the variances and covariance of $X$ and $Y$.

**4.69** If $\text{var}(X_1) = 5$, $\text{var}(X_2) = 4$, $\text{var}(X_3) = 7$, $\text{cov}(X_1, X_2) = 3$, $\text{cov}(X_1, X_3) = -2$, and $X_2$ and $X_3$ are independent, find the covariance of $Y_1 = X_1 - 2X_2 + 3X_3$ and $Y_2 = -2X_1 + 3X_2 + 4X_3$.

**4.70** With reference to Exercise 4.64, find $\text{cov}(Y, Z)$.

**4.71** With reference to Exercise 3.85, find the conditional mean and the conditional variance of $X$ given $Y = -1$.

**4.72** With reference to Exercise 3.87, find the conditional expectation of the random variable $U = Z^2$ given $X = 1$ and $Y = 2$.

**4.73** With reference to Exercise 3.90, find the conditional mean and the conditional variance of $Y$ given $X = \frac{1}{4}$.

**4.74** With reference to Example 3.22 and part (b) of Exercise 3.94, find the expected value of $X_2^2 X_3$ given $X_1 = \frac{1}{2}$.

**4.75** (a)  Show that the conditional distribution functon of the continuous random variable $X$ given $a < X \leqslant b$ is given by

$$F(x|a < X \leqslant b) = \begin{cases} 0 & \text{for } x \leqslant a \\ \dfrac{F(x) - F(a)}{F(b) - F(a)} & \text{for } a < x \leqslant b \\ 1 & \text{for } x > b \end{cases}$$

(b)  Differentiate the result of part (a) with respect to $x$ to find the conditional probability density of $X$ given $a < X \leqslant b$, and show that

$$E[u(X)|a < X \leqslant b] = \frac{\displaystyle\int_a^b u(x)f(x)\, dx}{\displaystyle\int_a^b f(x)\, dx}$$

## APPLICATIONS

**4.76** A quarter is bent so that the probabilities of heads and tails are 0.40 and 0.60. If it is tossed twice, what is the covariance of $Z$, the number of heads obtained on the first toss, and $W$, the total number of heads obtained in the two tosses of the coin?

**4.77** The inside diameter of a cylindrical tube is a random variable with a mean of 3 inches and a standard deviation of 0.02 inch, the thickness of the tube is a random variable with a mean of 0.3 inch and a standard deviation of 0.005 inch, and the two random variables are independent. Find the mean and the standard deviation of the outside diameter of the tube.

**4.78** The length of certain bricks is a random variable with a mean of 8 inches and a standard deviation of 0.1 inch, and the thickness of the mortar between two bricks is a random variable with a mean of 0.5 inch and a standard deviation of 0.03 inch. What is the mean and the standard deviation of the length of a wall made of 50 of these bricks laid side by side, if we can assume that all the random variables involved are independent?

**4.79** If heads is a success when we flip a coin, getting a six is a success when we roll a die, and getting an ace is a success when we draw a card from an ordinary deck of 52 playing cards, find the mean and the standard deviation of the total number of successes when we

(a)   flip a balanced coin, roll a balanced die, and then draw a card from a well-shuffled deck;

(b)   flip a balanced coin three times, roll a balanced die twice, and then draw a card from a well-shuffled deck.

**4.80** If we alternately flip a balanced coin and a coin which is loaded so that the probability of getting heads is 0.45, what are the mean and the standard deviation of the number of heads which we obtain in 10 flips of these coins?

**4.81** With reference to Exercise 3.80 and part (b) of Exericse 3.99, find the expected number of mathematics texts given that none of the statistics texts is selected.

**4.82** With reference to Exercise 3.103, by how much can a salesperson who spends $12 on gasoline expect to be reimbursed?

**4.83** The amount of time (in minutes) that an executive of a certain firm talks on the telephone is a random variable having the probability density

$$f(x) = \begin{cases} \dfrac{x}{4} & \text{for } 0 < x \leq 2 \\[2mm] \dfrac{4}{x^3} & \text{for } x > 2 \\[2mm] 0 & \text{elsewhere} \end{cases}$$

With reference to part (b) of Exercise 4.75, find the expected length of one of these telephone conversations that has lasted at least 1 minute.

## REFERENCES

Further information about the material in this chapter may be found in the more advanced mathematical statistics texts listed at the end of Chapter 3.

# 5

# *Special*
# *Probability Distributions*

## 5.1 INTRODUCTION

In this chapter we shall study some of the probability distributions that figure most prominently in statistical theory and applications. We shall also study their **parameters**, that is, the quantities that are constants for particular distributions but that can take on different values for different members of families of distributions of the same kind. The most common parameters are the lower moments, mainly $\mu$ and $\sigma^2$, and as we saw in the preceding chapter, there are essentially two ways in which they can be obtained: We can evaluate the necessary sums directly or we can work with moment-generating functions. Although it would seem logical to use in each case whichever method is simplest, we shall sometimes use both. In some instances this will be done because the results are needed later; in others it will merely serve to provide the reader with experience in the application of the respective mathematical techniques. Also, to keep the size of this chapter within bounds, many of the details are left as exercises.

## 5.2  THE DISCRETE UNIFORM DISTRIBUTION

If a random variable can take on $k$ different values with equal probability, we say that it has a **discrete uniform distribution**; symbolically,

---

**DEFINITION 5.1**  A random variable $X$ has a **discrete uniform distribution** and it is referred to as a discrete uniform random variable, if and only if its probability distribution is given by

$$f(x) = \frac{1}{k} \qquad \text{for } x = x_1, x_2, \ldots, x_k$$

where $x_i \neq x_j$ when $i \neq j$.

---

In accordance with Definitions 4.2 and 4.4, the mean and the variance of this distribution are $\mu = \sum_{i=1}^{k} x_i \cdot \frac{1}{k}$ and $\sigma^2 = \sum_{i=1}^{k} (x_i - \mu)^2 \cdot \frac{1}{k}$.

In the special case where $x_i = i$, the discrete uniform distribution becomes $f(x) = \frac{1}{k}$ for $x = 1, 2, \ldots, k$, and in this form it applies, for example, to the number of points we roll with a balanced die. The mean and the variance of this discrete uniform distribution, and its moment-generating function, are treated in Exercises 5.1 and 5.2.

## 5.3  THE BERNOULLI DISTRIBUTION

If an experiment has two possible outcomes, "success" and "failure," and their probabilities are, respectively, $\theta$ and $1 - \theta$, then the number of successes, 0 or 1, has a **Bernoulli distribution**; symbolically,

> **DEFINITION 5.2**   A random variable $X$ has a **Bernoulli distribution** and it is referred to as a Bernoulli random variable, if and only if its probability distribution is given by
>
> $$f(x; \theta) = \theta^x (1 - \theta)^{1-x} \qquad \text{for } x = 0, 1$$

Thus, $f(0; \theta) = 1 - \theta$ and $f(1; \theta) = \theta$ are combined into a single formula. Observe that we used the notation $f(x; \theta)$ to indicate explicitly that the Bernoulli distribution has the one parameter $\theta$. Since the Bernoulli distribution is a special case of the distribution of Section 5.4, we shall not discuss it here in any detail.

In connection with the Bernoulli distribution, a success may be getting heads with a balanced coin, it may be catching pneumonia, it may be passing (or failing) an examination, and it may be losing a race. This inconsistency is a carryover from the days when probability theory was applied only to games of chance (and one player's failure was the other's success). Also for this reason, we refer to an experiment to which the Bernoulli distribution applies as a **Bernoulli trial**, or simply a **trial**, and to sequences of such experiments as **repeated trials**.

## 5.4  THE BINOMIAL DISTRIBUTION

Repeated trials play a very important role in probability and statistics, especially when the number of trials is fixed, the parameter $\theta$ (the probability of a success) is the same for each trial, and the trials are all independent. As we shall see, there are several random variables that arise in connection with repeated trials. The one we shall study here concerns the total number of successes; others will be given in Section 5.5.

The theory which we shall discuss in this section has many applications; for instance, it applies if we want to know the probability of getting 5 heads in 12 flips of a coin, the probability that 7 of 10 persons will recover from a tropical disease, or the probability that 35 of 80 persons will respond to a mail-order solicitation. However, this is the case only if each of the 10 persons has the same chance of recovering from the disease and their recoveries are independent (say, they are treated by different doctors in different hospitals), and if the probability of getting a reply to the mail-order solicitation is the same for each of the 80 persons and there is independence (say, no two of them belong to the same household).

To derive a formula for the probability of getting "$x$ successes in $n$ trials" under the stated conditions, observe that the probability of getting $x$ successes and $n - x$ failures *in a specific order* is $\theta^x (1 - \theta)^{n-x}$. There is one factor $\theta$ for each success, one factor $1 - \theta$ for each failure, and the $x$ factors $\theta$ and $n - x$ factors

$1 - \theta$ are all multiplied together by virtue of the assumption of independence. Since this probability applies to any sequence of $n$ trials in which there are $x$ successes and $n - x$ failures, we have only to count how many sequences of this kind there are, and then multiply $\theta^x(1 - \theta)^{n-x}$ by that number. Clearly, the number of ways in which we can select the $x$ trials on which there is to be a success is $\binom{n}{x}$, and it follows that the desired probability for "$x$ successes in $n$ trials" is

$$\binom{n}{x}\theta^x(1 - \theta)^{n-x}.$$

> **DEFINITION 5.3**  A random variable $X$ has a **binomial distribution** and it is referred to as a binomial random variable, if and only if its probability distribution is given by
>
> $$b(x; n, \theta) = \binom{n}{x}\theta^x(1 - \theta)^{n-x} \qquad \text{for } x = 0, 1, 2, \ldots, n$$

Thus, the number of successes in $n$ trials is a random variable having a binomial distribution with the parameters $n$ and $\theta$. The name "binomial distribution" derives from the fact that the values of $b(x; n, \theta)$ for $x = 0, 1, 2, \ldots, n$ are the successive terms of the binomial expansion of $[(1 - \theta) + \theta]^n$; this shows also that the sum of the probabilities equals 1, as it should.

## EXAMPLE 5.1

Find the probability of getting five heads and seven tails in 12 flips of a balanced coin.

### Solution

Substituting $x = 5$, $n = 12$, and $\theta = \frac{1}{2}$ into the formula for the binomial distribution, we get

$$b(5; 12, \tfrac{1}{2}) = \binom{12}{5}(\tfrac{1}{2})^5(1 - \tfrac{1}{2})^{12-5}$$

and, looking up the value of $\binom{12}{5}$ in Table VII, we find that the result is $792(\tfrac{1}{2})^{12}$, or approximately 0.19.    ▲

## EXAMPLE 5.2

Find the probability that seven of 10 persons will recover from a tropical disease if we can assume independence and the probability is 0.80 that any one of them will recover from the disease.

*Solution*

Substituting $x = 7$, $n = 10$, and $\theta = 0.80$ into the formula for the binomial distribution, we get

$$b(7; 10, 0.80) = \binom{10}{7}(0.80)^7(1 - 0.80)^{10-7}$$

and, looking up the value of $\binom{10}{7}$ in Table VII, we find that the result is

$120(0.80)^7(0.20)^3$, or approximately 0.20.    ▲

If we tried to calculate the third probability asked for on page 185, the one concerning the responses to the mail-order solicitation, by substituting $x = 35$, $n = 80$, and, say, $\theta = 0.15$, into the formula for the binomial distribution, we would find that this requires a prohibitive amount of work. In actual practice, binomial probabilities are rarely calculated directly, for they are tabulated extensively for various values of $\theta$ and $n$, and there exists an abundance of computer software yielding binomial probabilities as well as corresponding cumulative probabilities

$$B(x; n, \theta) = \sum_{k=0}^{x} b(k; n, \theta)$$

upon simple commands. An example of such a printout (with somewhat different notation) is shown in Figure 5.1.

In the past, the National Bureau of Standards table and the book by H. G. Romig have been widely used; they are listed among the references at the end of this chapter. Also, Table I at the end of the book gives the values of $b(x; n, \theta)$ to four decimal places for $n = 1$ to $n = 20$ and $\theta = 0.05, 0.10, 0.15, \ldots, 0.45, 0.50$. To use this table when $\theta$ is greater than 0.50, we refer to the identity

**THEOREM 5.1**

$$b(x; n, \theta) = b(n - x; n, 1 - \theta)$$

```
MTB > BINOMIAL N=10 P=0.63

BINOMIAL PROBABILITIES FOR N = 10 AND P = .630000

        K                  P(X = K)              P(X LESS OR = K)
        0                   .0000                     .0000
        1                   .0008                     .0009
        2                   .0063                     .0071
        3                   .0285                     .0356
        4                   .0849                     .1205
        5                   .1734                     .2939
        6                   .2461                     .5400
        7                   .2394                     .7794
        8                   .1529                     .9323
        9                   .0578                     .9902
       10                   .0098                    1.0000
```

**Figure 5.1** Computer printout of binomial probabilities for $n = 10$ and $\theta = 0.63$.

which the reader will be asked to prove in part (a) of Exercise 5.5. For instance, to find $b(11; 18, 0.70)$, we look up $b(7; 18, 0.30)$ and get 0.1376. Also, there are several ways in which binomial probabilities can be approximated when $n$ is large; one of these will be mentioned in Section 5.7 and another in Section 6.6.

Let us now find formulas for the mean and the variance of the binomial distribution.

---

**THEOREM 5.2**  The mean and the variance of the binomial distribution are

$$\mu = n\theta \quad \text{and} \quad \sigma^2 = n\theta(1 - \theta)$$

---

***Proof.***

$$\mu = \sum_{x=0}^{n} x \cdot \binom{n}{x} \theta^x (1 - \theta)^{n-x}$$

$$= \sum_{x=1}^{n} \frac{n!}{(x - 1)!(n - x)!} \theta^x (1 - \theta)^{n-x}$$

where we omitted the term corresponding to $x = 0$, which is 0, and canceled the $x$ against the first factor of $x! = x(x - 1)!$ in the denominator of $\binom{n}{x}$.

Then, factoring out the factor $n$ in $n! = n(n - 1)!$ and one factor $\theta$, we get

$$\mu = n\theta \cdot \sum_{x=1}^{n} \binom{n-1}{x-1} \theta^{x-1}(1 - \theta)^{n-x}$$

and, letting $y = x - 1$ and $m = n - 1$, this becomes

$$\mu = n\theta \cdot \sum_{y=0}^{m} \binom{m}{y} \theta^{y}(1 - \theta)^{m-y} = n\theta$$

since the last summation is the sum of all the values of a binomial distribution with the parameters $m$ and $\theta$, and hence equal to 1.

To find expressions for $\mu_2'$ and $\sigma^2$, let us make use of the fact that $E(X^2) = E[X(X - 1)] + E(X)$ and first evaluate $E[X(X - 1)]$. Duplicating for all practical purposes the steps used before, we thus get

$$E[X(X - 1)] = \sum_{x=0}^{n} x(x - 1)\binom{n}{x} \theta^{x}(1 - \theta)^{n-x}$$

$$= \sum_{x=2}^{n} \frac{n!}{(x - 2)!(n - x)!} \theta^{x}(1 - \theta)^{n-x}$$

$$= n(n - 1)\theta^2 \cdot \sum_{x=2}^{n} \binom{n-2}{x-2} \theta^{x-2}(1 - \theta)^{n-x}$$

and, letting $y = x - 2$ and $m = n - 2$, this becomes

$$E[X(X - 1)] = n(n - 1)\theta^2 \cdot \sum_{y=0}^{m} \binom{m}{y} \theta^{y}(1 - \theta)^{m-y}$$

$$= n(n - 1)\theta^2$$

Therefore,

$$\mu_2' = E[X(X - 1)] + E(X) = n(n - 1)\theta^2 + n\theta$$

and, finally,

$$\sigma^2 = \mu_2' - \mu^2$$

$$= n(n - 1)\theta^2 + n\theta - n^2\theta^2$$

$$= n\theta(1 - \theta) \qquad \blacktriangledown$$

An alternative proof of this theorem, requiring much less algebraic detail, is suggested in Exercise 5.6.

It should not have come as a surprise that the mean of the binomial distribution is given by the product $n\theta$. After all, if a balanced coin is flipped 200 times, we expect (in the sense of a mathematical expectation) $200 \cdot \frac{1}{2} = 100$ heads and 100 tails; similarly, if a balanced die is rolled 240 times we expect $240 \cdot \frac{1}{6} = 40$ sixes, and if the probability is 0.80 that a person shopping at a department store will make a purchase, we would expect $400(0.80) = 320$ of 400 persons shopping at the department store to make a purchase.

The formula for the variance of the binomial distribution, being a measure of variation, has many important applications, but to emphasize its significance let us consider the random variable $Y = \dfrac{X}{n}$, where $X$ is a random variable having a binomial distribution with the parameters $n$ and $\theta$. This random variable is the proportion of successes in $n$ trials, and in Exercise 5.6 the reader will be asked to prove the following result.

---

**THEOREM 5.3** If $X$ has a binomial distribution with the parameters $n$ and $\theta$ and $Y = \dfrac{X}{n}$, then

$$E(Y) = \theta \quad \text{and} \quad \sigma_Y^2 = \frac{\theta(1 - \theta)}{n}$$

---

Now, if we apply Chebyshev's theorem with $k\sigma = c$ (see Exercise 4.40), we can assert that *for any positive constant c the probability is at least*

$$1 - \frac{\theta(1 - \theta)}{nc^2}$$

*that the proportion of successes in n trials falls between $\theta - c$ and $\theta + c$.* Hence, *when $n \to \infty$, the probability approaches 1 that the proportion of successes will differ from $\theta$ by less than any arbitrary constant c.* This result is called a **law of large numbers**, and it should be observed that it applies to the proportion of successes, not to their actual number. It is a fallacy to suppose that when $n$ is large, the number of successes must necessarily be close to $n\theta$.

Since the moment-generating function of the binomial distribution is easy to obtain, let us find it and use it to verify the results of Theorem 5.2.

**THEOREM 5.4** The moment-generating function of the binomial distribution is given by

$$M_X(t) = [1 + \theta(e^t - 1)]^n$$

***Proof.*** By Definitions 4.6 and 5.3, we get

$$M_X(t) = \sum_{x=0}^{n} e^{xt} \binom{n}{x} \theta^x (1 - \theta)^{n-x}$$

$$= \sum_{x=0}^{n} \binom{n}{x} (\theta e^t)^x (1 - \theta)^{n-x}$$

and by Theorem 1.9 this summation is easily recognized as the binomial expansion of $[\theta e^t + (1 - \theta)]^n = [1 + \theta(e^t - 1)]^n$. ▼

If we differentiate $M_X(t)$ twice with respect to $t$, we get

$$M_X'(t) = n\theta e^t [1 + \theta(e^t - 1)]^{n-1}$$

$$M_X''(t) = n\theta e^t [1 + \theta(e^t - 1)]^{n-1} + n(n - 1)\theta^2 e^{2t} [1 + \theta(e^t - 1)]^{n-2}$$

$$= n\theta e^t (1 - \theta + n\theta e^t)[1 + \theta(e^t - 1)]^{n-2}$$

and, upon substituting $t = 0$, we get $\mu_1' = n\theta$ and $\mu_2' = n\theta(1 - \theta + n\theta)$. Thus, $\mu = n\theta$ and $\sigma^2 = \mu_2' - \mu^2 = n\theta(1 - \theta + n\theta) - (n\theta)^2 = n\theta(1 - \theta)$, which agrees with the formulas given in Theorem 5.2.

From the work of this section it may seem easier to find the moments of the binomial distribution with the moment-generating function than to evaluate them directly, but it should be apparent that the differentiation becomes fairly involved if we want to determine, say, $\mu_3'$ or $\mu_4'$. Actually, there exists yet an easier way of determining the moments of the binomial distribution; it is based on its **factorial moment-generating function**, which is explained in Exercise 5.12.

## EXERCISES

**5.1** If $X$ has the discrete uniform distribution $f(x) = \frac{1}{k}$ for $x = 1, 2, \ldots, k$, show that

(a) its mean is $\mu = \dfrac{k + 1}{2}$;

(b)  its variance is $\sigma^2 = \dfrac{k^2 - 1}{12}$.

(*Hint*: Refer to the appendix at the end of the book.)

**5.2** If $X$ has the discrete uniform distribution $f(x) = \frac{1}{k}$ for $x = 1, 2, \ldots, k$, show that its moment-generating function is given by

$$M_X(t) = \frac{e^t(1 - e^{kt})}{k(1 - e^t)}$$

Also find the mean of this distribution by evaluating $\lim_{t \to 0} M_X'(t)$, and compare the result with that obtained in Exercise 5.1.

**5.3** We did not study the Bernoulli distribution in any detail in Section 5.3, because it can be looked upon as a binomial distribution with $n = 1$. Show that for the Bernoulli distribution $\mu_r' = \theta$ for $r = 1, 2, 3, \ldots$, by

(a)  evaluating the sum $\sum\limits_{x=0}^{1} x^r \cdot f(x; \theta)$;

(b)  letting $n = 1$ in the moment-generating function of the binomial distribution and examining its Maclaurin's series.

**5.4** Use the result of the preceding exercise to show that for the Bernoulli distribution,

(a)  $\alpha_3 = \dfrac{1 - 2\theta}{\sqrt{\theta(1 - \theta)}}$, where $\alpha_3$ is the measure of skewness defined in Exercise 4.34;

(b)  $\alpha_4 = \dfrac{1 - 3\theta(1 - \theta)}{\theta(1 - \theta)}$, where $\alpha_4$ is the measure of peakedness defined in Exercise 4.35.

**5.5** Verify that

(a)  $b(x; n, \theta) = b(n - x; n, 1 - \theta)$.

Also show that if $B(x; n, \theta) = \sum\limits_{k=0}^{x} b(k; n, \theta)$ for $x = 0, 1, 2, \ldots, n$, then

(b)  $b(x; n, \theta) = B(x; n, \theta) - B(x - 1; n, \theta)$;

(c)  $b(x; n, \theta) = B(n - x; n, 1 - \theta) - B(n - x - 1; n, 1 - \theta)$;

(d)  $B(x; n, \theta) = 1 - B(n - x - 1; n, 1 - \theta)$.

**5.6** An alternative proof of Theorem 5.2 may be based on the fact that if $X_1, X_2, \ldots,$ and $X_n$ are independent random variables having the same Bernoulli distribution with the parameter $\theta$, then $Y = X_1 + X_2 + \cdots + X_n$ is a random variable having the binomial distribution with the parameters $n$ and $\theta$.

(a)  Verify directly (that is, without making use of the fact that the Bernoulli distribution is a special case of the binomial distribution) that the mean and the variance of the Bernoulli distribution are $\mu = \theta$ and $\sigma^2 = \theta(1 - \theta)$.

(b)  Based on Theorem 4.14 and its corollary on pages 174 and 175, show that if $X_1, X_2, \ldots ,$ and $X_n$ are independent random variables having the same Bernoulli distribution with the parameter $\theta$ and $Y = X_1 + X_2 + \cdots + X_n$, then

$$E(Y) = n\theta \quad \text{and} \quad \text{var}(Y) = n\theta(1 - \theta)$$

**5.7**  Prove Theorem 5.3.

**5.8**  When calculating all the values of a binomial distribution, the work can usually be simplified by first calculating $b(0; n, \theta)$ and then using the recursion formula

$$b(x + 1; n, \theta) = \frac{\theta(n - x)}{(x + 1)(1 - \theta)} \cdot b(x; n, \theta)$$

Verify this formula and use it to calculate the values of the binomial distribution with $n = 7$ and $\theta = 0.25$.

**5.9**  Use the recursion formula of the preceding exercise to show that for $\theta = \frac{1}{2}$, the binomial distribution has

(a)  a maximum at $x = \dfrac{n}{2}$ when $n$ is even;

(b)  maxima at $x = \dfrac{n - 1}{2}$ and $x = \dfrac{n + 1}{2}$ when $n$ is odd.

**5.10**  If $X$ is a binomial random variable, for what value of $\theta$ is the probability $b(x; n, \theta)$ a maximum?

**5.11**  In the proof of Theorem 5.2 we determined the quantity $E[X(X - 1)]$, called the second **factorial moment**. In general, the $r$th factorial moment of $X$ is given by

$$\mu'_{(r)} = E[X(X - 1)(X - 2) \cdot \ldots \cdot (X - r + 1)]$$

Express $\mu'_2$, $\mu'_3$, and $\mu'_4$ in terms of factorial moments.

**5.12**  The **factorial moment-generating function** of a discrete random variable $X$ is given by

$$F_X(t) = E(t^X) = \sum_x t^x \cdot f(x)$$

Show that the $r$th derivative of $F_X(t)$ with respect to $t$ at $t = 1$ is $\mu'_{(r)}$, the $r$th factorial moment defined in Exercise 5.11.

**5.13** With reference to Exercise 5.12, find the factorial moment-generating function of

(a) the Bernoulli distribution and show that $\mu'_{(1)} = \theta$ and $\mu'_{(r)} = 0$ for $r > 1$;

(b) the binomial distribution and use it to find $\mu$ and $\sigma^2$.

**5.14** If we let $a = -\mu$ in the first part of Theorem 4.10, where $\mu$ is the mean of $X$, we get

$$M_Y(t) = M_{X-\mu}(t) = e^{-\mu t} \cdot M_X(t)$$

(a) Show that the $r$th derivative of $M_{X-\mu}(t)$ with respect to $t$ at $t = 0$ gives the $r$th moment about the mean of $X$.

(b) Find such a generating function for moments about the mean of the binomial distribution, and verify that the second derivative at $t = 0$ is $n\theta(1 - \theta)$.

**5.15** Use the result of part (b) of the preceding exercise to show that for the binomial distribution

$$\alpha_3 = \frac{1 - 2\theta}{\sqrt{n\theta(1 - \theta)}}$$

where $\alpha_3$ is the measure of skewness defined in Exercise 4.34. What can we conclude about the skewness of the binomial distribution when

(a) $\theta = \frac{1}{2}$;

(b) $n$ is large?

## APPLICATIONS

**5.16** A multiple-choice test consists of eight questions and three answers to each question (of which only one is correct). If a student answers each question by rolling a balanced die and checking the first answer if he gets a 1 or 2, the second answer if he gets a 3 or 4, and the third answer if he gets a 5 or 6, what is the probability that he will get exactly four correct answers?

**5.17** An automobile safety engineer claims that 1 in 10 automobile accidents is due to driver fatigue. Using the formula for the binomial distribution and rounding to four decimals, what is the probability that at least 3 of 5 automobile accidents are due to driver fatigue?

**5.18** If 40 percent of the mice used in an experiment will become very aggressive within 1 minute after having been administered an experimental drug, find the probability that exactly six of 15 mice which have been administered the drug will become very aggressive within 1 minute, using

(a) the formula for the binomial distribution;

(b) Table I.

**5.19** In a certain city, incompatibility is given as the legal reason in 70 percent of all divorce cases. Find the probability that five of the next six divorce cases filed in this city will claim incompatibility as the reason, using

(a) the formula for the binomial distribution;

(b) Table I.

**5.20** A social scientist claims that only 50 percent of all high school seniors capable of doing college work actually go to college. Assuming that this claim is true, use Table I to find the probabilities that among 18 high school seniors capable of doing college work

(a) exactly 10 will go to college;

(b) at least 10 will go to college;

(c) at most eight will go to college.

**5.21** Suppose that the probability is 0.63 that a car stolen in a certain Western city will be recovered. Use the computer printout of Figure 5.1 to find the probability that at least eight of 10 cars stolen in this city will be recovered, using

(a) the values in the P(X = K) column;

(b) the values in the P(X LESS OR = K) column.

**5.22** With reference to the preceding exercise and the computer printout of Figure 5.1, find the probability that among 10 cars stolen in the given city anywhere from three to five will be recovered, using

(a) the values in the P(X = K) column;

(b) the values in the P(X LESS OR = K) column.

**5.23** With reference to Exercise 5.18, suppose that the percentage had been 42 instead of 40. Use a suitable table or a computer printout of the binomial distribution with $n = 15$ and $\theta = 0.42$ to rework both parts of that exercise.

**5.24** With reference to Exercise 5.20, suppose that the percentage had been 51 instead of 50. Use a suitable table or a computer printout of the binomial distribution with $n = 18$ and $\theta = 0.51$ to rework the three parts of that exercise.

**5.25** In planning the operation of a new school, one school board member claims that four out of five newly hired teachers will stay with the school for more than a year, while another school board member claims that it would be correct to say three out of five. In the past, the two board members have been about equally reliable in their predictions, so that in the absence of any other information we would assign their judgments equal weight. If one or the other has to be right, what probabilities would we assign to their claims if it were found that 11 of 12 newly hired teachers stayed with the school for more than a year?

**5.26** Use Chebyshev's theorem and Theorem 5.3 to verify that the probability is at least $\frac{35}{36}$ that

(a) in 900 flips of a balanced coin the proportion of heads will be between 0.40 and 0.60;

      (b)   in 10,000 flips of a balanced coin the proportion of heads will be between 0.47 and 0.53;

      (c)   in 1,000,000 flips of a balanced coin the proportion of heads will be betwen 0.497 and 0.503.

Note that this serves to illustrate the law of large numbers.

## 5.5 THE NEGATIVE BINOMIAL AND GEOMETRIC DISTRIBUTIONS

In connection with repeated Bernoulli trials, we are sometimes interested in the number of the trial on which the $k$th success occurs. For instance, we may be interested in the probability that the tenth child exposed to a contagious disease will be the third to catch it, the probability that the fifth person to hear a rumor will be the first one to believe it, or the probability that a burglar will be caught for the second time on his or her eighth job.

If the $k$th success is to occur on the $x$th trial, there must be $k - 1$ successes on the first $x - 1$ trials, and the probability for this is

$$b(k - 1; x - 1, \theta) = \binom{x - 1}{k - 1} \theta^{k-1}(1 - \theta)^{x-k}$$

The probability of a success on the $x$th trial is $\theta$, and the probability that the $k$th success occurs on the $x$th trial is, therefore,

$$\theta \cdot b(k - 1; x - 1, \theta) = \binom{x - 1}{k - 1} \theta^{k}(1 - \theta)^{x-k}$$

---

**DEFINITION 5.4** A random variable $X$ has a **negative binomial distribution**, and it is referred to as a negative binomial random variable, if and only if

$$b^*(x; k, \theta) = \binom{x - 1}{k - 1} \theta^{k}(1 - \theta)^{x-k}$$

for $x = k, k + 1, k + 2, \ldots$ .

---

Thus, the number of the trial on which the $k$th success occurs is a random variable having a negative binomial distribution with the parameters $k$ and $\theta$. The name "negative binomial distribution" derives from the fact that the values of $b^*(x; k, \theta)$ for $x = k, k + 1, k + 2, \ldots$ , are the successive terms of the binomial expansion of $\left(\dfrac{1}{\theta} - \dfrac{1 - \theta}{\theta}\right)^{-k}$.[†] In the literature of statistics, negative binomial distributions are also referred to as **binomial waiting-time distributions** or as **Pascal distributions**.

## EXAMPLE 5.3

If the probability is 0.40 that a child exposed to a certain contagious disease will catch it, what is the probability that the tenth child exposed to the disease will be the third to catch it?

### Solution

Substituting $x = 10$, $k = 3$, and $\theta = 0.40$ into the formula for the negative binomial distribution, we get

$$b^*(10; 3, 0.40) = \binom{9}{2}(0.40)^3(0.60)^7$$

$$= 0.0645 \quad \blacktriangle$$

When a table of binomial probabilities is available, the determination of negative binomial probabilities can generally be simplified by making use of the identity

---

**THEOREM 5.5**

$$b^*(x; k, \theta) = \frac{k}{x} \cdot b(k; x, \theta)$$

---

The reader will be asked to verify this theorem in Exercise 5.29.

## EXAMPLE 5.4

Use Theorem 5.5 and Table I to rework Example 5.3.

---

[†] Binomial expansions with negative exponents are explained in the book by W. Feller listed among the references at the end of Chapter 2.

*Solution*

Substituting $x = 10$, $k = 3$, and $\theta = 0.40$ into the formula of Theorem 5.5, we get

$$b^*(10; 3, 0.40) = \tfrac{3}{10} \cdot b(3; 10, 0.40)$$

$$= \tfrac{3}{10} (0.2150)$$

$$= 0.0645 \quad \blacktriangle$$

Moments of the negative binomial distribution may be obtained by proceeding as in the proof of Theorem 5.2; for the mean and the variance we get

---

**THEOREM 5.6**  The mean and the variance of the negative binomial distribution are

$$\mu = \frac{k}{\theta} \quad \text{and} \quad \sigma^2 = \frac{k}{\theta}\left(\frac{1}{\theta} - 1\right)$$

---

as the reader will be asked to verify in Exercise 5.30.

Since the negative binomial distribution with $k = 1$ has many important applications, it is given a special name; it is called the **geometric distribution**.

---

**DEFINITION 5.5**  A random variable $X$ has a **geometric distribution**, and it is referred to as a geometric random variable, if and only if its probability distribution is given by

$$g(x; \theta) = \theta(1 - \theta)^{x-1} \quad \text{for } x = 1, 2, 3, \ldots$$

---

## EXAMPLE 5.5

If the probability is 0.75 that an applicant for a driver's license will pass the road test on any given try, what is the probability that an applicant will finally pass the test on the fourth try?

*Solution*

Substituting $x = 4$ and $\theta = 0.75$ into the formula for the geometric distribution, we get

$$g(4; 0.75) = 0.75(1 - 0.75)^{4-1}$$
$$= 0.75(0.25)^3$$
$$= 0.0117$$

Of course, this result is based on the assumption that the trials are all independent, and there may be some question here about its validity.    ▲

## 5.6 THE HYPERGEOMETRIC DISTRIBUTION

In Chapter 2 we used sampling with and without replacement to illustrate the multiplication rules for independent and dependent events. To obtain a formula analogous to that of the binomial distribution which applies to sampling without replacement, in which case the trials are not independent, let us consider a set of $N$ elements of which $k$ are looked upon as successes and the other $N - k$ as failures. As in connection with the binomial distribution, we are interested in the probability of getting $x$ successes in $n$ trials, but now we are choosing, without replacement, $n$ of the $N$ elements contained in the set.

There are $\binom{k}{x}$ ways of choosing $x$ of the $k$ successes and $\binom{N-k}{n-x}$ ways of

choosing $n - x$ of the $N - k$ failures, and, hence, $\binom{k}{x}\binom{N-k}{n-x}$ ways of choosing

$x$ successes and $n - x$ failures. Since there are $\binom{N}{n}$ ways of choosing $n$ of the

$N$ elements in the set, and we shall assume that they are all equally likely (which is what we mean when we say that the selection is random), it follows from Theorem 2.2 that the probability of "$x$ successes in $n$ trials" is $\binom{k}{x}\binom{N-k}{n-x} \Big/ \binom{N}{n}$.

---

**DEFINITION 5.6** A random variable $X$ has a **hypergeometric distribution**, and it is referred to as a hypergeometric random variable, if and only if its probability distribution is given by

$$h(x; n, N, k) = \frac{\binom{k}{x}\binom{N-k}{n-x}}{\binom{N}{n}} \qquad \begin{array}{l} \text{for } x = 0, 1, 2, \ldots, n, \\ x \leq k \text{ and } n - x \leq N - k \end{array}$$

---

Thus, for sampling without replacement, the number of successes in $n$ trials is a random variable having a hypergeometric distribution with the parameters $n$, $N$, and $k$.

## EXAMPLE 5.6

As part of an air-pollution survey, an inspector decides to examine the exhaust of six of a company's 24 trucks. If four of the company's trucks emit excessive amounts of pollutants, what is the probability that none of them will be included in the inspector's sample?

### Solution

Substituting $x = 0$, $n = 6$, $N = 24$, and $k = 4$ into the formula for the hypergeometric distribution, we get

$$h(0; 6, 24, 4) = \frac{\binom{4}{0}\binom{20}{6}}{\binom{24}{6}}$$

$$= 0.2880 \qquad \blacktriangle$$

The method by which we find the mean and the variance of the hypergeometric distribution is very similar to that employed in the proof of Theorem 5.2.

---

**THEOREM 5.7**   The mean and the variance of the hypergeometric distribution are

$$\mu = \frac{nk}{N} \quad \text{and} \quad \sigma^2 = \frac{nk(N - k)(N - n)}{N^2(N - 1)}$$

---

***Proof.***   To determine the mean, let us directly evaluate the sum

$$\mu = \sum_{x=0}^{n} x \cdot \frac{\binom{k}{x}\binom{N - k}{n - x}}{\binom{N}{n}}$$

$$= \sum_{x=1}^{n} \frac{k!}{(x-1)!(k-x)!} \cdot \frac{\binom{N-k}{n-x}}{\binom{N}{n}}$$

where we omitted the term corresponding to $x = 0$, which is 0, and canceled the $x$ against the first factor of $x! = x(x-1)!$ in the denominator of $\binom{k}{x}$.

Then, factoring out $k \Big/ \binom{N}{n}$, we get

$$\mu = \frac{k}{\binom{N}{n}} \cdot \sum_{x=1}^{n} \binom{k-1}{x-1}\binom{N-k}{n-x}$$

and, letting $y = x - 1$ and $m = n - 1$, this becomes

$$\mu = \frac{k}{\binom{N}{n}} \cdot \sum_{y=0}^{m} \binom{k-1}{y}\binom{N-k}{m-y}$$

Finally, using Theorem 1.12, we get

$$\mu = \frac{k}{\binom{N}{n}} \cdot \binom{N-1}{m} = \frac{k}{\binom{N}{n}} \cdot \binom{N-1}{n-1} = \frac{nk}{N}$$

To obtain the formula for $\sigma^2$, we proceed as in the proof of Theorem 5.2, namely, by first evaluating $E[X(X-1)]$ and then making use of the fact that $E(X^2) = E[X(X-1)] + E(X)$. Leaving it to the reader to show that

$$E[X(X-1)] = \frac{k(k-1)n(n-1)}{N(N-1)}$$

in Exercise 5.38, we thus get

$$\sigma^2 = \frac{k(k-1)n(n-1)}{N(N-1)} + \frac{nk}{N} - \left(\frac{nk}{N}\right)^2$$

$$= \frac{nk(N-k)(N-n)}{N^2(N-1)} \qquad \blacktriangledown$$

Since the moment-generating function of the hypergeometric distribution is fairly complicated, it will not be treated in this book. Details may be found, however, in the book by M. G. Kendall and A. Stuart listed among the references at the end of Chapter 3.

When $N$ is large and $n$ is relatively small compared to $N$ (the usual rule of thumb is that $n$ should not exceed 5 percent of $N$), there is not much difference between sampling with replacement and sampling without replacement, and the formula for the binomial distribution with the parameters $n$ and $\theta = \dfrac{k}{N}$ may be used to approximate hypergeometric probabilities.

## EXAMPLE 5.7

Among the 120 applicants for a job, only 80 are actually qualified. If five of the applicants are randomly selected for an in-depth interview, find the probability that only two of the five will be qualified for the job by using

- (a)  the formula for the hypergeometric distribution;
- (b)  the formula for the binomial distribution with $\theta = \frac{80}{120}$ as an approximation.

### Solution

- (a)  Substituting $x = 2$, $n = 5$, $N = 120$, and $k = 80$ into the formula for the hypergeometric distribution, we get

$$h(2; 5, 120, 80) = \frac{\dbinom{80}{2}\dbinom{40}{3}}{\dbinom{120}{5}}$$

$$= 0.164$$

rounded to three decimals;

- (b)  substituting $x = 2$, $n = 5$, and $\theta = \frac{80}{120} = \frac{2}{3}$ into the formula for the binomial distribution, we get

$$b(2; 5, \tfrac{2}{3}) = \binom{5}{2}\left(\frac{2}{3}\right)^{2}\left(1 - \frac{2}{3}\right)^{3}$$

$$= 0.165$$

rounded to three decimals. As can be seen from these results, the approximation is very close.    ▲

## 5.7 THE POISSON DISTRIBUTION

When $n$ is large, the calculation of binomial probabilities with the formula of Definition 5.3 will usually involve a prohibitive amount of work. For instance, to calculate the probability that 18 of 3,000 persons watching a parade on a very hot summer day will suffer from heat exhaustion, we first have to determine $\binom{3,000}{18}$, and if the probability is 0.005 that any one of the 3,000 persons watching the parade will suffer from heat exhaustion, we also have to calculate the value of $(0.005)^{18}(0.995)^{2,982}$.

In this section we shall present a probability distribution which can be used to approximate binomial probabilities of this kind. Specifically, we shall investigate the limiting form of the binomial distribution when $n \to \infty$, $\theta \to 0$, while $n\theta$ remains constant. Letting this constant be $\lambda$, that is, $n\theta = \lambda$ and, hence, $\theta = \dfrac{\lambda}{n}$, we can write

$$b(x; n, \theta) = \binom{n}{x}\left(\frac{\lambda}{n}\right)^{x}\left(1 - \frac{\lambda}{n}\right)^{n-x}$$

$$= \frac{n(n - 1)(n - 2)\cdot\ldots\cdot(n - x + 1)}{x!}\left(\frac{\lambda}{n}\right)^{x}\left(1 - \frac{\lambda}{n}\right)^{n-x}$$

Then, if we divide one of the $x$ factors $n$ in $\left(\dfrac{\lambda}{n}\right)^{x}$ into each factor of the product $n(n - 1)(n - 2)\cdot\ldots\cdot(n - x + 1)$ and write

$$\left(1 - \frac{\lambda}{n}\right)^{n-x} \quad \text{as} \quad \left[\left(1 - \frac{\lambda}{n}\right)^{-n/\lambda}\right]^{-\lambda}\left(1 - \frac{\lambda}{n}\right)^{-x}$$

we obtain

$$\frac{1\left(1 - \frac{1}{n}\right)\left(1 - \frac{2}{n}\right)\cdot\ldots\cdot\left(1 - \frac{x-1}{n}\right)}{x!} (\lambda)^x \left[\left(1 - \frac{\lambda}{n}\right)^{-n/\lambda}\right]^{-\lambda}\left(1 - \frac{\lambda}{n}\right)^{-x}$$

Finally, if we let $n \to \infty$ while $x$ and $\lambda$ remain fixed, we find that

$$1\left(1 - \frac{1}{n}\right)\left(1 - \frac{2}{n}\right)\cdot\ldots\cdot\left(1 - \frac{x-1}{n}\right) \to 1$$

$$\left(1 - \frac{\lambda}{n}\right)^{-x} \to 1$$

$$\left(1 - \frac{\lambda}{n}\right)^{-n/\lambda} \to e$$

and, hence, that the limiting distribution becomes

$$p(x; \lambda) = \frac{\lambda^x e^{-\lambda}}{x!} \qquad \text{for } x = 0, 1, 2, \ldots$$

---

**DEFINITION 5.7**  A random variable $X$ has a **Poisson distribution,** and it is referred to as a Poisson random variable, if and only if its probability distribution is given by

$$p(x; \lambda) = \frac{\lambda^x e^{-\lambda}}{x!} \qquad \text{for } x = 0, 1, 2, \ldots$$

---

Thus, in the limit when $n \to \infty$, $\theta \to 0$, and $n\theta = \lambda$ remains constant, the number of successes is a random variable having a Poisson distribution with the parameter $\lambda$. This distribution is named after the French mathematician Simeon Poisson (1781–1840). In general, the Poisson distribution will provide a good approximation to binomial probabilities when $n \geqslant 20$ and $\theta \leqslant 0.05$. When $n \geqslant 100$ and $n\theta < 10$, the approximation will generally be excellent.

To get some idea about the closeness of the Poisson approximation to the binomial distribution, consider the computer printout of Figure 5.2, which shows one above the other, the binomial distribution with $n = 150$ and $\theta = 0.05$ and the Poisson distribution with $\lambda = 150(0.05) = 7.5$.

```
MTB > BINOMIAL N=150 P=0.05

   BINOMIAL PROBABILITIES FOR N = 150 AND P = .050000

        K              P(X = K)              P(X LESS OR = K)
        0               .0005                    .0005
        1               .0036                    .0041
        2               .0141                    .0182
        3               .0366                    .0548
        4               .0708                    .1256
        5               .1088                    .2344
        6               .1384                    .3729
        7               .1499                    .5228
        8               .1410                    .6638
        9               .1171                    .7809
       10               .0869                    .8678
       11               .0582                    .9260
       12               .0355                    .9615
       13               .0198                    .9813
       14               .0102                    .9915
       15               .0049                    .9964
       16               .0022                    .9986
       17               .0009                    .9995
       18               .0003                    .9998
       19               .0001                    .9999




MTB > POISSON MU=7.5

   POISSON PROBABILITIES FOR MEAN =  7.500

        K              P(X = K)              P(X LESS OR = K)
        0               .0006                    .0006
        1               .0041                    .0047
        2               .0156                    .0203
        3               .0389                    .0591
        4               .0729                    .1321
        5               .1094                    .2414
        6               .1367                    .3782
        7               .1465                    .5246
        8               .1373                    .6620
        9               .1144                    .7764
       10               .0858                    .8622
       11               .0585                    .9208
       12               .0366                    .9573
       13               .0211                    .9784
       14               .0113                    .9897
       15               .0057                    .9954
       16               .0026                    .9980
       17               .0012                    .9992
       18               .0005                    .9997
       19               .0002                    .9999
       20               .0001                   1.0000
```

**Figure 5.2**  Computer printout of the binomial distribution with $n = 150$ and $\theta = 0.05$ and the Poisson distribution with $\lambda = 7.5$.

## EXAMPLE 5.8

Use Figure 5.2 to determine the value of $x$ (from 5 to 15) for which the error is greatest when we use the Poisson distributin with $\lambda = 7.5$ to approximate the binomial distribution with $n = 150$ and $\theta = 0.05$.

### Solution

Calculating the differences corresponding to $x = 5$, $x = 6$, ... , $x = 15$, we get 0.0006, $-0.0017$, $-0.0034$, $-0.0037$, $-0.0027$, $-0.0011$, 0.0003, 0.0011, 0.0013, 0.0011, and 0.0008. Thus, the maximum error (numerically) is $-0.0037$, and it corresponds to $x = 8$. ▲

The examples which follow illustrate the Poisson approximation to the binomial distribution.

## EXAMPLE 5.9

If 2 percent of the books bound at a certain bindery have defective bindings, use the Poisson approximation to the binomial distribution to determine the probability that five of 400 books bound by this bindery will have defective bindings.

### Solution

Substituting $x = 5$, $\lambda = 400(0.02) = 8$, and $e^{-8} = 0.00034$ (from Table VIII at the end of the book) into the formula of Definition 5.7, we get

$$p(5; 8) = \frac{8^5 \cdot e^{-8}}{5!} = \frac{(32,768)(0.00034)}{120} = 0.093 \quad ▲$$

In actual practice, Poisson probabilities are seldom obtained by direct substitution into the formula of Definition 5.7. Sometimes we refer to tables of Poisson probabilities, such as Table II at the end of this book or more extensive tables in handbooks of statistical tables, but more often than not, nowadays, we refer to suitable computer software. The use of tables or computers is of special importance when we are concerned with probabilities relating to several values of $x$.

## EXAMPLE 5.10

Records show that the probability is 0.00005 that a car will have a flat tire while crossing a certain bridge. Use the Poisson distribution to approximate the binomial probabilities that among 10,000 cars crossing this bridge,

(a)   exactly two will have a flat tire;

(b)   at most two will have a flat tire.

### Solution

(a)   Referring to Table II, we find that for $x = 2$ and $\lambda = 10,000(0.00005)$ = 0.5 the Poisson probability is 0.0758.

(b)   Referring to Table II, we find that for $x = 0$, 1, and 2, and $\lambda = 0.5$, the Poisson probabilities are 0.6065, 0.3033, and 0.0758. Thus, the probability that at most two of 10,000 cars crossing the bridge will have a flat tire is

$$0.6065 + 0.3033 + 0.0758 = 0.9856 \quad \blacktriangle$$

## EXAMPLE 5.11

Use Figure 5.3 to rework the preceding example.

### Solution

(a)   Reading off the value for $K = 2$ in the $P(X = K)$ column, we get 0.0758.

(b)   Here we could add the values for $K = 0$, $K = 1$, and $K = 2$ in the $P(X = K)$ column, or we could read the value for $K = 2$ in the P(X LESS OR = K) column, getting 0.9856.    $\blacktriangle$

Having derived the Poisson distribution as a limiting form of the binomial distribution, we can obtain formulas for its mean and its variance by applying the same limiting conditions ($n \to \infty$, $\theta \to 0$, and $n\theta = \lambda$ remains constant) to the

```
MTB > POISSON MU=.5

POISSON PROBABILITIES FOR MEAN =    .500

     K                P(X = K)            P(X LESS OR = K)
     0                 .6065                   .6065
     1                 .3033                   .9098
     2                 .0758                   .9856
     3                 .0126                   .9982
     4                 .0016                   .9998
     5                 .0002                  1.0000
```

**Figure 5.3**   Computer printout of the Poisson distribution with $\lambda = 0.5$.

mean and the variance of the binomial distribution. For the mean we get $\mu = n\theta$ $= \lambda$ and for the variance we get $\sigma^2 = n\theta(1 - \theta) = \lambda(1 - \theta)$, which approaches $\lambda$ when $\theta \rightarrow 0$.

---

**THEOREM 5.8** The mean and the variance of the Poisson distribution are given by

$$\mu = \lambda \quad \text{and} \quad \sigma^2 = \lambda$$

---

These results can also be obtained by directly evaluating the necessary summations (see Exercise 5.44) or by working with the moment-generating function given in the following theorem.

---

**THEOREM 5.9** The moment-generating function of the Poisson distribution is given by

$$M_X(t) = e^{\lambda(e^t - 1)}$$

---

***Proof.*** By Definitions 4.6 and 5.7,

$$M_X(t) = \sum_{x=0}^{\infty} e^{xt} \cdot \frac{\lambda^x e^{-\lambda}}{x!} = e^{-\lambda} \cdot \sum_{x=0}^{\infty} \frac{(\lambda e^t)^x}{x!}$$

where $\sum_{x=0}^{\infty} \frac{(\lambda e^t)^x}{x!}$ can be recognized as the Maclaurin's series of $e^z$ with $z = \lambda e^t$. Thus,

$$M_X(t) = e^{-\lambda} \cdot e^{\lambda e^t} = e^{\lambda(e^t - 1)} \qquad \blacktriangledown$$

Then, if we differentiate $M_X(t)$ twice with respect to $t$, we get

$$M_X'(t) = \lambda e^t e^{\lambda(e^t - 1)}$$

$$M_X''(t) = \lambda e^t e^{\lambda(e^t - 1)} + \lambda^2 e^{2t} e^{\lambda(e^t - 1)}$$

so that $\mu_1' = M_X'(0) = \lambda$ and $\mu_2' = M_X''(0) = \lambda + \lambda^2$. Thus, $\mu = \lambda$ and $\sigma^2 = \mu_2' - \mu^2 = (\lambda + \lambda^2) - \lambda^2 = \lambda$, which agrees with Theorem 5.8.

Although the Poisson distribution has been derived as a limiting form of the binomial distribution, it has many applications which have no direct connection with binomial distributions. For example, the Poisson distribution can serve as a model for the number of successes that occur during a given time interval or in a specified region when (1) the numbers of successes occurring in nonoverlapping time intervals or regions are independent; (2) the probability of a single success occurring in a very short time interval or in a very small region is proportional to the length of the time interval or the size of the region; and (3) the probability of more than one success occurring in such a short time interval or falling in such a small region is negligible. Hence, a Poisson distribution might describe the number of telephone calls per hour received by an office, the number of typing errors per page, or the number of bacteria in a given culture, when the average number of successes, $\lambda$, for the given time interval or specified region is known.

## EXAMPLE 5.12

The average number of trucks arriving on any one day at a truck depot in a certain city is known to be 12. What is the probability that on a given day fewer than nine trucks will arrive at this depot?

### Solution

Let $X$ be the number of trucks arriving on a given day. Then, using Table II with $\lambda = 12$, we get

$$P(X < 9) = \sum_{x=0}^{8} p(x; 12) = 0.1550 \quad \blacktriangle$$

If, in a situation where the preceding conditions apply, successes occur at a mean rate of $\alpha$ per *unit* time or per *unit* region, then the number of successes in an interval of $t$ units of time or $t$ units of the specified region is a Poisson random variable with the mean $\lambda = \alpha t$ (see Exercise 5.42). Therefore, the number of successes, $X$, in a time interval of length $t$ units or a region of size $t$ units has the Poisson distribution

$$p(x; \alpha t) = \frac{e^{-\alpha t}(\alpha t)^x}{x!} \quad \text{for } x = 0, 1, 2, \ldots$$

## EXAMPLE 5.13

A certain kind of sheet metal has, on the average, five defects per 10 square feet. If we assume a Poisson distribution, what is the probability that a 15-square-foot sheet of the metal will have at least six defects?

*Solution*

Let $X$ denote the number of defects in a 15-square-foot sheet of the metal. Then, since the unit of area is 10 square feet, we have

$$\lambda = \alpha t = (5)(1.5) = 7.5$$

and

$$P(X \geq 6) = 1 - P(X \leq 5) = 1 - 0.2414 = 0.7586$$

according to the computer printout shown in Figure 5.2.   ▲

*EXERCISES*

**5.27** The negative binomial distribution is sometimes defined in a different way as the distribution of the number of failures that precede the $k$th success. If the $k$th success occurs on the $x$th trial, it must be preceded by $x - k$ failures. Thus, find the distribution of $Y = X - k$, where $X$ has the distribution of Definition 5.4.

**5.28** With reference to the preceding exercise, find expressions for $\mu_Y$ and $\sigma_Y^2$.

**5.29** Prove Theorem 5.5.

**5.30** Prove Theorem 5.6 by first determining $E(X)$ and $E[X(X + 1)]$.

**5.31** Show that the moment-generating function of the geometric distribution is given by

$$M_X(t) = \frac{\theta e^t}{1 - e^t(1 - \theta)}$$

**5.32** Use the moment-generating function derived in the preceding exercise to show that for the geometric distribution, $\mu = \dfrac{1}{\theta}$ and $\sigma^2 = \dfrac{1 - \theta}{\theta^2}$.

**5.33** Differentiating with respect to $\theta$ the expressions on both sides of the equation

$$\sum_{x=1}^{\infty} \theta(1 - \theta)^{x-1} = 1$$

show that the mean of the geometric distribution is given by $\mu = \dfrac{1}{\theta}$. Then, differentiating again with respect to $\theta$, show that $\mu_2' = \dfrac{2 - \theta}{\theta^2}$ and, hence,

$$\sigma^2 = \frac{1 - \theta}{\theta^2}.$$

**5.34** If $X$ is a random variable having a geometric distribution, show that

$$P(X = x + n|X > n) = P(X = x)$$

**5.35** If the probability is $f(x)$ that a product fails the $x$th time it is being used, that is, on the $x$th trial, then its **failure rate** at the $x$th trial is the probability that it will fail on the $x$th trial given that it has not failed on the first $x - 1$ trials; symbolically, it is given by

$$Z(x) = \frac{f(x)}{1 - F(x - 1)}$$

where $F(x)$ is the value of the corresponding distribution function at $x$. Show that if $X$ is a geometric random variable, its failure rate is constant and equal to $\theta$.

**5.36** A variation of the binomial distribution arises when the $n$ trials are all independent, but the probability of a success on the $i$th trial is $\theta_i$, and these probabilities are not all equal. If $X$ is the number of successes obtained under these conditions in $n$ trials, show that

(a)  $\mu_X = n\theta$, where $\theta = \frac{1}{n} \cdot \sum\limits_{i=1}^{n} \theta_i$;

(b)  $\sigma_X^2 = n\theta(1 - \theta) - n\sigma_\theta^2$, where $\theta$ is as defined in part (a) and $\sigma_\theta^2 = \frac{1}{n} \cdot \sum\limits_{i=1}^{n} (\theta_i - \theta)^2$.

**5.37** When calculating all the values of a hypergeometric distribution, the work can often be simplified by first calculating $h(0; n, N, k)$ and then using the recursion formula

$$h(x + 1; n, N, k) = \frac{(n - x)(k - x)}{(x + 1)(N - k - n + x + 1)} \cdot h(x; n, N, k)$$

Verify this formula and use it to calculate the values of the hypergeometric distribution with $n = 4$, $N = 9$, and $k = 5$.

**5.38** Verify the expression given for $E[X(X - 1)]$ in the proof of Theorem 5.7.

**5.39** Show that if we let $\theta = \dfrac{k}{N}$ in Theorem 5.7, the mean and the variance of the hypergeometric distribution can be written as $\mu = n\theta$ and $\sigma^2 = n\theta(1 - \theta) \cdot \dfrac{N - n}{N - 1}$. How do these results tie in with the discussion on page 202?

**5.40** When calculating all the values of a Poisson distribution, the work can often be simplified by first calculating $p(0; \lambda)$ and then using the recursion formula

$$p(x + 1; \lambda) = \frac{\lambda}{x + 1} \cdot p(x; \lambda)$$

Verify this formula and use it and $e^{-2} = 0.1353$ to verify the values given in Table II for $\lambda = 2$.

**5.41** Approximate the binomial probability $b(3; 100, 0.10)$ by using
(a) the formula for the binomial distribution and logarithms;
(b) Table II.

**5.42** Suppose that $f(x, t)$ is the probability of getting $x$ successes during a time interval of length $t$ when (i) the probability of a success during a very small time interval from $t$ to $t + \Delta t$ is $\alpha \cdot \Delta t$, (ii) the probability of more than one success during such a time interval is negligible, and (iii) the probability of a success during such a time interval does not depend on what happened prior to time $t$.

(a) Show that under these conditions

$$f(x, t + \Delta t) = f(x, t)[1 - \alpha \cdot \Delta t] + f(x - 1, t)\alpha \cdot \Delta t$$

and, hence, that

$$\frac{d[f(x, t)]}{dt} = \alpha[f(x - 1, t) - f(x, t)]$$

(b) Show by direct substitution that a solution of this infinite system of differential equations (there is one for each value of $x$) is given by the Poisson distribution with $\lambda = \alpha t$.

**5.43** Use repeated integration by parts to show that

$$\sum_{y=0}^{x} \frac{\lambda^y e^{-\lambda}}{y!} = \frac{1}{x!} \cdot \int_{\lambda}^{\infty} t^x e^{-t} \, dt$$

This result is important because values of the distribution function of a Poisson random variable may, thus, be obtained by referring to a table of incomplete gamma functions.

**5.44** Derive the formulas for the mean and the variance of the Poisson distribution by first evaluating $E(X)$ and $E[X(X - 1)]$.

**5.45** Show that if the limiting conditions $n \to \infty$, $\theta \to 0$, while $n\theta$ remains constant, are applied to the moment-generating function of the binomial distribution, we get the moment-generating function of the Poisson distribution.

[*Hint*: Make use of the fact that $\lim\limits_{n \to \infty} \left(1 + \dfrac{z}{n}\right)^n = e^z$.]

**5.46** Use Theorem 5.9 to show that for the Poisson distribution $\alpha_3 = \dfrac{1}{\sqrt{\lambda}}$, where $\alpha_3$ is the measure of skewness defined in Exercise 4.34.

**5.47** Differentiating with respect to $\lambda$ the expressions on both sides of the equation

$$\mu_r = \sum_{x=0}^{\infty} (x - \lambda)^r \cdot \frac{\lambda^x e^{-\lambda}}{x!}$$

derive the following recursion formula for the moments about the mean of the Poisson distribution:

$$\mu_{r+1} = \lambda \left[ r\mu_{r-1} + \frac{d\mu_r}{d\lambda} \right]$$

for $r = 1, 2, 3, \ldots$ . Also, use this recursion formula and the fact that $\mu_0 = 1$ and $\mu_1 = 0$ to find $\mu_2$, $\mu_3$, and $\mu_4$, and verify the formula given for $\alpha_3$ in Exercise 5.46.

**5.48** Use Theorem 5.9 to find the moment-generating function of $Y = X - \lambda$, where $X$ is a random variable having the Poisson distribution with the parameter $\lambda$, and use it to verify that $\sigma_X^2 = \lambda$.

## APPLICATIONS

**5.49** If the probability is 0.75 that a person will believe a rumor about the transgressions of a certain politician, find the probabilities that

(a)   the eighth person to hear the rumor will be the fifth to believe it;

(b)   the fifteenth person to hear the rumor will be the tenth to believe it.

**5.50** If the probabilities of having a male or female child are both 0.50, find the probabilities that

(a)   a family's fourth child is their first son;

(b)   a family's seventh child is their second daughter;

(c)   a family's tenth child is their fourth or fifth son.

**5.51** An expert sharpshooter misses a target 5 percent of the time. Find the probability that she will miss the target for the second time on the fifteenth shot using

(a)   the formula for the negative binomial distribution;

(b)   Theorem 5.5 and Table I.

**5.52** When taping a television commercial, the probability is 0.30 that a certain actor will get his lines straight on any one take. What is the probability that he will get his lines straight for the first time on the sixth take?

**5.53** In a "torture test" a light switch is turned on and off until it fails. If the probability is 0.001 that the switch will fail any time it is turned on or off, what is the probability that the switch will not fail during the first 800 times it is turned on or off? Assume that the conditions underlying the geometric distribution are met and use logarithms.

**5.54** Adapt the formula of Theorem 5.5 so that it can be used to express geometric probabilities in terms of binomial probabilities, and use the formula and Table I to

(a) verify the result of Example 5.5;

(b) rework Exercise 5.52.

**5.55** A quality control engineer inspects a random sample of two hand-held calculators from each incoming lot of size 18, and accepts the lot if they are both in good working condition; otherwise, the entire lot is inspected with the cost charged to the vendor. What are the probabilities that such a lot will be accepted without further inspection if it contains

(a) four calculators that are not in good working condition;

(b) eight calculators that are not in good working condition;

(c) 12 calculators that are not in good working condition?

**5.56** Among the 16 applicants for a job, ten have college degrees. If three of the applicants are randomly chosen for interviews, what are the probabilities that

(a) none has a college degree;

(b) one has a college degree;

(c) two have college degrees

(d) all three have college degrees?

**5.57** Find the mean and the variance of the hypergeometric distribution with $n = 3$, $N = 16$, and $k = 10$, using

(a) the results of the preceding exercise;

(b) the formulas of Theorem 5.7.

**5.58** What is the probability that an IRS auditor will catch only two income tax returns with illegitimate deductions, if she randomly selects five returns from among 15 returns, of which nine contain illegitimate deductions?

**5.59** Check in each case whether the condition for the binomial approximation to the hypergeometric distribution is satisfied:

(a) $N = 200$ and $n = 12$;

(b) $N = 500$ and $n = 20$;

(c) $N = 640$ and $n = 30$.

**5.60** A shipment of 80 burglar alarms contains four that are defective. If three of these are randomly selected and shipped to a customer, find the probability that the customer will get exactly one bad unit using

(a)   the formula of the hypergeometric distribution;
(b)   the binomial distribution as an approximation.

**5.61**  Among the 300 employees of a company, 240 are union members, whereas the others are not. If six of the employees are chosen by lot to serve on a committee that administers the pension fund, find the probability that four of the six will be union members using

(a)   the formula for the hypergeometric distribution;
(b)   the binomial distribution as an approximation.

**5.62**  A panel of 300 persons chosen for jury duty includes 30 under 25 years of age. Since the jury of 12 persons chosen from this panel to judge a narcotics violation does not include anyone under 25 years of age, the youthful defendant's attorney complains that this jury is not really representative. Indeed, he argues, if the selection were random, the probability of having one of the 12 jurors under 25 years of age should be *many times* the probability of having none of them under 25 years of age. Actually, what is the ratio of these two probabilities?

**5.63**  Check in each case whether the values of $n$ and $\theta$ satisfy the rule of thumb for a good approximation, an excellent approximation, or neither, when we want to use the Poisson distribution to approximate binomial probabilities.

(a)   $n = 125$ and $\theta = 0.10$;
(b)   $n = 25$ and $\theta = 0.04$;
(c)   $n = 120$ and $\theta = 0.05$;
(d)   $n = 40$ and $\theta = 0.06$.

**5.64**  With reference to Example 5.8, determine the value of $x$ (from 5 to 15) for which the percentage error is greatest when we use the Poisson distribution with $\lambda = 7.5$ to approximate the binomial distribution with $n = 150$ and $\theta = 0.05$.

**5.65**  Is is known from experience that 1.4 percent of the calls received by a switchboard are wrong numbers. Use the Poisson approximation to the binomial distribution to determine the probability that among 150 calls received by the switchboard two are wrong numbers.

**5.66**  Records show that the probability is 0.0012 that a person will get food poisoning spending a day at a certain state fair. Use the Poisson approximation to the binomial distribution to find the probability that among 1,000 persons attending the fair, at most two will get food poisoning.

**5.67**  In a given city, 4 percent of all licensed drivers will be involved in at least one car accident in any given year. Use the Poisson approximation to the binomial distribution to determine the probability that among 150 licensed drivers randomly chosen in this city

(a)   only five will be involved in at least one accident in any given year;

(b)  at most three will be involved in at least one accident in any given year.

**5.68**  With reference to Example 5.13 and the computer printout of Figure 5.2, find the probability that a 15-square-foot sheet of the metal will have anywhere from eight to twelve defects, using

(a)  the values in the P(X = K) column;
(b)  the values in the P(X LESS OR = K) column.

**5.69**  The number of complaints which a dry-cleaning establishment receives per day is a random variable having a Poisson distribution with $\lambda = 3.3$. Use the formula for the Poisson distribution to find the probability that it will receive only two complaints on any given day.

**5.70**  The number of monthly breakdowns of a computer is a random variable having a Poisson distribution with $\lambda = 1.8$. Use the formula for the Poisson distribution to find the probabilities that this computer will function for a month

(a)  without a breakdown;
(b)  with only one breakdown.

**5.71**  Use Table II to verify the results of Exercise 5.70.

**5.72**  In a certain desert region the number of persons who become seriously ill each year from eating a certain poisonous plant is a random variable having a Poisson distribution with $\lambda = 5.2$. Use Table II to find the probabilities of

(a)  three such illnesses in a given year;
(b)  at least 10 such illnesses in a given year;
(c)  anywhere from four to six such illnesses in a given year.

**5.73**  In the inspection of a fabric produced in continuous rolls, the number of imperfections per yard is a random variable having the Poisson distribution with $\lambda = 0.25$. Find the probability that 2 yards of the fabric will have at most one imperfection using

(a)  Table II;
(b)  the computer printout of Figure 5.3.

## 5.8  THE MULTINOMIAL DISTRIBUTION

An immediate generalization of the binomial distribution arises when each trial has more than two possible outcomes, the probabilities of the respective outcomes are the same for each trial, and the trials are all independent. This would be the case, for instance, when persons interviewed by an opinion poll are asked whether they are for a candidate, against her, or undecided, or when samples of manufactured products are rated excellent, above average, average, or inferior.

To treat this kind of problem in general, let us consider the case where there are $n$ independent trials permitting $k$ mutually exclusive outcomes whose respective probabilities are $\theta_1$, $\theta_2$, ... , $\theta_k$ $\left( \text{with} \sum_{i=1}^{k} \theta_i = 1 \right)$. Referring to the outcomes as being of the first kind, the second kind, ... , and the $k$th kind, we shall be interested in the probability of getting $x_1$ outcomes of the first kind, $x_2$ outcomes of the second kind, ... , and $x_k$ outcomes of the $k$th kind $\left( \text{with} \sum_{i=1}^{k} x_i = n \right)$.

Proceeding as in the derivation of the formula for the binomial distribution, we first find that the probability of getting $x_1$ outcomes of the first kind, $x_2$ outcomes of the second kind, ... , and $x_k$ outcomes of the $k$th kind *in a specific order* is $\theta_1^{x_1} \cdot \theta_2^{x_2} \cdot \ldots \cdot \theta_k^{x_k}$. To get the corresponding probability for that many outcomes of each kind *in any order*, we shall have to multiply the probability for any specific order by

$$\binom{n}{x_1, x_2, \ldots, x_k} = \frac{n!}{x_1! \cdot x_2! \cdot \ldots \cdot x_k!}$$

according to Theorem 1.8.

---

**DEFINITION 5.8**  The random variables $X_1$, $X_2$, ... , $X_n$ have a **multinomial distribution** and they are referred to as multinomial random variables, if and only if their joint probability distribution is given by

$$f(x_1, x_2, \ldots, x_k; n, \theta_1, \theta_2, \ldots, \theta_k) = \binom{n}{x_1, x_2, \ldots, x_k} \cdot \theta_1^{x_1} \cdot \theta_2^{x_2} \cdot \ldots \cdot \theta_k^{x_k}$$

for $x_i = 0, 1, \ldots, n$ for each $i$, where $\sum_{i=1}^{k} x_i = n$ and $\sum_{i=1}^{k} \theta_i = 1$.

---

Thus, the numbers of outcomes of the different kinds are random variables having the multinomial distribution with the parameters $n$, $\theta_1$, $\theta_2$, ... , and $\theta_k$. The name "multinomial" derives from the fact that for various values of the $x_i$, the probabilities equal corresponding terms of the multinomial expansion of $(\theta_1 + \theta_2 + \ldots + \theta_k)^n$.

## EXAMPLE 5.14

A certain city has three television stations. During prime time on Saturday nights, Channel 12 has 50 percent of the viewing audience, Channel 10 has 30 percent of the viewing audience, and Channel 3 has 20 percent of the viewing audience. Find the probability that among eight television viewers in that city, randomly chosen on a Saturday night, five will be watching Channel 12, two will be watching Channel 10, and one will be watching Channel 3.

### Solution

Substituting $x_1 = 5$, $x_2 = 2$, $x_3 = 1$, $\theta_1 = 0.50$, $\theta_2 = 0.30$, $\theta_3 = 0.20$, and $n = 8$ into the formula of Definition 5.8, we get

$$f(5, 2, 1; 8, 0.50, 0.30, 0.20) = \frac{8!}{5! \cdot 2! \cdot 1!} (0.50)^5 (0.30)^2 (0.20)$$

$$= 0.0945 \quad \blacktriangle$$

## 5.9  THE MULTIVARIATE HYPERGEOMETRIC DISTRIBUTION

Just as the hypergeometric distribution takes the place of the binomial distribution for sampling without replacement, there also exists a multivariate distribution analogous to the multinomial distribution which applies to sampling without replacement. To derive its formula, let us consider a set of $N$ elements, of which $a_1$ are elements of the first kind, $a_2$ are elements of the second kind, ... , and $a_k$ are elements of the $k$th kind, such that $\sum_{i=1}^{k} a_i = N$. As in connection with the multinomial distribution, we are interested in the probability of getting $x_1$ elements (outcomes) of the first kind, $x_2$ elements of the second kind, ... , and $x_k$ elements of the $k$th kind, but now we are choosing, without replacement, $n$ of the $N$ elements of the set.

There are $\binom{a_1}{x_1}$ ways of choosing $x_1$ of the $a_1$ elements of the first kind, $\binom{a_2}{x_2}$ ways of choosing $x_2$ of the $a_2$ elements of the second kind, ... , and $\binom{a_k}{x_k}$ ways of choosing $x_k$ of the $a_k$ elements of the $k$th kind, and, hence, $\binom{a_1}{x_1}\binom{a_2}{x_2} \cdot \ldots \cdot \binom{a_k}{x_k}$ ways of choosing the required $\sum_{i=1}^{k} x_i = n$ elements. Since there are $\binom{N}{n}$ ways of choosing $n$ of the $N$ elements in the set and we assume that they are all equally likely (which is what we mean when we say that the selection is random), it follows that the desired probability is given by

$$\binom{a_1}{x_1}\binom{a_2}{x_2} \cdot \ldots \cdot \binom{a_k}{x_k} \bigg/ \binom{N}{n}.$$

**DEFINITION 5.9**  The random variables $X_1, X_2, \ldots, X_k$ have a **multivariate hypergeometric distribution** and they are referred to as multivariate hypergeometric random variables, if and only if their joint probability distribution is given by

$$f(x_1, x_2, \ldots, x_k; n, a_1, a_2, \ldots, a_k) = \frac{\binom{a_1}{x_1}\binom{a_2}{x_2} \cdot \ldots \cdot \binom{a_k}{x_k}}{\binom{N}{n}}$$

for $x_i = 0, 1, \ldots, n$ and $x_i \leq a_i$ for each $i$, where $\sum_{i=1}^{k} x_i = n$ and $\sum_{i=1}^{k} a_i = N$.

Thus, the joint distribution of the random variables under consideration, namely the distribution of the numbers of outcomes of the different kinds, is a multivariate hypergeometric distribution with the parameters $n, a_1, a_2, \ldots,$ and $a_k$.

## EXAMPLE 5.15

A panel of prospective jurors includes six married men, three single men, seven married women, and four single women. If the selection is random, what is the probability that a jury will consist of four married men, one single man, five married women and two single women?

### Solution

Substituting $x_1 = 4$, $x_2 = 1$, $x_3 = 5$, $x_4 = 2$, $a_1 = 6$, $a_2 = 3$, $a_3 = 7$, $a_4 = 4$, $N = 20$, and $n = 12$ into the formula of Definition 5.9, we get

$$f(4, 1, 5, 2; 12, 6, 3, 7, 4) = \frac{\binom{6}{4}\binom{3}{1}\binom{7}{5}\binom{4}{2}}{\binom{20}{12}}$$

$$= 0.0450 \quad \blacktriangle$$

## EXERCISES

**5.74** If $X_1, X_2, \ldots, X_k$ have the multinomial distribution of Definition 5.8, show that the mean of the marginal distribution of $X_i$ is $n\theta_i$ for $i = 1, 2, \ldots, k$.

**5.75** If $X_1, X_2, \ldots, X_k$ have the multinomial distribution of Definition 5.8, show that the covariance of $X_i$ and $X_j$ is $-n\theta_i\theta_j$ for $i = 1, 2, \ldots, k$, $j = 1, 2, \ldots, k$, and $i \neq j$.

## APPLICATIONS

**5.76** The probabilities are 0.40, 0.50, and 0.10 that, in city driving, a certain kind of compact car will average less than 22 miles per gallon, from 22 to 26 miles per gallon, or more than 26 miles per gallon. Find the probability that among ten such cars tested, three will average less than 22 miles per gallon, six will average from 22 to 26 miles per gallon, and one will average more than 26 miles per gallon.

**5.77** Suppose that the probabilities are 0.60, 0.20, 0.10, and 0.10 that a state income tax return will be filled out correctly, that it will contain only errors favoring the taxpayer, that it will contain only errors favoring the state, or that it will contain both kinds of errors. What is the probability that among 12 such income tax returns randomly chosen for audit, five will be filled out correctly, four will contain only errors favoring the taxpayer, two will contain only errors favoring the state, and one will contain both kinds of errors?

**5.78** According to the Mendelian theory of heredity, if plants with round yellow seeds are crossbred with plants with wrinkled green seeds, the probabilities of getting a plant that produces round yellow seeds, wrinkled yellow seeds, round green seeds, or wrinkled green seeds are, respectively, $\frac{9}{16}, \frac{3}{16}, \frac{3}{16}$, and $\frac{1}{16}$. What is the probability that among nine plants thus obtained there will be four that produce round yellow seeds, two that produce wrinkled yellow seeds, three that produce round green seeds, and none that produce wrinkled green seeds?

**5.79** If 18 defective glass bricks include 10 that have cracks but no discoloration, five that have discoloration but no cracks, and three that have cracks and discoloration, what is the probability that among six of the bricks (chosen at random for further checks) three will have cracks but no discoloration, one will have discoloration but no cracks, and two will have cracks and discoloration?

**5.80** Among 25 silver dollars struck in 1903 there are 15 from the Philadelphia mint, seven from the New Orleans mint, and three from the San Francisco mint. If five of these silver dollars are picked at random, find the probabilities of getting

(a)   four from the Philadelphia mint and one from the New Orleans mint;

(b)   three from the Phildelphia mint and one from each of the other two mints.

## REFERENCES

Useful information about various special probability distributions may be found in

DERMAN, C., GLESER, L., and OLKIN, I., *Probability Models and Applications*. New York: Macmillan Publishing Co., Inc., 1980,

HASTINGS, N. A. J., and PEACOCK, J. B., *Statistical Distributions*. London: Butterworth & Co. Ltd., 1975,

and

JOHNSON, N. L., and KOTZ, S., *Discrete Distributions*. Boston: Houghton Mifflin Company, 1969.

Binomial probabilities for $n = 2$ to $n = 49$ may be found in

*Tables of the Binomial Probability Distribution,* National Bureau of Standards Applied Mathematics Series No. 6, Washington, D.C.: U.S. Government Printing Office, 1950,

and for $n = 50$ to $n = 100$ in

ROMIG, H. G., *50–100 Binomial Tables*. New York: John Wiley & Sons, Inc., 1953.

The most widely used table of Poisson probabilities is

MOLINA, E. C., *Poisson's Exponential Binomial Limit*. Melbourne, Fla.: Robert E. Krieger Publishing Company, 1973 Reprint.

# 6

# *Special Probability Densities*

## 6.1 INTRODUCTION

In this chapter we shall study some of the probability densities which figure most prominently in statistical theory and in applications. In addition to the ones given in the text, several others are introduced in the exercises following Section 6.4, and three probability densities that are of basic importance in the theory of sampling will be taken up in Chapter 8. As in Chapter 5, we shall derive parameters and moment-generating functions, again leaving some of the details as exercises.

## 6.2  THE UNIFORM DENSITY

The probability densities of Examples 3.8 and 3.11 are special cases of the **uniform density**, whose graph may be pictured as in Figure 3.6.

---

**DEFINITION 6.1**  A random variable has a **uniform density**, and it is referred to as a continuous uniform random variable, if and only if its probability density is given by

$$f(x) = \begin{cases} \dfrac{1}{\beta - \alpha} & \text{for } \alpha < x < \beta \\ 0 & \text{elsewhere} \end{cases}$$

---

The parameters $\alpha$ and $\beta$ of this probability density are real constants with $\alpha < \beta$. In Exercise 6.2 the reader will be asked to verify that

---

**THEOREM 6.1**  The mean and the variance of the uniform density are given by

$$\mu = \frac{\alpha + \beta}{2} \quad \text{and} \quad \sigma^2 = \frac{1}{12}(\beta - \alpha)^2$$

---

Although the uniform density has some direct applications, one of which will be discussed in Example 7.8, its main value is that, due to its simplicity, it lends itself readily to the task of illustrating various aspects of statistical theory.

## 6.3  THE GAMMA, EXPONENTIAL, AND CHI-SQUARE DISTRIBUTIONS

Some of the examples and exercises of Chapters 3 and 4 dealt with random variables having probability densities of the form

$$f(x) = \begin{cases} kx^{\alpha-1}e^{-x/\beta} & \text{for } x > 0 \\ 0 & \text{elsewhere} \end{cases}$$

where $\alpha > 0$, $\beta > 0$, and $k$ must be such that the total area under the curve is equal to 1. To evaluate $k$, we first make the substitution $y = \dfrac{x}{\beta}$, which yields

$$\int_0^\infty kx^{\alpha-1}e^{-x/\beta}\, dx = k\beta^\alpha \int_0^\infty y^{\alpha-1}e^{-y}\, dy$$

The integral thus obtained depends on $\alpha$ alone, and it defines the well-known **gamma function**

$$\Gamma(\alpha) = \int_0^\infty y^{\alpha-1}e^{-y}\, dy \qquad \text{for } \alpha > 0$$

which is treated in detail in most advanced calculus texts. Integrating by parts, which is left to the reader in Exercise 6.7, we find that the gamma function satisfies the recursion formula

$$\Gamma(\alpha) = (\alpha - 1) \cdot \Gamma(\alpha - 1)$$

for $\alpha > 1$, and since

$$\Gamma(1) = \int_0^\infty e^{-y}\, dy = 1$$

it follows by repeated application of the recursion formula that $\Gamma(\alpha) = (\alpha - 1)!$ when $\alpha$ is a positive integer. Also, an important special value is $\Gamma(\frac{1}{2}) = \sqrt{\pi}$, as the reader will be asked to verify in Exercise 6.9.

Returning now to the problem of evaluating $k$, we equate the integral we obtained to 1, getting

$$\int_0^\infty kx^{\alpha-1}e^{-x/\beta}\, dx = k\beta^\alpha \Gamma(\alpha) = 1$$

and, hence,

$$k = \frac{1}{\beta^\alpha \Gamma(\alpha)}$$

This leads to the following definition of the **gamma distribution**.

---

**DEFINITION 6.2**  A random variable $X$ has a **gamma distribution**, and it is referred to as a gamma random variable, if and only if its probability density is given by

$$f(x) = \begin{cases} \dfrac{1}{\beta^{\alpha}\,\Gamma(\alpha)}\, x^{\alpha-1}e^{-x/\beta} & \text{for } x > 0 \\ 0 & \text{elsewhere} \end{cases}$$

where $\alpha > 0$ and $\beta > 0$.

When $\alpha$ is not a positive integer, the value of $\Gamma(\alpha)$ will have to be looked up in a special table. To give the reader some idea about the shape of the graphs of gamma densities, those for several special values of $\alpha$ and $\beta$ are shown in Figure 6.1.

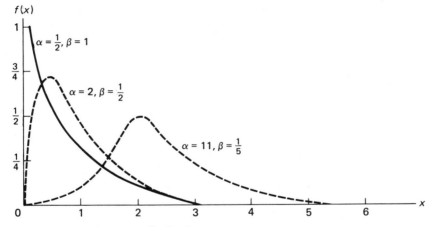

**Figure 6.1**  Graphs of gamma distributions.

Some special cases of the gamma distribution play important roles in statistics; for instance, for $\alpha = 1$ and $\beta = \theta$ we get

**DEFINITION 6.3**  A random variable $X$ has an **exponential distribution**, and it is referred to as an exponential random variable, if and only if its probability density is given by

$$f(x) = \begin{cases} \dfrac{1}{\theta}\, e^{-x/\theta} & \text{for } x > 0 \\ 0 & \text{elsewhere} \end{cases}$$

where $\theta > 0$.

To show how an exponential distribution might arise in practice, let us refer to the situation described in Exercise 5.42, where we were interested in the probability of getting $x$ successes during a time interval of length $t$ when (i) the probability of a success during a very small time interval from $t$ to $t + \Delta t$ is $\alpha \cdot \Delta t$, (ii) the probability of more than one success during such a time interval is negligible, and (iii) the probability of a success during such a time interval does not depend on what happened prior to time $t$. In that exercise, we showed that the number of successes is a value of the discrete random variable $X$ having the Poisson distribution with $\lambda = \alpha t$. Now let us determine the probability density of the continuous random variable $Y$, the **waiting time** until the first success. Clearly,

$$F(y) = P(Y \leqslant y) = 1 - P(Y > y)$$

$$= 1 - P(0 \text{ successes in a time interval of length } y)$$

$$= 1 - p(0; \alpha y)$$

$$= 1 - \frac{e^{-\alpha y}(\alpha y)^0}{0!}$$

$$= 1 - e^{-\alpha y} \qquad \text{for } y > 0$$

and $F(y) = 0$ for $y \leqslant 0$. Having thus found the distribution function of $Y$, we find that differentiation with respect to $y$ yields

$$f(y) = \begin{cases} \alpha e^{-\alpha y} & \text{for } y > 0 \\ 0 & \text{elsewhere} \end{cases}$$

which is the exponential distribution with $\theta = \dfrac{1}{\alpha}$.

The exponential distribution applies not only to the occurrence of the first success in a **Poisson process**, which is what we call a situation like that described in Exercise 5.42, but by virtue of condition (iii) (see Exercise 6.16), it applies also to the waiting times between successes.

## EXAMPLE 6.1

At a certain location on highway I-10, the number of cars exceeding the speed limit by more than 10 miles per hour in half an hour is a random variable having a Poisson distribution with $\lambda = 8.4$. What is the probability of a waiting time less than 5 minutes between cars exceeding the speed limit by more than 10 miles per hour?

*Solution*

Using half an hour as the unit of time, we have $\alpha = \lambda = 8.4$. Therefore, the waiting time is a random variable having an exponential distribution with $\theta = \frac{1}{8.4}$, and since 5 minutes is $\frac{1}{6}$ of the unit of time, we find that the desired probability is

$$\int_0^{1/6} 8.4e^{-8.4x}\,dx = -e^{-8.4x}\Big|_0^{1/6} = -e^{-1.4} + 1$$

which is approximately 0.75.    ▲

Another special case of the gamma distribution arises when $\alpha = \dfrac{\nu}{2}$ and $\beta = 2$, where $\nu$ is the lowercase Greek letter *nu*.

---

**DEFINITION 6.4**  A random variable $X$ has a **chi-square distribution**, and it is referred to as a chi-square random variable, if and only if its probability density is given by

$$f(x) = \begin{cases} \dfrac{1}{2^{\nu/2}\Gamma(\nu/2)}\, x^{\frac{\nu-2}{2}} e^{-\frac{x}{2}} & \text{for } x > 0 \\ 0 & \text{elsewhere} \end{cases}$$

---

The parameter $\nu$ is referred to as the **number of degrees of freedom**, or simply the **degrees of freedom**. The chi-square distribution plays a very important role in sampling theory, and it is discussed in some detail in Chapter 8.

To derive formulas for the mean and the variance of the gamma distribution and, hence, the exponential and chi-square distributions, let us first prove the following theorem.

---

**THEOREM 6.2**  The $r$th moment about the origin of the gamma distribution is given by

$$\mu_r' = \frac{\beta^r \Gamma(\alpha + r)}{\Gamma(\alpha)}$$

---

**Proof.**   By Definition 4.2,

$$\mu_r' = \int_0^\infty x^r \cdot \frac{1}{\beta^\alpha \Gamma(\alpha)}\, x^{\alpha-1} e^{-x/\beta}\, dx = \frac{\beta^r}{\Gamma(\alpha)} \cdot \int_0^\infty y^{\alpha+r-1} e^{-y}\, dy$$

where we let $y = \dfrac{x}{\beta}$. Since the integral on the right is $\Gamma(r + \alpha)$ according to the definition of the gamma function on page 224, this completes the proof.    ▼

Using this theorem, let us now derive the following results about the gamma distribution.

---

**THEOREM 6.3**  The mean and the variance of the gamma distribution are given by

$$\mu = \alpha\beta \quad \text{and} \quad \sigma^2 = \alpha\beta^2$$

---

**Proof.**   From Theorem 6.2 with $r = 1$ and $r = 2$, we get

$$\mu_1' = \frac{\beta\Gamma(\alpha + 1)}{\Gamma(\alpha)} = \alpha\beta$$

and

$$\mu_2' = \frac{\beta^2 \Gamma(\alpha + 2)}{\Gamma(\alpha)} = \alpha(\alpha + 1)\beta^2$$

so that $\mu = \alpha\beta$ and $\sigma^2 = \alpha(\alpha + 1)\beta^2 - (\alpha\beta)^2 = \alpha\beta^2$.    ▼

Substituting into these formulas $\alpha = 1$ and $\beta = \theta$ for the exponential distribution, and $\alpha = \dfrac{\nu}{2}$ and $\beta = 2$ for the chi-square distribution, we get

---

**COROLLARY 1**  The mean and the variance of the exponential distribution are given by

$$\mu = \theta \quad \text{and} \quad \sigma^2 = \theta^2$$

---

**COROLLARY 2**   The mean and the variance of the chi-square distribution are given by

$$\mu = \nu \quad \text{and} \quad \sigma^2 = 2\nu$$

---

For future reference, let us give here also the moment-generating function of the gamma distribution.

---

**THEOREM 6.4**   The moment-generating function of the gamma distribution is given by

$$M_X(t) = (1 - \beta t)^{-\alpha}$$

---

The reader will be asked to prove this result, and use it to find some of the lower moments, in Exercises 6.12 and 6.13.

## 6.4 THE BETA DISTRIBUTION

The uniform density $f(x) = 1$ for $0 < x < 1$ and $f(x) = 0$ elsewhere is a special case of the **beta distribution**, which is defined in the following way.

---

**DEFINITION 6.5**   A random variable $X$ has a **beta distribution**, and it is referred to as a beta random variable, if and only if its probability density is given by

$$f(x) = \begin{cases} \dfrac{\Gamma(\alpha + \beta)}{\Gamma(\alpha) \cdot \Gamma(\beta)} x^{\alpha - 1}(1 - x)^{\beta - 1} & \text{for } 0 < x < 1 \\ 0 & \text{elsewhere} \end{cases}$$

where $\alpha > 0$ and $\beta > 0$.

---

In recent years, the beta distribution has found important applications in **Bayesian inference**, where parameters are looked upon as random variables, and there is a need for a fairly "flexible" probability density for the parameter $\theta$ of the binomial distribution, which takes on nonzero values only on the interval from 0 to 1. By "flexible" we mean that the probability density can take on a great variety of different shapes, as the reader will be asked to verify for the beta distribution in Exercise 6.27. This use of the beta distribution is discussed in Chapter 10.

We shall not prove here that the total area under the curve of the beta distribution, like that of any probability density, is equal to 1, but in the proof of the theorem that follows, we shall make use of the fact that

$$\int_0^1 \frac{\Gamma(\alpha + \beta)}{\Gamma(\alpha) \cdot \Gamma(\beta)} x^{\alpha-1}(1 - x)^{\beta-1} \, dx = 1$$

and, hence, that

$$\int_0^1 x^{\alpha-1}(1 - x)^{\beta-1} \, dx = \frac{\Gamma(\alpha) \cdot \Gamma(\beta)}{\Gamma(\alpha + \beta)}$$

This integral defines the **beta function**, whose values are denoted $B(\alpha, \beta)$; in other words, $B(\alpha, \beta) = \dfrac{\Gamma(\alpha) \cdot \Gamma(\beta)}{\Gamma(\alpha + \beta)}$. Detailed discussion of the beta function may be found in any textbook on advanced calculus.

---

**THEOREM 6.5**   The mean and the variance of the beta distribution are given by

$$\mu = \frac{\alpha}{\alpha + \beta} \quad \text{and} \quad \sigma^2 = \frac{\alpha\beta}{(\alpha + \beta)^2(\alpha + \beta + 1)}$$

---

*Proof.*   By definition,

$$\mu = \frac{\Gamma(\alpha + \beta)}{\Gamma(\alpha) \cdot \Gamma(\beta)} \cdot \int_0^1 x \cdot x^{\alpha-1}(1 - x)^{\beta-1} \, dx$$

$$= \frac{\Gamma(\alpha + \beta)}{\Gamma(\alpha) \cdot \Gamma(\beta)} \cdot \frac{\Gamma(\alpha + 1) \cdot \Gamma(\beta)}{\Gamma(\alpha + \beta + 1)}$$

$$= \frac{\alpha}{\alpha + \beta}$$

where we recognized the integral as $B(\alpha + 1, \beta)$ and made use of the fact that $\Gamma(\alpha + 1) = \alpha \cdot \Gamma(\alpha)$ and $\Gamma(\alpha + \beta + 1) = (\alpha + \beta) \cdot \Gamma(\alpha + \beta)$. Similar steps, which will be left to the reader in Exercise 6.28, yield

$$\mu_2' = \frac{(\alpha + 1)\alpha}{(\alpha + \beta + 1)(\alpha + \beta)}$$

and it follows that

$$\sigma^2 = \frac{(\alpha + 1)\alpha}{(\alpha + \beta + 1)(\alpha + \beta)} - \left(\frac{\alpha}{\alpha + \beta}\right)^2$$

$$= \frac{\alpha\beta}{(\alpha + \beta)^2(\alpha + \beta + 1)} \quad \blacktriangledown$$

## EXERCISES

**6.1** Show that if a random variable has a uniform density with the parameters $\alpha$ and $\beta$, the probability that it will take on a value less than $\alpha + p(\beta - \alpha)$ is equal to $p$.

**6.2** Prove Theorem 6.1.

**6.3** If a random variable $X$ has a uniform density with the parameters $\alpha$ and $\beta$, find its distribution function.

**6.4** Show that if a random variable has a uniform density with the parameters $\alpha$ and $\beta$, the $r$th moment about the mean equals

(a)   0 when $r$ is odd;

(b)   $\dfrac{1}{r + 1}\left(\dfrac{\beta - \alpha}{2}\right)^r$ when $r$ is even.

**6.5** Use the results of Exercise 6.4 to find $\alpha_3$ and $\alpha_4$ for the uniform density with the parameters $\alpha$ and $\beta$.

**6.6** A random variable is said to have a **Cauchy distribution** if its density is given by

$$f(x) = \frac{\dfrac{\beta}{\pi}}{(x - \alpha)^2 + \beta^2} \qquad \text{for } -\infty < x < \infty$$

Show that for this distribution $\mu_1'$ and $\mu_2'$ do not exist.

**6.7** Use integration by parts to show that $\Gamma(\alpha) = (\alpha - 1) \cdot \Gamma(\alpha - 1)$ for $\alpha > 1$.

**6.8** Perform a suitable change of variable to show that the integral defining the gamma function can be written as

$$\Gamma(\alpha) = 2^{1-\alpha} \cdot \int_0^\infty z^{2\alpha-1} e^{-\frac{1}{2}z^2} \, dz \qquad \text{for } \alpha > 0$$

**6.9** Using the form of the gamma function of Exercise 6.8, we can write

$$\Gamma(\tfrac{1}{2}) = \sqrt{2} \int_0^\infty e^{-\frac{1}{2}z^2} \, dz$$

and, hence,

$$[\Gamma(\tfrac{1}{2})]^2 = 2\left\{\int_0^\infty e^{-\frac{1}{2}x^2} \, dx\right\}\left\{\int_0^\infty e^{-\frac{1}{2}y^2} \, dy\right\} = 2\int_0^\infty \int_0^\infty e^{-\frac{1}{2}(x^2+y^2)} \, dx \, dy$$

Change to polar coordinates to evaluate this double integral, and thus show that $\Gamma(\tfrac{1}{2}) = \sqrt{\pi}$.

**6.10** Find the probabilities that the value of a random variable will exceed 4, if it has a gamma distribution with
   (a)   $\alpha = 2$ and $\beta = 3$;
   (b)   $\alpha = 3$ and $\beta = 4$.

**6.11** Show that a gamma distribution with $\alpha > 1$ has a relative maximum at $x = \beta(\alpha - 1)$. What happens when $0 < \alpha < 1$ and when $\alpha = 1$?

**6.12** Prove Theorem 6.4, making the substitution $y = x\left(\dfrac{1}{\beta} - t\right)$ in the integral defining $M_X(t)$.

**6.13** Expand the moment-generating function of the gamma distribution as a binomial series, and read off the values of $\mu_1'$, $\mu_2'$, $\mu_3'$, and $\mu_4'$.

**6.14** Use the results of Exercise 6.13 to find $\alpha_3$ and $\alpha_4$ for the gamma distribution.

**6.15** Show that if a random variable has an exponential density with the parameter $\theta$, the probability that it will take on a value less than $-\theta \cdot \ln(1 - p)$ is equal to $p$ for $0 \leqslant p < 1$.

**6.16** If $X$ has an exponential distribution, show that

$$P(X \geqslant t + T | X \geqslant T) = P(X \geqslant t)$$

This property of an exponential random variable parallels that of a geometric random variable given in Exercise 5.34.

**6.17** If $X$ is a random variable having an exponential distribution with the parameter $\theta$, use Theorems 4.10 and 6.4 to find the moment-generating function of the random variable $Y = X - \theta$.

**6.18** With reference to Exercise 6.17, use the fact that the moments of $Y$ about the origin are the corresponding moments of $X$ about the mean, find $\alpha_3$ and $\alpha_4$ for the exponential distribution with the parameter $\theta$.

**6.19** Show that if $\nu > 2$, the chi-square distribution has a relative maximum at $x = \nu - 2$. What happens when $\nu = 2$ or $0 < \nu < 2$?

**6.20** A random variable $X$ has a **Rayleigh distribution** if and only if its probability density is given by

$$f(x) = \begin{cases} 2\alpha x e^{-\alpha x^2} & \text{for } x > 0 \\ 0 & \text{elsewhere} \end{cases}$$

where $\alpha > 0$. Show that for this distribution

(a)  $\mu = \dfrac{1}{2} \sqrt{\dfrac{\pi}{\alpha}}$;

(b)  $\sigma^2 = \dfrac{1}{\alpha}\left(1 - \dfrac{\pi}{4}\right)$.

**6.21** A random variable $X$ has a **Pareto distribution** if and only if its probability density is given by

$$f(x) = \begin{cases} \dfrac{\alpha}{x^{\alpha+1}} & \text{for } x > 1 \\ 0 & \text{elsewhere} \end{cases}$$

where $\alpha > 0$. Show that $\mu'_r$ exists only if $r < \alpha$.

**6.22** With reference to Exercise 6.21, show that for the Pareto distribution $\mu = \dfrac{\alpha}{\alpha - 1}$ provided $\alpha > 1$.

**6.23** A random variable $X$ has a **Weibull distribution** if and only if its probability density is given by

$$f(x) = \begin{cases} kx^{\beta-1}e^{-\alpha x^\beta} & \text{for } x > 0 \\ 0 & \text{elsewhere} \end{cases}$$

where $\alpha > 0$ and $\beta > 0$.

(a)   Express $k$ in terms of $\alpha$ and $\beta$.

(b)   Show that $\mu = \alpha^{-1/\beta}\Gamma\left(1 + \dfrac{1}{\beta}\right)$.

Note that Weibull distributions with $\beta = 1$ are exponential distributions.

**6.24** If the random variable $T$ is the time to failure of a commercial product and the values of its probability density and distribution function at time $t$ are $f(t)$ and $F(t)$, then its failure rate at time $t$ (see also Exercise 5.35) is given by $\dfrac{f(t)}{1 - F(t)}$. Thus, the failure rate at time $t$ is the probability density of failure at time $t$ given that failure does not occur prior to time $t$.

(a)   Show that if $T$ has an exponential distribution, the failure rate is constant.

(b)   Show that if $T$ has a Weibull distribution (see Exercise 6.23), the failure rate is given by $\alpha\beta t^{\beta-1}$.

**6.25** Verify that the integral of the beta density, from $-\infty$ to $\infty$, equals 1 for

(a)   $\alpha = 2$ and $\beta = 4$;

(b)   $\alpha = 3$ and $\beta = 3$.

**6.26** Show that if $\alpha > 1$ and $\beta > 1$, the beta density has a relative maximum at

$$x = \frac{\alpha - 1}{\alpha + \beta - 2}.$$

**6.27** Sketch the graphs of the beta densities having

(a)   $\alpha = 2$ and $\beta = 2$;

(b)   $\alpha = \frac{1}{2}$ and $\beta = 1$;

(c)   $\alpha = 2$ and $\beta = \frac{1}{2}$;

(d)   $\alpha = 2$ and $\beta = 5$.

[*Hint*: To evaluate $\Gamma(\frac{3}{2})$ and $\Gamma(\frac{5}{2})$ make use of the recursion formula $\Gamma(\alpha) = (\alpha - 1) \cdot \Gamma(\alpha - 1)$ and the result of Exercise 6.9.]

**6.28** Verify the expression given for $\mu_2'$ in the proof of Theorem 6.5.

**6.29** Show that the parameters of the beta distribution can be expressed as follows in terms of the mean and the variance of this distribution:

(a)   $\alpha = \mu\left[\dfrac{\mu(1 - \mu)}{\sigma^2} - 1\right]$;

(b)   $\beta = (1 - \mu)\left[\dfrac{\mu(1 - \mu)}{\sigma^2} - 1\right]$.

**6.30** Karl Pearson, one of the founders of modern statistics, showed that the differential equation

$$\frac{1}{f(x)} \cdot \frac{d[f(x)]}{dx} = \frac{d - x}{a + bx + cx^2}$$

yields (for appropriate values of the constants $a$, $b$, $c$, and $d$) most of the important distributions of statistics. Verify that the differential equation gives

(a)   the gamma distribution when $a = c = 0$, $b > 0$, and $d > -b$;

(b)   the exponential distribution when $a = c = d = 0$ and $b > 0$;

(c)   the beta distribution when $a = 0$, $b = -c$, $\dfrac{d - 1}{b} < 1$, and $\dfrac{d}{b} > -1$.

## APPLICATIONS

**6.31** A point $D$ is chosen on the line $AB$, whose midpoint is $C$ and whose length is $a$. If $X$, the distance from $D$ to $A$, is a random variable having the uniform density with $\alpha = 0$ and $\beta = a$, what is the probability that $AD$, $BD$, and $AC$ will form a triangle?

**6.32** In certain experiments, the error made in determining the density of a substance is a random variable having a uniform density with $\alpha = -0.015$ and $\beta = 0.015$. Find the probabilities that such an error will

(a)   be between $-0.002$ and $0.003$;

(b)   exceed $0.005$ in absolute value.

**6.33** If a company employs $n$ salespersons, its gross sales in thousands of dollars may be regarded as a random variable having a gamma distribution with $\alpha = 80\sqrt{n}$ and $\beta = 2$. If the sales cost is \$8,000 per salesperson, how many salespersons should the company employ to maximize the expected profit?

**6.34** In a certain city, the daily consumption of electric power, in millions of kilowatt-hours, can be treated as a random variable having a gamma distribution with $\alpha = 3$ and $\beta = 2$. If the power plant of this city has a daily capacity of 12 million kilowatt-hours, what is the probability that this power supply will be inadequate on any given day?

**6.35** The mileage (in thousands of miles) which car owners get with a certain kind of radial tire is a random variable having an exponential distribution with $\theta = 40$. Find the probabilities that one of these tires will last

(a)   at least 20,000 miles;

(b)   at most 30,000 miles.

**6.36** The amount of time that a watch will run without having to be reset is a random variable having an exponential distribution with $\theta = 120$ days. Find the probabilities that such a watch will

(a)   have to be reset in less than 24 days;

(b)   not have to be reset in at least 180 days.

**6.37** The number of planes arriving per day at a small private airport is a random variable having a Poisson distribution with $\lambda = 28.8$. What is the probability that the time between two such arrivals is at least 1 hour?

**6.38** The number of bad checks which a bank receives during a 5-hour business day is a Poisson random variable with $\lambda = 2$. What is the probability that it will not receive a bad check on any one day during the first 2 hours of business?

**6.39** A certain kind of appliance requires repairs on the average once every 2 years. Assuming that the times between repairs are exponentially distributed, what is the probability that such an appliance will work at least 3 years without requiring repairs?

**6.40** If the annual proportion of erroneous income tax returns filed with the IRS can be looked upon as a random variable having a beta distribution with $\alpha = 2$ and $\beta = 9$, what is the probability that in any given year there will be fewer than 10 percent erroneous returns?

**6.41** If the annual proportion of new restaurants that fail in a given city may be looked upon as a random variable having a beta distribution with $\alpha = 1$ and $\beta = 4$, find

(a)  the mean of this distribution, namely, the annual proportion of new restaurants that can be expected to fail in the given city;

(b)  the probability that at least 25 percent of all new restaurants will fail in the given city in any one year.

**6.42** Suppose that the service life, in hours, of a semiconductor is a random variable having a Weibull distribution (see Exercise 6.23) with $\alpha = 0.025$ and $\beta = 0.500$.

(a)  How long can such a semiconductor be expected to last?

(b)  What is the probability that such a semiconductor will still be in operating condition after 4,000 hours?

## 6.5  THE NORMAL DISTRIBUTION

The **normal distribution**, which we shall study in this section, is in many ways the cornerstone of modern statistical theory. It was investigated first in the eighteenth century when scientists observed an astonishing degree of regularity in errors of measurement. They found that the patterns (distributions) which they observed could be closely approximated by continuous curves which they referred to as "normal curves of errors" and attributed to the laws of chance. The mathematical properties of such normal curves were first studied by Abraham de Moivre (1667–1745), Pierre Laplace (1749–1827), and Karl Gauss (1777–1855).

---

**DEFINITION 6.6**  A random variable $X$ has a **normal distribution**, and it is referred to as a normal random variable, if and only if its probability density is given by

$$n(x; \mu, \sigma) = \frac{1}{\sigma\sqrt{2\pi}} e^{-\frac{1}{2}\left(\frac{x-\mu}{\sigma}\right)^2} \qquad \text{for } -\infty < x < \infty$$

where $\sigma > 0$.

---

The graph of a normal distribution, shaped like the cross section of a bell, is shown in Figure 6.2.

The notation which we used here is similar to that used in connection with some of the probability distributions of Chapter 5—it shows explicitly that the two parameters of the normal distribution are $\mu$ and $\sigma$. It remains to be shown, however, that the parameter $\mu$ is, in fact, $E(X)$ and that the parameter $\sigma$ is, in fact, the square root of var($X$), where $X$ is a random variable having the normal distribution with these two parameters.

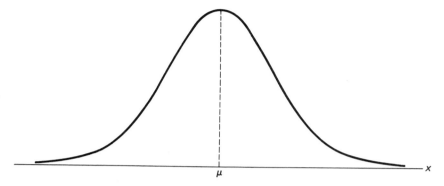

**Figure 6.2**  Graph of normal distribution.

First, though, let us show that the formula of Definition 6.6 can serve as a probability density. Since the values of $n(x; \mu, \sigma)$ are evidently positive so long as $\sigma > 0$, we must show that the total area under the curve is equal to 1. Integrating from $-\infty$ to $\infty$ and making the substitution $z = \dfrac{x - \mu}{\sigma}$, we get

$$\int_{-\infty}^{\infty} \frac{1}{\sigma\sqrt{2\pi}} \, e^{-\frac{1}{2}\left(\frac{x-\mu}{\sigma}\right)^2} \, dx = \frac{1}{\sqrt{2\pi}} \int_{-\infty}^{\infty} e^{-\frac{1}{2}z^2} \, dz = \frac{2}{\sqrt{2\pi}} \int_{0}^{\infty} e^{-\frac{1}{2}z^2} \, dz$$

Then, since the integral on the right equals $\dfrac{\Gamma(\frac{1}{2})}{\sqrt{2}} = \dfrac{\sqrt{\pi}}{\sqrt{2}}$ according to Exercise 6.9,

it follows that the total area under the curve is equal to $\dfrac{2}{\sqrt{2\pi}} \cdot \dfrac{\sqrt{\pi}}{\sqrt{2}} = 1$.

Next let us show that

---

**THEOREM 6.6**  The moment-generating function of the normal distribution is given by

$$M_X(t) = e^{\mu t + \frac{1}{2}\sigma^2 t^2}$$

---

***Proof.***  By definition,

$$M_X(t) = \int_{-\infty}^{\infty} e^{xt} \cdot \frac{1}{\sigma\sqrt{2\pi}} \, e^{-\frac{1}{2}\left(\frac{x-\mu}{\sigma}\right)^2} \, dx$$

$$= \frac{1}{\sigma\sqrt{2\pi}} \cdot \int_{-\infty}^{\infty} e^{-\frac{1}{2\sigma^2}[-2xt\sigma^2 + (x-\mu)^2]} \, dx$$

and if we complete the square, that is, use the identity

$$-2xt\sigma^2 + (x - \mu)^2 = [x - (\mu + t\sigma^2)]^2 - 2\mu t\sigma^2 - t^2\sigma^4$$

we get

$$M_X(t) = e^{\mu t + \frac{1}{2}t^2\sigma^2} \left\{ \frac{1}{\sigma\sqrt{2\pi}} \cdot \int_{-\infty}^{\infty} e^{-\frac{1}{2}\left[\frac{x - (\mu + t\sigma^2)}{\sigma}\right]^2} \, dx \right\}$$

Since the quantity inside the braces is the integral, from $-\infty$ to $\infty$, of a normal density with the parameters $\mu + t\sigma^2$ and $\sigma$, and hence equal to 1, it follows that

$$M_X(t) = e^{\mu t + \frac{1}{2}\sigma^2 t^2} \qquad \blacktriangledown$$

We are now ready to verify that the parameters $\mu$ and $\sigma$ in Definition 6.6 are, indeed, the mean and the standard deviation of the normal distribution. Twice differentiating $M_X(t)$ with respect to $t$, we get

$$M_X'(t) = (\mu + \sigma^2 t) \cdot M_X(t)$$

and

$$M_X''(t) = [(\mu + \sigma^2 t)^2 + \sigma^2] \cdot M_X(t)$$

so that $M_X'(0) = \mu$ and $M_X''(0) = \mu^2 + \sigma^2$. Thus, $E(X) = \mu$ and var$(X) = (\mu^2 + \sigma^2) - \mu^2 = \sigma^2$.

Since the normal distribution plays a basic role in statistics and its density cannot be integrated directly, its areas have been tabulated for the special case where $\mu = 0$ and $\sigma = 1$.

---

**DEFINITION 6.7** The normal distribution with $\mu = 0$ and $\sigma = 1$ is referred to as the **standard normal distribution**.

---

The entries in Table III, represented by the shaded area of Figure 6.3, are the values of

$$\int_0^z \frac{1}{\sqrt{2\pi}} e^{-\frac{1}{2}x^2} \, dx$$

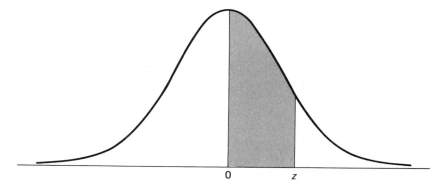

**Figure 6.3**  Tabulated areas under the standard normal distribution.

namely, the probabilities that a random variable having the standard normal distribution will take on a value on the interval from 0 to $z$, for $z$ = 0.00, 0.01, 0.02, ... , 3.08, and 3.09, and also $z$ = 4.0, $z$ = 5.0, and $z$ = 6.0. By virtue of the symmetry of the normal distribution about its mean, it is unnecessary to extend Table III to negative values of $z$.

## EXAMPLE 6.2

Find the probabilities that a random variable having the standard normal distribution will take on a value

(a)   less than 1.72;
(b)   less than −0.88;
(c)   between 1.30 and 1.75;
(d)   between −0.25 and 0.45.

*Solution*

(a)   We look up the entry corresponding to $z$ = 1.72 in Table III, add 0.5000 (see Figure 6.4), and get 0.4573 + 0.5000 = 0.9573.

(b)   We look up the entry corresponding to $z$ = 0.88 in Table III, subtract it from 0.5000 (see Figure 6.4), and get 0.5000 − 0.3106 = 0.1894.

(c)   We look up the entries corresponding to $z$ = 1.75 and $z$ = 1.30 in Table III, subtract the second from the first (see Figure 6.4), and get 0.4599 − 0.4032 = 0.0567.

(d)   We look up the entries corresponding to $z$ = 0.25 and $z$ = 0.45 in Table III, add them (see Figure 6.4), and get 0.0987 + 0.1736 = 0.2723.   ▲

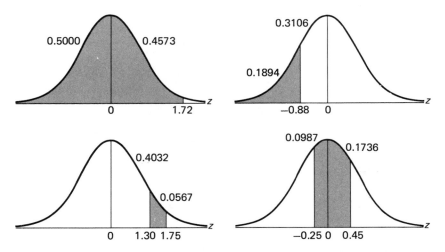

**Figure 6.4**  Diagrams for Example 6.2.

Occasionally, we are required to find a value of $z$ corresponding to a specified probability that falls between values listed in Table III. In that case, for convenience, we always choose the $z$ value corresponding to the tabular value that comes closest to the specified probability. However, if the given probability falls midway between tabular values, we shall choose for $z$ the value falling midway between the corresponding values of $z$.

## EXAMPLE 6.3

With reference to Table III, find the values of $z$ which correspond to entries of

(a)  0.3512;
(b)  0.2533.

*Solution*

(a)  Since 0.3512 falls between 0.3508 and 0.3531 corresponding to $z = $ 1.04 and $z = 1.05$, and since 0.3512 is closer to 0.3508 than 0.3531, we choose $z = 1.04$.
(b)  Since 0.2533 falls midway between 0.2517 and 0.2549 corresponding to $z = 0.68$ and $z = 0.69$, we choose $z = 0.685$.    ▲

To determine probabilities relating to random variables having normal distributions other than the standard normal distribution, we make use of the following theorem.

**THEOREM 6.7**  If $X$ has a normal distribution with the mean $\mu$ and the standard deviation $\sigma$, then

$$Z = \frac{X - \mu}{\sigma}$$

has the standard normal distribution.

***Proof.***    Since the relationship between the values of $X$ and $Z$ is linear, $Z$ must take on a value between $z_1 = \dfrac{x_1 - \mu}{\sigma}$ and $z_2 = \dfrac{x_2 - \mu}{\sigma}$ when $X$ takes on a value between $x_1$ and $x_2$. Hence, we can write

$$P(x_1 < X < x_2) = \frac{1}{\sqrt{2\pi}\,\sigma} \int_{x_1}^{x_2} e^{-\frac{1}{2}\left(\frac{x-\mu}{\sigma}\right)^2} dx$$

$$= \frac{1}{\sqrt{2\pi}} \int_{z_1}^{z_2} e^{-\frac{1}{2}z^2} dz$$

$$= \int_{z_1}^{z_2} n(z; 0, 1)\, dz$$

$$= P(z_1 < Z < z_2)$$

where $Z$ is seen to be a random varible having the standard normal distribution.    ▼

Thus, to use Table III in connection with any random variable having a normal distribution, we simply perform the change of scale $z = \dfrac{x - \mu}{\sigma}$.

## EXAMPLE 6.4

Suppose that the amount of cosmic radiation to which a person is exposed when flying by jet across the United States is a random variable having a normal distribution with a mean of 4.35 mrem and a standard deviation of 0.59 mrem. What is the probability that a person will be exposed to more than 5.20 mrem of cosmic radiation on such a flight?

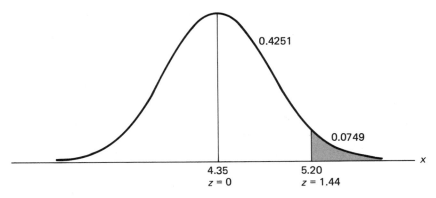

**Figure 6.5**  Diagram for Example 6.4.

*Solution*

Looking up the entry corresponding to $z = \dfrac{5.20 - 4.35}{0.59} = 1.44$ in Table III and subtracting it from 0.5000 (see Figure 6.5), we get $0.5000 - 0.4251 = 0.0749$.  ▲

## 6.6  THE NORMAL APPROXIMATION TO THE BINOMIAL DISTRIBUTION

The normal distribution is sometimes introduced as a continuous distribution which provides a close approximation to the binomial distribution when $n$, the number of trials, is very large and $\theta$, the probability of a success on an individual trial, is close to $\frac{1}{2}$. Figure 6.6 shows the histograms of binomial distributions with $\theta = \frac{1}{2}$ and $n = 2, 5, 10,$ and 25, and it can be seen that with increasing $n$ these distributions approach the symmetrical bell-shaped pattern of the normal distribution.

To provide a theoretical foundation for this argument, let us first prove the following theorem.

---

**THEOREM 6.8**  If $X$ is a random variable having a binomial distribution with the parameters $n$ and $\theta$, then the moment-generating function of

$$Z = \frac{X - n\theta}{\sqrt{n\theta(1 - \theta)}}$$

approaches that of the standard normal distribution when $n \to \infty$.

---

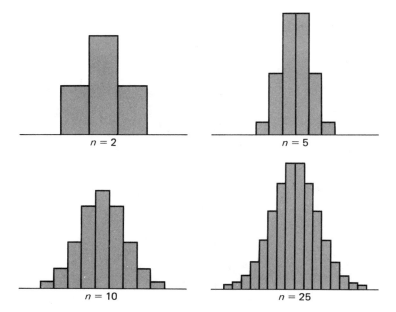

**Figure 6.6**  Binomial distributions with $\theta = \frac{1}{2}$.

***Proof.***   Making use of Theorems 4.10 and 5.4, we can write

$$M_Z(t) = M_{\frac{X-\mu}{\sigma}}(t) = e^{-\mu t/\sigma} \cdot [1 + \theta(e^{t/\sigma} - 1)]^n$$

where $\mu = n\theta$ and $\sigma = \sqrt{n\theta(1 - \theta)}$. Then, taking logarithms and substituting the Maclaurin's series of $e^{t/\sigma}$, we get

$$\ln M_{\frac{X-\mu}{\sigma}}(t) = -\frac{\mu t}{\sigma} + n \cdot \ln[1 + \theta(e^{t/\sigma} - 1)]$$

$$= -\frac{\mu t}{\sigma} + n \cdot \ln\left[1 + \theta\left\{\frac{t}{\sigma} + \frac{1}{2}\left(\frac{t}{\sigma}\right)^2 + \frac{1}{6}\left(\frac{t}{\sigma}\right)^3 + \cdots\right\}\right]$$

and, using the infinite series $\ln(1 + x) = x - \frac{1}{2}x^2 + \frac{1}{3}x^3 - \cdots$, which converges for $|x| < 1$, to expand this logarithm, it follows that

$$\ln M_{\frac{X-\mu}{\sigma}}(t) = -\frac{\mu t}{\sigma} + n\theta \left[\frac{t}{\sigma} + \frac{1}{2}\left(\frac{t}{\sigma}\right)^2 + \frac{1}{6}\left(\frac{t}{\sigma}\right)^3 + \cdots \right]$$

$$-\frac{n\theta^2}{2}\left[\frac{t}{\sigma} + \frac{1}{2}\left(\frac{t}{\sigma}\right)^2 + \frac{1}{6}\left(\frac{t}{\sigma}\right)^3 + \cdots \right]^2$$

$$+\frac{n\theta^3}{3}\left[\frac{t}{\sigma} + \frac{1}{2}\left(\frac{t}{\sigma}\right)^2 + \frac{1}{6}\left(\frac{t}{\sigma}\right)^3 + \cdots \right]^3 - \cdots$$

Collecting powers of $t$, we obtain

$$\ln M_{\frac{X-\mu}{\sigma}}(t) = \left(-\frac{\mu}{\sigma} + \frac{n\theta}{\sigma}\right)t + \left(\frac{n\theta}{2\sigma^2} - \frac{n\theta^2}{2\sigma^2}\right)t^2$$

$$+ \left(\frac{n\theta}{6\sigma^3} - \frac{n\theta^2}{2\sigma^3} + \frac{n\theta^3}{3\sigma^3}\right)t^3 + \cdots$$

$$= \frac{1}{\sigma^2}\left(\frac{n\theta - n\theta^2}{2}\right)t^2 + \frac{n}{\sigma^3}\left(\frac{\theta - 3\theta^2 + 2\theta^3}{6}\right)t^3 + \cdots$$

since $\mu = n\theta$. Then, substituting $\sigma = \sqrt{n\theta(1-\theta)}$, we find that

$$\ln M_{\frac{X-\mu}{\sigma}}(t) = \frac{1}{2}t^2 + \frac{n}{\sigma^3}\left(\frac{\theta - 3\theta^2 + 2\theta^3}{6}\right)t^3 + \cdots$$

where for $r > 2$ the coefficient of $t^r$ is a constant times $\dfrac{n}{\sigma^r}$, which approaches 0 when $n \to \infty$, It follows that

$$\lim_{n \to \infty} \ln M_{\frac{X-\mu}{\sigma}}(t) = \tfrac{1}{2}t^2$$

and since the limit of a logarithm equals the logarithm of the limit (provided the two limits exist), we conclude that

$$\lim_{n \to \infty} M_{\frac{X-\mu}{\sigma}}(t) = e^{\frac{1}{2}t^2}$$

which is the moment-generating function of Theorem 6.6 with $\mu = 0$ and $\sigma = 1$.    ▼

This completes the proof of Theorem 6.8, but have we shown that when $n \to \infty$ the distribution of $Z$, the **standardized** binomial random variable, approaches the standard normal distribution? Not quite. To this end, we must refer to two theorems which we shall state here without proof:

1.  *There is a one-to-one correspondence between moment-generating functions and probability distributions (densities) when the former exist.*

2.  *If the moment-generating function of one random variable approaches that of another random variable, then the distribution (density) of the first random variable approaches that of the second random variable under the same limiting conditions.*

Strictly speaking, our results apply when $n \to \infty$, but the normal distribution is often used to approximate binomial probabilities even when $n$ is fairly small. A good rule of thumb is to use this approximation only when $n\theta$ and $n(1 - \theta)$ are both greater than 5.

# EXAMPLE 6.5

Use the normal approximation to the binomial distribution to determine the probability of getting six heads and 10 tails in 16 flips of a balanced coin.

## Solution

To find this approximation, we must use the **continuity correction** according to which each nonnegative integer $k$ is represented by the interval from $k - \frac{1}{2}$ to $k + \frac{1}{2}$. With reference to Figure 6.7, we must thus determine the area under the curve between 5.5 and 6.5, and since $\mu = 16 \cdot \frac{1}{2} = 8$ and $\sigma = \sqrt{16 \cdot \frac{1}{2} \cdot \frac{1}{2}} = 2$, we must find the area between

$$z = \frac{5.5 - 8}{2} = -1.25 \quad \text{and} \quad z = \frac{6.5 - 8}{2} = -0.75$$

The entries in Table III corresponding to $z = 1.25$ and $z = 0.75$ are 0.3944 and 0.2734, and we find that the normal approximation to the probability of "six heads and 10 tails" is $0.3944 - 0.2734 = 0.1210$. Since the corresponding value in Table I is 0.1222, we find that the error of the approximation is $-0.0012$, and that the percentage error is $\frac{0.0012}{0.1222} \cdot 100 = 0.98\%$ in absolute value.    ▲

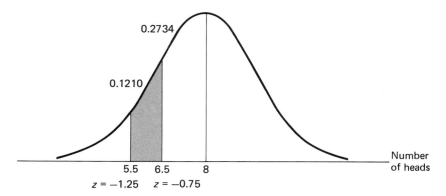

**Figure 6.7**  Diagram for Example 6.5.

The normal approximation to the binomial distribution used to be applied quite extensively, particularly in approximating probabilities associated with large sets of values of binomial random variables. Nowadays, most of this work is done with computers, and we have mentioned the relationship between the binomial and normal distributions primarily because of its theoretical applications. It forms the basis for many of the statistical procedures discussed in Chapters 11, 13, and 16.

### EXERCISES

**6.43** Show that the normal distribution has
  (a)   a relative maximum at $x = \mu$;
  (b)   inflection points at $x = \mu - \sigma$ and $x = \mu + \sigma$.

**6.44** Show that the differential equation of Exercise 6.30 with $b = c = 0$ and $a > 0$ yields a normal distribution.

**6.45** In the proof of Theorem 6.6 we twice differentiated the moment-generating function of the normal distribution with respect to $t$ to show that $E(X) = \mu$ and $\text{var}(X) = \sigma^2$. Differentiating twice more and using the formula of Exercise 4.33, find expressions for $\mu_3$ and $\mu_4$.

**6.46** If $X$ is a random variable having a normal distribution with the mean $\mu$ and the standard deviation $\sigma$, find the moment-generating function of $Y = X - c$, where $c$ is a constant, and use it to rework the preceding exercise.

**6.47** Use the results of Exercise 6.45 to show that $\alpha_3 = 0$ and $\alpha_4 = 3$ for normal distributions, where $\alpha_3$ and $\alpha_4$ are as defined in Exercises 4.34 and 4.35.

**6.48** If $X$ is a random variable having a normal distribution with the mean $\mu$ and the standard deviation $\sigma$, use the third part of Theorem 4.10 and Theorem 6.6 to show that the moment-generating function of

$$Z = \frac{X - \mu}{\sigma}$$

is the moment-generating function of the standard normal distribution. Note that, together with the two theorems on page 245, this proves Theorem 6.7.

**6.49** If $X$ is a random variable having the standard normal distribution and $Y = X^2$, show that $\text{cov}(X, Y) = 0$ even though $X$ and $Y$ are evidently not independent.

**6.50** Use the Maclaurin's series expansion of the moment-generating function of the standard normal distribution to show that

(a)  $\mu_r = 0$ when $r$ is odd;

(b)  $\mu_r = \dfrac{r!}{2^{r/2}\left(\dfrac{r}{2}\right)!}$ when $r$ is even.

**6.51** If we let $K_X(t) = \ln M_{X-\mu}(t)$, the coefficient of $\dfrac{t^r}{r!}$ in the Maclaurin's series of $K_X(t)$ is called the **$r$th cumulant**, and it is denoted by $\kappa_r$. Equating coefficients of like powers, show that

(a)  $\kappa_2 = \mu_2$;
(b)  $\kappa_3 = \mu_3$;
(c)  $\kappa_4 = \mu_4 - 3\mu_2^2$.

**6.52** With reference to the preceding exercise, show that for normal distributions $\kappa_2 = \sigma^2$ and all other cumulants are zero.

**6.53** Show that if $X$ is a random variable having the Poisson distribution with the parameter $\lambda$ and $\lambda \to \infty$, then the moment-generating function of

$$Z = \frac{X - \lambda}{\sqrt{\lambda}}$$

namely, that of a standardized Poisson random variable, approaches the moment-generating function of the standard normal distribution.

**6.54** Show that when $\alpha \to \infty$ and $\beta$ remains constant, the moment-generating function of a standardized gamma random variable approaches the moment-generating function of the standard normal distribution.

## APPLICATIONS

**6.55** If $Z$ is a random variable having the standard normal distribution, find the probabilities that it will take on a value

(a)  greater than 1.14;
(b)  greater than $-0.36$;

(c)   between $-0.46$ and $-0.09$;

(d)   between $-0.58$ and $1.12$.

**6.56** If $Z$ is a random variable having the standard normal distribution, find

(a)   $P(Z < 1.33)$;

(b)   $P(Z \leqslant -0.79)$;

(c)   $P(0.55 < Z < 1.22)$;

(d)   $P(-1.90 \leqslant Z \leqslant 0.44)$.

**6.57** Find $z$ if the standard-normal-curve area

(a)   between 0 and $z$ is 0.4726;

(b)   to the left of $z$ is 0.9868;

(c)   to the right of $z$ is 0.1314;

(d)   between $-z$ and $z$ is 0.8502.

**6.58** If $Z$ is a random variable having the standard normal distribution, find the respective values $z_1$, $z_2$, $z_3$, and $z_4$ such that

(a)   $P(0 < Z < z_1) = 0.4306$;

(b)   $P(Z \geqslant z_2) = 0.7704$;

(c)   $P(Z > z_3) = 0.2912$;

(d)   $P(-z_4 \leqslant Z < z_4) = 0.9700$.

**6.59** If $X$ is a random variable having a normal distribution, what are the probabilities of getting a value

(a)   within one standard deviation of the mean;

(b)   within two standard deviations of the mean;

(c)   within three standard deviations of the mean;

(d)   within four standard deviations of the mean?

**6.60** If $z_\alpha$ is defined by

$$\int_{z_\alpha}^{\infty} n(z; 0, 1)\, dz = \alpha$$

find its values for

(a)   $\alpha = 0.05$;

(b)   $\alpha = 0.025$;

(c)   $\alpha = 0.01$;

(d)   $\alpha = 0.005$.

**6.61** Suppose that during periods of transcendental meditation the reduction of a person's oxygen consumption is a random variable having a normal distribution with $\mu = 37.6$ cc per minute and $\sigma = 4.6$ cc per minute. Find the probabilities that during a period of transcendental meditation a person's oxygen consumption will be reduced by

(a)   at least 44.5 cc per minute;

(b)   at most 35.0 cc per minute;

(c)   anywhere from 30.0 to 40.0 cc per minute.

**6.62** In a photographic process, the developing time of prints may be looked upon as a random variable having the normal distribution with $\mu = 15.40$ seconds and $\sigma = 0.48$ second. Find the probabilities that the time it takes to develop one of the prints will be

(a)   at least 16.00 seconds;
(b)   at most 14.20 seconds;
(c)   anywhere from 15.00 to 15.80 seconds.

**6.63** Suppose that the actual amount of instant coffee which a filling machine puts into "6-ounce" jars is a random variable having a normal distribution with $\sigma = 0.05$ ounce. If only 3 percent of the jars are to contain less than 6 ounces of coffee, what must be the mean fill of these jars?

**6.64** A random variable has a normal distribution with $\sigma = 10$. If the probability that the random variable will take on a value less than 82.5 is 0.8212, what is the probability that it will take on a value greater than 58.3?

**6.65** Check in each case whether the normal approximation to the binomial distribution may be used according to the rule of thumb on page 245.

(a)   $n = 16$ and $\theta = 0.20$;
(b)   $n = 65$ and $\theta = 0.10$;
(c)   $n = 120$ and $\theta = 0.98$.

**6.66** Suppose we want to use the normal approximation to the binomial distribution to determine $b(1; 150, 0.05)$.

(a)   Based on the rule of thumb on page 245, would we be justified in using the approximation?
(b)   Make the approximation and round to four decimals.
(c)   If a computer printout shows that $b(1; 150, 0.05) = 0.0036$ rounded to four decimals, what is the percentage error of the approximation obtained in part (b)?

This serves to illustrate that the rule of thumb is just that and no more; making approximations like this also requires a good deal of professional judgment.

**6.67** With reference to Exercise 6.66, show that the Poisson distribution would have yielded a better approximation.

**6.68** Use the normal approximation to the binomial distribution to determine (to four decimals) the probability of getting seven heads and seven tails in 14 flips of a balanced coin. Also refer to Table I to find the error of this approximation.

**6.69** If 23 percent of all patients with high blood pressure have bad side effects from a certain kind of medicine, use the normal approximation to find the probability that among 120 patients with high blood pressure treated with this medicine more than 32 will have bad side effects.

**6.70** If the probability is 0.20 that a certain bank will refuse a loan application, use the normal approximation to determine (to three decimals) the probability that the bank will refuse at most 40 of 225 loan applications.

**6.71** To illustrate the law of large numbers (see also Exercise 5.26), use the normal approximation to the binomial distribution to determine the probabilities that the proportion of heads will be anywhere from 0.49 to 0.51 when a balanced coin is flipped

(a) 100 times;

(b) 1,000 times;

(c) 10,000 times.

## 6.7 THE BIVARIATE NORMAL DISTRIBUTION

Among multivariate densities, of special importance is the **multivariate normal distribution**, which is a generalization of the normal distribution in one variable. As it is best (indeed, virtually necessary) to present this distribution in matrix notation, we shall give here only the **bivariate** case; discussions of the general case are listed among the references at the end of this chapter.

---

**DEFINITION 6.8** A pair of random variables $X$ and $Y$ have a **bivariate normal distribution**, and they are referred to as jointly normally distributed random variables, if and only if their joint probability density is given by

$$f(x, y) = \frac{e^{-\frac{1}{2(1-\rho^2)}\left[\left(\frac{x-\mu_1}{\sigma_1}\right)^2 - 2\rho\left(\frac{x-\mu_1}{\sigma_1}\right)\left(\frac{y-\mu_2}{\sigma_2}\right) + \left(\frac{y-\mu_2}{\sigma_2}\right)^2\right]}}{2\pi\sigma_1\sigma_2\sqrt{1-\rho^2}}$$

for $-\infty < x < \infty$ and $-\infty < y < \infty$, where $\sigma_1 > 0$, $\sigma_2 > 0$, and $-1 < \rho < 1$.

---

To study this joint distribution, let us first show that the parameters $\mu_1$, $\mu_2$, $\sigma_1$, and $\sigma_2$ are the means and the standard deviations of the two random variables $X$ and $Y$. To begin with, we integrate on $y$ from $-\infty$ to $\infty$, getting

$$g(x) = \frac{e^{-\frac{1}{2(1-\rho^2)}\left(\frac{x-\mu_1}{\sigma_1}\right)^2}}{2\pi\sigma_1\sigma_2\sqrt{1-\rho^2}} \int_{-\infty}^{\infty} e^{-\frac{1}{2(1-\rho^2)}\left[\left(\frac{y-\mu_2}{\sigma_2}\right)^2 - 2\rho\left(\frac{x-\mu_1}{\sigma_1}\right)\left(\frac{y-\mu_2}{\sigma_2}\right)\right]} \, dy$$

for the marginal density of $X$. Then, temporarily making the substitution $u =$

$\dfrac{x - \mu_1}{\sigma_1}$ to simplify the notation and changing the variable of integration by letting

$v = \dfrac{y - \mu_2}{\sigma_2}$, we obtain

$$g(x) = \frac{e^{-\frac{1}{2(1-\rho^2)}u^2}}{2\pi\sigma_1\sqrt{1-\rho^2}} \int_{-\infty}^{\infty} e^{-\frac{1}{2(1-\rho^2)}(v^2 - 2\rho u v)}\, dv$$

After completing the square by letting $v^2 - 2\rho u v = (v - \rho u)^2 - \rho^2 u^2$ and collecting terms, this becomes

$$g(x) = \frac{e^{-\frac{1}{2}u^2}}{\sigma_1\sqrt{2\pi}} \left\{ \frac{1}{\sqrt{2\pi}\sqrt{1-\rho^2}} \int_{-\infty}^{\infty} e^{-\frac{1}{2}\left(\frac{v - \rho u}{\sqrt{1-\rho^2}}\right)^2}\, dv \right\}$$

Finally, identifying the quantity in parentheses as the integral of a normal density from $-\infty$ to $\infty$ and, hence, equalling 1, we get

$$g(x) = \frac{e^{-\frac{1}{2}u^2}}{\sigma_1\sqrt{2\pi}} = \frac{1}{\sigma_1\sqrt{2\pi}} e^{-\frac{1}{2}\left(\frac{x-\mu_1}{\sigma_1}\right)^2}$$

for $-\infty < x < \infty$. It follows by inspection that the marginal density of $X$ is a normal distribution with the mean $\mu_1$ and the standard deviation $\sigma_1$, and by symmetry that the marginal density of $Y$ is a normal distribution with the mean $\mu_2$ and the standard deviation $\sigma_2$.

So far as the parameter $\rho$ is concerned, where $\rho$ is the lowercase Greek letter *rho*, it is called the **correlation coefficient**, and the necessary integration will show that $\operatorname{cov}(X, Y) = \rho\sigma_1\sigma_2$. Thus, the parameter $\rho$ measures how the two random variables $X$ and $Y$ vary together, and its significance will be discussed further in Chapter 14.

When we deal with a pair of random variables having a bivariate normal distribution, their conditional densities are also of importance; so, let us prove the following theorem.

---

**THEOREM 6.9** If $X$ and $Y$ have a bivariate normal distribution, the conditional density of $Y$ given $X = x$ is a normal distribution with the mean

$$\mu_{Y|x} = \mu_2 + \rho\frac{\sigma_2}{\sigma_1}(x - \mu_1)$$

and the variance

$$\sigma^2_{Y|x} = \sigma^2_2(1 - \rho^2)$$

and the conditional density of $X$ given $Y = y$ is a normal distribution with the mean

$$\mu_{X|y} = \mu_1 + \rho \frac{\sigma_1}{\sigma_2}(y - \mu_2)$$

and the variance

$$\sigma^2_{X|y} = \sigma^2_1(1 - \rho^2)$$

**Proof.**   Writing $w(y|x) = \dfrac{f(x, y)}{g(x)}$ in accordance with Definition 3.13 and letting $u = \dfrac{x - \mu_1}{\sigma_1}$ and $v = \dfrac{y - \mu_2}{\sigma_2}$ to simplify the notation, we get

$$w(y|x) = \frac{\dfrac{1}{2\pi\sigma_1\sigma_2\sqrt{1 - \rho^2}} e^{-\frac{1}{2(1-\rho^2)}[u^2 - 2\rho uv + v^2]}}{\dfrac{1}{\sqrt{2\pi}\sigma_1} e^{-\frac{1}{2}u^2}}$$

$$= \frac{1}{\sqrt{2\pi}\sigma_2\sqrt{1 - \rho^2}} e^{-\frac{1}{2(1-\rho^2)}[v^2 - 2\rho uv + \rho^2 u^2]}$$

$$= \frac{1}{\sqrt{2\pi}\sigma_2\sqrt{1 - \rho^2}} e^{-\frac{1}{2}\left[\frac{v - \rho u}{\sqrt{1 - \rho^2}}\right]^2}$$

Then, expressing this result in terms of the original variables, we obtain

$$w(y|x) = \frac{1}{\sigma_2\sqrt{2\pi}\sqrt{1 - \rho^2}} e^{-\frac{1}{2}\left[\frac{y - \left\{\mu_2 + \rho\frac{\sigma_2}{\sigma_1}(x - \mu_1)\right\}}{\sigma_2\sqrt{1 - \rho^2}}\right]^2}$$

for $-\infty < y < \infty$, and it can be seen by inspection that this is a normal density with the mean $\mu_{Y|x} = \mu_2 + \rho \dfrac{\sigma_2}{\sigma_1} (x - \mu_1)$ and the variance $\sigma_{Y|x}^2 = \sigma_2^2(1 - \rho^2)$. The corresponding results for the conditional density of $X$ given $Y = y$ follow by symmetry.    ▼

The bivariate normal distribution has many important properties, some statistical and some purely mathematical. Among the former, there is the following property, which the reader will be asked to prove in Exercise 6.72.

---

**THEOREM 6.10**  If two random variables have a bivariate normal distribution, they are independent if and only if $\rho = 0$.

---

In connection with this, if $\rho = 0$ the random variables are said to be **uncorrelated**.

Also, we have shown that for two random variables having a bivariate normal distribution the two marginal densities are normal, but the converse is not necessarily true. In other words, the marginal distributions may both be normal without the joint distribution being a bivariate normal distribution. For instance, if the bivariate density of $X$ and $Y$ is given by

$$f^*(x, y) = \begin{cases} 2f(x, y) & \text{inside squares 2 and 4 of Figure 6.8} \\ 0 & \text{inside squares 1 and 3 of Figure 6.8} \\ f(x, y) & \text{elsewhere} \end{cases}$$

where $f(x, y)$ is the value of the bivariate normal density with $\mu_1 = 0$, $\mu_2 = 0$,

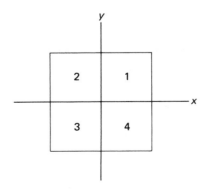

**Figure 6.8**  Sample space for the bivariate density given by $f^*(x, y)$.

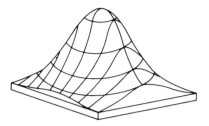

**Figure 6.9**  Bivariate normal surface.

and $\rho = 0$ at $(x, y)$, it is easy to see that the marginal densities of $X$ and $Y$ are normal even though their joint density is not a bivariate normal distribution.

Many interesting properties of the bivariate normal density are obtained by studying the **bivariate normal surface**, pictured in Figure 6.9, whose equation is $z = f(x, y)$, where $f(x, y)$ is the value of the bivariate normal density at $(x, y)$. As the reader will be asked to verify in the exercises that follow, the bivariate normal surface has a maximum at $(\mu_1, \mu_2)$, any plane parallel to the $z$-axis intersects the surface in a curve having the shape of a normal distribution, and any plane parallel to the $xy$-plane which intersects the surface intersects it in an ellipse called a **contour of constant probability density**. When $\rho = 0$ and $\sigma_1 = \sigma_2$, the contours of constant probability density are circles, and it is customary to refer to the corresponding joint density as a **circular normal distribution**.

### EXERCISES

**6.72** To prove Theorem 6.10, show that if $X$ and $Y$ have a bivariate normal distribution, then
   (a)  their independence implies that $\rho = 0$;
   (b)  $\rho = 0$ implies that they are independent.

**6.73** Show that any plane perpendicular to the $xy$-plane intersects the bivariate normal surface in a curve having the shape of a normal distribution.

**6.74** If the exponent of $e$ of a bivariate normal density is

$$\frac{-1}{102}\,[(x + 2)^2 - 2.8(x + 2)(y - 1) + 4(y - 1)^2]$$

   find
   (a)  $\mu_1$, $\mu_2$, $\sigma_1$, $\sigma_2$, and $\rho$;
   (b)  $\mu_{Y|x}$ and $\sigma_{Y|x}^2$.

**6.75** If the exponent of $e$ of a bivariate normal density is

$$-\frac{1}{54}(x^2 + 4y^2 + 2xy + 2x + 8y + 4)$$

find $\sigma_1$, $\sigma_2$, and $\rho$, given that $\mu_1 = 0$ and $\mu_2 = -1$.

**6.76** If $X$ and $Y$ have the bivariate normal distribution with $\mu_1 = 2$, $\mu_2 = 5$, $\sigma_1 = 3$, $\sigma_2 = 6$, and $\rho = \frac{2}{3}$, find $\mu_{Y|1}$ and $\sigma_{Y|1}$.

**6.77** If $X$ and $Y$ have a bivariate normal distribution, and $U = X + Y$ and $V = X - Y$, find an expression for the correlation coefficient of $U$ and $V$.

**6.78** If $X$ and $Y$ have a bivariate normal distribution, it can be shown that their joint moment-generating function (see Exercise 4.63) is given by

$$M_{X,Y}(t_1, t_2) = E(e^{t_1 X + t_2 Y})$$

$$= e^{t_1 \mu_1 + t_2 \mu_2 + \frac{1}{2}(\sigma_1^2 t_1^2 + 2\rho\sigma_1\sigma_2 t_1 t_2 + \sigma_2^2 t_2^2)}$$

Verify that

(a)   the first partial derivative of this function with respect to $t_1$ at $t_1 = 0$ and $t_2 = 0$ is $\mu_1$;

(b)   the second partial derivative with respect to $t_1$ at $t_1 = 0$ and $t_2 = 0$ is $\sigma_1^2 + \mu_1^2$;

(c)   the second partial derivative with respect to $t_1$ and $t_2$ at $t_1 = 0$ and $t_2 = 0$ is $\rho\sigma_1\sigma_2 + \mu_1\mu_2$.


## APPLICATIONS

**6.79** The center of a target is taken as the origin of a rectangular system of co-ordinates, with reference to which the point of impact of a missile has the coordinates $X$ and $Y$. If $X$ and $Y$ have a bivariate normal density with $\mu_1 = 0$, $\mu_2 = 0$, $\sigma_1 = 120$ feet, $\sigma_2 = 120$ feet, and $\rho = 0$, find the probabilities that the point of impact will be

(a)   inside a square with sides of 180 feet, whose center is at the origin and whose sides are parallel to the coordinate axes;

(b)   inside a circle with a radius of 75 feet with its center at the origin.

**6.80** If $X$ and $Y$ have the circular normal distribution with $\mu_1 = \mu_2 = 0$ and $\sigma_1 = \sigma_2 = 12$, find

(a)   the probability of getting a point $(x, y)$ inside the circle $x^2 + y^2 = 36$;

(b)   the value of $c$ for which the probability of getting a point $(x, y)$ inside the circle $x^2 + y^2 = c^2$ is 0.80.

**6.81** Suppose that $X$ and $Y$, the height and weight of certain animals, have a bivariate normal distribution with $\mu_1 = 18$ inches, $\mu_2 = 15$ pounds, $\sigma_1 = 3$ inches, $\sigma_2 = 2$ pounds, and $\rho = 0.75$. Find

(a)   the expected weight of one of these animals that is 17 inches tall;

(b)   the expected height of one of these animals that weighs 20 pounds.

## REFERENCES

Useful information about various special probability densities, in outline form, may be found in

DERMAN, C., GLESER, L., and OLKIN, I., *Probability Models and Applications*. New York: Macmillan Publishing Co., Inc., 1980,

HASTINGS, N. A. J., and PEACOCK, J. B., *Statistical Distributions*. London: Butterworth & Co. Ltd., 1975,

and

JOHNSON, N. L., and KOTZ, S., *Continuous Univariate Distributions*, Vols. 1 and 2. Boston: Houghton Mifflin Company, 1970.

A direct proof that the standardized binomial distribution approaches the standard normal distribution when $n \to \infty$ is given in

KEEPING, E. S., *Introduction to Statistical Inference*. Princeton, N.J.: D. Van Nostrand Co., Inc., 1962.

A detailed treatment of the mathematical and statistical properties of the bivariate normal surface may be found in

YULE, G. U., and KENDALL, M. G., *An Introduction to the Theory of Statistics*, 14th ed. New York: Hafner Publishing Co., Inc., 1950.

The multivariate normal distribution is treated in matrix notation in

BICKEL, P. J., and DOKSUM, K. A., *Mathematical Statistics: Basic Ideas and Selected Topics*. San Francisco: Holden-Day, Inc., 1977,

HOGG, R. V., and CRAIG, A. T., *Introduction to Mathematical Statistics*, 4th ed. New York: Macmillan Publishing Co., Inc., 1978,

LINDGREN, B. W., *Statistical Theory*, 3rd ed. New York: Macmillan Publishing Co., Inc., 1976.

# 7

# *Functions of Random Variables*

## 7.1 INTRODUCTION

In this chapter we shall concern ourselves with the problem of finding the probability distributions or densities of **functions of one or more random variables**. That is, given a set of random variables $X_1, X_2, \ldots, X_n$ and their joint probability distribution or density, we shall be interested in finding the probability distribution or density of some random variable $Y = u(X_1, X_2, \ldots, X_n)$. This means that the values of $Y$ are related to those of the $X$'s by means of the equation $y = u(x_1, x_2, \ldots, x_n)$.

Several methods are available for solving this kind of problem. The ones we shall discuss in the next four sections are called the **distribution function technique**, the **transformation technique**, and the **moment-generating function technique**. Although all three methods can be used in some situations, in most

**257**

problems one technique will be preferable (easier to use than the others). This is true, for example, in some instances where the function in question is linear in the random variables $X_1, X_2, \ldots, X_n$, and the moment-generating function technique yields the simplest derivations.

The various techniques we shall discuss in this chapter will be used again in Chapter 8 to derive several distributions that are of fundamental importance in statistical inference.

## 7.2 DISTRIBUTION FUNCTION TECHNIQUE

A straightforward method of obtaining the probability density of a function of continuous random variables consists of first finding its distribution function and then its probability density by differentiation. Thus, if $X_1, X_2, \ldots, X_n$ are continuous random variables with a given joint probability density, the probability density of $Y = u(X_1, X_2, \ldots, X_n)$ is obtained by first determining an expression for the probability

$$F(y) = P(Y \leq y) = P[u(X_1, X_2, \ldots, X_n) \leq y]$$

and then differentiating to get

$$f(y) = \frac{dF(y)}{dy}$$

according to Theorem 3.6.

## EXAMPLE 7.1

If the probability density of $X$ is given by

$$f(x) = \begin{cases} 6x(1 - x) & \text{for } 0 < x < 1 \\ 0 & \text{elsewhere} \end{cases}$$

find the probability density of $Y = X^3$.

### Solution

Letting $G(y)$ denote the value of the distribution function of $Y$ at $y$, we can write

$$G(y) = P(Y \leq y)$$
$$= P(X^3 \leq y)$$

$$= P(X \leq y^{1/3})$$

$$= \int_0^{y^{1/3}} 6x(1 - x)\, dx$$

$$= 3y^{2/3} - 2y$$

and, hence,

$$g(y) = 2(y^{-1/3} - 1)$$

for $0 < y < 1$; elsewhere, $g(y) = 0$. In Exercise 7.20 the reader will be asked to verify this result by a different technique.    ▲

## EXAMPLE 7.2

If $Y = |X|$, show that

$$g(y) = \begin{cases} f(y) + f(-y) & \text{for } y > 0 \\ 0 & \text{elsewhere} \end{cases}$$

where $f(x)$ is the value of the probability density of $X$ at $x$ and $g(y)$ is the value of the probability density of $Y$ at $y$. Also, use this result to find the probability density of $Y = |X|$ when $X$ has the standard normal distribution.

### Solution

For $y > 0$ we have

$$G(y) = P(Y \leq y)$$

$$= P(|X| \leq y)$$

$$= P(-y \leq X \leq y)$$

$$= F(y) - F(-y)$$

and, upon differentiation,

$$g(y) = f(y) + f(-y)$$

Also, since $|x|$ cannot be negative, $g(y) = 0$ for $y < 0$. Arbitrarily letting $g(0) = 0$, we can thus write

$$g(y) = \begin{cases} f(y) + f(-y) & \text{for } y > 0 \\ 0 & \text{elsewhere} \end{cases}$$

If $X$ has the standard normal distribution and $Y = |X|$, it follows that

$$g(y) = n(y; 0, 1) + n(-y; 0, 1)$$
$$= 2n(y; 0, 1)$$

for $y > 0$ and $g(y) = 0$ elsewhere. An important application of this result may be found in Example 7.9.    ▲

## EXAMPLE 7.3

If the joint density of $X_1$ and $X_2$ is given by

$$f(x_1, x_2) = \begin{cases} 6e^{-3x_1 - 2x_2} & \text{for } x_1 > 0, x_2 > 0 \\ 0 & \text{elsewhere} \end{cases}$$

find the probability density of $Y = X_1 + X_2$.

### Solution

Integrating the joint density over the shaded region of Figure 7.1, we get

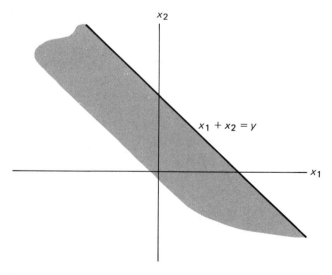

**Figure 7.1**  Diagram for Example 7.3.

$$F(y) = \int_0^y \int_0^{y-x_2} 6e^{-3x_1-2x_2} \, dx_1 \, dx_2$$

$$= 1 + 2e^{-3y} - 3e^{-2y}$$

and, differentiating with respect to $y$, we obtain

$$f(y) = 6(e^{-2y} - e^{-3y})$$

for $y > 0$; elsewhere, $f(y) = 0$.    ▲

## EXERCISES

**7.1**  If the probability density of $X$ is given by

$$f(x) = \begin{cases} 2xe^{-x^2} & \text{for } x > 0 \\ 0 & \text{elsewhere} \end{cases}$$

and $Y = X^2$, find
  (a)   the distribution function of $Y$;
  (b)   the probability density of $Y$.

**7.2**  If $X$ has an exponential distribution with the parameter $\theta$, use the distribution function technique to find the probability density of the random variable $Y = \ln X$.

**7.3**  If $X$ has the uniform density with the parameters $\alpha = 0$ and $\beta = 1$, use the distribution function technique to find the probability density of the random variable $Y = \sqrt{X}$.

**7.4**  If the joint probability density of $X$ and $Y$ is given by

$$f(x, y) = \begin{cases} 4xye^{-(x^2+y^2)} & \text{for } x > 0, y > 0 \\ 0 & \text{elsewhere} \end{cases}$$

and $Z = \sqrt{X^2 + Y^2}$, find
  (a)   the distribution function of $Z$;
  (b)   the probability density of $Z$.

**7.5**  If $X_1$ and $X_2$ are independent random variables having exponential densities with the parameters $\theta_1$ and $\theta_2$, use the distribution function technique to find the probability density of $Y = X_1 + X_2$ when
  (a)   $\theta_1 \neq \theta_2$;
  (b)   $\theta_1 = \theta_2$.
  (Example 7.3 is a special case of this with $\theta_1 = \frac{1}{3}$ and $\theta_2 = \frac{1}{2}$.)

**7.6**  With reference to the two random variables of Exercise 7.5, show that if $\theta_1 = \theta_2 = 1$, the random variable

$$Z = \frac{X_1}{X_1 + X_2}$$

has the uniform density with $\alpha = 0$ and $\beta = 1$.

**7.7**  Let $X_1$ and $X_2$ be independent random variables having the uniform density with $\alpha = 0$ and $\beta = 1$. Referring to Figure 7.2, find expressions for the distribution function of $Y = X_1 + X_2$ for

(a)  $y \leqslant 0$;
(b)  $0 < y < 1$;
(c)  $1 < y < 2$;
(d)  $y \geqslant 2$.

Also find the probability density of $Y$.

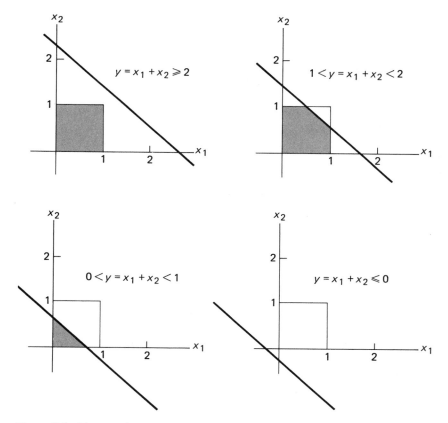

**Figure 7.2**  Diagram for Exercise 7.7.

**7.8** If the joint density of $X$ and $Y$ is given by

$$f(x, y) = \begin{cases} e^{-(x+y)} & \text{for } x > 0, y > 0 \\ 0 & \text{elsewhere} \end{cases}$$

and $Z = \dfrac{X + Y}{2}$, find the probability density of $Z$ by the distribution function technique.

## APPLICATIONS

**7.9** In Exercise 3.84, the price of a certain commodity (in dollars) and its total sales (in 10,000 units) were denoted by $P$ and $S$. Use the joint density given in that exercise and the distribution function technique to find the probability density of $V = SP$, the total amount of money (in \$10,000 units) that is spent on this commodity.

**7.10** With reference to Exercise 3.49, find the probability density of the average mileage of two such tires. Assume independence.

**7.11** In Exercise 3.103, $X$ is the amount of money (in dollars) that a salesperson spends on gasoline and $Y$ is the amount of money for which he or she is reimbursed. Use the joint probability density given in that exercise and the distribution function technique to find the probability density of the random variable $Z = X - Y$, the amount of money for which he or she is not reimbursed.

**7.12** Let $X$ be the amount of premium gasoline (in 1,000 gallons) that a service station has in its tanks at the beginning of a day, and $Y$ the amount which the service station sells during that day. If the joint density of $X$ and $Y$ is given by

$$f(x, y) = \begin{cases} \dfrac{1}{200} & \text{for } 0 < y < x < 20 \\ 0 & \text{elsewhere} \end{cases}$$

use the distribution function technique to find the probability density of the amount that the service station has left in its tanks at the end of the day.

**7.13** The percentages of copper and iron in a certain kind of ore are, respectively, $X_1$ and $X_2$. If the joint density of these two random variables is given by

$$f(x_1, x_2) = \begin{cases} \frac{3}{11}(5x_1 + x_2) & \text{for } x_1 > 0, x_2 > 0, \text{ and } x_1 + 2x_2 < 2 \\ 0 & \text{elsewhere} \end{cases}$$

use the distribution function technique to find the probability density of $Y = X_1 + X_2$. Also find $E(Y)$, the expected total percentage of copper and iron in the ore.

## 7.3 TRANSFORMATION TECHNIQUE: ONE VARIABLE

Let us show how the probability distribution or density of a function of a random variable can be determined without first getting its distribution function. In the discrete case there is no real problem so long as the relationship between the values of $X$ and $Y = u(X)$ is one-to-one; all we have to do is make the appropriate substitution.

### EXAMPLE 7.4

If $X$ is the number of heads obtained in four tosses of a balanced coin, find the probability distribution of $Y = \dfrac{1}{1 + X}$.

#### Solution

Using the formula for the binomial distribution with $n = 4$ and $\theta = \frac{1}{2}$, we find that the probability distribution of $X$ is given by

| $x$ | 0 | 1 | 2 | 3 | 4 |
|-----|---|---|---|---|---|
| $f(x)$ | $\frac{1}{16}$ | $\frac{4}{16}$ | $\frac{6}{16}$ | $\frac{4}{16}$ | $\frac{1}{16}$ |

Then, using the relationship $y = \dfrac{1}{1 + x}$ to substitute values of $Y$ for values of $X$, we find that the probability distribution of $Y$ is given by

| $y$ | 1 | $\frac{1}{2}$ | $\frac{1}{3}$ | $\frac{1}{4}$ | $\frac{1}{5}$ |
|-----|---|---|---|---|---|
| $g(y)$ | $\frac{1}{16}$ | $\frac{4}{16}$ | $\frac{6}{16}$ | $\frac{4}{16}$ | $\frac{1}{16}$ |

If we had wanted to make the substitution directly in the formula for the binomial distribution with $n = 4$ and $\theta = \frac{1}{2}$, we could have substituted $x = \dfrac{1}{y} - 1$ for $x$ in

$$f(x) = \binom{4}{x}\left(\frac{1}{2}\right)^4 \qquad \text{for } x = 0, 1, 2, 3, 4$$

getting

$$g(y) = f\left(\frac{1}{y} - 1\right) = \binom{4}{\frac{1}{y} - 1}\left(\frac{1}{2}\right)^4 \qquad \text{for } y = 1, \tfrac{1}{2}, \tfrac{1}{3}, \tfrac{1}{4}, \tfrac{1}{5} \qquad \blacktriangle$$

Note that in the preceding example the probabilities remained unchanged; the only difference is that in the result they are associated with the various values of $Y$ instead of the corresponding values of $X$. That is all there is to the **transformation** (or **change-of-variable**) **technique** in the discrete case so long as the relationship is one-to-one. If the relationship is not one-to-one, we may proceed as in the following example.

## EXAMPLE 7.5

With reference to Example 7.4, find the probability distribution of the random variable $Z = (X - 2)^2$.

### Solution

Calculating the probabilities $h(z)$ associated with the various values of $Z$, we get

$$h(0) = f(2) = \tfrac{6}{16}$$

$$h(1) = f(1) + f(3) = \tfrac{4}{16} + \tfrac{4}{16} = \tfrac{8}{16}$$

$$h(4) = f(0) + f(4) = \tfrac{1}{16} + \tfrac{1}{16} = \tfrac{2}{16}$$

and, hence,

| $z$ | 0 | 1 | 4 |
|-----|---|---|---|
| $h(z)$ | $\tfrac{3}{8}$ | $\tfrac{4}{8}$ | $\tfrac{1}{8}$ |

$\blacktriangle$

To perform a transformation of variable in the continuous case, we shall assume that the function given by $y = u(x)$ is differentiable and either increasing or decreasing for all values within the range of $X$ for which $f(x) \neq 0$, so that the inverse function, given by $x = w(y)$, exists for all the corresponding values of $y$ and is differentiable except where $u'(x) = 0$.[†] Under these conditions, we can prove the following theorem.

---

[†]To avoid points where $u'(x)$ might be 0, we generally do not include the endpoints of the intervals for which probability densities are nonzero. This is the practice which we have followed and shall continue to follow throughout this book.

**THEOREM 7.1** Let $f(x)$ be the value of the probability density of the continuous random variable $X$ at $x$. If the function given by $y = u(x)$ is differentiable and either increasing or decreasing for all values within the range of $X$ for which $f(x) \neq 0$, then, for these values of $x$, the equation $y = u(x)$ can be uniquely solved for $x$ to give $x = w(y)$, and for the corresponding values of $y$ the probability density of $Y = u(X)$ is given by

$$g(y) = f[w(y)] \cdot |w'(y)| \qquad \text{provided } u'(x) \neq 0$$

Elsewhere, $g(y) = 0$.

*Proof.* First let us prove the case where the function given by $y = u(x)$ is increasing. As can be seen from Figure 7.3, $X$ must take on a value between $w(a)$ and $w(b)$ when $Y$ takes on a value between $a$ and $b$. Hence

$$P(a < Y < b) = P[w(a) < X < w(b)]$$

$$= \int_{w(a)}^{w(b)} f(x)\, dx$$

$$= \int_{a}^{b} f[w(y)]w'(y)\, dy$$

where we performed the change of variable $y = u(x)$, or equivalently $x = w(y)$, in the integral. In accordance with Definition 3.4, the integrand gives the probability density of $Y$ so long as $w'(y)$ exists, and we can write

$$g(y) = f[w(y)]w'(y)$$

When the function given by $y = u(x)$ is decreasing, it can be seen from Figure 7.3 that $X$ must take on a value between $w(b)$ and $w(a)$ when $Y$ takes on a value between $a$ and $b$. Hence,

$$P(a < Y < b) = P[w(b) < X < w(a)]$$

$$= \int_{w(b)}^{w(a)} f(x)\, dx$$

$$= \int_{b}^{a} f[w(y)]w'(y)\, dy$$

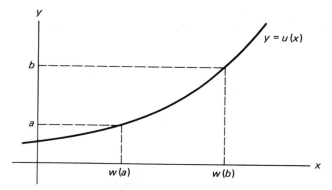

Increasing function

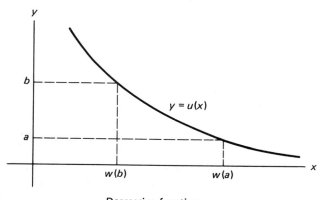

Decreasing function

**Figure 7.3**  Diagrams for proof of Theorem 7.1.

$$= -\int_a^b f[w(y)]w'(y)\,dy$$

where we performed the same change of variable as before, and it follows that

$$g(y) = -f[w(y)]w'(y)$$

Since $w'(y) = \dfrac{dx}{dy} = \dfrac{1}{\dfrac{dy}{dx}}$ is positive when the function given by $y = u(x)$ is increasing, and $-w'(y)$ is positive when the function given by $y = u(x)$ is decreasing, we can combine the two cases by writing

$$g(y) = f[w(y)] \cdot |w'(y)| \qquad \blacktriangledown$$

## EXAMPLE 7.6

If $X$ has the exponential distribution given by

$$f(x) = \begin{cases} e^{-x} & \text{for } x > 0 \\ 0 & \text{elsewhere} \end{cases}$$

find the probability density of the random variable $Y = \sqrt{X}$.

*Solution*

The equation $y = \sqrt{x}$, relating the values of $X$ and $Y$, has the unique inverse $x = y^2$, which yields $w'(y) = \dfrac{dx}{dy} = 2y$. Therefore,

$$g(y) = e^{-y^2}|2y| = 2ye^{-y^2}$$

for $y > 0$ in accordance with Theorem 7.1. Since the probability of getting a value of $Y$ less than or equal to 0, like the probability of getting a value of $X$ less than or equal to 0, is zero, it follows that the probability density of $Y$ is given by

$$g(y) = \begin{cases} 2ye^{-y^2} & \text{for } y > 0 \\ 0 & \text{elsewhere} \end{cases}$$

Note that this is the Weibull distribution of Exercise 6.23 with $\alpha = 1$ and $\beta = 2$.   $\blacktriangle$

The two diagrams of Figure 7.4 illustrate what happened in this example when we transformed from $X$ to $Y$. As in the discrete case (for instance, Example 7.4), the probabilities remain the same, but they pertain to different values (intervals of values) of the respective random variables. In the diagram on the left, the 0.35 probability pertains to the event that $X$ will take on a value on the interval from 1 to 4, and in the diagram on the right, the 0.35 probability pertains to the event that $Y$ will take on a value on the interval from 1 to 2.

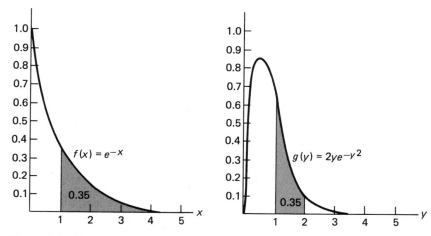

**Figure 7.4**  Diagrams for Example 7.6.

## EXAMPLE 7.7

If the double arrow of Figure 7.5 is spun so that the random variable $\Theta$ has the uniform density

$$f(\theta) = \begin{cases} \dfrac{1}{\pi} & \text{for } -\dfrac{\pi}{2} < \theta < \dfrac{\pi}{2} \\ 0 & \text{elsewhere} \end{cases}$$

determine the probability density of $X$, the abscissa of the point on the $x$-axis to which the arrow will point.

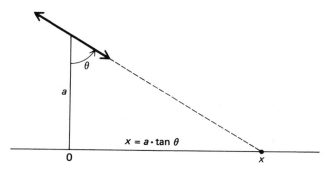

**Figure 7.5**  Diagram for Example 7.7.

*Solution*

As is apparent from the diagram, the relationship between $x$ and $\theta$ is given by $x = a \cdot \tan \theta$, so that

$$\frac{d\theta}{dx} = \frac{a}{a^2 + x^2}$$

and it follows that

$$g(x) = \frac{1}{\pi} \cdot \left| \frac{a}{a^2 + x^2} \right|$$

$$= \frac{1}{\pi} \cdot \frac{a}{a^2 + x^2} \qquad \text{for } -\infty < x < \infty$$

according to Theorem 7.1. Note that this is a special case of the Cauchy distribution of Exercise 6.6.    ▲

## EXAMPLE 7.8

If $F(x)$ is the value of the distribution function of the continuous random variable $X$ at $x$, find the probability density of $Y = F(X)$.

*Solution*

As can be seen from Figure 7.6, the value of $Y$ corresponding to any particular value of $X$ is given by the area under the curve, namely, the area

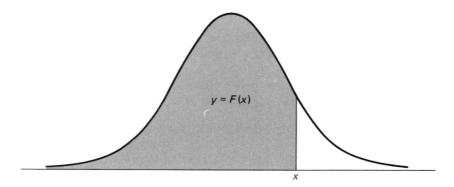

**Figure 7.6**   Diagram for Example 7.8.

under the graph of the density of $X$ to the left of $x$. Differentiating $y = F(x)$ with respect to $x$, we get

$$\frac{dy}{dx} = F'(x) = f(x)$$

and, hence,

$$\frac{dx}{dy} = \frac{1}{\dfrac{dy}{dx}} = \frac{1}{f(x)}$$

provided $f(x) \neq 0$. It follows from Theorem 7.1 that

$$g(y) = f(x) \cdot \left| \frac{1}{f(x)} \right| = 1$$

for $0 < y < 1$, and we can say that $y$ has the uniform density with $\alpha = 0$ and $\beta = 1$.   ▲

The transformation that we performed in this example is called the **probability integral transformation**. The result is not only of theoretical importance, but it facilitates the **simulation** of observed values of continuous random variables. A reference to how this is done, especially in connection with the normal distribution, is given on page 290.

When the conditions underlying Theorem 7.1 are not met, we can be in serious difficulties, and we may have to use the method of Section 7.2 or a generalization of Theorem 7.1 referred to among the references on page 290; sometimes, there is an easy way out, as in the following example.

## EXAMPLE 7.9

If $X$ has the standard normal distribution, find the probability density of $Z = X^2$.

### Solution

Since the function given by $z = x^2$ is decreasing for negative values of $x$ and increasing for positive values of $x$, the conditions of Theorem 7.1 are not met. However, the transformation from $X$ to $Z$ can be made in two steps: First we find the probability density of $Y = |X|$, and then we find the probability density of $Z = Y^2 \; (= X^2)$.

So far as the first step is concerned, we already studied the transformation $Y = |X|$ in Example 7.2; in fact, we showed that if $X$ has the standard normal distribution, then $Y = |X|$ has the probability density

$$g(y) = 2n(y; 0, 1) = \frac{2}{\sqrt{2\pi}} e^{-\frac{1}{2}y^2}$$

for $y > 0$, and $g(y) = 0$ elsewhere. For the second step, the function given by $z = y^2$ is increasing for $y > 0$; that is, for all values of $Y$ for which $g(y) \neq 0$. Thus, we can use Theorem 7.1, and since

$$\frac{dy}{dz} = \frac{1}{2} z^{-\frac{1}{2}}$$

we get

$$h(z) = \frac{2}{\sqrt{2\pi}} e^{-\frac{1}{2}z} \left| \frac{1}{2} z^{-\frac{1}{2}} \right|$$

$$= \frac{1}{\sqrt{2\pi}} z^{-\frac{1}{2}} e^{-\frac{1}{2}z}$$

for $z > 0$, and $h(z) = 0$ elsewhere. Observe that since $\Gamma(\frac{1}{2}) = \sqrt{\pi}$, the distribution we have arrived at for $Z$ is a chi-square distribution (see Definition 6.4) with $\nu = 1$. ▲

## 7.4 TRANSFORMATION TECHNIQUE: SEVERAL VARIABLES

The method of the preceding section can also be used to find the distribution of a random variable which is a function of two or more random variables. Suppose, for instance, that we are given the joint distribution of two random variables $X_1$ and $X_2$ and that we want to determine the probability distributon or the probability density of the random variable $Y = u(X_1, X_2)$. If the relationship between $y$ and $x_1$ with $x_2$ held constant or the relationship between $y$ and $x_2$ with $x_1$ held constant permits, we can proceed in the discrete case as in Example 7.4 to find the joint distribution of $Y$ and $X_2$ or that of $X_1$ and $Y$ and then sum on the values of the other random variable to get the marginal distribution of $Y$. In the continuous case, we first use Theorem 7.1 with the transformation formula written as

$$g(y, x_2) = f(x_1, x_2) \cdot \left| \frac{\partial x_1}{\partial y} \right|$$

or as

$$g(x_1, y) = f(x_1, x_2) \cdot \left| \frac{\partial x_2}{\partial y} \right|$$

where $f(x_1, x_2)$ and the partial derivative must be expressed in terms of $y$ and $x_2$, or $x_1$ and $y$. Then we integrate out the other variable to get the marginal density of $Y$.

## EXAMPLE 7.10

If $X_1$ and $X_2$ are independent random variables having Poisson distributions with the parameters $\lambda_1$ and $\lambda_2$, find the probability distribution of the random variable $Y = X_1 + X_2$.

### Solution

Since $X_1$ and $X_2$ are independent, their joint distribution is given by

$$f(x_1, x_2) = \frac{e^{-\lambda_1}(\lambda_1)^{x_1}}{x_1!} \cdot \frac{e^{-\lambda_2}(\lambda_2)^{x_2}}{x_2!}$$

$$= \frac{e^{-(\lambda_1+\lambda_2)}(\lambda_1)^{x_1}(\lambda_2)^{x_2}}{x_1!x_2!}$$

for $x_1 = 0, 1, 2, \ldots$ , and $x_2 = 0, 1, 2, \ldots$ . Since $y = x_1 + x_2$ and, hence, $x_1 = y - x_2$, we can substitute $y - x_2$ for $x_1$, getting

$$g(y, x_2) = \frac{e^{-(\lambda_1+\lambda_2)}(\lambda_2)^{x_2}(\lambda_1)^{y-x_2}}{x_2!(y - x_2)!}$$

for $y = 0, 1, 2, \ldots$ , and $x_2 = 0, 1, \ldots, y$, for the joint distribution of $Y$ and $X_2$. Then, summing on $x_2$ from 0 to $y$, we get

$$h(y) = \sum_{x_2=0}^{y} \frac{e^{-(\lambda_1+\lambda_2)}(\lambda_2)^{x_2}(\lambda_1)^{y-x_2}}{x_2!(y - x_2)!}$$

$$= \frac{e^{-(\lambda_1+\lambda_2)}}{y!} \cdot \sum_{x_2=0}^{y} \frac{y!}{x_2!(y - x_2)!}(\lambda_2)^{x_2}(\lambda_1)^{y-x_2}$$

after factoring out $e^{-(\lambda_1+\lambda_2)}$ and multiplying and dividing by $y!$. Identifying the summation at which we arrived as the binomial expansion of $(\lambda_1 + \lambda_2)^y$, we finally get

$$h(y) = \frac{e^{-(\lambda_1 + \lambda_2)}(\lambda_1 + \lambda_2)^y}{y!} \qquad \text{for } y = 0, 1, 2, \ldots$$

and we have, thus, shown that the sum of two independent random variables having Poisson distributions with the parameters $\lambda_1$ and $\lambda_2$ has a Poisson distribution with the parameter $\lambda = \lambda_1 + \lambda_2$.  ▲

## EXAMPLE 7.11

If the joint probability density of $X_1$ and $X_2$ is given by

$$f(x_1, x_2) = \begin{cases} e^{-(x_1 + x_2)} & \text{for } x_1 > 0, x_2 > 0 \\ 0 & \text{elsewhere} \end{cases}$$

find the probability density of $Y = \dfrac{X_1}{X_1 + X_2}$.

### Solution

Since $y$ decreases when $x_2$ increases and $x_1$ is held constant, we can use Theorem 7.1 (as modified on pages 272 and 273) to find the joint density of $X_1$ and $Y$. Since $y = \dfrac{x_1}{x_1 + x_2}$ yields $x_2 = x_1 \cdot \dfrac{1 - y}{y}$ and, hence,

$$\frac{\partial x_2}{\partial y} = -\frac{x_1}{y^2}$$

it follows that

$$g(x_1, y) = e^{-x_1/y} \left| -\frac{x_1}{y^2} \right| = \frac{x_1}{y^2} \cdot e^{-x_1/y}$$

for $x_1 > 0$ and $0 < y < 1$. Finally, integrating out $x_1$ and changing the variable of integration to $u = x_1/y$, we get

$$h(y) = \int_0^\infty \frac{x_1}{y^2} \cdot e^{-x_1/y} \, dx_1$$

$$= \int_0^\infty u \cdot e^{-u} \, du$$

$$= \Gamma(2)$$

$$= 1$$

for $0 < y < 1$, and $h(y) = 0$ elsewhere. Thus, the random variable $Y$ has the uniform density with $\alpha = 0$ and $\beta = 1$. (Note that in Exercise 7.6 the reader was asked to show this by the distribution function technique.)     ▲

The preceding example could also have been worked by a general method where we begin with the joint distribution of two random variables $X_1$ and $X_2$ and determine the joint distribution of two new random variables $Y_1 = u_1(X_1, X_2)$ and $Y_2 = u_2(X_1, X_2)$. Then we can find the marginal distribution of $Y_1$ or $Y_2$ by summation or integration.

This method is used mainly in the continuous case, where we need the following theorem, which is a direct generalization of Theorem 7.1.

---

**THEOREM 7.2**  Let $f(x_1, x_2)$ be the value of the joint probability density of the continuous random variables $X_1$ and $X_2$ at $(x_1, x_2)$. If the functions given by $y_1 = u_1(x_1, x_2)$ and $y_2 = u_2(x_1, x_2)$ are partially differentiable with respect to both $x_1$ and $x_2$ and represent a one-to-one transformation for all values within the range of $X_1$ and $X_2$ for which $f(x_1, x_2) \neq 0$, then for these values of $x_1$ and $x_2$, the equations $y_1 = u_1(x_1, x_2)$ and $y_2 = u_2(x_1, x_2)$ can be uniquely solved for $x_1$ and $x_2$ to give $x_1 = w_1(y_1, y_2)$ and $x_2 = w_2(y_1, y_2)$, and for the corresponding values of $y_1$ and $y_2$ the joint probability density of $Y_1 = u_1(X_1, X_2)$ and $Y_2 = u_2(X_1, X_2)$ is given by

$$g(y_1, y_2) = f[w_1(y_1, y_2), w_2(y_1, y_2)] \cdot |J|$$

Here, $J$, called the **Jacobian** of the transformation, is the determinant

$$J = \begin{vmatrix} \dfrac{\partial x_1}{\partial y_1} & \dfrac{\partial x_1}{\partial y_2} \\ \dfrac{\partial x_2}{\partial y_1} & \dfrac{\partial x_2}{\partial y_2} \end{vmatrix}$$

Elsewhere, $g(y_1, y_2) = 0$.

---

We shall not prove this theorem, but information about Jacobians and their applications can be found in most textbooks on advanced calculus. There they are used mainly in connection with multiple integrals, say, when we want to change from rectangular coordinates to polar coordinates or from rectangular coordinates to spherical coordinates.

## EXAMPLE 7.12

With reference to the random variables $X_1$ and $X_2$ of Example 7.11, find

(a)   the joint density of $Y_1 = X_1 + X_2$ and $Y_2 = \dfrac{X_1}{X_1 + X_2}$;

(b)   the marginal density of $Y_2$.

### Solution

(a)   Solving $y_1 = x_1 + x_2$ and $y_2 = \dfrac{x_1}{x_1 + x_2}$ for $x_1$ and $x_2$, we get $x_1 = y_1 y_2$ and $x_2 = y_1(1 - y_2)$, and it follows that

$$J = \begin{vmatrix} y_2 & y_1 \\ 1 - y_2 & -y_1 \end{vmatrix} = -y_1$$

Since the transformation is one-to-one, mapping the region $x_1 > 0$ and $x_2 > 0$ in the $x_1 x_2$-plane into the region $y_1 > 0$ and $0 < y_2 < 1$ in the $y_1 y_2$-plane, we can use Theorem 7.2 and it follows that

$$g(y_1, y_2) = e^{-y_1} \cdot |-y_1| = y_1 e^{-y_1}$$

for $y_1 > 0$ and $0 < y_2 < 1$; elsewhere, $g(y_1, y_2) = 0$.

(b)   Using the joint density obtained in part (a) and integrating out $y_1$, we get

$$h(y_2) = \int_0^\infty g(y_1, y_2)\, dy_1$$

$$= \int_0^\infty y_1 e^{-y_1}\, dy_1$$

$$= \Gamma(2)$$

$$= 1$$

for $0 < y_2 < 1$; elsewhere, $h(y_2) = 0$. Note that this result agrees with that obtained on page 274.    ▲

## EXAMPLE 7.13

If the joint density of $X_1$ and $X_2$ is given by

$$f(x_1, x_2) = \begin{cases} 1 & \text{for } 0 < x_1 < 1, 0 < x_2 < 1 \\ 0 & \text{elsewhere} \end{cases}$$

find

(a)  the joint density of $Y = X_1 + X_2$ and $Z = X_2$;

(b)  the marginal density of $Y$.

Note that in Exercise 7.7 the reader was asked to work the same problem by the distribution function technique.

*Solution*

(a)  Solving $y = x_1 + x_2$ and $z = x_2$ for $x_1$ and $x_2$, we get $x_1 = y - z$ and $x_2 = z$, so that

$$J = \begin{vmatrix} 1 & -1 \\ 0 & 1 \end{vmatrix} = 1$$

Since the transformation is one-to-one, mapping the region $0 < x_1 < 1$ and $0 < x_2 < 1$ in the $x_1x_2$-plane into the region $z < y < z + 1$ and $0 < z < 1$ in the $yz$-plane (see Figure 7.7), we can use Theorem 7.2 and we get

$$g(y, z) = 1 \cdot |1| = 1$$

for $z < y < z + 1$ and $0 < z < 1$; elsewhere, $g(y, z) = 0$.

(b)  Integrating out $z$ separately for $y \leq 0$, $0 < y < 1$, $1 < y < 2$, and $y \geq 2$, we get

$$h(y) = \begin{cases} 0 & \text{for } y \leq 0 \\ \displaystyle\int_0^y 1 \, dz = y & \text{for } 0 < y < 1 \\ \displaystyle\int_{y-1}^1 1 \, dz = 2 - y & \text{for } 1 < y < 2 \\ 0 & \text{for } y \geq 2 \end{cases}$$

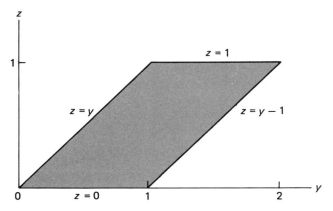

**Figure 7.7**  Transformed sample space for Example 7.13.

and to make the density function continuous, we let $h(1) = 1$. We have thus shown that the sum of the given random variables has the **triangular probability density**, whose graph is shown in Figure 7.8.    ▲

So far we have considered here only functions of two random variables, but the method based on Theorem 7.2 can easily be generalized to functions of three or more random variables. For instance, if we are given the joint probability density of three random variables $X_1$, $X_2$, and $X_3$ and we want to find the joint probability density of the random variables $Y_1 = u_1(X_1, X_2, X_3)$, $Y_2 = u_2(X_1, X_2, X_3)$, and $Y_3 = u_3(X_1, X_2, X_3)$, the general approach is the same, but the Jacobian is now the $3 \times 3$ determinant

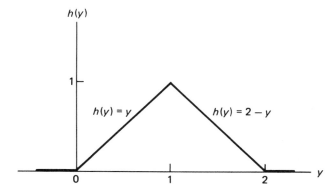

**Figure 7.8**  Triangular probability density.

$$J = \begin{vmatrix} \dfrac{\partial x_1}{\partial y_1} & \dfrac{\partial x_1}{\partial y_2} & \dfrac{\partial x_1}{\partial y_3} \\[2mm] \dfrac{\partial x_2}{\partial y_1} & \dfrac{\partial x_2}{\partial y_2} & \dfrac{\partial x_2}{\partial y_3} \\[2mm] \dfrac{\partial x_3}{\partial y_1} & \dfrac{\partial x_3}{\partial y_2} & \dfrac{\partial x_3}{\partial y_3} \end{vmatrix}$$

Once we have determined the joint probability density of the three new random variables, we can find the marginal density of any two of the random variables, or any one, by integration.

## EXAMPLE 7.14

If the joint probability density of $X_1$, $X_2$, and $X_3$ is given by

$$f(x_1, x_2, x_3) = \begin{cases} e^{-(x_1 + x_2 + x_3)} & \text{for } x_1 > 0, x_2 > 0, x_3 > 0 \\ 0 & \text{elsewhere} \end{cases}$$

find

(a)  the joint density of $Y_1 = X_1 + X_2 + X_3$, $Y_2 = X_2$, and $Y_3 = X_3$;
(b)  the marginal density of $Y_1$.

### Solution

(a)  Solving the system of equations $y_1 = x_1 + x_2 + x_3$, $y_2 = x_2$, and $y_3 = x_3$ for $x_1$, $x_2$, and $x_3$, we get $x_1 = y_1 - y_2 - y_3$, $x_2 = y_2$, and $x_3 = y_3$. It follows that

$$J = \begin{vmatrix} 1 & -1 & -1 \\ 0 & 1 & 0 \\ 0 & 0 & 1 \end{vmatrix} = 1$$

and, since the transformation is one-to-one, that

$$g(y_1, y_2, y_3) = e^{-y_1} \cdot |1|$$
$$= e^{-y_1}$$

for $y_2 > 0$, $y_3 > 0$, and $y_1 > y_2 + y_3$; elsewhere, $g(y_1, y_2, y_3) = 0$.

(b)  Integrating out $y_2$ and $y_3$, we get

$$h(y_1) = \int_0^{y_1} \int_0^{y_1 - y_3} e^{-y_1} dy_2 \, dy_3$$

$$= \frac{1}{2} y_1^2 \cdot e^{-y_1}$$

for $y_1 > 0$; $h(y_1) = 0$ elsewhere. Observe that we have shown that the sum of three independent random variables having the gamma distribution with $\alpha = 1$ and $\beta = 1$ is a random variable having the gamma distribution with $\alpha = 3$ and $\beta = 1$.   ▲

As the reader will find in Exercise 7.47, it would have been easier to obtain the result of part (b) of Example 7.14 by using the method based on Theorem 1 as modified on pages 272 and 273.

## EXERCISES

**7.14** If $X$ has a hypergeometric distribution with $k = 3$, $N = 6$, and $n = 2$, find the probability distribution of $Y$, the number of successes minus the number of failures.

**7.15** With reference to Exercise 7.14, find the probability distribution of the random variable $Z = (X - 1)^2$.

**7.16** If $X$ has a binomial distribution with $n = 3$ and $\theta = \frac{1}{3}$, find the probability distributions of

(a) $Y = \dfrac{X}{1 + X}$;

(b) $U = (X - 1)^4$.

**7.17** If $X$ has a geometric distribution with $\theta = \frac{1}{3}$, find the formula for the probability distribution of the random variable $Y = 4 - 5X$.

**7.18** If $X$ is the total we roll with a pair of dice, for which the probability distribution is given on page 86, find the probability distribution of the remainder we get when the values of $X$ are divided by 3.

**7.19** Use the transformation-of-variable technique to prove Theorem 6.7.

**7.20** Use the transformation technique to rework Example 7.1.

**7.21** If $X = \ln Y$ has a normal distribution with the mean $\mu$ and the standard deviation $\sigma$, find the probability density of $Y$, which is said to have the **log-normal distribution**.

**7.22** If $F(x)$ is the value of the distribution function of $X$ at $x$, then the **median** of $X$, denoted by $\tilde{\mu}$, is such that $F(\tilde{\mu}) = \frac{1}{2}$. With reference to the preceding exercise, show that $\tilde{\mu} = e^{\mu}$.

**7.23** With reference to Exercise 7.21, show that the log-normal distribution has a relative maximum at $e^{\mu - \sigma^2}$.

**7.24** If the probability density of $X$ is given by

$$f(x) = \begin{cases} \dfrac{x}{2} & \text{for } 0 < x < 2 \\ 0 & \text{elsewhere} \end{cases}$$

find the probability density of $Y = X^3$. Also plot the graphs of the probability densities of $X$ and $Y$ and indicate the respective areas under the curves which represent $P(\frac{1}{2} < X < 1)$ and $P(\frac{1}{8} < Y < 1)$.

**7.25** If the probability density of $X$ is given by

$$f(x) = \begin{cases} \dfrac{kx^3}{(1 + 2x)^6} & \text{for } x > 0 \\ 0 & \text{elsewhere} \end{cases}$$

where $k$ is an appropriate constant, find the probability density of the random variable $Y = \dfrac{2X}{1 + 2X}$. Identify the distribution of $Y$, and thus determine the value of $k$.

**7.26** If $X$ has a uniform density with $\alpha = 0$ and $\beta = 1$, show that the random variable $Y = -2 \cdot \ln X$ has a gamma distribution. What are its parameters?

**7.27** If $X$ has a uniform density with $\alpha = 0$ and $\beta = 1$, show that $Y = X^{-1/\alpha}$ with $\alpha > 0$ has the Pareto distribution of Exercise 6.21.

**7.28** Consider the random variable $X$ with the probability density

$$f(x) = \begin{cases} \dfrac{3x^2}{2} & \text{for } -1 < x < 1 \\ 0 & \text{elsewhere} \end{cases}$$

(a)  Use the result of Example 7.2 to find the probability density of $Y = |X|$.
(b)  Find the probability density of $Z = X^2 \; (= Y^2)$.

**7.29** Consider the random variable $X$ with the uniform density having $\alpha = 0$ and $\beta = 1$.

(a)  Use the result of Example 7.2 to find the probability density of $Y = |X|$.
(b)  Find the probability density of $Z = X^4 \; (= Y^4)$.

**7.30** If the joint probability distribution of $X_1$ and $X_2$ is given by

$$f(x_1, x_2) = \dfrac{x_1 x_2}{36}$$

for $x_1 = 1, 2, 3$, and $x_2 = 1, 2, 3$, find

(a) the probability distribution of $X_1 X_2$;
(b) the probability distribution of $X_1 / X_2$.

**7.31** With reference to Exercise 7.30, find

(a) the joint distribution of $Y_1 = X_1 + X_2$ and $Y_2 = X_1 - X_2$;
(b) the marginal distribution of $Y_1$.

**7.32** If the joint probability distribution of $X$ and $Y$ is given by

$$f(x, y) = \frac{(x - y)^2}{7}$$

for $x = 1, 2$, and $y = 1, 2, 3$, find

(a) the joint distribution of $U = X + Y$ and $V = X - Y$;
(b) the marginal distribution of $U$.

**7.33** If $X_1, X_2$, and $X_3$ have the multinomial distribution (see Definition 5.6) with $n = 2$, $\theta_1 = \frac{1}{4}$, $\theta_2 = \frac{1}{3}$, and $\theta_3 = \frac{5}{12}$, find the joint probability distribution of $Y_1 = X_1 + X_2$, $Y_2 = X_1 - X_2$, and $Y_3 = X_3$.

**7.34** With reference to Example 3.12, find

(a) the probability distribution of $U = X + Y$;
(b) the probability distribution of $V = XY$;
(c) the probability distribution of $W = X - Y$.

**7.35** If $X_1$ and $X_2$ are independent random variables having binomial distributions with the respective parameters $n_1$ and $\theta$ and $n_2$ and $\theta$, show that $Y = X_1 + X_2$ has the binomial distribution with the parameters $n_1 + n_2$ and $\theta$. (*Hint:* Use Theorem 1.12.)

**7.36** If $X_1$ and $X_2$ are independent random variables having the geometric distribution with the parameter $\theta$, show that $Y = X_1 + X_2$ is a random variable having the negative binomial distribution with the parameters $\theta$ and $k = 2$.

**7.37** If $X$ and $Y$ are independent random variables having the standard normal distribution, show that the random variable $Z = X + Y$ is also normally distributed. (*Hint:* Complete the square in the exponent.) What are the mean and the variance of this normal distribution?

**7.38** Consider two random variables $X$ and $Y$ with the joint probability density

$$f(x, y) = \begin{cases} 12xy(1 - y) & \text{for } 0 < x < 1, 0 < y < 1 \\ 0 & \text{elsewhere} \end{cases}$$

Find the probability density of $Z = XY^2$ by using Theorem 7.1 (as modified on pages 272 and 273) to determine the joint probability density of $Y$ and $Z$ and then integrating out $y$.

**7.39** Rework Exercise 7.38 by using Theorem 7.2 to determine the joint probability density of $Z = XY^2$ and $U = Y$ and then finding the marginal density of $Z$.

**7.40** Consider two independent random variables $X_1$ and $X_2$ having the same Cauchy distribution

$$f(x) = \frac{1}{\pi(1 + x^2)} \qquad \text{for } -\infty < x < \infty$$

Find the probability density of $Y_1 = X_1 + X_2$ by using Theorem 7.1 (as modified on pages 272 and 273) to determine the joint probability density of $X_1$ and $Y_1$ and then integrating out $x_1$. Also, identify the distribution of $Y_1$.

**7.41** Rework Exercise 7.40 by using Theorem 7.2 to determine the joint probability density of $Y_1 = X_1 + X_2$ and $Y_2 = X_1 - X_2$ and then finding the marginal density of $Y_1$.

**7.42** Consider two random variables $X$ and $Y$ whose joint probability density is given by

$$f(x, y) = \begin{cases} \frac{1}{2} & \text{for } x > 0, y > 0, x + y < 2 \\ 0 & \text{elsewhere} \end{cases}$$

Find the probability density of $U = Y - X$ by using Theorem 7.1 as modified on pages 272 and 273.

**7.43** Rework Exercise 7.42 by using Theorem 7.2 to determine the joint probability density of $U = Y - X$ and $V = X$ and then finding the marginal density of $U$.

**7.44** Let $X_1$ and $X_2$ be two continuous random variables having the joint probability density

$$f(x_1, x_2) = \begin{cases} 4x_1x_2 & \text{for } 0 < x_1 < 1, 0 < x_2 < 1 \\ 0 & \text{elsewhere} \end{cases}$$

Find the joint probability density of $Y_1 = X_1^2$ and $Y_2 = X_1X_2$.

**7.45** Let $X$ and $Y$ be two continuous random variables having the joint probability density

$$f(x, y) = \begin{cases} 24xy & \text{for } 0 < x < 1, 0 < y < 1, x + y < 1 \\ 0 & \text{elsewhere} \end{cases}$$

Find the joint probability density of $Z = X + Y$ and $W = X$.

**7.46** Let $X$ and $Y$ be two independent random variables having identical gamma distributions.

(a) Find the joint probability density of the random variables $U = \dfrac{X}{X+Y}$ and $V = X + Y$.

(b) Find and identify the marginal density of $U$.

**7.47** On pages 272 and 273 we indicated that the method of transformation based on Theorem 7.1 can be generalized so that it applies also to random variables that are functions of two or more random variables. So far we have used this method only for functions of two random variables, but when there are three, for example, we introduce the new random variable in place of one of the original random variables, and then we eliminate (by summation or integration) the other two random variables with which we began. Use this method to rework Example 7.14.

**7.48** In Example 7.13 we found the probability density of the sum of two independent random variables having the uniform density with $\alpha = 0$ and $\beta = 1$. Given a third random variable $X_3$, which has the same uniform density and is independent of both $X_1$ and $X_2$, show that if $U = Y + X_3 = X_1 + X_2 + X_3$, then

(a) the joint probability density of $U$ and $Y$ is given by

$$g(u, y) = \begin{cases} y & \text{for Regions I and II of Figure 7.9} \\ 2 - y & \text{for Regions III and IV of Figure 7.9} \\ 0 & \text{elsewhere} \end{cases}$$

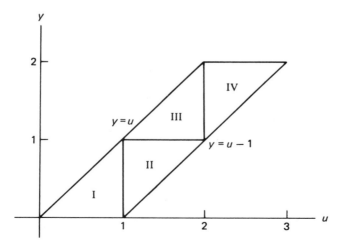

**Figure 7.9** Diagram for Exercise 7.48.

(b)   the probability density of $U$ is given by

$$
h(u) = \begin{cases}
0 & \text{for } u \leq 0 \\
\frac{1}{2} u^2 & \text{for } 0 < u < 1 \\
\frac{1}{2} u^2 - \frac{3}{2}(u-1)^2 & \text{for } 1 < u < 2 \\
\frac{1}{2} u^2 - \frac{3}{2}(u-1)^2 + \frac{3}{2}(u-2)^2 & \text{for } 2 < u < 3 \\
0 & \text{for } u \geq 3
\end{cases}
$$

Note that if we let $h(1) = h(2) = \frac{1}{2}$, this will make the probability density of $U$ continuous.

### APPLICATIONS

**7.49** According to the Maxwell-Boltzmann law of theoretical physics, the probability density of $V$, the velocity of a gas molecule, is

$$
f(v) = \begin{cases}
kv^2 e^{-\beta v^2} & \text{for } v > 0 \\
0 & \text{elsewhere}
\end{cases}
$$

where $\beta$ depends on its mass and the absolute temperature and $k$ is an appropriate constant. Show that the kinetic energy $E = \frac{1}{2} mV^2$ is a random variable having a gamma distribution.

**7.50** With reference to Exercise 3.82, find the probability density of the distance between the point of impact and the center of the target.

**7.51** With reference to Exercise 3.83, find the probability density of the random variable $Z = \dfrac{X + Y}{2}$, which is the average of the two proportions of correct answers a student will get on the two aptitude tests.

**7.52** With reference to Exercise 3.84, use Theorem 7.2 to find the joint probability density of the two random variables $V = SP$ and $W = P$, and then find the marginal density of $V$.

## 7.5 MOMENT-GENERATING FUNCTION TECHNIQUE

Moment-generating functions can play an important role in determining the probability distribution or density of a function of random variables when the function is a linear combination of $n$ *independent* random variables. We shall illustrate this technique here when such a linear combination is, in fact, the sum of $n$ independent random variables, leaving it to the reader to generalize it in Exercises 7.57 and 7.58.

The method is based on the theorem that the moment-generating function of the sum of $n$ independent random variables equals the product of their moment-generating functions, namely,

---

**THEOREM 7.3**  If $X_1$, $X_2$, . . . , and $X_n$ are independent random variables and $Y = X_1 + X_2 + \cdots + X_n$, then

$$M_Y(t) = \prod_{i=1}^{n} M_{X_i}(t)$$

where $M_{X_i}(t)$ is the value of the moment-generating function of $X_i$ at $t$.

---

***Proof.***  Making use of the fact that the random variables are independent and, hence,

$$f(x_1, x_2, \ldots, x_n) = f_1(x_1) \cdot f_2(x_2) \cdot \ldots \cdot f_n(x_n)$$

according to Definition 3.14, we can write

$$
\begin{aligned}
M_Y(t) &= E(e^{Yt}) \\
&= E[e^{(X_1 + X_2 + \cdots + X_n)t}] \\
&= \int_{-\infty}^{\infty} \cdots \int_{-\infty}^{\infty} e^{(x_1 + x_2 + \cdots + x_n)t} f(x_1, x_2, \ldots, x_n) \, dx_1 \, dx_2 \cdots dx_n \\
&= \int_{-\infty}^{\infty} e^{x_1 t} f_1(x_1) \, dx_1 \cdot \int_{-\infty}^{\infty} e^{x_2 t} f_2(x_2) \, dx_2 \cdots \int_{-\infty}^{\infty} e^{x_n t} f_n(x_n) \, dx_n \\
&= \prod_{i=1}^{n} M_{X_i}(t)
\end{aligned}
$$

which proves the theorem for the continuous case. To prove it for the discrete case, we have only to replace all of the integrals by sums.    ▼

Note that if we want to use Theorem 7.3 to find the probability distribution or the probability density of the random variable $Y = X_1 + X_2 + \cdots + X_n$, we must be able to identify whatever probability distribution or density corresponds to $M_Y(t)$. Then, we must rely on the first of the two theorems we gave on page 245, namely, the uniqueness theorem about the correspondence between moment-generating functions and probability distributions or densities.

## EXAMPLE 7.15

Find the probability distribution of the sum of $n$ independent random variables $X_1, X_2, \ldots, X_n$ having Poisson distributions with the respective parameters $\lambda_1, \lambda_2, \ldots, \lambda_n$.

### Solution

By Theorem 5.9 we have

$$M_{X_i}(t) = e^{\lambda_i(e^t - 1)}$$

and, hence, for $Y = X_1 + X_2 + \cdots + X_n$ we obtain

$$M_Y(t) = \prod_{i=1}^{n} e^{\lambda_i(e^t - 1)} = e^{(\lambda_1 + \lambda_2 + \cdots + \lambda_n)(e^t - 1)}$$

which can readily be identified as the moment-generating function of the Poisson distribution with the parameter $\lambda = \lambda_1 + \lambda_2 + \cdots + \lambda_n$. Thus, the distribution of the sum of $n$ independent random variables having Poisson distributions with the parameters $\lambda_i$ is a Poisson distribution with the parameter $\lambda = \lambda_1 + \lambda_2 + \cdots + \lambda_n$. Note that in Example 7.10 we proved this for $n = 2$.    ▲

## EXAMPLE 7.16

If $X_1, X_2, \ldots, X_n$ are independent random variables having exponential distributions with the same parameter $\theta$, find the probability density of the random variable $Y = X_1 + X_2 + \cdots + X_n$.

### Solution

Since the exponential distribution is a gamma distribution with $\alpha = 1$ and $\beta = \theta$, we have

$$M_{X_i}(t) = (1 - \theta t)^{-1}$$

by Theorem 6.4, and hence

$$M_Y(t) = \prod_{i=1}^{n} (1 - \theta t)^{-1} = (1 - \theta t)^{-n}$$

according to the second of the special rules for products in the appendix at the end of the book. Identifying the moment-generating function of $Y$ as that of a gamma distribution with $\alpha = n$ and $\beta = \theta$, we conclude that the distribution of the sum of $n$ independent random variables having exponential distributions with the same parameter $\theta$ is a gamma distribution with the parameters $\alpha = n$ and $\beta = \theta$. Note that this agrees with the result of Example 7.14, where we showed that the sum of three independent random variables having exponential distributions with the parameter $\theta = 1$ has a gamma distribution with $\alpha = 3$ and $\beta = 1$.   ▲

Theorem 7.3 also provides an easy and elegant way of deriving the moment-generating function of the binomial distribution. Suppose that $X_1, X_2, \ldots, X_n$ are independent random variables having the same Bernoulli distribution $f(x; \theta) = \theta^x (1 - \theta)^{1-x}$ for $x = 0, 1$. By Definition 4.6, we thus have

$$M_{X_i}(t) = e^{0 \cdot t}(1 - \theta) + e^{1 \cdot t}\theta = 1 + \theta(e^t - 1)$$

so that Theorem 7.3 yields

$$M_Y(t) = \prod_{i=1}^{n} [1 + \theta(e^t - 1)] = [1 + \theta(e^t - 1)]^n$$

This moment-generating function is readily identified as that of the binomial distribution with the parameters $n$ and $\theta$. Of course, $Y = X_1 + X_2 + \cdots + X_n$ is the total number of successes in $n$ trials, since $X_1$ is the number of successes on the first trial, $X_2$ is the number of successes on the second trial, $\ldots$, and $X_n$ is the number of successes on the $n$th trial. As we shall see later, this is a fruitful way of looking at the binomial distribution.

## EXERCISES

**7.53** Use the moment-generating function technique to rework Exercise 7.35.

**7.54** Find the moment-generating function of the negative binomial distribution by making use of the fact that if $k$ independent random variables have geometric distributions with the same parameter $\theta$, their sum is a random variable having the negative binomial distribution with the parameters $\theta$ and $k$. (*Hint*: Use the result of Exercise 5.31.)

**7.55** If $n$ independent random variables have the same gamma distribution with the parameters $\alpha$ and $\beta$, find the moment-generating function of their sum and, if possible, identify its distribution.

**7.56** If $n$ independent random variables $X_i$ have normal distributions with the means $\mu_i$ and the standard deviations $\sigma_i$, find the moment-generating function of their sum and identify the corresponding distribution, its mean, and its variance.

**7.57** Prove the following generalization of Theorem 7.3: If $X_1, X_2, \ldots$, and $X_n$ are independent random variables and $Y = a_1X_1 + a_2X_2 + \cdots + a_nX_n$, then

$$M_Y(t) = \prod_{i=1}^{n} M_{X_i}(a_i t)$$

where $M_{X_i}(t)$ is the value of the moment-generating function of $X_i$ at $t$.

**7.58** Use the result of the preceding exercise to show that if $n$ independent random variables $X_i$ have normal distributions with the means $\mu_i$ and the standard deviations $\sigma_i$, then $Y = a_1X_1 + a_2X_2 + \cdots + a_nX_n$ has a normal distribution. What are the mean and the variance of this distribution?

## APPLICATIONS

**7.59** A lawyer has an unlisted number on which she receives on the average 2.1 calls every half hour and a listed number on which she receives on the average 10.9 calls every half hour. If it can be assumed that the number of calls she receives on these phones are independent random variables having Poisson distributions, what are the probabilities that in half an hour she will receive altogether

(a)   14 calls;
(b)   at most six calls?

**7.60** In a newspaper ad, a car dealer lists a 1986 Chrysler Le Baron, a 1985 Ford Escort, and a 1987 Buick Skylark. If the numbers of inquiries he will get about these cars may be regarded as independent random variables having Poisson distributions with the parameters $\lambda_1 = 3.6$, $\lambda_2 = 5.8$, and $\lambda_3 = 4.6$, what are the probabilities that altogether he will receive

(a)   fewer than 10 inquiries about these cars;
(b)   anywhere from 15 to 20 inquiries about these cars;
(c)   at least 18 inquiries about these cars?

**7.61** With reference to Exercise 7.60, what is the probability that the car dealer will receive six inquiries about the 1985 Ford Escort and eight inquiries about the other two cars?

**7.62** If the number of complaints a dry-cleaning establishment receives per day is a random variable having the Poisson distribution with $\lambda = 3.3$, what are the probabilities that it will receive

(a)   two complaints on any given day;
(b)   five complaints altogether on any two given days;
(c)   at least 12 complaints altogether on any three given days?

**7.63** The number of fish a person catches per hour at Woods Canyon Lake is a random variable having the Poisson distribution with $\lambda = 1.6$. What are the probabilities that a person fishing there will catch

    (a)   four fish in 2 hours;

    (b)   at least two fish in 3 hours;

    (c)   at most three fish in 4 hours?

**7.64** If the number of minutes it takes a service station attendant to balance a tire is a random variable having an exponential distribution with the parameter $\theta = 5$, what are the probabilities that the attendant will take

    (a)   less than 8 minutes to balance two tires;

    (b)   at least 12 minutes to balance three tires?

**7.65** If the number of minutes a doctor spends with a patient is a random variable having an exponential distribution with the parameter $\theta = 9$, what are the probabilities that it will take the doctor at least 20 minutes to treat

    (a)   one patient;

    (b)   two patients;

    (c)   three patients?

## REFERENCES

The use of the probability integral transformation in problems of simulation is discussed in

FREUND, J. E., MILLER, I., and JOHNSON, R. A., *Probability and Statistics for Engineers*, 4th ed. Englewood Cliffs, N.J.: Prentice Hall, Inc., 1990.

A generalization of Theorem 7.1, which applies when the interval within the range of $X$ for which $f(x) \neq 0$ can be partitioned into $k$ subintervals so that the conditions of Theorem 7.1 apply separately for each of the subintervals, may be found in

WALPOLE, R. E., and MYERS, R. H., *Probability and Statistics for Engineers and Scientists*, 4th ed. New York: Macmillan Publishing Company, Inc., 1989.

More detailed and more advanced treatments of the material in this chapter are given in many advanced texts on mathematical statistics; for instance, in

HOGG, R. V., and CRAIG, A. T., *Introduction to Mathematical Statistics*, 4th ed. New York: Macmillan Publishing Company, Inc., 1978,

ROUSSAS, G. G., *A First Course in Mathematical Statistics*. Reading, Mass.: Addison-Wesley Publishing Company, Inc., 1973,

WILKS, S. S., *Mathematical Statistics*. New York: John Wiley & Sons, Inc., 1962.

# 8

# *Sampling Distributions*

## 8.1  INTRODUCTION

Statistics concerns itself mainly with conclusions and predictions resulting from chance outcomes that occur in carefully planned experiments or investigations. In the finite case, these chance outcomes constitute a subset, or **sample**, of measurements or observations from a larger set of values called the **population**. In the continuous case they are usually values of identically distributed random variables, whose distribution we refer to as the **population distribution**, or the **infinite population** sampled. The word "infinite" implies that there is, logically speaking, no limit to the number of values we could observe.

All these terms are used here somewhat unconventionally. If a scientist must choose and then weigh five of 40 guinea pigs as part of an experiment, a layman might say that the ones she selects constitute the sample. This is how the term "sample" is used in everyday language. In statistics, it is preferable to look upon the weights of the five guinea pigs as a sample from the population which consists of the weights of all 40 guinea pigs. In this way, the population as well as the sample consists of numbers. Also, suppose that, to estimate the average useful life of a certain kind of transistor, an engineer selects ten of these transistors, tests them over a period of time, and records for each one the time to failure. If these times to failure are values of random variables having an exponential distribution with the parameter $\theta$, we say that they constitute a sample from this exponential population.

As can well be imagined, not all samples lend themselves to valid generalizations about the populations from which they came. In fact, most of the methods of inference discussed in this book are based on the assumption that we are dealing with **random samples**. In practice, we often deal with random samples from populations that are finite, but large enough to be treated as if they were infinite. Thus, most statistical theory and most of the methods we shall discuss apply to samples from infinite populations, and we shall begin here with a definition of random samples from infinite populations. Random samples from finite populations will be treated later in Section 8.3.

---

**DEFINITION 8.1** If $X_1, X_2, \ldots, X_n$ are independent and identically distributed random variables, we say that they constitute a **random sample** from the infinite population given by their common distribution.

---

If $f(x_1, x_2, \ldots, x_n)$ is the value of the joint distribution of such a set of random variables at $(x_1, x_2, \ldots, x_n)$, we can write

$$f(x_1, x_2, \ldots, x_n) = \prod_{i=1}^{n} f(x_i)$$

where $f(x_i)$ is the value of the population distribution at $x_i$. Observe that Definition 8.1 and the subsequent discussion apply also to sampling with replacement from finite populations; sampling without replacement from finite populations is discussed on pages 298 and 299.

Statistical inferences are usually based on **statistics**, that is, on random variables that are functions of a set of random variables $X_1, X_2 \ldots, X_n$, constituting a random sample. Typical of what we mean by "statistic" are the **sample mean** and the **sample variance**.

**DEFINITION 8.2**  If $X_1, X_2, \ldots, X_n$ constitute a random sample, then

$$\overline{X} = \frac{\sum_{i=1}^{n} X_i}{n}$$

is called the **sample mean** and

$$S^2 = \frac{\sum_{i=1}^{n} (X_i - \overline{X})^2}{n - 1}$$

is called the **sample variance**.[†]

As they are given here, these definitions apply only to random samples, but the sample mean and the sample variance can, similarly, be defined for any set of random variables $X_1, X_2, \ldots, X_n$.

It is common practice also to apply the terms "random sample," "statistic," "sample mean," and "sample variance" to the values of the random variables instead of the random variables themselves. Intuitively, this makes more sense and it conforms with colloquial usage. Thus, we might calculate

$$\bar{x} = \frac{\sum_{i=1}^{n} x_i}{n} \quad \text{and} \quad s^2 = \frac{\sum_{i=1}^{n} (x_i - \bar{x})^2}{n - 1}$$

for observed sample data and refer to these statistics as the sample mean and the sample variance. Here, the $x_i$, $\bar{x}$, and $s^2$ are values of the corresponding random variables $X_i$, $\overline{X}$, and $S^2$. Indeed, the formulas for $\bar{x}$ and $s^2$ are used even when we deal with any kind of data, not necessarily sample data, in which case we refer to $\bar{x}$ and $s^2$ simply as the mean and the variance.

It should be understood that we have introduced $\overline{X}$ and $S^2$ here merely as examples of statistics and that there are many other statistics that will be introduced later in this chapter and in subsequent chapters.

---

[†]The reason for dividing by $n - 1$ rather than the seemingly more logical choice, $n$, will be explained in Section 10.3.

## 8.2 THE DISTRIBUTION OF THE MEAN

Since statistics are random variables, their values will vary from sample to sample, and it is customary to refer to their distributions as **sampling distributions**. Most of the remainder of this chapter will be devoted to the sampling distributions of statistics which play important roles in applications.

First let us study some theory about the **sampling distribution of the mean**, making only some very general assumptions about the nature of the populations sampled.

---

**THEOREM 8.1** If $X_1, X_2, \ldots, X_n$ constitute a random sample from an infinite population with the mean $\mu$ and the variance $\sigma^2$, then

$$E(\overline{X}) = \mu \quad \text{and} \quad \mathrm{var}(\overline{X}) = \frac{\sigma^2}{n}$$

---

***Proof.*** Letting $Y = \overline{X}$ in Theorem 4.14 and, hence, setting $a_i = \frac{1}{n}$, we get

$$E(\overline{X}) = \sum_{i=1}^{n} \frac{1}{n} \cdot \mu = n \left( \frac{1}{n} \cdot \mu \right) = \mu$$

since $E(X_i) = \mu$. Then by the corollary of Theorem 4.14 we conclude that

$$\mathrm{var}(\overline{X}) = \sum_{i=1}^{n} \frac{1}{n^2} \cdot \sigma^2 = n \left( \frac{1}{n^2} \cdot \sigma^2 \right) = \frac{\sigma^2}{n} \quad \blacktriangledown$$

It is customary to write $E(\overline{X})$ as $\mu_{\overline{X}}$, $\mathrm{var}(\overline{X})$ as $\sigma_{\overline{X}}^2$, and refer to $\sigma_{\overline{X}}$ as the **standard error of the mean**. The formula for the standard error of the mean, $\sigma_{\overline{X}} = \dfrac{\sigma}{\sqrt{n}}$, shows that the standard deviation of the distribution of $\overline{X}$ decreases when $n$, the **sample size**, is increased. This means that when $n$ becomes larger and we actually have more information (the values of more random variables), we can expect values of $\overline{X}$ to be closer to $\mu$, the quantity they are intended to estimate. If we refer to Chebyshev's theorem as it is formulated in Exercise 4.40, we can express this formally in the following way.

**THEOREM 8.2**  For any positive constant $c$, the probability that $\overline{X}$ will take on a value between $\mu - c$ and $\mu + c$ is at least

$$1 - \frac{\sigma^2}{nc^2}$$

When $n \to \infty$, this probability approaches 1.

This result, called a **law of large numbers**, is primarily of theoretical interest. Of much more practical value is the **central limit theorem**, one of the most important theorems of statistics, which concerns the limiting distribution of the **standardized mean** of $n$ random variables when $n \to \infty$. We shall prove this theorem here only for the case where the $n$ random variables are a random sample from a population whose moment-generating function exists. More general conditions under which the theorem holds are given in Exercises 8.7 and 8.9, and the most general conditions under which it holds are referred to at the end of this chapter.

**THEOREM 8.3**  (Central limit theorem) If $X_1, X_2, \ldots, X_n$ constitute a random sample from an infinite population with the mean $\mu$, the variance $\sigma^2$, and the moment-generating function $M_X(t)$, then the limiting distribution of

$$Z = \frac{\overline{X} - \mu}{\sigma/\sqrt{n}}$$

as $n \to \infty$ is the standard normal distribution.

*Proof.*  First using the third part of Theorem 4.10 and then the second, we get

$$M_Z(t) = M_{\underset{\sigma/\sqrt{n}}{\overline{X} - \mu}}(t) = e^{-\sqrt{n}\mu t/\sigma} \cdot M_{\overline{X}}\left(\frac{\sqrt{n}t}{\sigma}\right)$$

$$= e^{-\sqrt{n}\mu t/\sigma} \cdot M_{n\overline{X}}\left(\frac{t}{\sigma\sqrt{n}}\right)$$

Since $n\overline{X} = X_1 + X_2 + \cdots + X_n$, it follows from Theorem 7.3 that

$$M_Z(t) = e^{-\sqrt{n}\mu t/\sigma} \cdot \left[ M_X\!\left( \frac{t}{\sigma\sqrt{n}} \right) \right]^n$$

and hence that

$$\ln M_Z(t) = -\frac{\sqrt{n}\mu t}{\sigma} + n \cdot \ln M_X\!\left( \frac{t}{\sigma\sqrt{n}} \right)$$

Expanding $M_X\!\left( \dfrac{t}{\sigma\sqrt{n}} \right)$ as a power series in $t$, we obtain

$$\ln M_Z(t) = -\frac{\sqrt{n}\mu t}{\sigma} + n \cdot \ln\left[ 1 + \mu_1' \frac{t}{\sigma\sqrt{n}} + \mu_2' \frac{t^2}{2\sigma^2 n} + \mu_3' \frac{t^3}{6\sigma^3 n\sqrt{n}} + \cdots \right]$$

where $\mu_1'$, $\mu_2'$, and $\mu_3'$ are the moments about the origin of the population distribution, namely, those of the original random variables $X_i$.

If $n$ is sufficiently large, we can use the expansion of $\ln(1 + x)$ as a power series in $x$ (as on page 243), getting

$$\begin{aligned}
\ln M_Z(t) = -\frac{\sqrt{n}\mu t}{\sigma} + n\Bigg\{ & \left[ \mu_1' \frac{t}{\sigma\sqrt{n}} + \mu_2' \frac{t^2}{2\sigma^2 n} + \mu_3' \frac{t^3}{6\sigma^3 n\sqrt{n}} + \cdots \right] \\
& -\frac{1}{2}\left[ \mu_1' \frac{t}{\sigma\sqrt{n}} + \mu_2' \frac{t^2}{2\sigma^2 n} + \mu_3' \frac{t^3}{6\sigma^3 n\sqrt{n}} + \cdots \right]^2 \\
& +\frac{1}{3}\left[ \mu_1' \frac{t}{\sigma\sqrt{n}} + \mu_2' \frac{t^2}{2\sigma^2 n} + \mu_3' \frac{t^3}{6\sigma^3 n\sqrt{n}} + \cdots \right]^3 - \cdots \Bigg\}
\end{aligned}$$

Then, collecting powers of $t$, we obtain

$$\begin{aligned}
\ln M_Z(t) = & \left( -\frac{\sqrt{n}\mu}{\sigma} + \frac{\sqrt{n}\mu_1'}{\sigma} \right)t + \left( \frac{\mu_2'}{2\sigma^2} - \frac{\mu_1'^2}{2\sigma^2} \right)t^2 \\
& + \left( \frac{\mu_3'}{6\sigma^3\sqrt{n}} - \frac{\mu_1' \cdot \mu_2'}{2\sigma^3\sqrt{n}} + \frac{\mu_1'^3}{3\sigma^3\sqrt{n}} \right)t^3 + \cdots
\end{aligned}$$

and since $\mu_1' = \mu$ and $\mu_2' - \mu_1' = \sigma^2$, this reduces to

$$\ln M_Z(t) = \frac{1}{2} t^2 + \left( \frac{\mu_3'}{6} - \frac{\mu_1'\mu_2'}{2} + \frac{\mu_1'^3}{3} \right) \frac{t^3}{\sigma^3\sqrt{n}} + \cdots$$

Finally, observing that the coefficient of $t^3$ is a constant times $\dfrac{1}{\sqrt{n}}$ and in general for $r \geq 2$ the coefficient of $t^r$ is a constant times $\dfrac{1}{\sqrt{n^{r-2}}}$, we get

$$\lim_{n \to \infty} \ln M_Z(t) = \tfrac{1}{2}t^2$$

and hence

$$\lim_{n \to \infty} M_Z(t) = e^{\frac{1}{2}t^2}$$

since the limit of a logarithm equals the logarithm of the limit (provided these limits exist). Identifying the limiting moment-generating function at which we have arrived as that of the standard normal distribution, we need only the two theorems stated on page 245 to complete the proof of Theorem 8.3.    ▼

Sometimes, the central limit theorem is interpreted incorrectly as implying that the distribution of $\overline{X}$ approaches a normal distribution when $n \to \infty$. This is incorrect because $\text{var}(\overline{X}) \to 0$ when $n \to \infty$; on the other hand, the central limit theorem does justify approximating the distribution of $\overline{X}$ with a normal distribution having the mean $\mu$ and the variance $\dfrac{\sigma^2}{n}$ when $n$ is large. In practice, this approximation is used when $n \geq 30$ regardless of the actual shape of the population sampled. For smaller values of $n$ the approximation is questionable, but see Theorem 8.4.

## EXAMPLE 8.1

A soft-drink vending machine is set so that the amount of drink dispensed is a random variable with a mean of 200 milliliters and a standard deviation of 15 milliliters. What is the probability that the average (mean) amount dispensed in a random sample of size 36 is at least 204 milliliters?

### Solution

According to Theorem 8.1, the distribution of $\overline{X}$ has the mean $\mu_{\overline{x}} = 200$ and the standard deviation $\sigma_{\overline{x}} = \dfrac{15}{\sqrt{36}} = 2.5$, and according to the central limit theorem, this distribution is approximately normal. Since

$z = \dfrac{204 - 200}{2.5} = 1.6$, it follows from Table III that $P(\overline{X} \geq 204) = P(Z \geq 1.6) = 0.5000 - 0.4452 = 0.0548$.   ▲

It is of interest to note that when the population we are sampling is normal, the distribution of $\overline{X}$ is a normal distribution regardless of the size of $n$.

---

**THEOREM 8.4**   If $\overline{X}$ is the mean of a random sample of size $n$ from a normal population with the mean $\mu$ and the variance $\sigma^2$, its sampling distribution is a normal distribution with the mean $\mu$ and the variance $\sigma^2/n$.

---

***Proof.***   According to Theorems 4.10 and 7.3 we can write

$$M_{\overline{X}}(t) = \left[ M_X\left(\frac{t}{n}\right) \right]^n$$

and since the moment-generating function of a normal distribution with the mean $\mu$ and the variance $\sigma^2$ is given by

$$M_X(t) = e^{\mu t + \frac{1}{2}\sigma^2 t^2}$$

according to Theorem 6.6, we get

$$M_{\overline{X}}(t) = \left[ e^{\mu \cdot \frac{t}{n} + \frac{1}{2}\left(\frac{t}{n}\right)^2 \sigma^2} \right]^n$$

$$= e^{\mu t + \frac{1}{2}t^2\left(\frac{\sigma^2}{n}\right)}$$

This moment-generating function is readily seen to be that of a normal distribution with the mean $\mu$ and the variance $\sigma^2/n$, and to complete the proof of Theorem 8.4 we have only to refer to the two theorems on page 245.   ▲

## 8.3 THE DISTRIBUTION OF THE MEAN: FINITE POPULATIONS

If an experiment consists of selecting one or more values from a finite set of numbers $\{c_1, c_2, \ldots, c_N\}$, this set is referred to as a **finite population of size $N$**. In the definition which follows, it will be assumed that we are sampling without replacement from a finite population of size $N$.

**DEFINITION 8.3** If $X_1$ is the first value drawn from a finite population of size $N$, $X_2$ is the second value drawn, ... , $X_n$ is the $n$th value drawn, and the joint probability distribution of these $n$ random variables is given by

$$f(x_1, x_2, \ldots, x_n) = \frac{1}{N(N-1)\cdot \ldots \cdot (N-n+1)}$$

for each ordered $n$-tuple of values of these random variables, then $X_1$, $X_2$, and $X_n$ are said to constitute a **random sample** from the given finite population.

As in Definition 8.1, the random sample is a set of random variables, but here again it is common practice also to apply the term "random sample" to the values of the random variables, namely, to the actual numbers drawn.

From the joint probability distribution of Definition 8.3, it follows that the probability for each subset of $n$ of the $N$ elements of the finite population (regardless of the order in which the values are drawn) is

$$\frac{n!}{N(N-1)\cdot \ldots \cdot (N-n+1)} = \frac{1}{\binom{N}{n}}$$

This is often given as an alternative definition or as a criterion for the selection of a random sample of size $n$ from a finite population of size $N$: *Each of the* $\binom{N}{n}$ *possible samples must have the same probability.*

It also follows from the joint probability distribution of Definition 8.3 that the marginal distribution of $X_r$ is given by

$$f(x_r) = \frac{1}{N} \qquad \text{for } x_r = c_1, c_2, \ldots, c_N$$

for $r = 1, 2, \ldots, n$, and we refer to the mean and the variance of this discrete uniform distribution as the mean and the variance of the finite population. Therefore,

**DEFINITION 8.4** The **mean** and the **variance** of the finite population $\{c_1, c_2, \ldots, c_N\}$ are

$$\mu = \sum_{i=1}^{N} c_i \cdot \frac{1}{N} \qquad \text{and} \qquad \sigma^2 = \sum_{i=1}^{N} (c_i - \mu)^2 \cdot \frac{1}{N}$$

Finally, it follows from the joint probability distribution of Definition 8.3 that the joint marginal distribution of any two of the random variables $X_1, X_2, \ldots, X_n$ is given by

$$g(x_r, x_s) = \frac{1}{N(N - 1)}$$

for each ordered pair of elements of the finite population. Thus, we can show that

---

**THEOREM 8.5**   If $X_r$ and $X_s$ are the $r$th and $s$th random variables of a random sample of size $n$ drawn from the finite population $\{c_1, c_2, \ldots, c_N\}$, then

$$\mathrm{cov}(X_r, X_s) = -\frac{\sigma^2}{N - 1}$$

---

***Proof.***   According to Definition 4.9,

$$\mathrm{cov}(X_r, X_s) = \sum_{\substack{i=1 \\ }}^{N} \sum_{\substack{j=1 \\ i \neq j}}^{N} \frac{1}{N(N - 1)} (c_i - \mu)(c_j - \mu)$$

$$= \frac{1}{N(N - 1)} \cdot \sum_{i=1}^{N} (c_i - \mu) \left[ \sum_{\substack{j=1 \\ j \neq i}}^{N} (c_j - \mu) \right]$$

and since $\displaystyle\sum_{\substack{j=1 \\ j \neq i}}^{N} (c_j - \mu) = \sum_{j=1}^{N} (c_j - \mu) - (c_i - \mu) = -(c_i - \mu)$, we get

$$\mathrm{cov}(X_r, X_s) = -\frac{1}{N(N - 1)} \cdot \sum_{i=1}^{N} (c_i - \mu)^2$$

$$= -\frac{1}{N - 1} \cdot \sigma^2 \qquad \blacktriangledown$$

Making use of all these results, let us now prove the following theorem, which, for random samples from finite populations, corresponds to Theorem 8.1.

---

**THEOREM 8.6** If $\overline{X}$ is the mean of a random sample of size $n$ from a finite population of size $N$ with the mean $\mu$ and the variance $\sigma^2$, then

$$E(\overline{X}) = \mu \quad \text{and} \quad \text{var}(\overline{X}) = \frac{\sigma^2}{n} \cdot \frac{N-n}{N-1}$$

---

***Proof.*** Substituting $a_i = \dfrac{1}{N}$, $\text{var}(X_i) = \sigma^2$, and $\text{cov}(X_i, X_j) = -\dfrac{\sigma^2}{N-1}$ into the formula of Theorem 4.14, we get

$$E(\overline{X}) = \sum_{i=1}^{n} \frac{1}{n} \cdot \mu = \mu$$

and

$$\begin{aligned}
\text{var}(\overline{X}) &= \sum_{i=1}^{n} \frac{1}{n^2} \cdot \sigma^2 + 2 \cdot \sum\sum_{i<j} \frac{1}{n^2}\left(-\frac{\sigma^2}{N-1}\right) \\
&= \frac{\sigma^2}{n} + 2 \cdot \frac{n(n-1)}{2} \cdot \frac{1}{n^2}\left(-\frac{\sigma^2}{N-1}\right) \\
&= \frac{\sigma^2}{n} \cdot \frac{N-n}{N-1} \quad \blacktriangledown
\end{aligned}$$

It is of interest to note that the formulas we obtained for var($\overline{X}$) in Theorems 8.1 and 8.6 differ only by the **finite population correction factor** $\dfrac{N-n}{N-1}$.[†] Indeed, when $N$ is large compared to $n$, the difference between the two formulas for var($\overline{X}$) is usually negligible, and the formula $\sigma_{\overline{x}} = \dfrac{\sigma}{\sqrt{n}}$ is often used as an approximation when we are sampling from a large finite population. A general

---

[†]Since there are many problems in which we are interested in the standard deviation rather than the variance, the term "finite population correction factor" often refers to $\sqrt{\dfrac{N-n}{N-1}}$ instead of $\dfrac{N-n}{N-1}$. This does not matter, of course, so long as the usage is clearly understood.

rule of thumb is to use this approximation when the sample does not constitute more than 5 percent of the population.

## EXERCISES

**8.1** Use the corollary of Theorem 4.15 to show that if $X_1, X_2, \ldots, X_n$ constitute a random sample from an infinite population, then

$$\operatorname{cov}(X_r - \overline{X}, \overline{X}) = 0$$

for $r = 1, 2, \ldots, n$.

**8.2** Use Theorem 4.14 and its corollary to show that if $X_{11}, X_{12}, \ldots, X_{1n_1}, X_{21}, X_{22}, \ldots, X_{2n_2}$ are independent random variables, with the first $n_1$ constituting a random sample from an infinite population with the mean $\mu_1$ and the variance $\sigma_1^2$ and the other $n_2$ constituting a random sample from an infinite population with the mean $\mu_2$ and the variance $\sigma_2^2$, then

(a) $E(\overline{X}_1 - \overline{X}_2) = \mu_1 - \mu_2$;

(b) $\operatorname{var}(\overline{X}_1 - \overline{X}_2) = \dfrac{\sigma_1^2}{n_1} + \dfrac{\sigma_2^2}{n_2}$.

**8.3** With reference to Exercise 8.2, show that if the two samples come from normal populations, then $\overline{X}_1 - \overline{X}_2$ is a random variable having a normal distribution with the mean $\mu_1 - \mu_2$ and the variance $\dfrac{\sigma_1^2}{n_1} + \dfrac{\sigma_2^2}{n_2}$. (*Hint:* Proceed as in the proof of Theorem 8.4.)

**8.4** If $X_1, X_2, \ldots, X_n$ are independent random variables having identical Bernoulli distributions with the parameter $\theta$, then $\overline{X}$ is the proportion of successes in $n$ trials, which we denote by $\hat{\Theta}$. Verify that

(a) $E(\hat{\Theta}) = \theta$;

(b) $\operatorname{var}(\hat{\Theta}) = \dfrac{\theta(1 - \theta)}{n}$.

**8.5** If the first $n_1$ random variables of Exercise 8.2 have Bernoulli distributions with the parameter $\theta_1$ and the other $n_2$ random variables have Bernoulli distributions with the parameter $\theta_2$, show that in the notation of Exercise 8.4,

(a) $E(\hat{\Theta}_1 - \hat{\Theta}_2) = \theta_1 - \theta_2$;

(b) $\operatorname{var}(\hat{\Theta}_1 - \hat{\Theta}_2) = \dfrac{\theta_1(1 - \theta_1)}{n_1} + \dfrac{\theta_2(1 - \theta_2)}{n_2}$.

**8.6** Looking at binomial random variables as on page 288—namely, as sums of identically distributed independent Bernoulli random variables—and using the central limit theorem, prove Theorem 6.8.

**8.7** The following is a sufficient condition for the central limit theorem: If the random variables $X_1, X_2, \ldots, X_n$ are independent and uniformly bounded (that is, there exists a positive constant $k$ such that the probability is zero that any one of the random variables $X_i$ will take on a value greater than $k$ or less than $-k$), then if the variance of

$$Y_n = X_1 + X_2 + \cdots + X_n$$

becomes infinite when $n \to \infty$, the distribution of the standardized mean of the $X_i$ approaches the standard normal distribution. Show that this sufficient condition holds for a sequence of independent random variables $X_i$ having the respective probability distributions

$$f_i(x_i) = \begin{cases} \frac{1}{2} & \text{for } x_i = 1 - (\frac{1}{2})^i \\ \frac{1}{2} & \text{for } x_i = (\frac{1}{2})^i - 1 \end{cases}$$

**8.8** Consider the sequence of independent random variables $X_1$, $X_2$, $X_3$, ... , having the uniform densities

$$f_i(x_i) = \begin{cases} \dfrac{1}{2 - \dfrac{1}{i}} & \text{for } 0 < x_i < 2 - \dfrac{1}{i} \\ 0 & \text{elsewhere} \end{cases}$$

Use the sufficient condition of the preceding exercise to show that the central limit theorem holds.

**8.9** The following is a sufficient condition, the *Laplace-Liapounoff condition*, for the central limit theorem: If $X_1$, $X_2$, $X_3$, ... is a sequence of independent random variables, each having an absolute third moment

$$c_i = E(|X_i - \mu_i|^3)$$

and if

$$\lim_{n \to \infty} [\text{var}(Y_n)]^{-\frac{3}{2}} \cdot \sum_{i=1}^{n} c_i = 0$$

where $Y_n = X_1 + X_2 + \cdots + X_n$, then the distribution of the standardized mean of the $X_i$ approaches the standard normal distribution when $n \to \infty$. Use this condition to show that the central limit theorem holds for the sequence of random variables of Exercise 8.7.

**8.10** Use the condition of Exercise 8.9 to show that the central limit theorem holds for the sequence of random variables of Exercise 8.8.

**8.11** Explain why, when we sample with replacement from a finite population, the results of Theorem 8.1 apply rather than those of Theorem 8.6.

**8.12** Explain the results of Exercise 5.39 in the light of Theorem 8.6.

**8.13** If a random sample of size $n$ is selected from the finite population which consists of the integers $1, 2, \ldots, N$, show that

(a)   the mean of $\overline{X}$ is $\dfrac{N + 1}{2}$;

(b)   the variance of $\overline{X}$ is $\dfrac{(N + 1)(N - n)}{12n}$;

(c)   the mean and the variance of $Y = n \cdot \overline{X}$ are

$$E(Y) = \frac{n(N + 1)}{2} \quad \text{and} \quad \text{var}(Y) = \frac{n(N + 1)(N - n)}{12}$$

(*Hint*: Refer to the appendix at the end of the book or to the results of Exercise 5.1.)

**8.14** Find the mean and the variance of the finite population which consists of the 10 numbers 15, 13, 18, 10, 6, 21, 7, 11, 20, and 9.

**8.15** Show that the variance of the finite population $\{c_1, c_2, \ldots, c_N\}$ can be written as

$$\sigma^2 = \frac{\displaystyle\sum_{i=1}^{N} c_i^2}{N} - \mu^2$$

**8.16** Use the formula of Exercise 8.15 to recalculate the variance of the finite population of Exercise 8.14.

**8.17** Write a formula for the sample variance that is analogous to the formula of Exercise 8.15.

## APPLICATIONS

**8.18** How many different samples of size $n = 3$ can be drawn from a finite population of size

(a)   $N = 12$;

(b)   $N = 20$;

(c)   $N = 50$?

**8.19** What is the probability of each possible sample if

(a)   a random sample of size $n = 4$ is to be drawn from a finite population of size $N = 12$;

(b)   a random sample of size $n = 5$ is to be drawn from a finite population of size $N = 22$?

**8.20** If a random sample of size $n = 3$ is drawn from a finite population of size $N = 50$, what is the probability that a particular element of the population will be included in the sample?

**8.21** For random samples from an infinite population, what happens to the standard error of the mean if the sample size is

(a)  increased from 30 to 120;
(b)  increased from 80 to 180;
(c)  decreased from 450 to 50;
(d)  decreased from 250 to 40?

**8.22** Find the value of the finite population correction factor $\dfrac{N - n}{N - 1}$ for

(a)  $n = 5$ and $N = 200$;
(b)  $n = 50$ and $N = 300$;
(c)  $n = 200$ and $N = 800$.

**8.23** A random sample of size $n = 100$ is taken from an infinite population with the mean $\mu = 75$ and the variance $\sigma^2 = 256$. If we use Chebyshev's theorem, with what probability can we assert that the value we obtain for $\overline{X}$ will fall between 67 and 83?

**8.24** Use the central limit theorem instead of Chebyshev's theorem to rework Exercise 8.23.

**8.25** A random sample of size $n = 81$ is taken from an infinite population with the mean $\mu = 128$ and the standard deviation $\sigma = 6.3$. With what probability can we assert that the value we obtain for $\overline{X}$ will not fall between 126.6 and 129.4 if we use

(a)  Chebyshev's theorem;
(b)  the central limit theorem?

**8.26** Rework part (b) of Exercise 8.25, assuming that the population is not infinite but finite and of size $N = 400$.

**8.27** A random sample of size 64 is taken from a normal population with $\mu = 51.4$ and $\sigma = 6.8$. What is the probability that the mean of the sample will

(a)  exceed 52.9;
(b)  fall between 50.5 and 52.3;
(c)  be less than 50.6?

**8.28** A random sample of size 100 is taken from a normal population with $\sigma = 25$. What is the probability that the mean of the sample will differ from the mean of the population by 3 or more either way?

**8.29** Independent random samples of size 400 are taken from each of two populations having equal means and the standard deviations $\sigma_1 = 20$ and $\sigma_2 = 30$. Using Chebyshev's theorem and the result of Exercise 8.2, what can we assert with a probability of at least 0.99 about the value we will get for $\overline{X}_1 - \overline{X}_2$? (By "independent" we mean that the samples satisfy the conditions of Exercise 8.2.)

**8.30** Assume that the two populations of Exercise 8.29 are normal and use the result of Exercise 8.3 to find $k$ such that

$$P(-k < \overline{X}_1 - \overline{X}_2 < k) = 0.99$$

**8.31** Independent random samples of size $n_1 = 30$ and $n_2 = 50$ are taken from two normal populations having the means $\mu_1 = 78$ and $\mu_2 = 75$, and the variances $\sigma_1^2 = 150$ and $\sigma_2^2 = 200$. Use the results of Exercise 8.3 to find the probability that the mean of the first sample will exceed that of the second sample by at least 4.8.

**8.32** The actual proportion of families in a certain city who own, rather than rent, their home is 0.70. If 84 families in this city are interviewed at random and their responses to the question whether or not they own their home are looked upon as values of independent random variables having identical Bernoulli distributions with the parameter $\theta = 0.70$, with what probability can we assert that the value we obtain for the sample proportion $\hat{\Theta}$ will fall between 0.64 and 0.76, using the result of Exercise 8.4 and

(a) Chebyshev's theorem;

(b) the central limit theorem?

**8.33** The actual proportion of men who favor a certain tax proposal is 0.40 and the corresponding proportion for women is 0.25; $n_1 = 500$ men and $n_2 = 400$ women are interviewed at random and their individual responses are looked upon as the values of independent random variables having Bernoulli distributions with the respective parameters $\theta_1 = 0.40$ and $\theta_2 = 0.25$. What can we assert, according to Chebyshev's theorem, with a probability of at least 0.9375 about the value we will get for $\hat{\Theta}_1 - \hat{\Theta}_2$, the difference between the two sample proportions of favorable responses? Use the result of Exercise 8.5.

## 8.4 THE CHI-SQUARE DISTRIBUTION

In Example 7.9 we showed that if $X$ has the standard normal distribution, then $X^2$ has the special gamma distribution to which we referred as the **chi-square distribution**, and this accounts for the important role that the chi-square distribution plays in problems of sampling from normal populations. The chi-square distribution is often denoted by "$\chi^2$ distribution," where $\chi$ is the lowercase Greek letter *chi*. We also use $\chi^2$ for values of random variables having chi-square distributions, but we shall refrain from denoting the corresponding random variables by $X^2$, where X is the capital Greek letter *chi*. This avoids having to reiterate in each case whether $X$ is a random variable with values $x$ or a random variable with values $\chi$.

To review some of the results of Section 6.3, a random variable $X$ has the chi-square distribution with $\nu$ degrees of freedom if its probability density is given by

$$f(x) = \begin{cases} \dfrac{1}{2^{\nu/2}\Gamma(\nu/2)} x^{\frac{\nu-2}{2}} e^{-x/2} & \text{for } x > 0 \\ 0 & \text{elsewhere} \end{cases}$$

The mean and the variance of the chi-square distribution with $\nu$ degrees of freedom are $\nu$ and $2\nu$, and its moment-generating function is given by

$$M_X(t) = (1 - 2t)^{-\nu/2}$$

The chi-square distribution has several important mathematical properties, which are given in Theorems 8.7 through 8.10. First, let us formally state the result of Example 7.9, which we referred to above.

---

**THEOREM 8.7**  If $X$ has the standard normal distribution, then $X^2$ has the chi-square distribution with $\nu = 1$ degree of freedom.

---

More generally, let us show that

---

**THEOREM 8.8**  If $X_1, X_2, \ldots, X_n$ are independent random variables having standard normal distributions, then

$$Y = \sum_{i=1}^{n} X_i^2$$

has the chi-square distribution with $\nu = n$ degrees of freedom.

---

*Proof.*  Using the moment-generating function given above with $\nu = 1$ and Theorem 8.7, we find that

$$M_{X_i^2}(t) = (1 - 2t)^{-\frac{1}{2}}$$

and it follows by Theorem 7.3 that

$$M_Y(t) = \prod_{i=1}^{n} (1 - 2t)^{-\frac{1}{2}} = (1 - 2t)^{-\frac{n}{2}}$$

This moment-generating function is readily identified as that of the chi-square distribution with $\nu = n$ degrees of freedom. ▼

Two further properties of the chi-square distribution are given in the two theorems that follow; the reader will be asked to prove them in Exercises 8.34 and 8.35.

---

**THEOREM 8.9** If $X_1, X_2, \ldots, X_n$ are independent random variables having chi-square distributions with $\nu_1, \nu_2, \ldots, \nu_n$ degrees of freedom, then

$$Y = \sum_{i=1}^{n} X_i$$

has the chi-square distribution with $\nu_1 + \nu_2 + \cdots + \nu_n$ degrees of freedom.

---

**THEOREM 8.10** If $X_1$ and $X_2$ are independent random variables, $X_1$ has a chi-square distribution with $\nu_1$ degrees of freedom, and $X_1 + X_2$ has a chi-square distribution with $\nu > \nu_1$ degrees of freedom, then $X_2$ has a chi-square distribution with $\nu - \nu_1$ degrees of freedom.

---

The chi-square distribution has many important applications, of which several are discussed in Chapters 10 through 13. Foremost, there are those based, directly or indirectly, on the following theorem.

---

**THEOREM 8.11** If $\overline{X}$ and $S^2$ are the mean and the variance of a random sample of size $n$ from a normal population with the mean $\mu$ and the standard deviation $\sigma$, then

1. $\overline{X}$ and $S^2$ are independent;
2. the random variable $\dfrac{(n-1)S^2}{\sigma^2}$ has a chi-square distribution with $n - 1$ degrees of freedom.

---

***Proof.*** Since a detailed proof of part 1 would go beyond the scope of this text, we shall assume the independence of $\overline{X}$ and $S^2$ in our proof of part 2. In addition to the references to proofs of part 1 at the end of this chapter, Exercise 8.45 outlines the major steps of a somewhat simpler proof based on the idea of a conditional moment-generating function, and in Exercise 8.44 the reader will be asked to prove the independence of $\overline{X}$ and $S^2$ for the special case where $n = 2$.

To prove part 2, we begin with the identity

$$\sum_{i=1}^{n} (X_i - \mu)^2 = \sum_{i=1}^{n} (X_i - \overline{X})^2 + n(\overline{X} - \mu)^2$$

which the reader will be asked to verify in Exercise 8.36. Now, if we divide each term by $\sigma^2$ and substitute $(n - 1)S^2$ for $\sum_{i=1}^{n} (X_i - \overline{X})^2$, it follows that

$$\sum_{i=1}^{n} \left( \frac{X_i - \mu}{\sigma} \right)^2 = \frac{(n - 1)S^2}{\sigma^2} + \left( \frac{\overline{X} - \mu}{\sigma/\sqrt{n}} \right)^2$$

With regard to the three terms of this identity, we know from Theorem 8.8 that the one on the left-hand side of the equation is a random variable having a chi-square distribution with $n$ degrees of freedom. Also, according to Theorems 8.4 and 8.7, the second term on the right-hand side of the equation is a random variable having a chi-square distribution with 1 degree of freedom. Now, since $\overline{X}$ and $S^2$ are assumed to be independent, it follows that the two terms on the right-hand side of the equation are independent, and we conclude by Theorem 8.10 that $\dfrac{(n - 1)S^2}{\sigma^2}$ is a random variable having a chi-square distribution with $n - 1$ degrees of freedom.  ▼

Since the chi-square distribution arises in many important applications, integrals of its density have been extensively tabulated. Table V at the end of this book contains values of $\chi^2_{\alpha,\nu}$ for $\alpha = 0.995, 0.99, 0.975, 0.95, 0.05,$ 0.025, 0.01, 0.005, and $\nu = 1, 2, \ldots, 30$, where $\chi^2_{\alpha,\nu}$ is such that the area to its right under the chi-square curve with $\nu$ degrees of freedom (see Figure 8.1) is equal to $\alpha$. That is, $\chi^2_{\alpha,\nu}$ is such that if $X$ is a random variable having a chi-square distribution with $\nu$ degrees of freedom, then

$$P(X \geqslant \chi^2_{\alpha,\nu}) = \alpha$$

When $\nu$ is greater than 30, Table V cannot be used and probabilities related to chi-square distributions are usually approximated with normal distributions, as in Exercise 8.39 or 8.42.

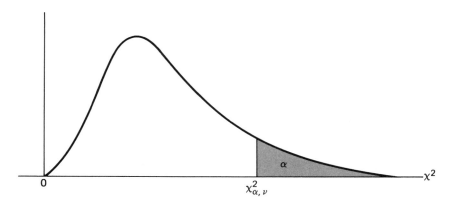

**Figure 8.1** Chi-square distribution.

## EXAMPLE 8.2

Suppose that the thickness of a part used in a semiconductor is its critical dimension, and that the process of manufacturing these parts is considered to be under control if the true variation among the thicknesses of the parts is given by a standard deviation not greater than $\sigma = 0.60$ thousandth of an inch. To keep a check on the process, random samples of size $n = 20$ are taken periodically, and it is regarded to be "out of control" if the probability that $S^2$ will take on a value greater than or equal to the observed sample value is 0.01 or less (even though $\sigma = 0.60$). What can one conclude about the process if the standard deviation of such a periodic random sample is $s = 0.84$ thousandth of an inch?

### Solution

The process will be declared "out of control" if $\dfrac{(n-1)s^2}{\sigma^2}$ with $n = 20$ and $\sigma = 0.60$ exceeds $\chi^2_{.01,19} = 36.191$. Since

$$\frac{(n-1)s^2}{\sigma^2} = \frac{19(0.84)^2}{(0.60)^2} = 37.24$$

exceeds 36.191, the process is declared out of control. Of course, it is assumed here that the sample may be regarded as a random sample from a normal population. ▲

## 8.5  THE *t* DISTRIBUTION

In Theorem 8.4 we showed that for random samples from a normal population with the mean $\mu$ and the variance $\sigma^2$, the random variable $\overline{X}$ has a normal distribution with the mean $\mu$ and the variance $\dfrac{\sigma^2}{n}$; in other words,

$$\frac{\overline{X} - \mu}{\sigma/\sqrt{n}}$$

has the standard normal distribution. This is an important result, but the major difficulty in applying it is that in most realistic applications the population standard deviation $\sigma$ is unknown. This makes it necessary to replace $\sigma$ with an estimate, usually with the value of the sample standard deviation $S$. Thus, the theory which follows leads to the exact distribution of $\dfrac{\overline{X} - \mu}{S/\sqrt{n}}$ for random samples from normal populations.

To derive this sampling distribution, let us first study the more general situation treated in the following theorem.

---

**THEOREM 8.12**  If $Y$ and $Z$ are independent random variables, $Y$ has a chi-square distribution with $\nu$ degrees of freedom, and $Z$ has the standard normal distribution, then the distribution of

$$T = \frac{Z}{\sqrt{Y/\nu}}$$

is given by

$$f(t) = \frac{\Gamma\left(\dfrac{\nu + 1}{2}\right)}{\sqrt{\pi\nu}\,\Gamma\left(\dfrac{\nu}{2}\right)} \cdot \left(1 + \frac{t^2}{\nu}\right)^{-\frac{\nu+1}{2}} \qquad \text{for } -\infty < t < \infty$$

and it is called the ***t* distribution** with $\nu$ degrees of freedom.

---

***Proof.***   Since $Y$ and $Z$ are independent, their joint probability density is given by

$$f(y, z) = \frac{1}{\sqrt{2\pi}} e^{-\frac{1}{2}z^2} \cdot \frac{1}{\Gamma\left(\frac{\nu}{2}\right)2^{\frac{\nu}{2}}} y^{\frac{\nu}{2}-1} e^{-\frac{y}{2}}$$

for $y > 0$ and $-\infty < z < \infty$, and $f(y, z) = 0$ elsewhere. Then, to use the change-of-variable technique of Section 7.3, we solve $t = \dfrac{z}{\sqrt{y/\nu}}$ for $z$ getting $z = t\sqrt{y/\nu}$ and, hence, $\dfrac{\partial z}{\partial t} = \sqrt{y/\nu}$. Thus, by Theorem 7.1, the joint density of $Y$ and $T$ is given by

$$g(y, t) = \begin{cases} \dfrac{1}{\sqrt{2\pi\nu}\Gamma\left(\dfrac{\nu}{2}\right)2^{\frac{\nu}{2}}} y^{\frac{\nu-1}{2}} e^{-\frac{y}{2}\left(1+\frac{t^2}{\nu}\right)} & \text{for } y > 0 \text{ and } -\infty < t < \infty \\ 0 & \text{elsewhere} \end{cases}$$

and, integrating out $y$ with the aid of the substitution $w = \dfrac{y}{2}\left(1 + \dfrac{t^2}{\nu}\right)$, we finally get

$$f(t) = \frac{\Gamma\left(\dfrac{\nu+1}{2}\right)}{\sqrt{\pi\nu}\Gamma\left(\dfrac{\nu}{2}\right)} \cdot \left(1 + \frac{t^2}{\nu}\right)^{-\frac{\nu+1}{2}} \qquad \text{for } -\infty < t < \infty \qquad \blacktriangledown$$

The $t$ distribution was introduced originally by W. S. Gosset, who published his scientific writings under the pen name "Student," since the company for which he worked, a brewery, did not permit publication by employees. Thus, the $t$ distribution is also known as the **Student-$t$ distribution**, or **Student's $t$ distribution**.

In view of its importance, the $t$ distribution has been tabulated extensively. Table IV, for example, contains values of $t_{\alpha,\nu}$ for $\alpha = 0.10, 0.05, 0.025, 0.01$, 0.005, and $\nu = 1, 2, \ldots, 29$, where $t_{\alpha,\nu}$ is such that the area to its right under the curve of the $t$ distribution with $\nu$ degrees of freedom (see Figure 8.2) is equal to $\alpha$. That is, $t_{\alpha,\nu}$ is such that if $T$ is a random variable having a $t$ distribution with $\nu$ degrees of freedom, then

$$P(T \geq t_{\alpha,\nu}) = \alpha$$

The table does not contain values of $t_{\alpha,\nu}$ for $\alpha > 0.50$, since the density is symmetrical about $t = 0$ and, hence, $t_{1-\alpha,\nu} = -t_{\alpha,\nu}$. When $\nu$ is 30 or more, proba-

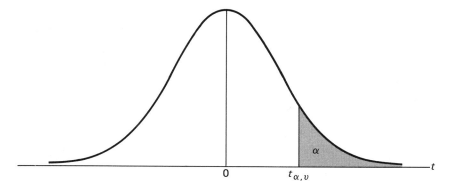

**Figure 8.2** *t* distribution.

bilities related to the *t* distribution are usually approximated with the use of normal distributions (see Exercise 8.49).

Among the many applications of the *t* distribution, some of which will be treated in Chapters 11 and 13, its major application (for which it was originally developed) is based on the following theorem.

---

**THEOREM 8.13** If $\overline{X}$ and $S^2$ are the mean and the variance of a random sample of size $n$ from a normal population with the mean $\mu$ and the variance $\sigma^2$, then

$$T = \frac{\overline{X} - \mu}{S/\sqrt{n}}$$

has the *t* distribution with $n - 1$ degrees of freedom.

---

***Proof.*** By Theorems 8.11 and 8.4, the random variables

$$Y = \frac{(n - 1)S^2}{\sigma^2} \quad \text{and} \quad Z = \frac{\overline{X} - \mu}{\sigma/\sqrt{n}}$$

have, respectively, a chi-square distribution with $n - 1$ degrees of freedom and the standard normal distribution. Since they are also independent by part 1 of Theorem 8.11, substitution into the formula for $T$ of Theorem 8.12 yields

$$T = \frac{\dfrac{\overline{X} - \mu}{\sigma/\sqrt{n}}}{\sqrt{S^2/\sigma^2}} = \frac{\overline{X} - \mu}{S/\sqrt{n}}$$

and this completes the proof.    ▼

## EXAMPLE 8.3

In 16 one-hour test runs, the gasoline consumption of an engine averaged 16.4 gallons with a standard deviation of 2.1 gallons. Test the claim that the average gasoline consumption of this engine is 12.0 gallons per hour.

### Solution

Substituting $n = 16$, $\mu = 12.0$, $\bar{x} = 16.4$, and $s = 2.1$ into the formula for $t$ in Theorem 8.13, we get

$$t = \frac{\bar{x} - \mu}{s/\sqrt{n}} = \frac{16.4 - 12.0}{2.1/\sqrt{16}} = 8.38$$

Since Table IV shows that for $\nu = 15$ the probability of getting a value of $T$ greater than 2.947 is 0.005, the probability of getting a value greater than 8 must be negligible. Thus, it would seem reasonable to conclude that the true average hourly gasoline consumption of the engine exceeds 12.0 gallons.    ▲

## 8.6  THE *F* DISTRIBUTION

Another distribution which plays an important role in connection with sampling from normal populations is the $F$ distribution, named after Sir Ronald A. Fisher, one of the most prominent statisticians of this century. Originally, it was studied as the sampling distribution of the ratio of two independent random variables with chi-square distributions, each divided by its respective degrees of freedom, and this is how we shall present it here.

> **THEOREM 8.14** If $U$ and $V$ are independent random variables having chi-square distributions with $\nu_1$ and $\nu_2$ degrees of freedom, then
>
> $$F = \frac{U/\nu_1}{V/\nu_2}$$

is a random variable having an **F distribution**, namely, a random variable whose probability density is given by

$$g(f) = \frac{\Gamma\left(\dfrac{\nu_1 + \nu_2}{2}\right)}{\Gamma\left(\dfrac{\nu_1}{2}\right)\Gamma\left(\dfrac{\nu_2}{2}\right)} \left(\frac{\nu_1}{\nu_2}\right)^{\frac{\nu_1}{2}} \cdot f^{\frac{\nu_1}{2}-1}\left(1 + \frac{\nu_1}{\nu_2}f\right)^{-\frac{1}{2}(\nu_1+\nu_2)}$$

$$0$$

for $f > 0$ and $g(f) = 0$ elsewhere.

***Proof.***   The joint density of $U$ and $V$ is given by

$$f(u, v) = \frac{1}{2^{\nu_1/2}\Gamma\left(\dfrac{\nu_1}{2}\right)} \cdot u^{\frac{\nu_1}{2}-1}e^{-\frac{u}{2}} \cdot \frac{1}{2^{\nu_2/2}\Gamma\left(\dfrac{\nu_2}{2}\right)} \cdot v^{\frac{\nu_2}{2}-1}e^{-\frac{v}{2}}$$

$$= \frac{1}{2^{(\nu_1+\nu_2)/2}\Gamma\left(\dfrac{\nu_1}{2}\right)\Gamma\left(\dfrac{\nu_2}{2}\right)} \cdot u^{\frac{\nu_1}{2}-1} v^{\frac{\nu_2}{2}-1} e^{-\frac{u+v}{2}}$$

for $u > 0$ and $v > 0$, and $f(u, v) = 0$ elsewhere. Then, to use the change-of-variable technique of Section 7.3, we solve

$$f = \frac{u/\nu_1}{v/\nu_2}$$

for $u$ getting $u = \dfrac{\nu_1}{\nu_2} \cdot vf$ and, hence, $\dfrac{\partial u}{\partial f} = \dfrac{\nu_1}{\nu_2} \cdot v$. Thus, by Theorem 7.1 the joint density of $F$ and $V$ is given by

$$g(f, v) = \frac{\left(\dfrac{\nu_1}{\nu_2}\right)^{\nu_1/2}}{2^{(\nu_1+\nu_2)/2}\Gamma\left(\dfrac{\nu_1}{2}\right)\Gamma\left(\dfrac{\nu_2}{2}\right)} \cdot f^{\frac{\nu_1}{2}-1} v^{\frac{\nu_1+\nu_2}{2}-1} e^{-\frac{v}{2}\left(\frac{\nu_1 f}{\nu_2}+1\right)}$$

for $f > 0$ and $v > 0$, and $g(f, v) = 0$ elsewhere. Now, integrating out $v$ by making the substitution $w = \dfrac{v}{2}\left(\dfrac{v_1 f}{v_2} + 1\right)$, we finally get

$$g(f) = \frac{\Gamma\left(\dfrac{v_1 + v_2}{2}\right)}{\Gamma\left(\dfrac{v_1}{2}\right)\Gamma\left(\dfrac{v_2}{2}\right)}\left(\dfrac{v_1}{v_2}\right)^{\frac{v_1}{2}} \cdot f^{\frac{v_1}{2}-1}\left(1 + \dfrac{v_1}{v_2}f\right)^{-\frac{1}{2}(v_1 + v_2)}$$

for $f > 0$, and $g(f) = 0$ elsewhere. ▼

In view of its importance, the $F$ distribution has been tabulated extensively. Table VI, for example, contains values of $f_{\alpha, v_1, v_2}$ for $\alpha = 0.05$ and $0.01$, and for various values of $v_1$ and $v_2$, where $f_{\alpha, v_1, v_2}$ is such that the area to its right under the curve of the $F$ distribution with $v_1$ and $v_2$ degrees of freedom (see Figure 8.3) is equal to $\alpha$. That is, $f_{\alpha, v_1, v_2}$ is such that

$$P(F \geq f_{\alpha, v_1, v_2}) = \alpha$$

Applications of Theorem 8.14 arise in problems in which we are interested in comparing the variances $\sigma_1^2$ and $\sigma_2^2$ of two normal populations; for instance, in problems in which we want to estimate the ratio $\dfrac{\sigma_1^2}{\sigma_2^2}$, or perhaps test whether $\sigma_1^2 = \sigma_2^2$. We base such inferences on **independent random samples** of size $n_1$ and $n_2$ from the two populations and Theorem 8.11, according to which

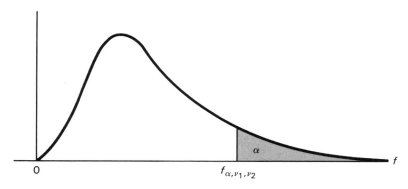

**Figure 8.3** *F* distribution.

$$\chi_1^2 = \frac{(n_1 - 1)s_1^2}{\sigma_1^2} \quad \text{and} \quad \chi_2^2 = \frac{(n_2 - 1)s_2^2}{\sigma_2^2}$$

are values of random variables having chi-square distributions with $n_1 - 1$ and $n_2 - 1$ degrees of freedom. By "independent random samples" we mean that the $n_1 + n_2$ random variables constituting the two random samples are all independent, so that the two chi-square random variables are independent and the substitution of their values for $U$ and $V$ in Theorem 8.14 yields the following result.

---

**THEOREM 8.15**  If $S_1^2$ and $S_2^2$ are the variances of independent random samples of size $n_1$ and $n_2$ from normal populations with the variances $\sigma_1^2$ and $\sigma_2^2$, then

$$F = \frac{S_1^2/\sigma_1^2}{S_2^2/\sigma_2^2} = \frac{\sigma_2^2 S_1^2}{\sigma_1^2 S_2^2}$$

is a random variable having an *F* distribution with $n_1 - 1$ and $n_2 - 1$ degrees of freedom.

---

In Chapter 11 we shall apply this theorem to the problem of estimating the ratio $\dfrac{\sigma_1^2}{\sigma_2^2}$ when these two population variances are unknown; also, in Chapter 13 we shall demonstrate how to test whether $\sigma_1^2 = \sigma_2^2$. Still other tests based on the *F* distribution are presented in the analysis-of-variance procedures of Chapter 15. Since all these applications are based on the ratios of sample variances, the *F* distribution is also known as the **variance-ratio distribution**.

## EXERCISES

**8.34**  Prove Theorem 8.9.

**8.35**  Prove Theorem 8.10.

**8.36**  Verify the identity

$$\sum_{i=1}^{n} (X_i - \mu)^2 = \sum_{i=1}^{n} (X_i - \overline{X})^2 + n(\overline{X} - \mu)^2$$

which we used in the proof of Theorem 8.11.

**8.37**  Use Theorem 8.11 to show that for random samples of size $n$ from a normal population with the variance $\sigma^2$, the sampling distribution of $S^2$ has the mean $\sigma^2$ and the variance $\dfrac{2\sigma^4}{n-1}$. (A general formula for the variance of $S^2$ for random samples from any population with finite second and fourth moments may be found in the book by H. Cramér listed among the references at the end of this chapter.)

**8.38** Show that if $X_1, X_2, \ldots, X_n$ are independent random variables having the chi-square distribution with $\nu = 1$ and $Y_n = X_1 + X_2 + \cdots + X_n$, then the limiting distribution of

$$Z = \frac{\dfrac{Y_n}{n} - 1}{\sqrt{2/n}}$$

as $n \to \infty$ is the standard normal distribution.

**8.39** Based on the result of Exercise 8.38, show that if $X$ is a random variable having a chi-square distribution with $\nu$ degrees of freedom and $\nu$ is large, the distribution of $\dfrac{X - \nu}{\sqrt{2\nu}}$ can be approximated with the standard normal distribution.

**8.40** Use the method of the preceding exercise to find the approximate value of the probability that a random variable having a chi-square distribution with $\nu = 50$ will take on a value greater than 68.0.

**8.41** If the range of $X$ is the set of all positive real numbers, show that for $k > 0$ the probability that $\sqrt{2X} - \sqrt{2\nu}$ will take on a value less than $k$ equals the probability that $\dfrac{X - \nu}{\sqrt{2\nu}}$ will take on a value less than $k + \dfrac{k^2}{2\sqrt{2\nu}}$.

**8.42** Use the results of Exercises 8.39 and 8.41 to show that if $X$ has a chi-square distribution with $\nu$ degrees of freedom, then for large $\nu$ the distribution of $\sqrt{2X} - \sqrt{2\nu}$ can be approximated with the standard normal distribution. Also, use this method of approximation to rework Exercise 8.40.

**8.43** Find the percentage errors of the approximations of Exercises 8.40 and 8.42, given that the actual value of the probability (rounded to five decimals) is 0.04596.

**8.44** (*Proof of the independence of $\overline{X}$ and $S^2$ for $n = 2$.*) If $X_1$ and $X_2$ are independent random variables having the standard normal distribution, show that

(a) the joint density of $X_1$ and $\overline{X}$ is given by

$$f(x_1, \bar{x}) = \frac{1}{\pi} \cdot e^{-\bar{x}^2} e^{-(x_1 - \bar{x})^2}$$

for $-\infty < x_1 < \infty$ and $-\infty < \bar{x} < \infty$;

(b)   the joint density of $U = |X_1 - \overline{X}|$ and $\overline{X}$ is given by

$$g(u, \bar{x}) = \frac{2}{\pi} \cdot e^{-(\bar{x}^2 + u^2)}$$

for $u > 0$ and $-\infty < \bar{x} < \infty$, since $f(x_1, \bar{x})$ is symmetrical about $\bar{x}$ for fixed $\bar{x}$;

(c)   $S^2 = 2(X_1 - \overline{X})^2 = 2U^2$;

(d)   the joint density of $\overline{X}$ and $S^2$ is given by

$$h(s^2, \bar{x}) = \frac{1}{\sqrt{\pi}} e^{-\bar{x}^2} \cdot \frac{1}{\sqrt{2\pi}} (s^2)^{-\frac{1}{2}} e^{-\frac{1}{2}s^2}$$

for $s^2 > 0$ and $-\infty < \bar{x} < \infty$, so that $\overline{X}$ and $S^2$ are independent.

**8.45** (*Proof of the independence of $\overline{X}$ and $S^2$*.) If $X_1, X_2, \ldots, X_n$ constitute a random sample from a normal population with the mean $\mu$ and the variance $\sigma^2$,

(a)   find the conditional density of $X_1$ given $X_2 = x_2, X_3 = x_3, \ldots, X_n = x_n$, and then set $X_1 = n\overline{X} - X_2 - \cdots - X_n$ and use the transformation technique to find the conditional density of $\overline{X}$ given $X_2 = x_2, X_3 = x_3, \ldots, X_n = x_n$;

(b)   find the joint density of $\overline{X}, X_2, X_3, \ldots, X_n$ by multiplying the conditional density of $\overline{X}$ obtained in part (a) by the joint density of $X_2, X_3, \ldots, X_n$, and show that

$$g(x_2, x_3, \ldots, x_n | \bar{x}) = \sqrt{n} \left( \frac{1}{\sigma \sqrt{2\pi}} \right)^{n-1} e^{-\frac{(n-1)s^2}{2\sigma^2}}$$

for $-\infty < x_i < \infty$, $i = 2, 3, \ldots, n$;

(c)   show that the conditional moment-generating function of $\dfrac{(n-1)S^2}{\sigma^2}$ given $\overline{X} = \bar{x}$ is

$$E\left[ e^{\frac{(n-1)S^2}{\sigma^2} \cdot t} \middle| \bar{x} \right] = (1 - 2t)^{-\frac{n-1}{2}} \qquad \text{for } t < \tfrac{1}{2}$$

Since this result is free of $\bar{x}$, it follows that $\overline{X}$ and $S^2$ are independent; it also shows that $\dfrac{(n-1)S^2}{\sigma^2}$ has a chi-square distribution with $n - 1$ degrees of freedom.

This proof, due to J. Shuster, is listed among the references at the end of this chapter.

**8.46** Use the transformation technique based on Theorem 7.2 to rework the proof of Theorem 8.12. (*Hint*: Let $t = \dfrac{z}{\sqrt{y/\nu}}$ and $u = y$.)

**8.47** Show that for $\nu > 2$ the variance of the $t$ distribution with $\nu$ degrees of freedom is $\dfrac{\nu}{\nu - 2}$. (*Hint*: Make the substitution $1 + \dfrac{t^2}{\nu} = \dfrac{1}{u}$.)

**8.48** Show that for the $t$ distribution with $\nu > 4$ degrees of freedom

(a)  $\mu_4 = \dfrac{3\nu^2}{(\nu - 2)(\nu - 4)}$;

(b)  $\alpha_4 = 3 + \dfrac{6}{\nu - 4}$.

(*Hint*: Make the substitution $1 + \dfrac{t^2}{\nu} = \dfrac{1}{u}$.)

**8.49** Use Stirling's formula of Exercise 1.6 to show that when $\nu \to \infty$, the $t$ distribution approaches the standard normal distribution.

**8.50** By what name did we refer to the $t$ distribution with $\nu = 1$ degree of freedom?

**8.51** Use the transformation technique based on Theorem 7.2 to rework the proof of Theorem 8.14. (*Hint*: Let $f = \dfrac{u/\nu_1}{v/\nu_2}$ and $w = v$.)

**8.52** Show that for $\nu_2 > 2$ the mean of the $F$ distribution is $\dfrac{\nu_2}{\nu_2 - 2}$, making use of the definition of $F$ in Theorem 8.14 and the fact that for a random variable $V$ having the chi-square distribution with $\nu_2$ degrees of freedom, $E\left(\dfrac{1}{V}\right) = \dfrac{1}{\nu_2 - 2}$.

**8.53** Verify that if $X$ has an $F$ distribution with $\nu_1$ and $\nu_2$ degrees of freedom and $\nu_2 \to \infty$, the distribution of $Y = \nu_1 X$ approaches the chi-square distribution with $\nu_1$ degrees of freedom.

**8.54** Verify that if $T$ has a $t$ distribution with $\nu$ degrees of freedom, then $X = T^2$ has an $F$ distribution with $\nu_1 = 1$ and $\nu_2 = \nu$ degrees of freedom.

**8.55** If $X$ has an $F$ distribution with $\nu_1$ and $\nu_2$ degrees of freedom, show that $Y = \dfrac{1}{X}$ has an $F$ distribution with $\nu_2$ and $\nu_1$ degrees of freedom.

**8.56** Use the result of the preceding exercise to show that

$$ f_{1-\alpha, \nu_1, \nu_2} = \dfrac{1}{f_{\alpha, \nu_2, \nu_1}} $$

**8.57** Verify that if $Y$ has a beta distribution with $\alpha = \dfrac{\nu_1}{2}$ and $\beta = \dfrac{\nu_2}{2}$, then

$$X = \frac{\nu_2 Y}{\nu_1(1-Y)}$$

has an $F$ distribution with $\nu_1$ and $\nu_2$ degrees of freedom.

**8.58** Show that the $F$ distribution with 4 and 4 degrees of freedom is given by

$$g(f) = \begin{cases} 6f(1+f)^{-4} & \text{for } f > 0 \\ 0 & \text{elsewhere} \end{cases}$$

and use this density to find the probability that for independent random samples of size $n = 5$ from normal populations with the same variance, $S_1^2/S_2^2$ will take on a value less than $\frac{1}{2}$ or greater than 2.

## APPLICATIONS

(In Exercises 8.60 through 8.65, refer to Tables IV, V, and VI.)

**8.59** Integrate the appropriate chi-square density to find the probability that the variance of a random sample of size 5 from a normal population with $\sigma^2 = 25$ will fall between 20 and 30.

**8.60** The claim that the variance of a normal population is $\sigma^2 = 25$ is to be rejected if the variance of a random sample of size 16 exceeds 54.668 or is less than 12.102. What is the probability that this claim will be rejected even though $\sigma^2 = 25$?

**8.61** The claim that the variance of a normal population is $\sigma^2 = 4$ is to be rejected if the variance of a random sample of size 9 exceeds 7.7535. What is the probability that this claim will be rejected even though $\sigma^2 = 4$?

**8.62** A random sample of size $n = 25$ from a normal population has the mean $\bar{x} = 47$ and the standard deviation $s = 7$. If we base our decision on the statistic of Theorem 8.13, can we say that the given information supports the conjecture that the mean of the population is $\mu = 42$?

**8.63** A random sample of size $n = 12$ from a normal population has the mean $\bar{x} = 27.8$ and the variance $s^2 = 3.24$. If we base our decision on the statistic of Theorem 8.13, can we say that the given information supports the claim that the mean of the population is $\mu = 28.5$?

**8.64** If $S_1$ and $S_2$ are the standard deviations of independent random samples of size $n_1 = 61$ and $n_2 = 31$ from normal populations with $\sigma_1^2 = 12$ and $\sigma_2^2 = 18$, find $P(S_1^2/S_2^2 > 1.16)$.

**8.65** If $S_1^2$ and $S_2^2$ are the variances of independent random samples of size $n_1 = 10$ and $n_2 = 15$ from normal populations with equal variances, find $P(S_1^2/S_2^2 < 4.03)$.

## 8.7   ORDER STATISTICS

Consider a random sample of size $n$ from an infinite population with a continuous density and suppose that we arrange the values of $X_1, X_2, \ldots,$ and $X_n$ according to size. If we look upon the smallest of the $x$'s as a value of the random variable $Y_1$, the next largest as a value of the random variable $Y_2$, the next largest after that as a value of the random variable $Y_3, \ldots,$ and the largest as a value of the random variable $Y_n$, we refer to these $Y$'s as **order statistics**. In particular, $Y_1$ is the first order statistic, $Y_2$ is the second order statistic, $Y_3$ is the third order statistic, and so on. (We are limiting this discussion to infinite populations with continuous densities so that there is zero probability that any two of the $x$'s will be alike.)

To be more explicit, consider the case where $n = 2$ and the relationship between the values of the $X$'s and the $Y$'s is

$$y_1 = x_1 \quad \text{and} \quad y_2 = x_2 \quad \text{when} \quad x_1 < x_2$$

$$y_1 = x_2 \quad \text{and} \quad y_2 = x_1 \quad \text{when} \quad x_2 < x_1$$

Similarly, for $n = 3$ the relationship between the values of the respective random variables is

$$y_1 = x_1, \quad y_2 = x_2, \quad \text{and} \quad y_3 = x_3, \quad \text{when} \quad x_1 < x_2 < x_3$$

$$y_1 = x_1, \quad y_2 = x_3, \quad \text{and} \quad y_3 = x_2, \quad \text{when} \quad x_1 < x_3 < x_2$$

$$\cdots \cdots \cdots \cdots \cdots \cdots \cdots \cdots \cdots \cdots \cdots$$

$$y_1 = x_3, \quad y_2 = x_2, \quad \text{and} \quad y_3 = x_1, \quad \text{when} \quad x_3 < x_2 < x_1$$

Let us now derive a formula for the probability density of the $r$th order statistic for $r = 1, 2, \ldots, n$.

**THEOREM 8.16**   For random samples of size $n$ from an infinite population which has the value $f(x)$ at $x$, the probability density of the $r$th order statistic $Y_r$ is given by

$$g_r(y_r) = \frac{n!}{(r-1)!(n-r)!} \left[ \int_{-\infty}^{y_r} f(x)\, dx \right]^{r-1} f(y_r) \left[ \int_{y_r}^{\infty} f(x)\, dx \right]^{n-r}$$

for $-\infty < y_r < \infty$.

***Proof.***    Suppose that the real axis is divided into three intervals, one from $-\infty$ to $y_r$, a second from $y_r$ to $y_r + h$ (where $h$ is a positive constant), and the third from $y_r + h$ to $\infty$. Since the population we are sampling has the value $f(x)$ at $x$, the probability that $r - 1$ of the sample values fall into the first interval, one falls into the second interval, and $n - r$ fall into the third interval is

$$\frac{n!}{(r-1)!\,1!\,(n-r)!} \left[ \int_{-\infty}^{y_r} f(x)\,dx \right]^{r-1} \left[ \int_{y_r}^{y_r+h} f(x)\,dx \right] \left[ \int_{y_r+h}^{\infty} f(x)\,dx \right]^{n-r}$$

according to the formula for the multinomial distribution. Using the law of the mean from calculus, we have

$$\int_{y_r}^{y_r+h} f(x)\,dx = f(\xi) \cdot h \qquad \text{where } y_r \leq \xi \leq y_r + h$$

and if we let $h \to 0$, we finally get

$$g_r(y_r) = \frac{n!}{(r-1)!\,(n-r)!} \left[ \int_{-\infty}^{y_r} f(x)\,dx \right]^{r-1} f(y_r) \left[ \int_{y_r}^{\infty} f(x)\,dx \right]^{n-r}$$

for $-\infty < y_r < \infty$ for the probability density of the $r$th order statistic.    ▼

In particular, the sampling distribution of $Y_1$, the smallest value in a random sample of size $n$, is given by

$$g_1(y_1) = n \cdot f(y_1) \left[ \int_{y_1}^{\infty} f(x)\,dx \right]^{n-1} \qquad \text{for } -\infty < y_1 < \infty$$

while the sampling distribution of $Y_n$, the largest value in a random sample of size $n$, is given by

$$g_n(y_n) = n \cdot f(y_n) \left[ \int_{-\infty}^{y_n} f(x)\,dx \right]^{n-1} \qquad \text{for } -\infty < y_n < \infty$$

Also, in a random sample of size $n = 2m + 1$ the **sample median** $\tilde{X}$ is $Y_{m+1}$, whose sampling distribution is given by

$$h(\tilde{x}) = \frac{(2m+1)!}{m!\,m!} \left[ \int_{-\infty}^{\tilde{x}} f(x)\,dx \right]^{m} f(\tilde{x}) \left[ \int_{\tilde{x}}^{\infty} f(x)\,dx \right]^{m} \qquad \text{for } -\infty < \tilde{x} < \infty$$

[For random samples of size $n = 2m$ the median is defined as $\frac{1}{2}(Y_m + Y_{m+1})$.]

In some instances it is possible to perform the integrations required to obtain the densities of the various order statistics; for other populations there may be no choice but to approximate these integrals by using numerical methods.

## EXAMPLE 8.4

Show that for random samples of size $n$ from an exponential population with the parameter $\theta$, the sampling distributions of $Y_1$ and $Y_n$ are given by

$$g_1(y_1) = \begin{cases} \dfrac{n}{\theta} \cdot e^{-ny_1/\theta} & \text{for } y_1 > 0 \\ 0 & \text{elsewhere} \end{cases}$$

and

$$g_n(y_n) = \begin{cases} \dfrac{n}{\theta} \cdot e^{-y_n/\theta}[1 - e^{-y_n/\theta}]^{n-1} & \text{for } y_n > 0 \\ 0 & \text{elsewhere} \end{cases}$$

and that, for random samples of size $n = 2m + 1$ from this kind of population, the sampling distribution of the median is given by

$$h(\tilde{x}) = \begin{cases} \dfrac{(2m + 1)!}{m!m!\theta} \cdot e^{-\tilde{x}(m+1)/\theta}[1 - e^{-\tilde{x}/\theta}]^m & \text{for } \tilde{x} > 0 \\ 0 & \text{elsewhere} \end{cases}$$

### Solution

The integrations required to obtain these results are straightforward, and they will be left to the reader in Exercise 8.66.  ▲

The following is an interesting result about the sampling distribution of the median which holds when the population density is continuous and nonzero at the **population median** $\tilde{\mu}$, which is such that $\int_{-\infty}^{\tilde{\mu}} f(x)\, dx = \frac{1}{2}$.

---

**THEOREM 8.17** For large $n$, the sampling distribution of the median for random samples of size $2n + 1$ is approximately normal with the mean $\tilde{\mu}$ and the variance $\dfrac{1}{8[f(\tilde{\mu})]^2 n}$.

---

A proof of this theorem is referred to on page 328. Note that for random samples of size $2n + 1$ from a normal population we have $\mu = \tilde{\mu}$, so that

$$f(\tilde{\mu}) = f(\mu) = \frac{1}{\sigma\sqrt{2\pi}}$$

and the variance of the median is approximately $\dfrac{\pi\sigma^2}{4n}$. If we compare this with the variance of the mean, which for random samples of size $2n + 1$ from an infinite population is $\dfrac{\sigma^2}{2n + 1}$, we find that for large samples from normal populations the mean is **more reliable** than the median; that is, the mean is subject to smaller chance fluctuations than the median.

## EXERCISES

**8.66** Verify the results of Example 8.4, namely, the sampling distributions of $Y_1$, $Y_n$, and $\bar{X}$ shown there for random samples from an exponential population.

**8.67** Find the sampling distributions of $Y_1$ and $Y_n$ for random samples of size $n$ from a continuous uniform population with $\alpha = 0$ and $\beta = 1$.

**8.68** Find the sampling distributon of the median for random samples of size $2m + 1$ from the population of Exercise 8.67.

**8.69** Find the mean and the variance of the sampling distribution of $Y_1$ for random samples of size $n$ from the population of Exercise 8.67.

**8.70** Find the sampling distributions of $Y_1$ and $Y_n$ for random samples of size $n$ from a population having the beta distribution with $\alpha = 3$ and $\beta = 2$.

**8.71** Find the sampling distribution of the median for random samples of size $2m + 1$ from the population of Exercise 8.70.

**8.72** Find the sampling distribution of $Y_1$ for random samples of size $n = 2$ taken
    (a)   without replacement from the finite population which consists of the first five positive integers;
    (b)   with replacement from the same population.
    (*Hint*: Enumerate all possibilities.)

**8.73** Duplicate the method used in the proof of Theorem 8.16 to show that the joint density of $Y_1$ and $Y_n$ is given by

$$g(y_1, y_n) = n(n - 1)f(y_1)f(y_n)\left[\int_{y_1}^{y_n} f(x)\,dx\right]^{n-2} \qquad \text{for } -\infty < y_1 < y_n < \infty$$

and $g(y_1, y_n) = 0$ elsewhere.

(a) Use this result to find the joint density of $Y_1$ and $Y_n$ for random samples of size $n$ from an exponential population.

(b) Use this result to find the joint density of $Y_1$ and $Y_n$ for the population of Exercise 8.67.

**8.74** With reference to part (b) of Exercise 8.73, find the covariance of $Y_1$ and $Y_n$.

**8.75** Use the formula for the joint density of $Y_1$ and $Y_n$ shown in Exercise 8.73 and the transformation technique of Section 7.4 to find an expression for the joint density of $Y_1$ and the **sample range** $R = Y_n - Y_1$.

**8.76** Use the result of the preceding exercise and that of part (a) of Exercise 8.73 to find the sampling distribution of $R$ for random samples of size $n$ from an exponential population.

**8.77** Use the result of Exercise 8.75 to find the sampling distribution of $R$ for random samples of size $n$ from the continuous uniform population of Exercise 8.67.

**8.78** Use the result of Exercise 8.77 to find the mean and the variance of the sampling distribution of $R$ for random samples of size $n$ from the continuous uniform population of Exercise 8.67.

**8.79** There are many problems, particularly in industrial applications, in which we are interested in the proportion of a population that lies between certain limits. Such limits are called **tolerance limits**. The following steps lead to the sampling distribution of the statistic $P$, which is the proportion of a population (having a continuous density) that lies between the smallest and the largest values of a random sample of size $n$.

(a) Use the formula for the joint density of $Y_1$ and $Y_n$ shown in Exercise 8.73 and the transformation technique of Section 7.4 to show that the joint density of $Y_1$ and $P$, whose values are given by

$$p = \int_{y_1}^{y_n} f(x)\, dx$$

is

$$h(y_1, p) = n(n - 1)f(y_1)p^{n-2}$$

(b) Use the result of part (a) and the transformation technique of Section 7.4 to show that the joint density of $P$ and $W$, whose values are given by

$$w = \int_{-\infty}^{y_1} f(x)\, dx$$

is

$$\varphi(w, p) = n(n - 1)p^{n-2}$$

for $w > 0$, $p > 0$, $w + p < 1$, and $\varphi(w, p) = 0$ elsewhere.

(c)  Use the result of part (b) to show that the marginal density of $P$ is given by

$$g(p) = \begin{cases} n(n - 1)p^{n-2}(1 - p) & \text{for } 0 < p < 1 \\ 0 & \text{elsewhere} \end{cases}$$

This is the desired density of the proportion of the population that lies between the smallest and the largest values of a random sample of size $n$, and it is of interest to note that it does not depend on the form of the population distribution.

**8.80** Use the result of Exercise 8.79 to show that for the random variable $P$ defined there,

$$E(P) = \frac{n - 1}{n + 1} \quad \text{and} \quad \text{var}(P) = \frac{2(n - 1)}{(n + 1)^2(n + 2)}$$

What can we conclude from this about the distribution of $P$ when $n$ is large?

## APPLICATIONS

**8.81** Find the probability that in a random sample of size $n = 4$ from the continuous uniform population of Exercise 8.67, the smallest value will be at least 0.20.

**8.82** Find the probability that in a random sample of size $n = 3$ from the beta population of Exercise 8.70, the largest value will be less than 0.90.

**8.83** Use the result of Exercise 8.77 to find the probability that the range of a random sample of size $n = 5$ from the given uniform population will be at least 0.75.

**8.84** Use the result of part (c) of Exercise 8.79 to find the probability that in a random sample of size $n = 10$ at least 80 percent of the population will lie between the smallest and largest values.

**8.85** Use the result of part (c) of Exercise 8.79 to set up an equation in $n$, whose solution will give the sample size that is required to be able to assert with probability $1 - \alpha$ that the proportion of the population contained between the smallest and largest sample values is at least $p$. Show that for $p = 0.90$ and $\alpha = 0.05$ this equation can be written as

$$(0.90)^{n-1} = \frac{1}{2n + 18}$$

This kind of equation is difficult to solve, but it can be shown that an approximate solution for $n$ is given by

$$\frac{1}{2} + \frac{1}{4} \cdot \frac{1+p}{1-p} \cdot \chi^2_{\alpha,4}$$

where $\chi^2_{\alpha,4}$ must be looked up in Table V. Use this method to find an approximate solution of the equation for $p = 0.90$ and $\alpha = 0.05$.

## REFERENCES

Necessary and sufficient conditions for the strongest form of the central limit theorem for independent random variables, the so-called *Lindeberg-Feller* conditions, are given in

FELLER, W., *An Introduction to Probability Theory and Its Applications*, Vol. I, 3rd ed. New York: John Wiley & Sons, Inc., 1968,

as well as in other advanced texts on probability theory.
Extensive tables of the normal, chi-square, $F$, and $t$ distributions may be found in

PEARSON, E. S., and HARTLEY, H. O., *Biometrika Tables for Statisticians*, Vol. I. New York: John Wiley & Sons, Inc., 1968.

A general formula for the variance of the sampling distribution of the second sample moment $M_2$ (which differs from $S^2$ only insofar as we divide by $n$ instead of $n-1$) is derived in

CRAMÉR, H., *Mathematical Methods of Statistics*. Princeton, N.J.: Princeton University Press, 1950,

and a proof of Theorem 8.17 is given in

WILKS, S. S., *Mathematical Statistics*. New York: John Wiley & Sons, Inc., 1962.

Proofs of the independence of $\overline{X}$ and $S^2$ for random samples from normal populations are given in many advanced texts on mathematical statistics. For instance, a proof based on moment-generating functions may be found in the above-mentioned book by S. S. Wilks, and a somewhat more elementary proof, illustrated for $n = 3$, may be found in

KEEPING, E. S., *Introduction to Statistical Inference*. Princeton, N.J.: D. Van Nostrand Co., Inc., 1962.

The proof outlined in Exercise 8.45 is given in

SHUSTER, J., "A Simple Method of Teaching the Independence of $\overline{X}$ and $S^2$." *The American Statistician*, Vol. 27, No. 1, 1973.

# 9

# Decision Theory

## 9.1 INTRODUCTION

In Chapter 4 we introduced the concept of a mathematical expectation to study expected values of random variables; in particular, the moments of their distributions. In applied situations, mathematical expectations are often used as a guide in choosing among alternatives, that is, in making decisions, because it is generally considered rational to select alternatives with the "most promising" mathematical expectations—the ones that maximize expected profits, minimize expected losses, maximize expected sales, minimize expected costs, and so on.

---

*Note:* The material in this chapter provides a unified approach to statistical inference. However, it is not a prerequisite for the classical approach to which we devote most of the remainder of this text and, hence, it may be omitted without loss of continuity.

Although this approach to decision making has great intuitive appeal, it is not without complications, for there are many problems in which it is difficult, if not impossible, to assign numerical values to the consequences of one's actions and to the probabilities of all eventualities.

## EXAMPLE 9.1

A manufacturer of leather goods must decide whether to expand his plant capacity now or wait at least another year. His advisors tell him that if he expands now and economic conditions remain good, there will be a profit of $164,000 during the next fiscal year; if he expands now and there is a recession, there will be a loss of $40,000; if he waits at least another year and economic conditions remain good, there will be a profit of $80,000; and if he waits at least another year and there is a recession, there will be a small profit of $8,000. What should the manufacturer decide to do, if he wants to minimize the expected loss during the next fiscal year and he feels that the odds are 2 to 1 that there will be a recession?

### Solution

Schematically, all these "payoffs" can be represented as in the following table, where the entries are the losses which correspond to the various possibilities, and, hence, gains are represented by negative numbers:

|  | *Expand now* | *Delay expansion* |
|---|---|---|
| *Economic conditions remain good* | $-164,000$ | $-80,000$ |
| *There is a recession* | $40,000$ | $-8,000$ |

We are working with losses here rather than profits to make this example fit the general scheme which we shall present in Sections 9.2 and 9.3.

Since the probabilities that economic conditions will remain good and that there will be a recession are, respectively, $\frac{1}{3}$ and $\frac{2}{3}$, the manufacturer's expected loss for the next fiscal year is

$$-164,000 \cdot \tfrac{1}{3} + 40,000 \cdot \tfrac{2}{3} = -28,000$$

if he expands his plant capacity now, and

$$-80,000 \cdot \tfrac{1}{3} + (-8,000) \cdot \tfrac{2}{3} = -32,000$$

if he waits at least another year. Since an expected profit (negative expected loss) of $32,000 is preferable to an expected profit (negative expected loss) of $28,000, it follows that the manufacturer should delay expanding the capacity of his plant.    ▲

The result at which we arrived in this example assumes that the values given in the table and also the odds for a recession are properly assessed. As the reader will be asked to show in Exercises 9.2 and 9.3, changes in these quantities can easily lead to different results.

## EXAMPLE 9.2

With reference to Example 9.1, suppose that the manufacturer has no idea about the odds that there will be a recession. What should he decide to do, if he is a confirmed pessimist?

### Solution

Being the kind of person who always expects the worst to happen, he might argue that if he expands his plant capacity now he could lose $40,000, if he delays expansion there would be a profit of at least $8,000, and, hence, that he will minimize the maximum loss (or maximize the minimum profit) if he waits at least another year.    ▲

The criterion used in this example is called the **minimax criterion**, and it is only one of many different criteria that can be used in this kind of situation. One such criterion, based on optimism rather than pessimism, is referred to in Exercise 9.7, and another, based on the fear of "losing out on a good deal," is referred to in Exercise 9.8.

## 9.2 THE THEORY OF GAMES

The examples of the preceding section may well have given the impression that the manufacturer is playing a game—a game between him and Nature (or call it fate or whatever "controls" whether there will be a recession). Each of the "players" has the choice of two moves: The manufacturer has the choice between actions $a_1$ and $a_2$ (to expand his plant capacity now or to delay expansion for at least a year) and Nature controls the choice between $\theta_1$ and $\theta_2$ (whether economic conditions are to remain good or whether there is to be a recession). Depending on the choice of their moves, there are the "payoffs" shown in the following table:

*Player A*
*(The Manufacturer)*

|  |  | $a_1$ | $a_2$ |
|---|---|---|---|
| *Player B* | $\theta_1$ | $L(a_1,\ \theta_1)$ | $L(a_2,\ \theta_1)$ |
| *(Nature)* | $\theta_2$ | $L(a_1,\ \theta_2)$ | $L(a_2,\ \theta_2)$ |

The amounts $L(a_1,\ \theta_1)$, $L(a_2,\ \theta_1)$, ... , are referred to as the values of the **loss function** which characterizes the particular "game"; in other words, $L(a_i,\ \theta_j)$ is the loss of Player $A$ (the amount he has to pay Player $B$) when he chooses alternative $a_i$ and Player $B$ chooses alternative $\theta_j$. Although it does not really matter, we shall assume here that these amounts are in dollars. In actual practice, they can also be expressed in terms of any goods or services, in units of utility (desirability or satisfaction), and even in terms of life or death (as in Russian roulette or in the conduct of a war).

The analogy we have drawn here is not really farfetched; the problem of Example 9.1 is typical of the kind of situation treated in the **theory of games**, a relatively new branch of mathematics which has stimulated considerable interest in recent years. This theory is not limited to parlor games, as its name might suggest, but it applies to any kind of competitive situation and, as we shall see, it has led to a unified approach to solving problems of statistical inference.

To introduce some of the basic concepts of the theory of games, let us begin by explaining what we mean by a **zero-sum two-person game**. In this term, "two-person" means that there are two players (or, more generally, two parties with conflicting interests), and "zero-sum" means that whatever one player loses the other player wins. Thus, in a zero-sum game there is no "cut for the house" as in professional gambling, and no capital is created or destroyed during the course of play. Of course, the theory of games also includes games which are neither zero-sum nor limited to two players, but as can well be imagined, such games are generally much more complicated. Exercise 9.19 is an example of a game which is not zero-sum.

Games are also classified according to the number of **strategies** (moves, choices, or alternatives) each player has at his disposal. For instance, if each player has to choose one of two alternatives (as in Example 9.1), we say that it is a $2 \times 2$ game; if one player has 3 possible moves while the other has 4, the game is $3 \times 4$ or $4 \times 3$, as the case may be. In this section we shall consider only **finite** games, that is, games in which each player has only a finite, or fixed, number of possible moves, but later we shall consider also games where each player has infinitely many moves.

It is customary in the theory of games to refer to the two players as Player $A$ and Player $B$ as we did in the table above, but the moves (choices, or alternatives) of Player $A$ are usually labeled I, II, III, ... , instead of $a_1$, $a_2$, $a_3$, ... , and those of Player $B$ are usually labeled 1, 2, 3, ... , instead of $\theta_1$, $\theta_2$, $\theta_3$, .... The **payoffs**, the amounts of money which change hands when the players choose their respective strategies, are usually shown in a table like that on page

332, which is referred to as a **payoff matrix** in the theory of games. (As before, positive payoffs represent losses of Player A and negative payoffs represent losses of Player B.) Let us also add that it is always assumed in the theory of games that each player must choose his strategy without knowing what his opponent is going to do, and that once a player has made his choice, it cannot be changed.

The objectives of the theory of games are to determine **optimum strategies** (namely, strategies which are most profitable to the respective players) and the corresponding payoff, which is called the **value** of the game.

## EXAMPLE 9.3

Given the 2 × 2 zero-sum two-person game

|  |  | Player A | |
|---|---|:---:|:---:|
|  |  | *I* | *II* |
| *Player B* | *1* | 7 | −4 |
|  | *2* | 8 | 10 |

find the optimum strategies of Players A and B and the value of the game.

### Solution

As can be seen by inspection, it would be foolish for Player B to choose Strategy 1, since Strategy 2 will yield more than Strategy 1 regardless of the choice made by Player A. In a situation like this we say that Strategy 1 is **dominated** by Strategy 2 (or that Strategy 2 **dominates** Strategy 1), and it stands to reason that any strategy which is dominated by another should be discarded. If we do this here, we find that Player B's optimum strategy is Strategy 2, the only one left, and that Player A's optimum strategy is Strategy I, since a loss of $8 is obviously preferable to a loss of $10. Also, the value of the game, the payoff corresponding to Strategies I and 2, is $8.    ▲

## EXAMPLE 9.4

Given the 3 × 2 zero-sum two-person game

|  |  | Player A | | |
|---|---|:---:|:---:|:---:|
|  |  | *I* | *II* | *III* |
| *Player B* | *1* | −4 | 1 | 7 |
|  | *2* | 4 | 3 | 5 |

find the optimum strategies of Players A and B and the value of the game.

*Solution*

In this game neither strategy of Player *B* dominates the other, but the third strategy of Player *A* is dominated by each of the other two—clearly, a profit of $4 or a loss of $1 is preferable to a loss of $7, and a loss of $4 or a loss of $3 is preferable to a loss of $5. Thus, we can discard the third column of the payoff matrix and study the 2 × 2 game

*Player A*

|            |       | *I*  | *II* |
|------------|-------|------|------|
|            | *1*   | −4   | 1    |
| *Player B* |       |      |      |
|            | *2*   | 4    | 3    |

where now Strategy 2 of Player *B* dominates Strategy 1. Thus, the optimum choice of Player *B* is Strategy 2, the optimum choice of Player *A* is Strategy II (since a loss of $3 is preferable to a loss of $4), and the value of the game is $3.   ▲

The process of discarding dominated strategies can be of great help in the solution of a game (that is, in finding optimum strategies and the value of the game), but it is the exception rather than the rule that it will lead to a complete solution. Dominances may not even exist, as is illustrated by the following 3 × 3 zero-sum two-person game:

*Player A*

|              |     | *I*  | *II* | *III* |
|--------------|-----|------|------|-------|
|              | *1* | −1   | 6    | −2    |
| *Player B*   | *2* | 2    | 4    | 6     |
|              | *3* | −2   | −6   | 12    |

So, we must look for other ways of arriving at optimum strategies. From the point of view of Player *A*, we might argue as follows: If he chooses Strategy I, the worst that can happen is that he loses $2; if he chooses Strategy II, the worst that can happen is that he loses $6; and if he chooses Strategy III, the worst that can happen is that he loses $12. Thus, he could minimize the maximum loss by choosing Strategy I.

Applying the same kind of argument to select a strategy for Player *B*, we find that if she chooses Strategy 1, the worst that can happen is that she loses $2; if she chooses Strategy 2, the worst that can happen is that she wins $2; and if she chooses Strategy 3, the worst that can happen is that she loses $6. Thus, she could minimize the maximum loss (or maximize the minimum gain, which is the same) by choosing Strategy 2.

The selection of Strategies I and 2, appropriately called **minimax strategies** (or strategies based on the **minimax criterion**), is really quite reasonable. By choosing Strategy I, Player A makes sure that his opponent can win at most $2, and by choosing Strategy 2, Player B makes sure that she will actually win this amount. This $2 is the value of the game, which means that the game favors Player B, but we could make it **equitable** by charging Player B $2 for the privilege of playing the game and giving the $2 to Player A.

A very important aspect of the minimax strategies I and 2 of this example is that they are completely "spyproof" in the sense that neither player can profit from knowing the other's choice. In our example, even if Player A announced publicly that he will choose Strategy I, it would still be best for Player B to choose Strategy 2, and if Player B announced publicly that she will choose Strategy 2, it would still be best for Player A to choose Strategy I. Unfortunately, not all games are spyproof.

# EXAMPLE 9.5

Show that the minimax strategies of Players A and B are not spyproof in the following game:

|  | | Player A | |
| --- | --- | :---: | :---: |
|  | | I | II |
| Player B | 1 | 8 | −5 |
|  | 2 | 2 | 6 |

## Solution

Player A can minimize his maximum loss by choosing Strategy II, and Player B can minimize her maximum loss by choosing Strategy 2. However, if Player A knew that Player B was going to base her choice on the minimax criterion, he could switch to Strategy I and thus reduce his loss from $6 to $2. Of course, if Player B discovered that Player A would try to outsmart her in this way, she could in turn switch to Strategy 1 and increase her gain to $8. In any case, the minimax strategies of the two players are not spyproof, thus leaving room for all sorts of trickery or deception.    ▲

There exists an easy way of determining for any given game whether minimax strategies are spyproof. What we have to look for are **saddle points**, namely, pairs of strategies for which the corresponding entry in the payoff matrix is the smallest value of its row and the greatest value of its column. In Example 9.5 there is no saddle point since the smallest value of each row is also the smallest value of its column. On the other hand, in the game of Example 9.3 there is a saddle point corresponding to Strategies I and 2 since 8, the smallest value of the second row, is the greatest value of the first column. Also, the 3 × 2 game of Example 9.4 has a saddle point corresponding to Strategies II and 2 since 3, the smallest value of the second row, is the greatest value of the second column, and the 3 × 3 game on page 334 has a saddle point corresponding to Strategies I and 2 since 2, the smallest value of the second row, is the greatest value of the first column. In general, if a game has a saddle point it is said to be **strictly determined**, and the strategies corresponding to the saddle point are spyproof (and, hence, optimum) minimax strategies. The fact that there can be more than one saddle point in a game is illustrated in Exercise 9.1; it also follows from this exercise that it does not matter in that case which of the saddle points is used to determine the optimum strategies of the two players.

If a game does not have a saddle point, minimax strategies are not spyproof, and each player can outsmart the other if he or she knows how the opponent will react in a given situation. To avoid this possibility, it suggests itself that each player should somehow mix up his or her behavior patterns intentionally, and the best way of doing this is by introducing an element of chance into the selection of strategies.

## EXAMPLE 9.6

With reference to the game of Example 9.5, suppose that Player $A$ uses a gambling device (dice, cards, numbered slips of paper, a table of random numbers) which leads to the choice of Strategy I with probability $x$, and to the choice of Strategy II with probability $1 - x$. Find the value of $x$ which will minimize Player $A$'s maximum expected loss.

### Solution

If Player $B$ chooses Strategy 1, Player $A$ can expect to lose

$$E = 8x - 5(1 - x)$$

dollars, and if Player $B$ chooses Strategy 2, Player $A$ can expect to lose

$$E = 2x + 6(1 - x)$$

dollars. Graphically, this situation is described in Figure 9.1, where we have plotted the lines whose equations are $E = 8x - 5(1 - x)$ and $E = 2x + 6(1 - x)$ for values of $x$ from 0 to 1.

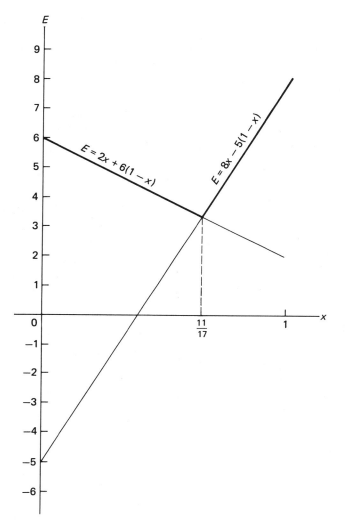

**Figure 9.1**  Diagram for Example 9.6.

Applying the minimax criterion to the expected losses of Player $A$, we find from Figure 9.1 that the greater of the two values of $E$ for any given value of $x$ is smallest where the two lines intersect, and to find the corresponding value of $x$ we have only to solve the equation

$$8x - 5(1 - x) = 2x + 6(1 - x)$$

which yields $x = \frac{11}{17}$. Thus, if Player $A$ uses eleven slips of paper numbered I and six slips of paper numbered II, shuffles them thoroughly, and then acts according to which kind he randomly draws, he will be holding his maximum expected loss down to $8 \cdot \frac{11}{17} - 5 \cdot \frac{6}{17} = 3\frac{7}{17}$, or \$3.41 to the nearest cent.   ▲

So far as Player $B$ of the preceding example is concerned, in Exercise 9.14 the reader will be asked to use a similar argument to show that Player $B$ will maximize her minimum gain (which is the same as minimizing her maximum loss) by choosing between Strategies 1 and 2 with respective probabilities of $\frac{4}{17}$ and $\frac{13}{17}$, and that she will thus assure for herself an expected gain of $3\frac{7}{17}$, or \$3.41 to the nearest cent. Incidentally, the \$3.41 to which Player $A$ can hold down his expected loss and Player $B$ can raise her expected gain is called the value of this game. Also, if a player's ultimate choice is thus left to chance, the overall strategy is referred to as **randomized** or **mixed**, whereas the original Strategies I, II, 1, and 2 are referred to as **pure**.

The examples of this section were all given without any "physical" interpretation because we were interested only in introducing some of the basic concepts of the theory of games. If we apply these methods to Example 9.1, we find that the "game" has a saddle point and that the manufacturer's minimax strategy is to delay expanding the capacity of his plant. Of course, this assumes, questionably so, that Nature (which controls whether there is going to be a recession) is a malevolent opponent. Also, it would seem that in a situation like this the manufacturer ought to have some idea about the chances for a recession and hence that the problem should be solved by the first method of Section 9.1.

## EXERCISE

**9.1**  If a zero-sum two-person game has a saddle point corresponding to the $i$th row and the $j$th column of the payoff matrix and another corresponding to the $k$th row and the $l$th column, show that

(a)  there are also saddle points corresponding to the $i$th row and the $l$th column of the payoff matrix and the $k$th row and the $j$th column;

(b)  the payoff must be the same for all four saddle points.

## APPLICATIONS

**9.2**  With reference to Example 9.1, what decision would minimize the manufacturer's expected loss if he felt that

(a)  the odds for a recession are 3 to 2;

(b)  the odds for a recession are 7 to 4?

**9.3**  With reference to Example 9.1, would the manufacturer's decision remain the same if

(a)   the $164,000 profit is replaced by a $200,000 profit and the odds are
      2 to 1 that there will be a recession;

(b)   the $40,000 loss is replaced by a $60,000 loss and the odds are 3 to
      2 that there will be a recession?

**9.4**   Ms. Cooper is planning to attend a convention in Honolulu, and she must
          send in her room reservation immediately. The convention is so large that
          the activities are held partly in Hotel $X$ and partly in Hotel $Y$, and Ms.
          Cooper does not know whether the particular session she wants to attend
          will be held at Hotel $X$ or Hotel $Y$. She is planning to stay only one night,
          which would cost her $66.00 at Hotel $X$ and $62.40 at Hotel $Y$, and it will
          cost her an extra $6.00 for cab fare if she stays at the wrong hotel.

(a)   If Ms. Cooper feels that the odds are 3 to 1 that the session she wants
      to attend will be held at Hotel $X$, where should she make her reser-
      vation so as to minimize her expected cost?

(b)   If Ms. Cooper feels that the odds are 5 to 1 that the session she wants
      to attend will be held at Hotel $X$, where should she make her reser-
      vation so as to minimize her expected cost?

**9.5**   A truck driver has to deliver a load of lumber to one of two construction
          sites, which are, respectively, 27 and 33 miles from the lumberyard, but he
          has misplaced the order telling him where the load of lumber should go.
          The two construction sites are 12 miles apart, and, to complicate matters,
          the telephone at the lumberyard is out of order. Where should he go first if
          he wants to minimize the distance he can expect to drive and he feels that

(a)   the odds are 5 to 1 that the lumber should go to the construction site
      which is 33 miles from the lumberyard;

(b)   the odds are 2 to 1 that the lumber should go to the construction site
      which is 33 miles from the lumberyard;

(c)   the odds are 3 to 1 that the lumber should go to the construction site
      which is 33 miles from the lumberyard?

**9.6**   Basing their decisions on pessimism as in Example 9.2, where should

(a)   Ms. Cooper of Exercise 9.4 make her reservation;
(b)   the truck driver of Exercise 9.5 go first?

**9.7**   Basing their decisions on optimism (that is, maximizing maximum gains or
          minimizing minimum losses), what decisions should be reached by

(a)   the manufacturer of Example 9.1;
(b)   Ms. Cooper of Exercise 9.4;
(c)   the truck driver of Exercise 9.5?

**9.8**   Suppose that the manufacturer of Example 9.1 is the kind of person who
          always worries about losing out on a good deal. For instance, he finds that
          if he delays expansion and economic conditions remain good, he will lose
          out by $84,000 (the difference between the $164,000 profit he would have
          made if he had decided to expand right away, and the $80,000 profit he

will actually make). Referring to this quantity as an **opportunity loss**, or **regret**, find

(a)    the opportunity losses that correspond to the other three possibilities;
(b)    the decision which would minimize the manufacturer's maximum loss of opportunity.

**9.9**    With reference to the definition of Exercise 9.8, find the decisions that will minimize the maximum opportunity loss of

(a)    Ms. Cooper of Exercise 9.4;
(b)    the truck driver of Exercise 9.5.

**9.10**    With reference to Example 9.1, suppose that the manufacturer has the option of hiring an infallible forecaster for $15,000 to find out for certain whether there will be a recession. Based on the original 2 to 1 odds that there will be a recession, would it be worthwhile for the manufacturer to spend this $15,000?

**9.11**    Each of the following is the payoff matrix (the payments Player A makes to Player B) for a zero-sum two-person game. Eliminate all dominated strategies and determine the optimum strategy for each player as well as the value of the game:

(a)

| 3 | −2 |
|---|---|
| 5 | 7 |

(b)

| 14 | 11 |
|---|---|
| 16 | −2 |

(c)

| −5 | 0 | 3 |
|---|---|---|
| −6 | −3 | −3 |
| −12 | −1 | 1 |

(d)

| 7 | 10 | 8 |
|---|---|---|
| 8 | 8 | 11 |
| 7 | 5 | 9 |

**9.12**    Each of the following is the payoff matrix of a zero-sum two-person game. Find the saddle point (or saddle points) and the value of each game:

(a)

| −1 | 5 | −2 |
|---|---|---|
| 0 | 3 | 1 |
| −2 | −4 | 5 |

(b)

| 3 | 2 | 4 | 9 |
|---|---|---|---|
| 4 | 4 | 4 | 3 |
| 5 | 6 | 5 | 6 |
| 5 | 7 | 5 | 9 |

**9.13**    A small town has two service stations, which share the town's market for gasoline. The owner of Station A is debating whether or not to give away free glasses to her customers as part of a promotional scheme, and the owner of Station $B$ is debating whether or not to give away free steak knives. They know (from similar situations elsewhere) that if Station $A$ gives away free glasses and Station $B$ does not give away free steak knives, Station $A$'s share of the market will increase by 6 percent; if Station $B$ gives away free steak knives and Station $A$ does not give away free glasses, Station $B$'s share of the market will increase by 8 percent; and if both stations give away the respective items, Station $B$'s share of the market will increase by 3 percent.

(a)   Present this information in the form of a payoff table, in which the entries are Station $A$'s losses in its share of the market.

(b)   Find optimum strategies for the owners of the two stations.

**9.14**  Verify the two probabilities $\frac{4}{17}$ and $\frac{13}{17}$, which we gave on page 338 for the randomized strategy of Player $B$.

**9.15**  The following is the payoff matrix of a $2 \times 2$ zero-sum two-person game:

| 3  | −4 |
|----|----|
| −3 | 1  |

(a)   What randomized strategy should Player $A$ use so as to minimize his maximum expected loss?

(b)   What randomized strategy should Player $B$ use so as to maximize her minimum expected gain?

(c)   What is the value of the game?

**9.16**  With reference to Exercise 9.4, what randomized strategy will minimize Ms. Cooper's maximum expected cost?

**9.17**  A country has two airfields with installations worth $2,000,000 and $10,000,000, respectively, of which it can defend only one against an attack by its enemy. The enemy, on the other hand, can attack only one of these airfields and take it successfully only if it is left undefended. Considering the "payoff" to the country to be the total value of the installations it holds after the attack, find the optimum strategy of the country as well as that of its enemy, and the value of the "game."

**9.18**  Two persons agree to play the following game: The first writes either 1 or 4 on a slip of paper and at the same time the second writes either 0 or 3 on another slip of paper. If the sum of the two numbers is odd, the first wins this amount in dollars; otherwise, the second wins $2.

(a)   Construct the payoff matrix in which the payoffs are the first person's losses.

(b)   What randomized decision procedure should the first person use so as to minimize her maximum expected loss?

(c)   What randomized decision procedure should the second person use so as to maximize his minimum expected gain?

**9.19**   There are two gas stations in a certain block, and the owner of the first station knows that if neither station lowers its prices, he can expect a net profit of $100 on any given day. If he lowers his prices while the other station does not, he can expect a net profit of $140; if he does not lower his prices but the other station does, he can expect a net profit of $70; and if both stations participate in this "price war," he can expect a net profit of $80. The owners of the two gas stations decide independently what prices to charge on any given day, and it is assumed that they cannot change their prices after they discover those charged by the other.

(a)   Should the owner of the first gas station charge his regular prices or should he lower them, if he wants to maximize his minimum net profit?

(b)   Assuming that the above profit figures apply also to the second gas station, how might the owners of the gas stations collude so that each could expect a net profit of $105?

Note that this "game" is not zero-sum, so that the possibility of collusion opens entirely new possibilities.

## 9.3   STATISTICAL GAMES

In statistical inference we base decisions about populations on sample data, and it is by no means farfetched to look upon such an inference as a game between Nature, which controls the relevant feature (or features) of the population, and the person (scientist, or statistician) who must arrive at some decision about Nature's choice. For instance, if we want to estimate the mean $\mu$ of a normal population on the basis of a random sample of size $n$, we could say that Nature has control over the "true" value of $\mu$. On the other hand, we might estimate $\mu$ in terms of the value of the sample mean or that of the sample median, and presumably there is some penalty or reward which depends on the size of our error.

In spite of the obvious similarity between this problem and the ones of the preceding section, there are essentially two features in which **statistical games** are different. First, there is the question which we already met when we tried to apply the theory of games to the decision problem of Example 9.1, namely, the question of whether it is reasonable to treat Nature as a malevolent opponent. Obviously not, but this does not simplify matters; if we could treat Nature as a rational opponent, we would know, at least, what to expect.

The other distinction is that in the games of Section 9.2 each player had to choose his strategy without any knowledge of what his opponent had done or was planning to do, whereas in a statistical game the statistician is supplied with sample data which provide him with some information about Nature's choice. This also complicates matters, but it merely amounts to the fact that we are dealing with more complicated kinds of games. To illustrate, let us consider the following decision problem: *We are told that a coin is either balanced with heads on one side and tails on the other or two-headed. We cannot inspect the coin, but we can flip it once and observe whether it comes up heads or tails. Then we must decide whether or not it is two-headed, keeping in mind that there is a penalty of $1 if our decision is wrong, and no penalty (or reward) if our decision is right.* If we ignored the fact that we can observe one flip of the coin, we could treat the problem as the following game:

*Player A (The Statistician)*

|  | | $a_1$ | $a_2$ |
|---|---|---|---|
| *Player B* | $\theta_1$ | $L(a_1, \theta_1) = 0$ | $L(a_2, \theta_1) = 1$ |
| *(Nature)* | $\theta_2$ | $L(a_1, \theta_2) = 1$ | $L(a_2, \theta_2) = 0$ |

which should remind the reader of the scheme on page 332. Now, $\theta_1$ is the "state of Nature" that the coin is two-headed, $\theta_2$ is the "state of Nature" that the coin is balanced with heads on one side and tails on the other, $a_1$ is the statistician's decision that the coin is two-headed, and $a_2$ is the statistician's decision that the coin is balanced with heads on one side and tails on the other. The entries in the table are the corresponding values of the given loss function.

Now let us consider also the fact that we (Player $A$, or the statistician) know what happened in the flip of the coin; that is, we know whether a random variable $X$ has taken on the value $x = 0$ (heads) or $x = 1$ (tails). Since we shall want to make use of this information in choosing between $a_1$ and $a_2$, we need a function, a **decision function**, which tells us what action to take when $x = 0$ and what action to take when $x = 1$. One possibility is to choose $a_1$ when $x = 0$ and $a_2$ when $x = 1$, and we can express this symbolically by writing

$$d_1(x) = \begin{cases} a_1 & \text{when } x = 0 \\ a_2 & \text{when } x = 1 \end{cases}$$

or more simply $d_1(0) = a_1$ and $d_1(1) = a_2$. The purpose of the subscript is to distinguish this decision function from others, for instance, from

$$d_2(0) = a_1 \quad \text{and} \quad d_2(1) = a_1$$

which tells us to choose $a_1$ regardless of the outcome of the experiment, from

$$d_3(0) = a_2 \quad \text{and} \quad d_3(1) = a_2$$

which tells us to choose $a_2$ regardless of the outcome of the experiment, and from

$$d_4(0) = a_2 \quad \text{and} \quad d_4(1) = a_1$$

which tells us to choose $a_2$ when $x = 0$ and $a_1$ when $x = 1$.

To compare the merits of all these decision functions, let us first determine the expected losses to which they lead for the various strategies of Nature, namely, the values of the **risk function**

$$R(d_i, \theta_j) = E\{L[d_i(X), \theta_j]\}$$

where the expectation is taken with respect to the random variable $X$. Since the probabilities for $x = 0$ and $x = 1$ are, respectively, 1 and 0 for $\theta_1$, and $\frac{1}{2}$ and $\frac{1}{2}$ for $\theta_2$, we get

$$R(d_1, \theta_1) = 1 \cdot L(a_1, \theta_1) + 0 \cdot L(a_2, \theta_1) = 1 \cdot 0 + 0 \cdot 1 = 0$$
$$R(d_1, \theta_2) = \tfrac{1}{2} \cdot L(a_1, \theta_2) + \tfrac{1}{2} \cdot L(a_2, \theta_2) = \tfrac{1}{2} \cdot 1 + \tfrac{1}{2} \cdot 0 = \tfrac{1}{2}$$
$$R(d_2, \theta_1) = 1 \cdot L(a_1, \theta_1) + 0 \cdot L(a_1, \theta_1) = 1 \cdot 0 + 0 \cdot 0 = 0$$
$$R(d_2, \theta_2) = \tfrac{1}{2} \cdot L(a_1, \theta_2) + \tfrac{1}{2} \cdot L(a_1, \theta_2) = \tfrac{1}{2} \cdot 1 + \tfrac{1}{2} \cdot 1 = 1$$
$$R(d_3, \theta_1) = 1 \cdot L(a_2, \theta_1) + 0 \cdot L(a_2, \theta_1) = 1 \cdot 1 + 0 \cdot 1 = 1$$
$$R(d_3, \theta_2) = \tfrac{1}{2} \cdot L(a_2, \theta_2) + \tfrac{1}{2} \cdot L(a_2, \theta_2) = \tfrac{1}{2} \cdot 0 + \tfrac{1}{2} \cdot 0 = 0$$
$$R(d_4, \theta_1) = 1 \cdot L(a_2, \theta_1) + 0 \cdot L(a_1, \theta_1) = 1 \cdot 1 + 0 \cdot 0 = 1$$
$$R(d_4, \theta_2) = \tfrac{1}{2} \cdot L(a_2, \theta_2) + \tfrac{1}{2} \cdot L(a_1, \theta_2) = \tfrac{1}{2} \cdot 0 + \tfrac{1}{2} \cdot 1 = \tfrac{1}{2}$$

where the values of the loss function were obtained from the table on page 343.

We have thus arrived at the following $4 \times 2$ zero-sum two-person game, in which the payoffs are the corresponding values of the risk function:

|  |  | $d_1$ | $d_2$ | $d_3$ | $d_4$ |
|---|---|---|---|---|---|
| *Player B* | $\theta_1$ | 0 | 0 | 1 | 1 |
| *(Nature)* | $\theta_2$ | $\frac{1}{2}$ | 1 | 0 | $\frac{1}{2}$ |

*Player A*
*(The Statistician)*

As can be seen by inspection, $d_2$ is dominated by $d_1$ and $d_4$ is dominated by $d_3$, so that $d_2$ and $d_4$ can be discarded—in decision theory we say that they are **inadmissible**. Actually, this should not come as a surprise, since in $d_2$ as well as $d_4$ we accept alternative $a_1$ (that the coin is two-headed) even though it came up tails.

This leaves us with the $2 \times 2$ zero-sum two-person game in which Player A has to choose between $d_1$ and $d_3$. It can easily be verified that if Nature is looked upon as a malevolent opponent, the optimum strategy is to randomize between $d_1$ and $d_3$ with respective probabilities of $\frac{2}{3}$ and $\frac{1}{3}$, and the value of the game (the expected risk) is $\frac{1}{3}$ of a dollar. If Nature is not looked upon as a malevolent opponent, some other criterion will have to be used for choosing between $d_1$ and $d_3$, and this will be discussed in the sections which follow. Incidentally, we formulated this problem with reference to a two-headed coin and an ordinary coin, but we could just as well have formulated it more abstractly as a decision problem in which we must decide on the basis of a single observation whether a random variable has the Bernoulli distribution with the parameter $\theta = 0$ or the parameter $\theta = \frac{1}{2}$.

To illustrate further the concepts of a loss function and a risk function, let us consider the following example, in which Nature as well as the statistician has a continuum of strategies.

## EXAMPLE 9.7

A random variable has the uniform density

$$f(x) = \begin{cases} \dfrac{1}{\theta} & \text{for } 0 < x < \theta \\ 0 & \text{elsewhere} \end{cases}$$

and we want to estimate the parameter $\theta$ (the "move" of Nature) on the basis of a single observation. If the decision function is to be of the form $d(x) = kx$, where $k \geqslant 1$, and the losses are proportional to the absolute value of the errors, that is,

$$L(kx, \theta) = c|kx - \theta|$$

where $c$ is a positive constant, find the value of $k$ which will minimize the risk.

### Solution

For the risk function we get

$$R(d, \theta) = \int_0^{\theta/k} c(\theta - kx) \cdot \frac{1}{\theta}\, dx + \int_{\theta/k}^{\theta} c(kx - \theta) \cdot \frac{1}{\theta}\, dx$$

$$= c\theta\left(\frac{k}{2} - 1 + \frac{1}{k}\right)$$

and there is nothing we can do about the factor $\theta$, but it can easily be verified that $k = \sqrt{2}$ will minimize $\dfrac{k}{2} - 1 + \dfrac{1}{k}$. Thus, if we actually took the observation and got $x = 5$, our estimate of $\theta$ would be $5\sqrt{2}$, or approximately 7.07.    ▲

## 9.4    DECISION CRITERIA

In Example 9.7 we were able to find a decision function which minimized the risk regardless of the true state of Nature (that is, regardless of the true value of the parameter $\theta$), but this is the exception rather than the rule. Had we not limited ourselves to decision functions of the form $d(x) = kx$, then the decision function given by $d(x) = \theta_1$ would be best when $\theta$ happens to equal $\theta_1$, the one given by $d(x) = \theta_2$ would be best when $\theta$ happens to equal $\theta_2, \ldots$, and it is obvious that there can be no decision function which is best for all values of $\theta$.

In general, we thus have to be satisfied with decision functions that are best only with respect to some criterion, and the two criteria which we shall study in this chapter are: (1) the **minimax criterion**, according to which we choose the decision function $d$ for which $R(d, \theta)$, maximized with respect to $\theta$, is a minimum; and (2) the **Bayes criterion**, according to which we choose the decision function $d$ for which the **Bayes risk** $E[R(d, \Theta)]$ is a minimum, where the expectation is taken with respect to $\Theta$. This requires that we look upon $\Theta$ as a random variable having a given distribution.

It is of interest to note that in the example of Section 9.1 we used both of these criteria. When we quoted odds for a recession, we assigned probabilities to the two states of Nature, $\theta_1$ and $\theta_2$, and when we suggested that the manufacturer minimize his expected loss, we suggested, in fact, that he use the Bayes criterion. Also, when we asked on page 331 what the manufacturer might do if he were a confirmed pessimist, we suggested that he would protect himself against the worst that can happen by using the minimax criterion.

## 9.5    THE MINIMAX CRITERION

If we apply the minimax criterion to the illustration of Section 9.3, dealing with the coin which is either two-headed or balanced with heads on one side and tails on the other, we find from the table on page 344 with $d_2$ and $d_4$ deleted that for $d_1$ the maximum risk is $\frac{1}{2}$, for $d_3$ the maximum risk is 1, and, hence, the one that minimizes the maximum risk is $d_1$.

## EXAMPLE 9.8

Use the minimax criterion to estimate the parameter $\theta$ of a binomial distribution on the basis of the random variable $X$, the observed number of successes in $n$ trials, when the decision function is of the form

$$d(x) = \frac{x + a}{n + b}$$

where $a$ and $b$ are constants, and the loss function is given by

$$L\left(\frac{x + a}{n + b}, \theta\right) = c\left(\frac{x + a}{n + b} - \theta\right)^2$$

where $c$ is a positive constant.

### Solution

The problem is to find the values of $a$ and $b$ which will minimize the corresponding risk function after it has been maximized with respect to $\theta$. After all, we have control over the choice of $a$ and $b$, while Nature (our presumed opponent) has control over the choice of $\theta$.

Since $E(X) = n\theta$ and $E(X^2) = n\theta(1 - \theta + n\theta)$, as we saw on page 189, it follows that

$$R(d, \theta) = E\left[c\left(\frac{X + a}{n + b} - \theta\right)^2\right]$$

$$= \frac{c}{(n + b)^2}[\theta^2(b^2 - n) + \theta(n - 2ab) + a^2]$$

and, using calculus, we could find the value of $\theta$ which maximizes this expression and then minimize $R(d, \theta)$ for this value of $\theta$ with respect to $a$ and $b$. This is not particularly difficult, but it is left to the reader in Exercise 9.23 as it involves some tedious algebraic detail.    ▲

To simplify the work in a problem of this kind, we can often use the **equalizer principle**, according to which (under fairly general conditions) the risk function of a minimax decision rule is a constant; for instance, it tells us that in Example 9.8 the risk function should not depend on the value of $\theta$.[†] To justify this principle, at least intuitively, observe that in Example 9.6 the minimax strategy of Player $A$ leads to an expected loss of \$3.41 regardless of whether Player $B$ chooses Strategy 1 or Strategy 2.

To make the risk function of Example 9.8 independent of $\theta$, the coefficients of $\theta$ and $\theta^2$ must both equal 0 in the expression for $R(d, \theta)$. This yields $b^2 - n$

---

[†]The exact conditions under which the equalizer principle holds are given in the book by T. S. Ferguson listed among the references at the end of this chapter.

= 0 and $n - 2ab = 0$, and, hence, $a = \frac{1}{2}\sqrt{n}$ and $b = \sqrt{n}$. Thus, the minimax decision function is given by

$$d(x) = \frac{x + \frac{1}{2}\sqrt{n}}{n + \sqrt{n}}$$

and if we actually obtained 39 successes in 100 trials, we would estimate the parameter $\theta$ of this binomial distribution as

$$d(39) = \frac{39 + \frac{1}{2}\sqrt{100}}{100 + \sqrt{100}} = 0.40$$

## 9.6 THE BAYES CRITERION

To apply the Bayes criterion in the illustration of Section 9.3, the one dealing with the coin which is either two-headed or balanced with heads on one side and tails on the other, we will have to assign probabilities to the two strategies of Nature, $\theta_1$ and $\theta_2$. If we assign $\theta_1$ and $\theta_2$, respectively, the probabilities $p$ and $1 - p$, it can be seen from the table on page 344 that for $d_1$ the Bayes risk is

$$0 \cdot p + \tfrac{1}{2} \cdot (1 - p) = \tfrac{1}{2} \cdot (1 - p)$$

and that for $d_3$ the Bayes risk is

$$1 \cdot p + 0 \cdot (1 - p) = p$$

It follows that the Bayes risk of $d_1$ is less than that of $d_3$ (and $d_1$ is to be preferred to $d_3$) when $p > \frac{1}{3}$, and that the Bayes risk of $d_3$ is less than that of $d_1$ (and $d_3$ is to be preferred to $d_1$) when $p < \frac{1}{3}$. When $p = \frac{1}{3}$, the two Bayes risks are equal, and we can use either $d_1$ or $d_3$.

## EXAMPLE 9.9

With reference to Example 9.7, suppose that the parameter of the uniform density is looked upon as a random variable with the probability density

$$h(\theta) = \begin{cases} \theta \cdot e^{-\theta} & \text{for } \theta > 0 \\ 0 & \text{elsewhere} \end{cases}$$

If there is no restriction on the form of the decision function and the loss function is quadratic, that is, its values are given by

$$L[d(x), \theta] = c\{d(x) - \theta\}^2$$

find the decision function which minimizes the Bayes risk.

## Solution

Since $\Theta$ is now a random variable, we look upon the original probability density as the conditional density

$$f(x|\theta) = \begin{cases} \dfrac{1}{\theta} & \text{for } 0 < x < \theta \\ 0 & \text{elsewhere} \end{cases}$$

and, letting $f(x, \theta) = f(x|\theta) \cdot h(\theta)$ in accordance with Definition 3.13, we get

$$f(x, \theta) = \begin{cases} e^{-\theta} & \text{for } 0 < x < \theta \\ 0 & \text{elsewhere} \end{cases}$$

As the reader will be asked to verify in Exercise 9.25, this yields

$$g(x) = \begin{cases} e^{-x} & \text{for } x > 0 \\ 0 & \text{elsewhere} \end{cases}$$

for the marginal density of $X$ and

$$\varphi(\theta|x) = \begin{cases} e^{x-\theta} & \theta > x \\ 0 & \text{elsewhere} \end{cases}$$

for the conditional density of $\Theta$ given $X = x$.

Now, the Bayes risk $E[R(d, \Theta)]$ which we shall want to minimize is given by the double integral

$$\int_0^\infty \left\{ \int_0^\theta c[d(x) - \theta]^2 f(x|\theta)\, dx \right\} h(\theta)\, d\theta$$

which can also be written as

$$\int_0^\infty \left\{ \int_x^\infty c[d(x) - \theta]^2 \varphi(\theta|x)\, d\theta \right\} g(x)\, dx$$

making use of the fact that $f(x|\theta) \cdot h(\theta) = \varphi(\theta|x) \cdot g(x)$ and changing the order of integration. To minimize this double integral, we must choose $d(x)$ for each $x$ so that the integral

$$\int_x^\infty c[d(x) - \theta]^2 \varphi(\theta|x) \, d\theta = \int_x^\infty c[d(x) - \theta]^2 e^{x-\theta} \, d\theta$$

is as small as possible. Differentiating with respect to $d(x)$ and putting the derivative equal to 0, we get

$$2ce^x \cdot \int_x^\infty [d(x) - \theta] e^{-\theta} \, d\theta = 0$$

This yields

$$d(x) \cdot \int_x^\infty e^{-\theta} \, d\theta - \int_x^\infty \theta e^{-\theta} \, d\theta = 0$$

and finally,

$$d(x) = \frac{\displaystyle\int_x^\infty \theta e^{-\theta} \, d\theta}{\displaystyle\int_x^\infty e^{-\theta} \, d\theta} = \frac{(x+1)e^{-x}}{e^{-x}} = x + 1$$

Thus, if the observation we get is $x = 5$ (as on page 346), this decision function gives the Bayes estimate $5 + 1 = 6$ for the parameter of the original uniform density. ▲

## EXERCISES

**9.20** With reference to the illustration on page 343, show that even if the coin is flipped $n$ times, there are only two admissible decision functions. Also construct a table showing the values of the risk function corresponding to these two decision functions and the two states of Nature.

**9.21** With reference to Example 9.7, show that if the losses are proportional to the squared errors instead of their absolute values, the risk function becomes

$$R(d, \theta) = \frac{c\theta^2}{3}(k^2 - 3k + 3)$$

and its minimum is at $k = \frac{3}{2}$.

**9.22** A statistician has to decide on the basis of a single observation whether the parameter $\theta$ of the density

$$f(x) = \begin{cases} \dfrac{2x}{\theta^2} & \text{for } 0 < x < \theta \\ 0 & \text{elsewhere} \end{cases}$$

equals $\theta_1$ or $\theta_2$, where $\theta_1 < \theta_2$. If he decides on $\theta_1$ when the observed value is less than the constant $k$, on $\theta_2$ when the observed value is greater than or equal to the constant $k$, and he is fined $C$ dollars for making the wrong decision, which value of $k$ will minimize the maximum risk?

**9.23** Finds the value of $\theta$ which maximizes the risk function of Example 9.8, and then find the values of $a$ and $b$ which minimize the risk function for that value of $\theta$. Compare the results with those given on page 348.

**9.24** If we assume in Example 9.8 that $\Theta$ is a random variable having a uniform density with $\alpha = 0$ and $\beta = 1$, show that the Bayes risk is given by

$$\frac{c}{(n + b)^2} \left[ \tfrac{1}{3}(b^2 - n) + \tfrac{1}{2}(n - 2ab) + a^2 \right]$$

Also show that this Bayes risk is a minimum when $a = 1$ and $b = 2$, so that the optimum Bayes decision rule is given by $d(x) = \dfrac{x + 1}{n + 2}$.

**9.25** Verify the results given on page 349 for the marginal density of $X$ and the conditional density of $\Theta$ given $X = x$.

**9.26** Suppose that we want to estimate the parameter $\theta$ of the geometric distribution on the basis of a single observation. If the loss function is given by

$$L[d(x), \theta] = c\{d(x) - \theta\}^2$$

and $\Theta$ is looked upon as a random variable having the uniform density $h(\theta) = 1$ for $0 < \theta < 1$ and $h(\theta) = 0$ elsewhere, duplicate the steps in Example 9.9 to show that

(a)   the conditional density of $\Theta$ given $X = x$ is

$$\varphi(\theta|x) = \begin{cases} x(x + 1)\theta(1 - \theta)^{x-1} & \text{for } 0 < \theta < 1 \\ 0 & \text{elsewhere} \end{cases}$$

(b)  the Bayes risk is minimized by the decision function

$$d(x) = \frac{2}{x + 2}$$

(*Hint*: Make use of the fact that the integral of any beta density is equal to 1.)

## APPLICATIONS

**9.27** A statistician has to decide on the basis of one observation whether the parameter $\theta$ of a Bernoulli distribution is $0$, $\frac{1}{2}$, or $1$; her loss in dollars (a penalty which is deducted from her fee) is 100 times the absolute value of her error.
  (a)  Construct a table showing the nine possible values of the loss function.
  (b)  List the nine possible decision functions and construct a table showing all the values of the corresponding risk function.
  (c)  Show that five of the decision functions are not admissible, and that according to the minimax criterion the remaining decision functions are all equally good.
  (d)  Which decision function is best according to the Bayes criterion, if the three possible values of the parameter $\theta$ are regarded as equally likely?

**9.28** A statistician has to decide on the basis of two observations whether the parameter $\theta$ of a binomial distribution is $\frac{1}{4}$ or $\frac{1}{2}$; his loss (a penalty which is deducted from his fee) is \$160 if he is wrong.
  (a)  Construct a table showing the four possible values of the loss function.
  (b)  List the eight possible decision functions and construct a table showing all the values of the corresponding risk function.
  (c)  Show that three of the decision functions are not admissible.
  (d)  Find the decision function which is best according to the minimax criterion.
  (e)  Find the decision function which is best according to the Bayes criterion, if the probabilities assigned to $\theta = \frac{1}{4}$ and $\theta = \frac{1}{2}$ are, respectively, $\frac{2}{3}$ and $\frac{1}{3}$.

**9.29** A manufacturer produces an item consisting of two components, which must both work for the item to function properly. The cost of returning one of the items to the manufacturer for repairs is $\alpha$ dollars, the cost of inspecting one of the components is $\beta$ dollars, and the cost of repairing a faulty component is $\varphi$ dollars. She can ship each item without inspection with the guarantee that it will be put into perfect working condition at her factory in case it does not work; she can inspect both components and repair them if necessary; or she can randomly select one of the components and ship the item with the original guarantee if it works, or repair it and also check the other component.

(a)  Construct a table showing the manufacturer's expected losses corresponding to her three "strategies" and the three "states" of Nature that 0, 1, or 2 of the components do not work.

(b)  What should the manufacturer do if $\alpha = \$25.00$, $\varphi = \$10.00$, and she wants to minimize her maximum expected losses?

(c)  What should the manufacturer do to minimize her Bayes risk if $\alpha = \$10.00$, $\beta = \$12.00$, $\varphi = \$30.00$, and she feels that the probabilities for 0, 1, and 2 defective components are, respectively, 0.70, 0.20, and 0.10?

## REFERENCES

Some fairly elementary material on the theory of games and decision theory can be found in

CHERNOFF, H., and MOSES, L. E., *Elementary Decision Theory*. Mineola, N.Y.: Dover Publications, Inc. (Republication of 1959 edition),

DRESHER, M., *Games of Strategy: Theory and Applications*. Englewood Cliffs, N.J.: Prentice Hall, Inc., 1961,

HAMBURG, M. *Statistical Analysis for Decision Making,* 4th ed. Orlando, Fla.: Harcourt Brace Jovanovich, 1988,

MCKINSEY, J. C. C., *Introduction to the Theory of Games*. New York: McGraw-Hill Book Company, 1952,

OWEN, G., *Game Theory*. Philadelphia: W. B. Saunders Company, 1968,

WILLIAMS, J. D., *The Compleat Strategyst*. New York: McGraw-Hill Book Company, 1954,

and more advanced treatments in

BICKEL, P. J., and DOKSUM, K. A., *Mathematical Statistics: Basic Ideas and Selected Topics*. San Francisco: Holden-Day, Inc., 1977,

FERGUSON, T. S., *Mathematical Statistics: A Decision Theoretic Approach*. New York: Academic Press, Inc., 1967,

WALD, A., *Statistical Decision Functions*. New York: John Wiley & Sons, Inc., 1950.

# 10

*Estimation:*
*Theory*

## 10.1 INTRODUCTION

Traditionally, problems of statistical inference are divided into **problems of estimation** and **tests of hypotheses**, though actually they are all decision problems and, hence, could be handled by the unified approach which we presented in the preceding chapter. The main difference between the two kinds of problems is that in problems of estimation we must determine the value of a parameter (or the values of several parameters) from a possible continuum of alternatives, whereas in tests of hypotheses we must decide whether to accept or reject a specific value or a set of specific values of a parameter (or those of several parameters).

When we use the value of a statistic to estimate a population parameter, we call this **point estimation**, and we refer to the value of the statistic as a **point estimate** of the parameter. For example, if we use a value of $\overline{X}$ to estimate the mean of a population, an observed sample proportion to estimate the parameter $\theta$ of a binomial population, or a value of $S^2$ to estimate a population variance, we are in each case using a point estimate of the parameter in question. These estimates are called point estimates because in each case a single number, or a single point on the real axis, is used to estimate the parameter.

Correspondingly, we refer to the statistics themselves as **point estimators**. For instance, $\overline{X}$ may be used as a point estimator of $\mu$, in which case $\bar{x}$ is a point estimate of this parameter. Similarly, $S^2$ may be used as a point estimator of $\sigma^2$, in which case $s^2$ is a point estimate of this parameter. Here we used the word "point" to distinguish between these estimators and estimates and the **interval estimators** and **interval estimates**, which we shall present in Chapter 11.

Since estimators are random variables, one of the key problems of point estimation is to study their sampling distributions. For instance, when we estimate the variance of a population on the basis of a random sample, we can hardly expect that the value of $S^2$ we get will actually equal $\sigma^2$, but it would be reassuring, at least, to know whether we can expect it to be close. Also, if we must decide whether to use a sample mean or a sample median to estimate the mean of a population, it would be important to know, among other things, whether $\overline{X}$ or $\tilde{X}$ is more likely to yield a value that is actually close.

Various statistical properties of estimators can, thus, be used to decide which estimator is most appropriate in a given situation, which will expose us to the smallest risk, which will give us the most information at the lowest cost, and so forth. The particular properties of estimators that we shall discuss in Sections 10.2 through 10.6 are **unbiasedness**, **minimum variance**, **efficiency**, **consistency**, **sufficiency**, and **robustness**.

## 10.2 UNBIASED ESTIMATORS

As we saw on page 346, perfect decision functions do not exist, and in connection with problems of estimation this means that there are no perfect estimators that always give the right answer. Thus, it would seem reasonable that an estimator should do so at least on the average—namely, that its expected value should equal the parameter which it is supposed to estimate. If this is the case, the estimator is said to be **unbiased**; otherwise, it is said to be **biased**. Formally,

---

**DEFINITION 10.1**  A statistic $\hat{\Theta}$ is an **unbiased estimator** of the parameter $\theta$ if and only if $E(\hat{\Theta}) = \theta$.

---

The following are some examples of unbiased as well as biased estimators.

## EXAMPLE 10.1

If $X$ has the binomial distribution with the parameters $n$ and $\theta$, show that the sample proportion, $\dfrac{X}{n}$, is an unbiased estimator of $\theta$.

*Solution*

Since $E(X) = n\theta$, it follows that

$$E\left(\frac{X}{n}\right) = \frac{1}{n} \cdot E(X) = \frac{1}{n} \cdot n\theta = \theta$$

and hence that $\dfrac{X}{n}$ is an unbiased estimator of $\theta$.    ▲

## EXAMPLE 10.2

Show that unless $\theta = \frac{1}{2}$, the minimax estimator of the binomial parameter $\theta$ on page 348 is biased.

*Solution*

Since $E(X) = n\theta$, it follows that

$$E\left(\frac{X + \frac{1}{2}\sqrt{n}}{n + \sqrt{n}}\right) = \frac{E(X + \frac{1}{2}\sqrt{n})}{n + \sqrt{n}} = \frac{n\theta + \frac{1}{2}\sqrt{n}}{n + \sqrt{n}}$$

and it can easily be seen that this quantity does not equal $\theta$ unless $\theta = \frac{1}{2}$.    ▲

## EXAMPLE 10.3

If $X_1, X_2, \ldots, X_n$ constitute a random sample from the population given by

$$f(x) = \begin{cases} e^{-(x-\delta)} & \text{for } x > \delta \\ 0 & \text{elsewhere} \end{cases}$$

show that $\overline{X}$ is a biased estimator of $\delta$.

*Solution*

Since the mean of the population is

$$\mu = \int_{\delta}^{\infty} x \cdot e^{-(x-\delta)} \, dx = 1 + \delta$$

it follows from Theorem 8.6 that $E(\overline{X}) = 1 + \delta \neq \delta$ and hence that $\overline{X}$ is a biased estimator of $\delta$.     ▲

When $\hat{\Theta}$ is a biased estimator of $\theta$, it may be of interest to know the extent of the **bias**, given by

$$b(\theta) = E(\hat{\Theta}) - \theta$$

Thus, for Example 10.2 the bias is

$$\frac{n\theta + \frac{1}{2}\sqrt{n}}{n + \sqrt{n}} - \theta = \frac{\frac{1}{2} - \theta}{\sqrt{n} + 1}$$

and it can be seen that it tends to be small when $\theta$ is close to $\frac{1}{2}$ and also when $n$ is large. Indeed, $\lim_{n\to\infty} b(\theta) = 0$ and we say that the estimator is **asymptotically unbiased**.

So far as Example 10.3 is concerned, the bias is $(1 + \delta) - \delta = 1$, but here there is something we can do about it. Since $E(\overline{X}) = 1 + \delta$, it follows that $E(\overline{X} - 1) = \delta$ and hence that $\overline{X} - 1$ is an unbiased estimator of $\delta$. The following is another example where a minor modification of an estimator leads to an estimator that is unbiased.

# EXAMPLE 10.4

If $X_1, X_2, \ldots, X_n$ constitute a random sample from a uniform population with $\alpha = 0$, show that the largest sample value (that is, the $n$th order statistic, $Y_n$) is a biased estimator of the parameter $\beta$. Also, modify this estimator of $\beta$ to make it unbiased.

*Solution*

Substituting into the formula for $g_n(y_n)$ on page 323, we find that the sampling distribution of $Y_n$ is given by

$$g_n(y_n) = n \cdot \frac{1}{\beta} \cdot \left( \int_0^{y_n} \frac{1}{\beta} \, dx \right)^{n-1}$$

$$= \frac{n}{\beta^n} \cdot y_n^{n-1}$$

for $0 < y_n < \beta$ and $g_n(y_n) = 0$ elsewhere, and hence that

$$E(Y_n) = \frac{n}{\beta^n} \cdot \int_0^\beta y_n^n \, dy_n$$

$$= \frac{n}{n+1} \cdot \beta$$

Thus, $E(Y_n) \neq \beta$ and the $n$th order statistic is a biased estimator of the parameter $\beta$. However, since

$$E\left( \frac{n+1}{n} \cdot Y_n \right) = \frac{n+1}{n} \cdot \frac{n}{n+1} \cdot \beta$$

$$= \beta$$

it follows that $\dfrac{n+1}{n}$ times the largest sample value is an unbiased estimator of the parameter $\beta$.  ▲

Having discussed unbiasedness as a desirable property of an estimator, we can now explain why we divided by $n - 1$ and not by $n$ when we defined the sample variance—it makes $S^2$ an unbiased estimator of $\sigma^2$ for random samples from infinite populations.

---

**THEOREM 10.1** If $S^2$ is the variance of a random sample from an infinite population with the finite variance $\sigma^2$, then $E(S^2) = \sigma^2$.

---

***Proof.***   By Definition 8.2,

$$E(S^2) = E\left[ \frac{1}{n-1} \cdot \sum_{i=1}^n (X_i - \overline{X})^2 \right]$$

$$= \frac{1}{n-1} \cdot E\left[ \sum_{i=1}^{n} \{(X_i - \mu) - (\overline{X} - \mu)\}^2 \right]$$

$$= \frac{1}{n-1} \cdot \left[ \sum_{i=1}^{n} E\{(X_i - \mu)^2\} - n \cdot E\{(\overline{X} - \mu)^2\} \right]$$

Then, since $E\{(X_i - \mu)^2\} = \sigma^2$ and $E\{(\overline{X} - \mu)^2\} = \dfrac{\sigma^2}{n}$, it follows that

$$E(S^2) = \frac{1}{n-1} \cdot \left[ \sum_{i=1}^{n} \sigma^2 - n \cdot \frac{\sigma^2}{n} \right] = \sigma^2 \qquad \blacktriangledown$$

Although $S^2$ is an unbiased estimator of the variance of an infinite population, it is not an unbiased estimator of the variance of a finite population, and in neither case is $S$ an unbiased estimator of $\sigma$. The bias of $S$ as an estimator of $\sigma$ is discussed, among others, in the book by E. S. Keeping listed among the references at the end of this chapter.

The discussion of the preceding paragraph illustrates one of the difficulties associated with the concept of unbiasedness. It may not be retained under functional transformations; that is, if $\hat{\Theta}$ is an unbiased estimator of $\theta$, it does not necessarily follow that $\omega(\hat{\Theta})$ is an unbiased estimator of $\omega(\theta)$. Another difficulty associated with the concept of unbiasedness is that unbiased estimators are not necessarily unique. For instance, in Example 10.6 we shall see that $\dfrac{n+1}{n} \cdot Y_n$ is not the only unbiased estimator of the parameter $\beta$ of Example 10.4, and in Exercise 10.8 we shall see that $\overline{X} - 1$ is not the only unbiased estimator of the parameter $\delta$ of Example 10.3.

## 10.3  EFFICIENCY

If we have to choose one of several unbiased estimators of a given parameter, we usually take the one whose sampling distribution has the smallest variance. We already mentioned this on page 325, where, comparing the sample median with the sample mean, we said that the estimator with the smaller variance is "more reliable."

To check whether a given unbiased estimator has the smallest possible variance, namely, whether it is a **minimum variance unbiased estimator** (also called a **best unbiased estimator**), we make use of the fact that if $\hat{\Theta}$ is an unbiased estimator of $\theta$, it can be shown under very general conditions (referred to on page 397) that the variance of $\hat{\Theta}$ must satisfy the inequality

$$\text{var}(\hat{\Theta}) \geq \frac{1}{n \cdot E\left[\left(\dfrac{\partial \ln f(X)}{\partial \theta}\right)^2\right]}$$

where $f(x)$ is the value of the population density at $x$ and $n$ is the size of the random sample. This inequality, the **Cramér-Rao inequality**, leads to the following result.

---

**THEOREM 10.2** If $\hat{\Theta}$ is an unbiased estimator of $\theta$ and

$$\text{var}(\hat{\Theta}) = \frac{1}{n \cdot E\left[\left(\dfrac{\partial \ln f(X)}{\partial \theta}\right)^2\right]}$$

then $\hat{\Theta}$ is a minimum variance unbiased estimator of $\theta$.

---

Here, the quantity in the denominator is referred to as the **information** about $\theta$ which is supplied by the sample (see also Exercise 10.19). Thus, the smaller the variance, the greater the information.

## EXAMPLE 10.5

Show that $\overline{X}$ is a minimum variance unbiased estimator of the mean $\mu$ of a normal population.

### Solution

Since

$$f(x) = \frac{1}{\sigma\sqrt{2\pi}} \cdot e^{-\frac{1}{2}\left(\frac{x-\mu}{\sigma}\right)^2} \qquad \text{for } -\infty < x < \infty$$

it follows that

$$\ln f(x) = -\ln \sigma\sqrt{2\pi} - \frac{1}{2}\left(\frac{x-\mu}{\sigma}\right)^2$$

so that

$$\frac{\partial \ln f(x)}{\partial \mu} = \frac{1}{\sigma}\left(\frac{x-\mu}{\sigma}\right)$$

and, hence,

$$E\left[\left(\frac{\partial \ln f(X)}{\partial \mu}\right)^2\right] = \frac{1}{\sigma^2}\cdot E\left[\left(\frac{X-\mu}{\sigma}\right)^2\right] = \frac{1}{\sigma^2}\cdot 1 = \frac{1}{\sigma^2}$$

Thus,

$$\frac{1}{n\cdot E\left[\left(\dfrac{\partial \ln f(X)}{\partial \mu}\right)^2\right]} = \frac{1}{n\cdot\dfrac{1}{\sigma^2}} = \frac{\sigma^2}{n}$$

and since $\overline{X}$ is unbiased and $\text{var}(\overline{X}) = \dfrac{\sigma^2}{n}$ according to Theorem 8.1, it follows that $\overline{X}$ is a minimum variance unbiased estimator of $\mu$.    ▲

It would be erroneous to conclude from this example that $\overline{X}$ is a minimum variance unbiased estimator of the mean of any population. Indeed, in Exercise 10.30 the reader will be asked to verify that this is not so for random samples of size $n = 3$ from the continuous uniform population with $\alpha = \theta - \frac{1}{2}$ and $\beta = \theta + \frac{1}{2}$.

As we have indicated, unbiased estimators of one and the same parameter are usually compared in terms of the size of their variances. If $\hat{\Theta}_1$ and $\hat{\Theta}_2$ are two unbiased estimators of the parameter $\theta$ of a given population and the variance of $\hat{\Theta}_1$ is less than the variance of $\hat{\Theta}_2$, we say that $\hat{\Theta}_1$ is **relatively more efficient** than $\hat{\Theta}_2$. Also, we use the ratio

$$\frac{\text{var}(\hat{\Theta}_1)}{\text{var}(\hat{\Theta}_2)}$$

as a measure of the efficiency of $\hat{\Theta}_2$ relative to $\hat{\Theta}_1$.

## EXAMPLE 10.6

In Example 10.4 we showed that if $X_1, X_2, \ldots, X_n$ constitute a random sample from a uniform population with $\alpha = 0$, then $\dfrac{n+1}{n}\cdot Y_n$ is an unbiased estimator of $\beta$.

(a)   Show that $2\overline{X}$ is also an unbiased estimator of $\beta$.

(b)   Compare the efficiency of these two estimators of $\beta$.

## Solution

(a)   Since the mean of the population is $\mu = \dfrac{\beta}{2}$ according to Theorem 6.1,

it follows from Theorem 8.1 that $E(\overline{X}) = \dfrac{\beta}{2}$ and hence that $E(2\overline{X}) =$

$\beta$. Thus, $2\overline{X}$ is an unbiased estimator of $\beta$.

(b)   First we must find the variances of the two estimators. Using the sampling distribution of $Y_n$ and the expression for $E(Y_n)$ given in Example 10.4, we get

$$E(Y_n^2) = \frac{n}{\beta^n} \cdot \int_0^\beta y_n^{n+1}\, dy_n = \frac{n}{n+2} \cdot \beta^2$$

and

$$\text{var}(Y_n) = \frac{n}{n+2} \cdot \beta^2 - \left(\frac{n}{n+1} \cdot \beta\right)^2$$

Leaving the details to the reader in Exercise 10.27, it can thus be shown that

$$\text{var}\left(\frac{n+1}{n} \cdot Y_n\right) = \frac{\beta^2}{n(n+2)}$$

Since the variance of the population is $\sigma^2 = \dfrac{\beta^2}{12}$ according to Theorem

6.1, it follows from Theorem 8.1 that $\text{var}(\overline{X}) = \dfrac{\beta^2}{12n}$ and hence that

$$\text{var}(2\overline{X}) = 4 \cdot \text{var}(\overline{X}) = \frac{\beta^2}{3n}$$

Therefore, the efficiency of $2\overline{X}$ relative to $\dfrac{n+1}{n} \cdot Y_n$ is given by

$$\frac{\text{var}\left(\frac{n+1}{n} \cdot Y_n\right)}{\text{var}(2\bar{X})} = \frac{\frac{\beta^2}{n(n+2)}}{\frac{\beta^2}{3n}} = \frac{3}{n+2}$$

and it can be seen that for $n > 1$ the estimator based on the $n$th order statistic is much more efficient than the other one. For $n = 10$, for example, the relative efficiency is only 25 percent, and for $n = 25$ it is only 11 percent.    ▲

# EXAMPLE 10.7

When the mean of a normal population is estimated on the basis of a random sample of size $2n + 1$, what is the efficiency of the median relative to the mean?

## Solution

From Theorem 8.1 we know that $\bar{X}$ is unbiased and that

$$\text{var}(\bar{X}) = \frac{\sigma^2}{2n+1}$$

So far as $\tilde{X}$ is concerned, it is unbiased by virtue of the symmetry of the normal distribution about its mean, and we know from the discussion following Theorem 8.17 that for large samples

$$\text{var}(\tilde{X}) = \frac{\pi\sigma^2}{4n}$$

Thus, for large samples, the efficiency of the median relative to the mean is approximately

$$\frac{\text{var}(\bar{X})}{\text{var}(\tilde{X})} = \frac{\frac{\sigma^2}{2n+1}}{\frac{\pi\sigma^2}{4n}} = \frac{4n}{\pi(2n+1)}$$

and the **asymptotic efficiency** of the median with respect to the mean is

$$\lim_{n \to \infty} \frac{4n}{\pi(2n+1)} = \frac{2}{\pi}$$

or about 64 percent.   ▲

The result of the preceding example may be interpreted as follows: For large samples, the mean requires only 64 percent as many observations as the median to estimate $\mu$ with the same reliability.

It is important to note that we have limited our discussion of relative efficiency to unbiased estimators. If we included biased estimators, we could always assure ourselves of an estimator with zero variance by letting its values equal the same constant regardless of the data which we may obtain. Therefore, if $\hat{\Theta}$ is not an unbiased estimator of a given parameter $\theta$, we judge its merits and make efficiency comparisons on the basis of the **mean square error** $E[(\hat{\Theta} - \theta)^2]$ instead of the variance of $\hat{\Theta}$.

## EXERCISES

**10.1** If $X_1, X_2, \ldots , X_n$ constitute a random sample from a population with the mean $\mu$, what condition must be imposed on the constants $a_1, a_2, \ldots , a_n$ so that

$$a_1X_1 + a_2X_2 + \cdots + a_nX_n$$

is an unbiased estimator of $\mu$?

**10.2** If $\hat{\Theta}_1$ and $\hat{\Theta}_2$ are unbiased estimators of the same parameter $\theta$, what condition must be imposed on the constants $k_1$ and $k_2$ so that

$$k_1\hat{\Theta}_1 + k_2\hat{\Theta}_2$$

is also an unbiased estimator of $\theta$?

**10.3** Use the formula for the sampling distribution of $\bar{X}$ on page 323 to show that for random samples of size $n = 3$ the median is an unbiased estimator of the parameter $\theta$ of a uniform population with $\alpha = \theta - \frac{1}{2}$ and $\beta = \theta + \frac{1}{2}$.

**10.4** Use the result of Example 8.4 to show that for random samples of size $n = 3$ the median is a biased estimator of the parameter $\theta$ of an exponential population.

**10.5** Given a random sample of size $n$ from a population which has the known mean $\mu$ and the finite variance $\sigma^2$, show that

$$\frac{1}{n} \cdot \sum_{i=1}^{n} (X_i - \mu)^2$$

is an unbiased estimator of $\sigma^2$.

**10.6** Use the results of Theorem 8.1 to show that $\bar{X}^2$ is an asymptotically unbiased estimator of $\mu^2$.

**10.7** Show that $\dfrac{X + 1}{n + 2}$ is a biased estimator of the binomial parameter $\theta$. Is this estimator asymptotically unbiased?

**10.8** With reference to Example 10.3, find an unbiased estimator of $\delta$ based on the smallest sample value (that is, on the first order statistic, $Y_1$).

**10.9** With reference to Example 10.4, find an unbiased estimator of $\beta$ based on the smallest sample value (that is, on the first order statistic, $Y_1$).

**10.10** If $X_1, X_2, \ldots, X_n$ constitute a random sample from a normal population with $\mu = 0$, show that

$$\sum_{i=1}^{n} \frac{X_i^2}{n}$$

is an unbiased estimator of $\sigma^2$.

**10.11** If $X$ is a random variable having the binomial distribution with the parameters $n$ and $\theta$, show that $n \cdot \dfrac{X}{n} \cdot \left(1 - \dfrac{X}{n}\right)$ is a biased estimator of the variance of $X$.

**10.12** If a random sample of size $n$ is taken without replacement from the finite population which consists of the positive integers $1, 2, \ldots, k$, show that

(a)   the sampling distribution of the $n$th order statistic, $Y_n$, is given by

$$f(y_n) = \frac{\binom{y_n - 1}{n - 1}}{\binom{k}{n}}$$

for $y_n = n, \ldots, k$;

(b)   $\dfrac{n + 1}{n} \cdot Y_n - 1$ is an unbiased estimator of $k$.

See also Exercise 10.35.

**10.13** Show that if $\hat{\Theta}$ is an unbiased estimator of $\theta$ and $\text{var}(\hat{\Theta}) \neq 0$, then $\hat{\Theta}^2$ is not an unbiased estimator of $\theta^2$.

**10.14** Show that the sample proportion $\dfrac{X}{n}$ is a minimum variance unbiased estimator of the binomial parameter $\theta$. $\left(Hint:\ \text{Treat}\ \dfrac{X}{n}\ \text{as the mean of a random sample of size } n \text{ from a Bernoulli population with the parameter } \theta.\right)$

**10.15** Show that the mean of a random sample of size $n$ is a minimum variance unbiased estimator of the parameter $\lambda$ of a Poisson population.

**10.16** If $\hat{\Theta}_1$ and $\hat{\Theta}_2$ are independent unbiased estimators of a given parameter $\theta$ and $\text{var}(\hat{\Theta}_1) = 3 \cdot \text{var}(\hat{\Theta}_2)$, find the constants $a_1$ and $a_2$ such that $a_1\hat{\Theta}_1 + a_2\hat{\Theta}_2$ is an unbiased estimator with minimum variance for such a linear combination.

**10.17** Show that the mean of a random sample of size $n$ from an exponential population is a minimum variance unbiased estimator of the parameter $\theta$.

**10.18** Show that for the unbiased estimator of Example 10.4, $\dfrac{n+1}{n} \cdot Y_n$, the Cramér-Rao inequality is not satisfied.

**10.19** The information about $\theta$ in a random sample of size $n$ is also given by

$$-n \cdot E\left[\frac{\partial^2 \ln f(X)}{\partial\theta^2}\right]$$

where $f(x)$ is the value of the population density at $x$, provided that the extremes of the region for which $f(x) \neq 0$ do not depend on $\theta$. The derivation of this formula takes the following steps:

(a)  Differentiating the expressions on both sides of

$$\int f(x)\, dx = 1$$

with respect to $\theta$, show that

$$\int \frac{\partial \ln f(x)}{\partial\theta} \cdot f(x)\, dx = 0$$

by interchanging the order of integration and differentiation.

(b)  Differentiating again with respect to $\theta$, show that

$$E\left[\left(\frac{\partial \ln f(X)}{\partial\theta}\right)^2\right] = -E\left[\frac{\partial^2 \ln f(X)}{\partial\theta^2}\right]$$

**10.20** Rework Example 10.5 using the alternative formula for the information given in Exercise 10.19.

**10.21** If $\overline{X}_1$ is the mean of a random sample of size $n$ from a normal population with the mean $\mu$ and the variance $\sigma_1^2$, $\overline{X}_2$ is the mean of a random sample of size $n$ from a normal population with the mean $\mu$ and the variance $\sigma_2^2$, and the two samples are independent, show that

(a)  $\omega \cdot \overline{X}_1 + (1 - \omega) \cdot \overline{X}_2$, where $0 \leqslant \omega \leqslant 1$, is an unbiased estimator of $\mu$;

(b)  the variance of this estimator is a minimum when

$$\omega = \frac{\sigma_2^2}{\sigma_1^2 + \sigma_2^2}$$

**10.22** With reference to Exercise 10.21, find the efficiency of the estimator of part (a) with $\omega = \frac{1}{2}$ relative to this estimator with

$$\omega = \frac{\sigma_2^2}{\sigma_1^2 + \sigma_2^2}$$

**10.23** If $\overline{X}_1$ and $\overline{X}_2$ are the means of independent random samples of size $n_1$ and $n_2$ from a normal population with the mean $\mu$ and the variance $\sigma^2$, show that the variance of the unbiased estimator

$$\omega \cdot \overline{X}_1 + (1 - \omega) \cdot \overline{X}_2$$

is a minimum when $\omega = \dfrac{n_1}{n_1 + n_2}$.

**10.24** With reference to Exercise 10.23, find the efficiency of the estimator with $\omega = \frac{1}{2}$ relative to the estimator with $\omega = \dfrac{n_1}{n_1 + n_2}$.

**10.25** If $X_1$, $X_2$, and $X_3$ constitute a random sample of size $n = 3$ from a normal population with the mean $\mu$ and the variance $\sigma^2$, find the efficiency of $\dfrac{X_1 + 2X_2 + X_3}{4}$ relative to $\dfrac{X_1 + X_2 + X_3}{3}$.

**10.26** If $X_1$ and $X_2$ constitute a random sample of size $n = 2$ from an exponential population, find the efficiency of $2Y_1$ relative to $\overline{X}$, where $Y_1$ is the first order statistic and $2Y_1$ and $\overline{X}$ are both unbiased estimators of the parameter $\theta$.

**10.27** Verify the result given for $\text{var}\left(\dfrac{n + 1}{n} \cdot Y_n\right)$ in Example 10.6.

**10.28** With reference to Example 10.3, we showed on page 357 that $\overline{X} - 1$ is an unbiased estimator of $\delta$, and in Exercise 10.8 the reader was asked to find another unbiased estimator of $\delta$ based on the smallest sample value. Find the efficiency of the first of these two estimators relative to the second.

**10.29** With reference to Exercise 10.12, show that $2\overline{X} - 1$ is also an unbiased estimator of $k$, and find the efficiency of this estimator relative to the one of part (b) of Exercise 10.12 for
(a)  $n = 2$;
(b)  $n = 3$.

**10.30** Since the variances of the mean and the midrange are not affected if the same constant is added to each observation, we can determine these variances for random samples of size 3 from the uniform population

$$f(x) = \begin{cases} 1 & \text{for } \theta - \tfrac{1}{2} < x < \theta + \tfrac{1}{2} \\ 0 & \text{elsewhere} \end{cases}$$

by referring instead to the uniform population

$$f(x) = \begin{cases} 1 & \text{for } 0 < x < 1 \\ 0 & \text{elsewhere} \end{cases}$$

(a)  Show that $E(X) = \tfrac{1}{2}$, $E(X^2) = \tfrac{1}{3}$, and $\text{var}(X) = \tfrac{1}{12}$ for this population, so that for a random sample of size $n = 3$, $\text{var}(\overline{X}) = \tfrac{1}{36}$.

(b)  Use the results of Exercises 8.67 and 8.73 (or derive the necessary densities and joint density) to show that for a random sample of size $n = 3$ from this population, the order statistics $Y_1$ and $Y_3$ have $E(Y_1) = \tfrac{1}{4}$, $E(Y_1^2) = \tfrac{1}{10}$, $E(Y_3) = \tfrac{3}{4}$, $E(Y_3^2) = \tfrac{3}{5}$, and $E(Y_1 Y_3) = \tfrac{1}{5}$, so that $\text{var}(Y_1) = \tfrac{3}{80}$, $\text{var}(Y_3) = \tfrac{3}{80}$, and $\text{cov}(Y_1, Y_3) = \tfrac{1}{80}$.

(c)  Use the results of part (b) and Theorem 4.14 to show that
$$E\left(\frac{Y_1 + Y_3}{2}\right) = \frac{1}{2} \text{ and } \text{var}\left(\frac{Y_1 + Y_3}{2}\right) = \frac{1}{40},$$
and hence that for random samples of size $n = 3$ from the given uniform population, the midrange is unbiased and more efficient than the mean.

**10.31** Show that if $\hat{\Theta}$ is a biased estimator of $\theta$, then

$$E[(\hat{\Theta} - \theta)^2] = \text{var}(\hat{\Theta}) + [b(\theta)]^2$$

**10.32** If $\hat{\Theta}_1 = \dfrac{X}{n}$, $\hat{\Theta}_2 = \dfrac{X + 1}{n + 2}$, and $\hat{\Theta}_3 = \tfrac{1}{3}$ are estimators of the parameter $\theta$ of a binomial population and $\theta = \tfrac{1}{2}$, for what values of $n$ is
(a)  the mean square error of $\hat{\Theta}_2$ less than the variance of $\hat{\Theta}_1$;
(b)  the mean square error of $\hat{\Theta}_3$ less than the variance of $\hat{\Theta}_1$?

### APPLICATIONS

**10.33** Random samples of size $n$ are taken from normal populations with the mean $\mu$ and the variances $\sigma_1^2 = 4$ and $\sigma_2^2 = 9$. If $\bar{x}_1 = 26.0$ and $\bar{x}_2 = 32.5$, estimate $\mu$ using the estimator of part (b) of Exercise 10.21.

**10.34** Independent random samples of size $n_1$ and $n_2$ are taken from a normal population with the mean $\mu$ and the variance $\sigma^2$. If $n_1 = 25$, $n_2 = 50$, $\bar{x}_1 = 27.6$, and $\bar{x}_2 = 38.1$, estimate $\mu$ using the estimator of Exercise 10.23.

**10.35** A country's military intelligence knows that an enemy built certain new tanks numbered serially from 1 to $k$. If three of these tanks are captured and their serial numbers are 210, 38, and 155, use the estimator of part (b) of Exercise 10.12 to estimate $k$.

## 10.4 CONSISTENCY

In the preceding section we assumed that the variance of an estimator, or its mean square error, is a good indication of its chance fluctuations. The fact that these measures may not provide good criteria for this purpose is illustrated by the following example: Suppose we want to estimate on the basis of one observation the parameter $\theta$ of the population given by

$$f(x) = \omega \cdot \frac{1}{\sigma\sqrt{2\pi}} \cdot e^{-\frac{1}{2}\left(\frac{x-\theta}{\sigma}\right)^2} + (1 - \omega) \cdot \frac{1}{\pi} \cdot \frac{1}{1 + (x - \theta)^2}$$

for $-\infty < x < \infty$ and $0 < \omega < 1$. Evidently, this population is a combination of a normal population with the mean $\theta$ and the variance $\sigma^2$ and a Cauchy population (see Exercise 6.6) with $\alpha = \theta$ and $\beta = 1$. Now, if $\omega$ is very close to 1, say, $\omega = 1 - 10^{-100}$, and $\sigma$ is very small, say, $\sigma = 10^{-100}$, the probability that a random variable having this distribution will take on a value which is very close to $\theta$, and hence is a very good estimate of $\theta$, is practically 1. Yet, since the variance of the Cauchy distribution does not exist, neither will the variance of this estimator.

The example of the preceding paragraph is a bit farfetched, but it suggests that we pay more attention to the probabilities with which estimators will take on values that are close to the parameters they are supposed to estimate. The reader may recall that we already touched upon this question in Sections 5.4 and 8.2. Basing our argument on Chebyshev's theorem, we showed on page 190 that when $n \to \infty$, the probability approaches 1 that the sample proportion $\dfrac{X}{n}$ will take on a value that differs from the binomial parameter $\theta$ by less than any arbitrary constant $c > 0$. Also using Chebyshev's theorem, we showed in Theorem 8.2 that when

$n \to \infty$, the probability approaches 1 that $\overline{X}$ will take on a value which differs from the mean of the population sampled by less than any arbitrary constant $c > 0$.

In both of these examples we were practically assured that, for large $n$, the estimators will take on values that are very close to the respective parameters. Formally, this concept of "closeness" is expressed by means of the following definition of **consistency**.

---

**DEFINITION 10.2**    The statistic $\hat{\Theta}$ is a **consistent estimator** of the parameter $\theta$ if and only if for each $c > 0$

$$\lim_{n \to \infty} P(|\hat{\Theta} - \theta| < c) = 1$$

---

Note that consistency is an **asymptotic property**, that is, a limiting property of an estimator. Informally, Definition 10.2 says that when $n$ is sufficiently large, we can be practically certain that the error made with a consistent estimator will be less than any small preassigned positive constant. The kind of convergence expressed by the limit in Definition 10.2 is generally called **convergence in probability**.

Based on Chebyshev's theorem, we have thus shown in Section 5.4 that $\dfrac{X}{n}$ is a consistent estimator of the binomial parameter $\theta$ and in Theorem 8.2 that $\overline{X}$ is a consistent estimator of the mean of a population with a finite variance. In practice, we can often judge whether an estimator is consistent by using the following sufficient condition, which, in fact, is an immediate consequence of Chebyshev's theorem.

---

**THEOREM 10.3**    If $\hat{\Theta}$ is an unbiased estimator of the parameter $\theta$ and $\text{var}(\hat{\Theta}) \to 0$ as $n \to \infty$, then $\hat{\Theta}$ is a consistent estimator of $\theta$.

---

## EXAMPLE 10.8

Show that for a random sample from a normal population, the sample variance $S^2$ is a consistent estimator of $\sigma^2$.

### Solution

Since $S^2$ is an unbiased estimator of $\sigma^2$ in accordance with Theorem 10.1, it remains to be shown that $\text{var}(S^2) \to 0$ as $n \to \infty$. Referring to the result of Exercise 8.37 (or to Theorem 8.11, on which this exercise is based), we find that for a random sample from a normal population,

$$\text{var}(S^2) = \frac{2\sigma^4}{n-1}$$

It follows that $\text{var}(S^2) \to 0$ as $n \to \infty$, and we have thus shown that $S^2$ is a consistent estimator of the variance of a normal population.    ▲

It is of interest to note that Theorem 10.3 also holds if we substitute "asymptotically unbiased" for "unbiased." This is illustrated by the following example.

## EXAMPLE 10.9

With reference to Example 10.3, show that the smallest sample value (that is, the first order statistic $Y_1$), is a consistent estimator of the parameter $\delta$.

### Solution

Substituting into the formula for $g_1(y_1)$ on page 323, we find that the sampling distribution of $Y_1$ is given by

$$g_1(y_1) = n \cdot e^{-(y_1-\delta)} \cdot \left[ \int_{y_1}^{\infty} e^{-(x-\delta)} \, dx \right]^{n-1}$$

$$= n \cdot e^{-n(y_1-\delta)}$$

for $y_1 > \delta$ and $g_1(y_1) = 0$ elsewhere. Based on this result it can easily be shown that $E(Y_1) = \delta + \dfrac{1}{n}$ and hence that $Y_1$ is an asymptotically unbiased estimator of $\delta$. Furthermore,

$$P(|Y_1 - \delta| < c) = P(\delta < Y_1 < \delta + c)$$

$$= \int_{\delta}^{\delta+c} n \cdot e^{-n(y_1-\delta)} \, dy_1$$

$$= 1 - e^{-nc}$$

Since $\lim_{n \to \infty} (1 - e^{-nc}) = 1$, it follows from Definition 10.2 that $Y_1$ is a consistent estimator of $\delta$.    ▲

As we indicated on page 370, Theorem 10.3 provides a sufficient condition for the consistency of an estimator. It is not a necessary condition because consistent estimators need not be unbiased, or even asymptotically unbiased. This is illustrated by Exercise 10.44.

## 10.5  SUFFICIENCY

An estimator $\hat{\Theta}$ is said to be **sufficient** if it utilizes all the information in a sample relevant to the estimation of $\theta$; that is, if all the knowledge about $\theta$ that can be gained from the individual sample values and their order can just as well be gained from the value of $\hat{\Theta}$ alone.

Formally, we can describe this property of an estimator by referring to the conditional probability distribution or density of the sample values given $\hat{\Theta} = \hat{\theta}$, which is given by

$$f(x_1, x_2, \ldots, x_n | \hat{\theta}) = \frac{f(x_1, x_2, \ldots, x_n, \hat{\theta})}{g(\hat{\theta})} = \frac{f(x_1, x_2, \ldots, x_n)}{g(\hat{\theta})}$$

If it depends on $\theta$, then particular values of $X_1, X_2, \ldots, X_n$ yielding $\hat{\Theta} = \hat{\theta}$ will be more probable for some values of $\theta$ than for others, and the knowledge of these sample values will help in the estimation of $\theta$. On the other hand, if it does not depend on $\theta$, then particular values of $X_1, X_2, \ldots, X_n$ yielding $\hat{\Theta} = \hat{\theta}$ will be just as likely for any value of $\theta$, and the knowledge of these sample values will be of no help in the estimation of $\theta$.

> **DEFINITION 10.3**  The statistic $\hat{\Theta}$ is a **sufficient estimator** of the parameter $\theta$ if and only if for each value of $\hat{\Theta}$ the conditional probability distribution or density of the random sample $X_1, X_2, \ldots, X_n$ given $\hat{\Theta} = \hat{\theta}$ is independent of $\theta$.

## EXAMPLE 10.10

If $X_1, X_2, \ldots, X_n$ constitute a random sample of size $n$ from a Bernoulli population, show that

$$\hat{\Theta} = \frac{X_1 + X_2 + \cdots + X_n}{n}$$

is a sufficient estimator of the parameter $\theta$.

*Solution*

By Definition 5.2,

$$f(x_i; \theta) = \theta^{x_i}(1 - \theta)^{1-x_i} \qquad \text{for } x_i = 0, 1$$

so that

$$f(x_1, x_2, \ldots, x_n) = \prod_{i=1}^{n} \theta^{x_i}(1 - \theta)^{1-x_i}$$

$$= \theta^{\sum_{i=1}^{n} x_i}(1 - \theta)^{n-\sum_{i=1}^{n} x_i}$$

$$= \theta^x(1 - \theta)^{n-x}$$

$$= \theta^{n\hat{\theta}}(1 - \theta)^{n-n\hat{\theta}}$$

for $x_i = 0$ or 1 and $i = 1, 2, \ldots, n$. Also, since

$$X = X_1 + X_2 + \cdots + X_n$$

is a binomial random variable with the parameters $\theta$ and $n$, its distribution is given by

$$b(x; n, \theta) = \binom{n}{x}\theta^x(1 - \theta)^{n-x}$$

and the transformation-of-variable technique of Section 7.3 yields

$$g(\hat{\theta}) = \binom{n}{n\hat{\theta}}\theta^{n\hat{\theta}}(1 - \theta)^{n-n\hat{\theta}} \qquad \text{for } \hat{\theta} = 0, \frac{1}{n}, \ldots, 1$$

Now, substituting into the formula for $f(x_1, x_2, \ldots, x_n | \hat{\theta})$ on page 372, we get

$$\frac{f(x_1, x_2, \ldots, x_n, \hat{\theta})}{g(\hat{\theta})} = \frac{f(x_1, x_2, \ldots, x_n)}{g(\hat{\theta})}$$

$$= \frac{\theta^{n\hat{\theta}}(1 - \theta)^{n-n\hat{\theta}}}{\binom{n}{n\hat{\theta}}\theta^{n\hat{\theta}}(1 - \theta)^{n-n\hat{\theta}}}$$

$$= \frac{1}{\binom{n}{n\hat{\theta}}}$$

$$= \frac{1}{\binom{n}{x}}$$

$$= \frac{1}{\binom{n}{x_1 + x_2 + \cdots + x_n}}$$

for $x_i = 0$ or $1$ and $i = 1, 2, \ldots, n$. Evidently, this does not depend on $\theta$ and we have shown, therefore, that $\hat{\Theta} = \dfrac{X}{n}$ is a sufficient estimator of $\theta$. ▲

### EXAMPLE 10.11

Show that $Y = \frac{1}{6}(X_1 + 2X_2 + 3X_3)$ is not a sufficient estimator of the Bernoulli parameter $\theta$.

#### Solution

Since we must show that

$$f(x_1, x_2, x_3 | y) = \frac{f(x_1, x_2, x_3, y)}{g(y)}$$

is not independent of $\theta$ for some values of $X_1$, $X_2$, and $X_3$, let us consider the case where $x_1 = 1$, $x_2 = 1$, and $x_3 = 0$. Thus, $y = \frac{1}{6}(1 + 2 \cdot 1 + 3 \cdot 0) = \frac{1}{2}$ and

$$f(1, 1, 0 | Y = \tfrac{1}{2}) = \frac{P(X_1 = 1, X_2 = 1, X_3 = 0, Y = \tfrac{1}{2})}{P(Y = \tfrac{1}{2})}$$

$$= \frac{f(1, 1, 0)}{f(1, 1, 0) + f(0, 0, 1)}$$

where

$$f(x_1, x_2, x_3) = \theta^{x_1 + x_2 + x_3}(1 - \theta)^{3 - (x_1 + x_2 + x_3)}$$

for $x_1 = 0$ or 1 and $i = 1, 2, 3$. Since $f(1, 1, 0) = \theta^2(1 - \theta)$ and $f(0, 0, 1) = \theta(1 - \theta)^2$, it follows that

$$f(1, 1, 0 | Y = \tfrac{1}{2}) = \frac{\theta^2(1 - \theta)}{\theta^2(1 - \theta) + \theta(1 - \theta)^2} = \theta$$

and it can be seen that this conditional probability depends on $\theta$. We have thus shown that $Y = \frac{1}{6}(X_1 + 2X_2 + 3X_3)$ is not a sufficient estimator of the parameter $\theta$ of a Bernoulli population.    ▲

Because it can be very tedious to check whether a statistic is a sufficient estimator of a given parameter based directly on Definition 10.3, it is usually easier to base it instead on the following **factorization theorem**.

---

**THEOREM 10.4**  The statistic $\hat{\Theta}$ is a sufficient estimator of the parameter $\theta$ if and only if the joint probability distribution or density of the random sample can be factored so that

$$f(x_1, x_2, \ldots, x_n; \theta) = g(\hat{\theta}, \theta) \cdot h(x_1, x_2, \ldots, x_n)$$

where $g(\hat{\theta}, \theta)$ depends only on $\hat{\theta}$ and $\theta$, and $h(x_1, x_2, \ldots, x_n)$ does not depend on $\theta$.

---

A proof of this theorem may be found in more advanced texts; see, for instance, the book by Hogg and Craig listed among the references at the end of this chapter. Here, let us illustrate the use of Theorem 10.4 by means of the following example.

## EXAMPLE 10.12

Show that $\overline{X}$ is a sufficient estimator of the mean $\mu$ of a normal population with the known variance $\sigma^2$.

### Solution

Making use of the fact that

$$f(x_1, x_2, \ldots, x_n; \mu) = \left(\frac{1}{\sigma\sqrt{2\pi}}\right)^n \cdot e^{-\frac{1}{2}\sum_{i=1}^{n}\left(\frac{x_i - \mu}{\sigma}\right)^2}$$

and that

$$\sum_{i=1}^{n} (x_i - \mu)^2 = \sum_{i=1}^{n} [(x_i - \bar{x}) - (\mu - \bar{x})]^2$$

$$= \sum_{i=1}^{n} (x_i - \bar{x})^2 + \sum_{i=1}^{n} (\bar{x} - \mu)^2$$

$$= \sum_{i=1}^{n} (x_i - \bar{x})^2 + n(\bar{x} - \mu)^2$$

we get

$$f(x_1, x_2, \ldots, x_n; \mu) = \left\{ \frac{\sqrt{n}}{\sigma \sqrt{2\pi}} \cdot e^{-\frac{1}{2}\left(\frac{\bar{x}-\mu}{\sigma/\sqrt{n}}\right)^2} \right\}$$

$$\times \left\{ \frac{1}{\sqrt{n}} \left(\frac{1}{\sigma \sqrt{2\pi}}\right)^{n-1} \cdot e^{-\frac{1}{2} \cdot \sum_{i=1}^{n} \left(\frac{x_i - \bar{x}}{\sigma}\right)^2} \right\}$$

where the first factor on the right-hand side depends only on the estimate $\bar{x}$ and the population mean $\mu$, and the second factor does not involve $\mu$. According to Theorem 10.4, it follows that $\bar{X}$ is a sufficient estimator of the mean $\mu$ of a normal population with the known variance $\sigma^2$.   ▲

Based on Definition 10.3 and Theorem 10.4, respectively, we have presented two ways of checking whether a statistic $\hat{\Theta}$ is a sufficient estimator of a given parameter $\theta$. As we already said, the factorization theorem usually leads to easier solutions, but if we want to show that a statistic $\hat{\Theta}$ is not a sufficient estimator of a given parameter $\theta$, it is nearly always easier to proceed with Definition 10.3. This was illustrated by Example 10.11.

Let us also mention the following important property of sufficient estimators. If $\hat{\Theta}$ is a sufficient estimator of $\theta$, then any single-valued function $Y = u(\hat{\Theta})$, not involving $\theta$, is also a sufficient estimator of $\theta$, and therefore of $u(\theta)$, provided $y = u(\hat{\theta})$ can be solved to give the single-valued inverse $\hat{\theta} = w(y)$. This follows directly from Theorem 10.4, since we can write

$$f(x_1, x_2, \ldots, x_n; \theta) = g[w(y), \theta] \cdot h(x_1, x_2, \ldots, x_n)$$

where $g[w(y), \theta]$ depends only on $y$ and $\theta$. If we apply this result to Example 10.10, where we showed that $\hat{\Theta} = \dfrac{X}{n}$ is a sufficient estimator of the Bernoulli parameter $\theta$, it follows that $X = X_1 + X_2 + \cdots + X_n$ is also a sufficient estimator of the mean $\mu = n\theta$ of a binomial population.

## 10.6 ROBUSTNESS

In recent years, special attention has been paid to a statistical property called **robustness**. It is indicative of the extent to which estimation procedures (and, as we shall see later, also other methods of inference) are adversely affected by violations of underlying assumptions. In other words, an estimator is said to be **robust** if its sampling distribution is not seriously affected by violations of assumptions. Such violations are often due to outliers caused by outright errors made, say, in reading instruments or recording the data, or by mistakes in experimental procedures. They may also pertain to the nature of the populations sampled or their parameters. For instance, when estimating the average useful life of a certain electronic component, we may think that we are sampling an exponential population, whereas actually we are sampling a Weibull population, or when estimating the average income of a certain age group, we may use a method based on the assumption that we are sampling a normal population, whereas actually the population (income distribution) is highly skewed. Also, when estimating the difference between the average weights of two kinds of frogs, the difference between the mean IQ's of two ethnic groups, and in general the difference $\mu_1 - \mu_2$ between the means of two populations, we may be assuming that the two populations have the same variance $\sigma^2$, whereas in reality $\sigma_1^2 \neq \sigma_2^2$.

As should be apparent, most questions of robustness are difficult to answer; indeed, much of the language used in the preceding paragraph is relatively imprecise. After all, what do we mean by "not seriously affected," and when we speak of violations of underlying assumptions, it should be clear that some violations are more serious than others. When it comes to questions of robustness, we are thus faced by all sorts of difficulties, mathematically and otherwise, and for the most part they can be resolved only by computer simulations. The subject of robustness will be mentioned again briefly in Section 16.1.

### EXERCISES

**10.36** Use Definition 10.2 to show that $Y_1$, the first order statistic, is a consistent estimator of the parameter $\alpha$ of a uniform population with $\beta = \alpha + 1$.

**10.37** With reference to Exercise 10.36, use Theorem 10.3 to show that $Y_1 - \dfrac{1}{n+1}$ is a consistent estimator of the parameter $\alpha$.

**10.38** With reference to the uniform population of Example 10.4, use the definition of consistency to show that $Y_n$, the $n$th order statistic, is a consistent estimator of the parameter $\beta$.

**10.39** If $X_1, X_2, \ldots, X_n$ constitute a random sample of size $n$ from an exponential population, show that $\overline{X}$ is a consistent estimator of the parameter $\theta$.

**10.40** With reference to Exercise 10.39, is $X_n$ a consistent estimator of the parameter $\theta$?

**10.41** Show that the estimator of Exercise 10.21 is consistent.

**10.42** Substituting "asymptotically unbiased" for "unbiased" in Theorem 10.3, show that $\dfrac{X + 1}{n + 2}$ is a consistent estimator of the binomial parameter $\theta$.

**10.43** Substituting "asymptotically unbiased" for "unbiased" in Theorem 10.3, use this theorem to rework Exercise 10.38.

**10.44** To show that an estimator can be consistent without being unbiased or even asymptotically unbiased, consider the following estimation procedure: To estimate the mean of a population with the finite variance $\sigma^2$, we first take a random sample of size $n$. Then, we randomly draw one of $n$ slips of paper numbered from 1 through $n$, and if the number we draw is 2, 3, $\ldots$, or $n$, we use as our estimator the mean of the random sample; otherwise, we use the estimate $n^2$. Show that this estimation procedure is

(a)    consistent;
(b)    neither unbiased nor asymptotically unbiased.

**10.45** If $X_1, X_2, \ldots, X_n$ constitute a random sample of size $n$ from an exponential population, show that $\overline{X}$ is a sufficient estimator of the parameter $\theta$.

**10.46** If $X_1$ and $X_2$ are independent random variables having binomial distributions with the parameters $\theta$ and $n_1$ and $\theta$ and $n_2$, show that $\dfrac{X_1 + X_2}{n_1 + n_2}$ is a sufficient estimator of $\theta$.

**10.47** With reference to the preceding exercise, is $\dfrac{X_1 + 2X_2}{n_1 + 2n_2}$ a sufficient estimator of $\theta$?

**10.48** With reference to Example 10.4, is the $n$th order statistic, $Y_n$, a sufficient estimator of the parameter $\beta$?

**10.49** If $X_1$ and $X_2$ constitute a random sample of size $n = 2$ from a Poisson population, show that the mean of the sample is a sufficient estimator of the parameter $\lambda$.

**10.50** If $X_1, X_2$, and $X_3$ constitute a random sample of size $n = 3$ from a Bernoulli population, show that $Y = X_1 + 2X_2 + X_3$ is not a sufficient estimator of $\theta$. (*Hint:* Consider special values of $X_1, X_2$, and $X_3$.)

**10.51** If $X_1, X_2, \ldots, X_n$ constitute a random sample of size $n$ from a geometric population, show that $Y = X_1 + X_2 + \cdots + X_n$ is a sufficient estimator of the parameter $\theta$.

**10.52** Show that the estimator of Exercise 10.5 is a sufficient estimator of the variance of a normal population with the known mean $\mu$.

## 10.7  THE METHOD OF MOMENTS

As we have seen in this chapter, there can be many different estimators of one and the same parameter of a population. Therefore, it would seem desirable to have some general method, or methods, that yield estimators with as many desirable properties as possible. In this section and in Section 10.8 we shall present two such methods, the **method of moments**, which is historically one of the oldest methods, and the **method of maximum likelihood**. Furthermore, **Bayesian estimation** will be treated briefly in Section 10.9 and another method, the **method of least squares**, will be taken up in Chapter 14.

The method of moments consists of equating the first few moments of a population to the corresponding moments of a sample, thus getting as many equations as are needed to solve for the unknown parameters of the population.

---

**DEFINITION 10.4**  The $k$th **sample moment** of a set of observations $x_1$, $x_2$, $\ldots$ , $x_n$ is the mean of their $k$th powers and it is denoted by $m_k'$; symbolically,

$$m_k' = \frac{\sum_{i=1}^{n} x_i^k}{n}$$

.

---

Thus, if a population has $r$ parameters, the method of moments consists of solving the system of equations

$$m_k' = \mu_k' \qquad k = 1, 2, \ldots, r$$

for the $r$ parameters.

## EXAMPLE 10.13

Given a random sample of size $n$ from a gamma population, use the method of moments to obtain formulas for estimating the parameters $\alpha$ and $\beta$.

### Solution

The system of equations we shall have to solve is

$$m_1' = \mu_1' \quad \text{and} \quad m_2' = \mu_2'$$

where $\mu_1' = \alpha\beta$ and $\mu_2' = \alpha(\alpha + 1)\beta^2$ according to Theorem 6.2. Thus,

$$m'_1 = \alpha\beta \quad \text{and} \quad m'_2 = \alpha(\alpha + 1)\beta^2$$

and, solving for $\alpha$ and $\beta$, we get the following formulas for estimating the two parameters of the gamma distribution:

$$\hat{\alpha} = \frac{(m'_1)^2}{m'_2 - (m'_1)^2} \quad \text{and} \quad \hat{\beta} = \frac{m'_2 - (m'_1)^2}{m'_1}$$

Since $m'_1 = \dfrac{\displaystyle\sum_{i=1}^{n} x_i}{n} = \bar{x}$ and $m'_2 = \dfrac{\displaystyle\sum_{i=1}^{n} x_i^2}{n}$, we can write

$$\hat{\alpha} = \frac{n\bar{x}^2}{\displaystyle\sum_{i=1}^{n}(x_i - \bar{x})^2} \quad \text{and} \quad \hat{\beta} = \frac{\displaystyle\sum_{i=1}^{n}(x_i - \bar{x})^2}{n\bar{x}}$$

in terms of the original observations.    ▲

In this example we were concerned with the parameters of a specific population. It is important to note, however, that when the parameters to be estimated are the moments of the population, then the method of moments can be used without any knowledge about the nature, or functional form, of the population.

## 10.8 THE METHOD OF MAXIMUM LIKELIHOOD

In two papers published early in this century, R. A. Fisher, the prominent statistician whom we already mentioned on page 314, proposed a general method of estimation called the **method of maximum likelihood**. He also demonstrated the advantages of this method by showing that it yields sufficient estimators whenever they exist and that maximum likelihood estimators are asymptotically minimum variance unbiased estimators.

To help understand the principle on which the method of maximum likelihood is based, suppose that four letters arrive in somebody's morning mail, but unfortunately one of them is misplaced before the recipient has a chance to open it. If, among the remaining three letters, two contain credit-card billings and the other one does not, what might be a good estimate of $k$, the total number of credit-card billings among the four letters received? Clearly, $k$ must be two or three, and if we assume that each letter had the same chance of being misplaced, we find that the probability of the observed data (two of the three remaining letters contain credit-card billings) is

$$\frac{\binom{2}{2}\binom{2}{1}}{\binom{4}{3}} = \frac{1}{2}$$

for $k = 2$ and

$$\frac{\binom{3}{2}\binom{1}{1}}{\binom{4}{3}} = \frac{3}{4}$$

for $k = 3$. Therefore, if we choose as our estimate of $k$ the value which maximizes the probability of getting the observed data, we obtain $k = 3$. We call this estimate a **maximum likelihood estimate**, and the method by which it was obtained, the method of maximum likelihood.

Thus, the essential feature of the method of maximum likelihood is that we look at the sample values and then choose as our estimates of the unknown parameters the values for which the probability or probability density of getting the sample values is a maximum. In what follows, we shall limit ourselves to the one-parameter case, but as we shall see in Example 10.17, the general idea applies also when there are several unknown parameters. In the discrete case, if the observed sample values are $x_1, x_2, \ldots, x_n$, the probability of getting them is

$$P(X_1 = x_1, X_2 = x_2, \ldots, X_n = x_n) = f(x_1, x_2, \ldots, x_n; \theta)$$

which is just the value of the joint probability distribution of the random variables $X_1, X_2, \ldots, X_n$ at $X_1 = x_1, X_2 = x_2, \ldots, X_n = x_n$. Since the sample values have been observed and are therefore fixed numbers, we regard $f(x_1, x_2, \ldots, x_n; \theta)$ as a value of a function of $\theta$, and we refer to this function as the **likelihood function**. An analogous definition applies when the random sample comes from a continuous population, but in that case $f(x_1, x_2, \ldots, x_n; \theta)$ is the value of the joint probability density of the random variables $X_1, X_2, \ldots, X_n$ at $X_1 = x_1, X_2 = x_2, \ldots, X_n = x_n$.

---

**DEFINITION 10.5** If $x_1, x_2, \ldots, x_n$ are the values of a random sample from a population with the parameter $\theta$, the **likelihood function** of the sample is given by

$$L(\theta) = f(x_1 x_2, \ldots, x_n; \theta)$$

for values of $\theta$ within a given domain. Here $f(x_1, x_2, \ldots, x_n; \theta)$ is the value of the joint probability distribution or the joint probability density of the random variables $X_1, X_2, \ldots, X_n$ at $X_1 = x_1, X_2 = x_2, \ldots, X_n = x_n$.

Thus, the method of maximum likelihood consists of maximizing the likelihood function with respect to $\theta$, and we refer to the value of $\theta$ which maximizes the likelihood function as the maximum likelihood estimate of $\theta$.

## EXAMPLE 10.14

Given $x$ "successes" in $n$ trials, find the maximum likelihood estimate of the parameter $\theta$ of the corresponding binomial distribution.

### Solution

To find the value of $\theta$ which maximizes

$$L(\theta) = \binom{n}{x} \theta^x (1 - \theta)^{n-x}$$

it will be convenient to make use of the fact that the value of $\theta$ which maximizes $L(\theta)$ will also maximize

$$\ln L(\theta) = \ln\binom{n}{x} + x \cdot \ln \theta + (n - x) \cdot \ln(1 - \theta)$$

Thus, we get

$$\frac{d[\ln L(\theta)]}{d\theta} = \frac{x}{\theta} - \frac{n - x}{1 - \theta}$$

and, equating this derivative to 0 and solving for $\theta$, we find that the likelihood function has a maximum at $\theta = \frac{x}{n}$. This is the maximum likelihood estimate of the binomial parameter $\theta$, and we refer to $\hat{\theta} = \frac{X}{n}$ as the corresponding **maximum likelihood estimator**.    ▲

## EXAMPLE 10.15

If $x_1, x_2, \ldots, x_n$ are the values of a random sample from an exponential population, find the maximum likelihood estimator of its parameter $\theta$.

*Solution*

Since the likelihood function is given by

$$L(\theta) = f(x_1, x_2, \ldots, x_n; \theta)$$

$$= \prod_{i=1}^{n} f(x_i; \theta)$$

$$= \left(\frac{1}{\theta}\right)^n \cdot e^{-\frac{1}{\theta}\left(\sum_{i=1}^{n} x_i\right)}$$

differentiation of $\ln L(\theta)$ with respect to $\theta$ yields

$$\frac{d[\ln L(\theta)]}{d\theta} = -\frac{n}{\theta} + \frac{1}{\theta^2} \cdot \sum_{i=1}^{n} x_i$$

Equating this derivative to zero and solving for $\theta$, we get the maximum likelihood estimate

$$\hat{\theta} = \frac{1}{n} \cdot \sum_{i=1}^{n} x_i = \bar{x}$$

Hence, the maximum likelihood estimator is $\hat{\Theta} = \overline{X}$.    ▲

Now let us consider an example in which straightforward differentiation cannot be used to find the maximum value of the likelihood function.

## EXAMPLE 10.16

If $x_1, x_2, \ldots, x_n$ are the values of a random sample of size $n$ from a uniform population with $\alpha = 0$ (as in Example 10.4), find the maximum likelihood estimator of $\beta$.

*Solution*

The likelihood function is given by

$$L(\beta) = \prod_{i=1}^{n} f(x_i; \beta) = \left(\frac{1}{\beta}\right)^n$$

for $\beta$ greater than or equal to the largest of the $x$'s and 0 otherwise. Since the value of this likelihood function increases as $\beta$ decreases, we must make $\beta$ as small as possible, and it follows that the maximum likelihood estimator of $\beta$ is $Y_n$, the $n$th order statistic.    ▲

Comparing the result of this example with that of Example 10.4, we find that maximum likelihood estimators need not be unbiased. The ones of Examples 10.14 and 10.15 were unbiased.

As we have already indicated, the method of maximum likelihood can also be used for the simultaneous estimation of several parameters of a given population. In that case we must find the values of the parameters that maximize the likelihood function.

## EXAMPLE 10.17

If $X_1, X_2, \ldots, X_n$ constitute a random sample of size $n$ from a normal population with the mean $\mu$ and the variance $\sigma^2$, find joint maximum likelihood estimates of these two parameters.

### Solution

Since the likelihood function is given by

$$L(\mu, \sigma^2) = \prod_{i=1}^{n} n(x_i; \mu, \sigma)$$

$$= \left(\frac{1}{\sigma\sqrt{2\pi}}\right)^n \cdot e^{-\frac{1}{2\sigma^2}\cdot\sum_{i=1}^{n}(x_i-\mu)^2}$$

partial differentiation of $\ln L(\mu, \sigma^2)$ with respect to $\mu$ and $\sigma^2$ yields

$$\frac{\partial[\ln L(\mu, \sigma^2)]}{\partial\mu} = \frac{1}{\sigma^2}\cdot\sum_{i=1}^{n}(x_i - \mu)$$

and

$$\frac{\partial[\ln L(\mu, \sigma^2)]}{\partial\sigma^2} = -\frac{n}{2\sigma^2} + \frac{1}{2\sigma^4}\cdot\sum_{i=1}^{n}(x_i - \mu)^2$$

Equating the first of these two partial derivatives to zero and solving for $\mu$, we get

$$\hat{\mu} = \frac{1}{n} \cdot \sum_{i=1}^{n} x_i = \bar{x}$$

and equating the second of these partial derivatives to zero and solving for $\sigma^2$ after substituting $\mu = \bar{x}$, we get

$$\hat{\sigma}^2 = \frac{1}{n} \cdot \sum_{i=1}^{n} (x_i - \bar{x})^2 \qquad \blacktriangle$$

It should be observed that we did not show that $\hat{\sigma}$ is a maximum likelihood estimate of $\sigma$, only that $\hat{\sigma}^2$ is a maximum likelihood estimate of $\sigma^2$. However, it can be shown (see reference at the end of this chapter) that maximum likelihood estimators have the **invariance property** that if $\hat{\Theta}$ is a maximum likelihood estimator of $\theta$ and the function given by $g(\theta)$ is continuous, then $g(\hat{\Theta})$ is also a maximum likelihood estimator of $g(\theta)$. It follows that

$$\hat{\sigma} = \sqrt{\frac{1}{n} \cdot \sum_{i=1}^{n} (x_i - \bar{x})^2}$$

which differs from $s$ in that we divide by $n$ instead of $n - 1$, is a maximum likelihood estimate of $\sigma$.

In Examples 10.14, 10.15, and 10.17 we maximized the logarithm of the likelihood function instead of the likelihood function itself, but this is by no means necessary. It just so happened that it was convenient in each case.

## EXERCISES

**10.53** If $X_1, X_2, \ldots, X_n$ constitute a random sample from a population with the mean $\mu$ and the variance $\sigma^2$, use the method of moments to find estimators for $\mu$ and $\sigma^2$.

**10.54** Given a random sample of size $n$ from an exponential population, use the method of moments to find an estimator of the parameter $\theta$.

**10.55** Given a random sample of size $n$ from a uniform population with $\alpha = 0$, find an estimator for $\beta$ by the method of moments.

**10.56** Given a random sample of size $n$ from a Poisson population, use the method of moments to obtain an estimator for the parameter $\lambda$.

**10.57** Given a random sample of size $n$ from a beta population with $\beta = 1$, use the method of moments to find a formula for estimating the parameter $\alpha$.

**10.58** If $X_1, X_2, \ldots, X_n$ constitute a random sample of size $n$ from a population given by

$$f(x; \theta) = \begin{cases} \dfrac{2(\theta - x)}{\theta^2} & \text{for } 0 < x < \theta \\ 0 & \text{elsewhere} \end{cases}$$

find an estimator for $\theta$ by the method of moments.

**10.59** If $X_1, X_2, \ldots, X_n$ constitute a random sample of size $n$ from a population given by

$$g(x; \theta) = \begin{cases} \dfrac{1}{\theta} \cdot e^{-\frac{x-\delta}{\theta}} & \text{for } x > \delta \\ 0 & \text{elsewhere} \end{cases}$$

find estimators for $\delta$ and $\theta$ by the method of moments. This distribution is sometimes referred to as the **two-parameter exponential distribution**, and for $\theta = 1$ it is the distribution of Example 10.3.

**10.60** Given a random sample of size $n$ from a continuous uniform population, use the method of moments to find formulas for estimating the parameters $\alpha$ and $\beta$.

**10.61** Consider $N$ independent random variables having identical binomial distributions with the parameters $\theta$ and $n = 3$. If $n_0$ of them take on the value 0, $n_1$ take on the value 1, $n_2$ take on the value 2, and $n_3$ take on the value 3, use the method of moments to find a formula for estimating $\theta$.

**10.62** Use the method of maximum likelihood to rework Exercise 10.56.

**10.63** Use the method of maximum likelihood to rework Exercise 10.57.

**10.64** If $X_1, X_2, \ldots, X_n$ constitute a random sample of size $n$ from a gamma population with $\alpha = 2$, use the method of maximum likelihood to find a formula for estimating $\beta$.

**10.65** Given a random sample of size $n$ from a normal population with the known mean $\mu$, find the maximum likelihood estimator for $\sigma$.

**10.66** If $X_1, X_2, \ldots, X_n$ constitute a random sample of size $n$ from a geometric population, find formulas for estimating its parameter $\theta$ by using
(a)   the method of moments;
(b)   the method of maximum likelihood.

**10.67** Given a random sample of size $n$ from a Rayleigh population (see Exercise 6.20), find an estimator for its parameter $\alpha$ by the method of maximum likelihood.

**10.68** Given a random sample of size $n$ from a Pareto population (see Exercise 6.21), use the method of maximum likelihood to find a formula for estimating its parameter $\alpha$.

**10.69** Use the method of maximum likelihood to rework Exercise 10.59.

**10.70** Use the method of maximum likelihood to rework Exercise 10.60.

**10.71** Use the method of maximum likelihood to rework Exercise 10.61.

**10.72** Given a random sample of size $n$ from a gamma population with the known parameter $\alpha$, find the maximum likelihood estimator for
   (a) $\beta$;
   (b) $\tau = (2\beta - 1)^2$.

**10.73** If $V_1, V_2, \ldots, V_n$ and $W_1, W_2, \ldots, W_n$ are independent random samples of size $n$ from normal populations with the means $\mu_1 = \alpha + \beta$ and $\mu_2 = \alpha - \beta$ and the common variance $\sigma^2 = 1$, find maximum likelihood estimators for $\alpha$ and $\beta$.

**10.74** If $V_1, V_2, \ldots, V_{n_1}$ and $W_1, W_2, \ldots, W_{n_2}$ are independent random samples of size $n_1$ and $n_2$ from normal populations with the means $\mu_1$ and $\mu_2$ and the common variance $\sigma^2$, find maximum likelihood estimators for $\mu_1$, $\mu_2$, and $\sigma^2$.

**10.75** Let $X_1, X_2, \ldots, X_n$ be a random sample of size $n$ from the uniform population given by

$$f(x; \theta) = \begin{cases} 1 & \text{for } \theta - \frac{1}{2} < x < \theta + \frac{1}{2} \\ 0 & \text{elsewhere} \end{cases}$$

Show that if $Y_1$ and $Y_n$ are the first and $n$th order statistic, any estimator $\hat{\theta}$ such that

$$Y_n - \tfrac{1}{2} \leqslant \hat{\theta} \leqslant Y_1 + \tfrac{1}{2}$$

can serve as a maximum likelihood estimator of $\theta$. This shows that maximum likelihood estimators need not be unique.

**10.76** With reference to Exercise 10.75, check whether the following estimators are maximum likelihood estimators of $\theta$:
   (a) $\frac{1}{2}(Y_1 + Y_n)$;
   (b) $\frac{1}{3}(Y_1 + 2Y_2)$.

## APPLICATIONS

**10.77** On 12 days selected at random, a city's consumption of electricity was 6.4, 4.5, 10.8, 7.2, 6.8, 4.9, 3.5, 16.3, 4.8, 7.0, 8.8, and 5.4 million kilowatt-hours. Assuming that these data may be looked upon as a random sample from a gamma population, use the estimators obtained in Example 10.13 to estimate the parameters $\alpha$ and $\beta$.

**10.78** The size of an animal population is sometimes estimated by the **capture-recapture method**. In this method, $n_1$ of the animals are captured in the area under consideration, tagged, and released. Later, $n_2$ of the animals are captured, $X$ of them are found to be tagged, and this information is used to estimate $N$, the total number of animals of the given kind in the area under consideration. If $n_1 = 3$ rare owls are captured in a section of a forest, tagged, and released, and later $n_2 = 4$ such owls are captured and only one of them is found to be tagged, estimate $N$ by the method of maximum likelihood. (*Hint:* Try $N = 9$, 10, 11, 12, 13, and 14.)

**10.79** Certain radial tires had useful lives of 35,200, 41,000, 44,700, 38,600, and 41,500 miles. Assuming that these data can be looked upon as a random sample from an exponential population, use the estimator obtained in Exercise 10.54 to estimate the parameter $\theta$.

**10.80** Among six measurements of the boiling point of a silicon compound, the size of the error was 0.07, 0.03, 0.14, 0.04, 0.08, and 0.03° C. Assuming that these data can be looked upon as a random sample from the population of Exercise 10.58, use the estimator obtained there by the method of moments to estimate the parameter $\theta$.

**10.81** Not counting the ones which failed immediately, certain light bulbs had useful lives of 415, 433, 489, 531, 466, 410, 479, 403, 562, 422, 475, and 439 hours. Assuming that these data can be looked upon as a random sample from a two-parameter exponential population, use the estimators obtained in Exercise 10.59 to estimate the parameters $\delta$ and $\theta$.

**10.82** Rework Exercise 10.81 using the estimators obtained in Exercise 10.69 by the method of maximum likelihood.

**10.83** Data collected over a number of years show that when a broker called a random sample of eight of her clients, she got a busy signal 6.5, 10.6, 8.1, 4.1, 9.3, 11.5, 7.3, and 5.7 percent of the time. Assuming that these figures can be looked upon as a random sample from a continuous uniform population, use the estimators obtained in Exercise 10.60 to estimate the parameters $\alpha$ and $\beta$.

**10.84** Rework Exercise 10.83 using the estimators obtained in Exercise 10.70.

**10.85** Every time Mr. Jones goes to the race track he bets on three races. In a random sample of 20 of his visits to the race track, he lost all his bets 11 times, won once seven times, and won twice on two occasions. If $\theta$ is the probability that he will win any one of his bets, estimate it by using the maximum likelihood estimator obtained in Exercise 10.71.

**10.86** In a random sample of the teachers in a large school district, their annual salaries were $23,900, $21,500, $26,400, $24,800, $33,600, $24,500, $29,200, $36,200, $22,400, $21,500, $28,300, $26,800, $31,400, $22,700, and $23,100. Assuming that these data can be looked upon as a random sample from a Pareto population, use the estimator obtained in Exercise 10.68 to estimate the parameter $\alpha$.

**10.87** On 20 very cold days, a farmer got her tractor started on the first, third, fifth, first, second, first, third, seventh, second, fourth, fourth, eighth, first, third, sixth, fifth, second, first, sixth, and second try. Assuming that these data can be looked upon as a random sample from a geometric population, estimate its parameter $\theta$ by either of the methods of Exercise 10.66.

**10.88** The IQ's of 10 teenagers belonging to one ethnic group are 98, 114, 105, 101, 123, 117, 106, 92, 110, and 108, whereas those of six teenagers belonging to another ethnic group are 122, 105, 99, 126, 114, and 108. Assuming that these data can be looked upon as independent random samples from normal populations with the means $\mu_1$ and $\mu_2$ and the common variance $\sigma^2$, estimate these parameters by means of the maximum likelihood estimators obtained in Exercise 10.74.

## 10.9  BAYESIAN ESTIMATION[†]

So far we have assumed in this chapter that the parameters which we want to estimate are unknown constants; in Bayesian estimation the parameters are looked upon as random variables having **prior distributions**, usually reflecting the strength of one's belief about the possible values they can assume. In Section 9.6, we already met a problem of Bayesian estimation—the parameter was that of a uniform density and its prior distribution was a gamma distribution.

The main problem of Bayesian estimation is that of combining prior feelings about a parameter with direct sample evidence, and in Example 9.9 we accomplished this by determining $\varphi(\theta|x)$, the conditional density of $\Theta$ given $X = x$. In contrast to the prior distribution of $\Theta$, this conditional distribution (which also reflects the direct sample evidence) is called the **posterior distribution** of $\Theta$. In general, if $h(\theta)$ is the value of the prior distribution of $\Theta$ at $\theta$ and we want to combine the information which it conveys with direct sample evidence about $\Theta$, for instance, the value of a statistic $W = u(X_1, X_2, \ldots, X_n)$, we determine the posterior distribution of $\Theta$ by means of the formula

$$\varphi(\theta|w) = \frac{f(\theta, w)}{g(w)} = \frac{h(\theta) \cdot f(w|\theta)}{g(w)}$$

Here $f(w|\theta)$ is the value of the sampling distribution of $W$ given $\Theta = \theta$ at $w$, $f(\theta, w)$ is the value of the joint distribution of $\Theta$ and $W$ at $\theta$ and $w$, and $g(w)$ is the value of the marginal distribution of $W$ at $w$. Note that the above formula for $\varphi(\theta|w)$ is, in fact, an extension of Bayes' theorem, Theorem 2.13, to the continuous case. Hence, the term "Bayesian estimation."

Once the posterior distribution of a parameter has been obtained, it can be used to make estimates as in Example 9.9, or it can be used to make probability

---

[†]Some of the concepts and language used in this section were introduced in Chapter 9, the optional chapter on decision theory.

statements about the parameter, as will be illustrated in Example 10.19. Although the method we have described has extensive applications, we shall limit our discussion here to inferences about the parameter $\Theta$ of a binomial population and the mean of a normal population; inferences about the parameter of a Poisson population are treated in Exercise 10.92.

---

**THEOREM 10.5** If $X$ is a binomial random variable and the prior distribution of $\Theta$ is a beta distribution with the parameters $\alpha$ and $\beta$, then the posterior distribution of $\Theta$ given $X = x$ is a beta distribution with the parameters $x + \alpha$ and $n - x + \beta$.

---

**Proof.** For $\Theta = \theta$ we have

$$f(x|\theta) = \binom{n}{x}\theta^x(1 - \theta)^{n-x} \qquad \text{for } x = 0, 1, 2, \ldots, n$$

$$h(\theta) = \begin{cases} \dfrac{\Gamma(\alpha + \beta)}{\Gamma(\alpha) \cdot \Gamma(\beta)} \cdot \theta^{\alpha-1}(1 - \theta)^{\beta-1} & \text{for } 0 < \theta < 1 \\ 0 & \text{elsewhere} \end{cases}$$

and, hence,

$$f(\theta, x) = \frac{\Gamma(\alpha + \beta)}{\Gamma(\alpha) \cdot \Gamma(\beta)} \cdot \theta^{\alpha-1}(1 - \theta)^{\beta-1} \times \binom{n}{x}\theta^x(1 - \theta)^{n-x}$$

$$= \binom{n}{x} \cdot \frac{\Gamma(\alpha + \beta)}{\Gamma(\alpha) \cdot \Gamma(\beta)} \cdot \theta^{x+\alpha-1}(1 - \theta)^{n-x+\beta-1}$$

for $0 < \theta < 1$ and $x = 0, 1, 2, \ldots, n$, and $f(\theta, x) = 0$ elsewhere. To obtain the marginal density of $X$, let us make use of the fact that the integral of the beta density from 0 to 1 equals 1, namely, that

$$\int_0^1 x^{\alpha-1}(1 - x)^{\beta-1} \, dx = \frac{\Gamma(\alpha) \cdot \Gamma(\beta)}{\Gamma(\alpha + \beta)}$$

Thus, we get

$$g(x) = \binom{n}{x} \cdot \frac{\Gamma(\alpha + \beta)}{\Gamma(\alpha) \cdot \Gamma(\beta)} \cdot \frac{\Gamma(\alpha + x) \cdot \Gamma(n - x + \beta)}{\Gamma(n + \alpha + \beta)}$$

for $x = 0, 1, \ldots, n$ and, hence,

$$\varphi(\theta|x) = \frac{\Gamma(n + \alpha + \beta)}{\Gamma(\alpha + x) \cdot \Gamma(n - x + \beta)} \cdot \theta^{x+\alpha-1}(1 - \theta)^{n-x+\beta-1}$$

for $0 < \theta < 1$, and $\varphi(\theta|x) = 0$ elsewhere. As can be seen by inspection, this is a beta density with the parameters $x + \alpha$ and $n - x + \beta$.    ▼

To make use of this theorem, let us refer to the result that (under very general conditions) the mean of the posterior distribution minimizes the Bayes risk when the loss function is quadratic, namely, when the loss function is given by

$$L[d(x), \theta] = c[d(x) - \theta]^2$$

where $c$ is a positive constant. Note that this is the loss function which we used in Example 9.9. Since the posterior distribution of $\Theta$ is a beta distribution with parameters $x + \alpha$ and $n - x + \beta$, it follows from Theorem 6.5 that

$$E(\Theta|x) = \frac{x + \alpha}{\alpha + \beta + n}$$

is a value of an estimator of $\theta$ which minimizes the Bayes risk when the loss function is quadratic and the prior distribution of $\Theta$ is of the given form.

## EXAMPLE 10.18

Find the mean of the posterior distribution as an estimate of the "true" probability of a success, if 42 successes are obtained in 120 binomial trials and the prior distribution of $\Theta$ is a beta distribution with $\alpha = \beta = 40$.

### Solution

Substituting $x = 42$, $n = 120$, $\alpha = 40$, and $\beta = 40$ into the above formula for $E(\Theta|x)$, we get

$$E(\Theta|42) = \frac{42 + 40}{40 + 40 + 120} = 0.41$$

Note that without knowledge of the prior distribution of $\Theta$, the minimum variance unbiased estimate of $\theta$ (see Exercise 10.14) would be the sample proportion

$$\hat{\theta} = \frac{x}{n} = \frac{42}{120} = 0.35    \blacktriangle$$

**THEOREM 10.6** If $\overline{X}$ is the mean of a random sample of size $n$ from a normal population with the known variance $\sigma^2$ and the prior distribution of M (capital Greek *mu*) is a normal distribution with the mean $\mu_0$ and the variance $\sigma_0^2$, then the posterior distribution of M given $\overline{X} = \bar{x}$ is a normal distribution with the mean $\mu_1$ and the variance $\sigma_1^2$, where

$$\mu_1 = \frac{n\bar{x}\sigma_0^2 + \mu_0\sigma^2}{n\sigma_0^2 + \sigma^2} \quad \text{and} \quad \frac{1}{\sigma_1^2} = \frac{n}{\sigma^2} + \frac{1}{\sigma_0^2}$$

***Proof.*** For M $= \mu$ we have

$$f(\bar{x}|\mu) = \frac{\sqrt{n}}{\sigma\sqrt{2\pi}} \cdot e^{-\frac{1}{2}\left(\frac{\bar{x}-\mu}{\sigma/\sqrt{n}}\right)^2} \qquad \text{for } -\infty < \bar{x} < \infty$$

according to Theorem 8.4, and

$$h(\mu) = \frac{1}{\sigma_0\sqrt{2\pi}} \cdot e^{-\frac{1}{2}\left(\frac{\mu-\mu_0}{\sigma_0}\right)^2} \qquad \text{for } -\infty < \mu < \infty$$

so that

$$\varphi(\mu|\bar{x}) = \frac{h(\mu) \cdot f(\bar{x}|\mu)}{g(\bar{x})}$$

$$= \frac{\sqrt{n}}{2\pi\sigma\sigma_0 g(\bar{x})} \cdot e^{-\frac{1}{2}\left(\frac{\bar{x}-\mu}{\sigma/\sqrt{n}}\right)^2 - \frac{1}{2}\left(\frac{\mu-\mu_0}{\sigma_0}\right)^2} \qquad \text{for } -\infty < \mu < \infty$$

Now, if we collect powers of $\mu$ in the exponent of $e$, we get

$$-\frac{1}{2}\left(\frac{n}{\sigma^2} + \frac{1}{\sigma_0^2}\right)\mu^2 + \left(\frac{n\bar{x}}{\sigma^2} + \frac{\mu_0}{\sigma_0^2}\right)\mu - \frac{1}{2}\left(\frac{n\bar{x}^2}{\sigma^2} + \frac{\mu_0^2}{\sigma_0^2}\right)$$

and if we let

$$\frac{1}{\sigma_1^2} = \frac{n}{\sigma^2} + \frac{1}{\sigma_0^2} \quad \text{and} \quad \mu_1 = \frac{n\bar{x}\sigma_0^2 + \mu_0\sigma^2}{n\sigma_0^2 + \sigma^2}$$

factor out $-\dfrac{1}{2\sigma_1^2}$, and complete the square, the exponent of $e$ in the expression for $\varphi(\mu|\bar{x})$ becomes

$$-\frac{1}{2\sigma_1^2}(\mu-\mu_1)^2 + R$$

where $R$ involves $n$, $\bar{x}$, $\mu_0$, $\sigma$, and $\sigma_0$, but not $\mu$. Thus, the posterior distribution of M becomes

$$\varphi(\mu|\bar{x}) = \frac{\sqrt{n}\cdot e^R}{2\pi\sigma\sigma_0 g(\bar{x})}\cdot e^{-\frac{1}{2\sigma_1^2}(\mu-\mu_1)^2} \qquad \text{for } -\infty < \mu < \infty$$

which is easily identified as a normal distribution with the mean $\mu_1$ and the variance of $\sigma_1^2$. Hence, it can be written as

$$\varphi(\mu|\bar{x}) = \frac{1}{\sigma_1\sqrt{2\pi}}\cdot e^{-\frac{1}{2}\left(\frac{\mu-\mu_1}{\sigma_1}\right)^2} \qquad \text{for } -\infty < \mu < \infty$$

where $\mu_1$ and $\sigma_1$ are defined above. Note that we did not have to determine $g(\bar{x})$ as it was absorbed in the constant in the final result.    ▼

## EXAMPLE 10.19

A distributor of soft-drink vending machines feels that in a supermarket one of his machines will sell on the average $\mu_0 = 738$ drinks per week. Of course, the mean will vary somewhat from market to market, and the distributor feels that this variation is measured by the standard deviation $\sigma_0 = 13.4$. So far as a machine placed in a particular market is concerned, the number of drinks sold will vary from week to week, and this variation is measured by the standard deviation $\sigma = 42.5$. If one of the distributor's machines put into a new supermarket averaged $\bar{x} = 692$ during the first 10 weeks, what is the probability (the distributor's personal probability) that for this market the value of M is actually between 700 and 720?

### Solution

Assuming that the population sampled is approximately normal and that it is reasonable to treat the prior distribution of M as a normal distribution with the mean $\mu_0$ and the standard deviation $\sigma_0 = 13.4$, we find that substitution into the two formulas of Theorem 10.6 yields

$$\mu_1 = \frac{10 \cdot 692(13.4)^2 + 738(42.5)^2}{10(13.4)^2 + (42.5)^2} = 715$$

and

$$\frac{1}{\sigma_1^2} = \frac{10}{(42.5)^2} + \frac{1}{(13.4)^2} = 0.0111$$

so that $\sigma_1^2 = 90.0$ and $\sigma_1 = 9.5$. Now, the answer to our question is given by the area of the shaded region of Figure 10.1, namely, the area under the standard normal curve between

$$z = \frac{700 - 715}{9.5} = -1.58 \quad \text{and} \quad z = \frac{720 - 715}{9.5} = 0.53$$

Thus, the probability that the value of M is between 700 and 720 is $0.4429 + 0.2019 = 0.6448$, or approximately 0.645.     ▲

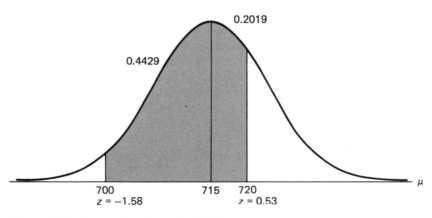

**Figure 10.1**  Diagram for Example 10.19.

## EXERCISES

**10.89** Making use of the results of Exercise 6.29, show that the mean of the posterior distribution of $\Theta$ given on page 391 can be written as

$$E(\Theta|x) = w \cdot \frac{x}{n} + (1 - w) \cdot \theta_0$$

namely, as a weighted mean of $\dfrac{x}{n}$ and $\theta_0$, where $\theta_0$ and $\sigma_0^2$ are the mean and the variance of the prior beta distribution of $\Theta$ and

$$w = \frac{n}{n + \dfrac{\theta_0(1 - \theta_0)}{\sigma_0^2} - 1}$$

**10.90** In Example 10.18 the prior distribution of the parameter $\Theta$ of the binomial distribution was a beta distribution with $\alpha = \beta = 40$. Use Theorem 6.5 to find the mean and the variance of this prior distribution and describe its shape.

**10.91** Show that the mean of the posterior distribution of M given in Theorem 10.6 can be written as

$$\mu_1 = w \cdot \bar{x} + (1 - w) \cdot \mu_0$$

namely, as a weighted mean of $\bar{x}$ and $\mu_0$, where

$$w = \frac{n}{n + \dfrac{\sigma^2}{\sigma_0^2}}$$

**10.92** If $X$ has a Poisson distribution and the prior distribution of its parameter $\Lambda$ (capital Greek *lambda*) is a gamma distribution with the parameters $\alpha$ and $\beta$, show that

(a)  the posterior distribution of $\Lambda$ given $X = x$ is a gamma distribution with the parameters $\alpha + x$ and $\dfrac{\beta}{\beta + 1}$;

(b)  the mean of the posterior distribution of $\Lambda$ is

$$\mu_1 = \frac{\beta(\alpha + x)}{\beta + 1}$$

## APPLICATIONS

**10.93** The output of a certain transistor production line is checked daily by inspecting a sample of 100 units. Over a long period of time, the process has maintained a yield of 80 percent, namely, a proportion defective of 20 percent, and the variation of the proportion defective from day to day is measured by a standard deviation of 0.04. If on a certain day the sample contains 38 defectives, find the mean of the posterior distribution of $\Theta$ as an estimate of that day's proportion defective. Assume that the prior distribution of $\Theta$ is a beta distribution.

**10.94** Records of a university (collected over many years) show that on the average 74 percent of all incoming freshmen have IQ's of at least 115. Of course, the percentage varies somewhat from year to year and this variation is measured by a standard deviation of 3 percent. If a sample check of 30 freshmen entering the university in 1991 showed that only 18 of them have IQ's of at least 115, estimate the true proportion of students with IQ's of at least 115 in that freshman class using

(a) only the prior information;
(b) only the direct information;
(c) the result of Exercise 10.89 to combine the prior information with the direct information.

**10.95** With reference to Example 10.19, find $P(712 < M < 725 | \bar{x} = 692)$.

**10.96** A history professor is making up a final examination which is to be given to a very large group of students. His feelings about the average grade they should get is expressed subjectively by a normal distribution with the mean $\mu_0 = 65.2$ and the standard deviation $\sigma_0 = 1.5$.

(a) What prior probability does the professor assign to the actual average grade being somewhere on the interval from 63.0 to 68.0?
(b) What posterior probability would he assign to this event if the examination is tried on a random sample of 40 students whose grades have a mean of 72.9 and a standard deviation of 7.4? Use $s = 7.4$ as an estimate of $\sigma$.

**10.97** An office manager feels that for a certain kind of business the daily number of incoming telephone calls is a random variable having a Poisson distribution, whose parameter has a prior gamma distribution with $\alpha = 50$ and $\beta = 2$. Being told that one such business had 112 incoming calls on a given day, what would be her estimate of that particular business' average daily number of incoming calls if she considers

(a) only the prior information;
(b) only the direct information;
(c) both kinds of information and the theory of Exercise 10.92?

## REFERENCES

Various properties of sufficient estimators are discussed in

LEHMANN, E. L., *Theory of Point Estimation.* New York: John Wiley & Sons, Inc., 1983,
WILKS, S. S., *Mathematical Statistics.* New York: John Wiley & Sons, Inc., 1962,

and a proof of Theorem 10.4 may be found in

HOGG, R. V., and CRAIG, A. T., *Introduction to Mathematical Statistics*, 4th ed. New York: Macmillan Publishing Co., Inc., 1978.

Important properties of maximum likelihood estimators are discussed in

KEEPING, E. S., *Introduction to Statistical Inference*. Princeton, N.J.: D. Van Nostrand
Co., Inc., 1962,

and a derivation of the Cramér-Rao inequality, as well as the most general conditions under
which it applies, may be found in

RAO, C. R., *Advanced Statistical Methods in Biometric Research*. New York: John Wiley
& Sons, Inc., 1952.

# 11

*Estimation:*
*Applications*

## 11.1 INTRODUCTION

In Chapter 10 we concerned ourselves with point estimation. Although this is a common way in which estimates are expressed, it leaves room for many questions. For instance, it does not tell us on how much information the estimate is based, nor does it tell us anything about the possible size of the error. Thus we might have to supplement a point estimate $\hat{\theta}$ of $\theta$ with the size of the sample and the value of var($\hat{\Theta}$), or with some other information about the sampling distribution

398

of $\hat{\Theta}$. As we shall see, this will enable us to appraise the possible size of the error.

Alternatively, we might use **interval estimation**. An interval estimate of $\theta$ is an interval of the form $\hat{\theta}_1 < \theta < \hat{\theta}_2$, where $\hat{\theta}_1$ and $\hat{\theta}_2$ are values of appropriate random variables $\hat{\Theta}_1$ and $\hat{\Theta}_2$. By "appropriate" we mean that

$$P(\hat{\Theta}_1 < \theta < \hat{\Theta}_2) = 1 - \alpha$$

for some specified probability $1 - \alpha$. For a specified value of $1 - \alpha$, we refer to $\hat{\theta}_1 < \theta < \hat{\theta}_2$ as a $(1 - \alpha)100\%$ **confidence interval** for $\theta$. Also, $1 - \alpha$ is called the **degree of confidence**, and the endpoints of the interval, $\hat{\theta}_1$ and $\hat{\theta}_2$, are called the lower and upper **confidence limits**. For instance, when $\alpha = 0.05$, the degree of confidence is 0.95 and we get a 95% confidence interval.

It should be understood that, like point estimates, interval estimates of a given parameter are not unique. This is illustrated by Exercises 11.2 and 11.3, and also in Section 11.2, where we show that, based on a single random sample, there are various confidence intervals for $\mu$, all having the same degree of confidence $1 - \alpha$. As was the case in point estimation, methods of interval estimation are judged by their various statistical properties. For instance, one desirable property is to have the length of a $(1 - \alpha)100\%$ confidence interval as short as possible; another desirable property is to have the expected length, $E(\hat{\Theta}_2 - \hat{\Theta}_1)$, as small as possible.

## 11.2 THE ESTIMATION OF MEANS

To illustrate how the possible size of errors can be appraised in point estimation, suppose that the mean of a random sample is to be used to estimate the mean of a normal population with the known variance $\sigma^2$. By Theorem 8.4, the sampling distribution of $\overline{X}$ for random samples of size $n$ from a normal population with the mean $\mu$ and the variance $\sigma^2$ is a normal distribution with $\mu_{\overline{X}} = \mu$ and $\sigma_{\overline{X}}^2 = \dfrac{\sigma^2}{n}$. Thus, we can write

$$P(|Z| < z_{\alpha/2}) = 1 - \alpha$$

where

$$Z = \frac{\overline{X} - \mu}{\sigma/\sqrt{n}}$$

and $z_{\alpha/2}$ is such that the integral of the standard normal density from $z_{\alpha/2}$ to $\infty$ equals $\alpha/2$ (see also Exercise 6.60). It follows that

$$P\left(|\overline{X} - \mu| < z_{\alpha/2} \cdot \frac{\sigma}{\sqrt{n}}\right) = 1 - \alpha$$

or, in words, that

---

**THEOREM 11.1**  If $\overline{X}$, the mean of a random sample of size $n$ from a normal population with the known variance $\sigma^2$, is to be used as an estimator of the mean of the population, the probability is $1 - \alpha$ that the error will be less than $z_{\alpha/2} \cdot \dfrac{\sigma}{\sqrt{n}}$.

---

# EXAMPLE 11.1

A team of efficiency experts intends to use the mean of a random sample of size $n = 150$ to estimate the average mechanical aptitude of assembly-line workers in a large industry (as measured by a certain standardized test). If, based on experience, the efficiency experts can assume that $\sigma = 6.2$ for such data, what can they assert with probability 0.99 about the maximum error of their estimate?

*Solution*

Substituting $n = 150$, $\sigma = 6.2$, and $z_{.005} = 2.575$ into the expression for the maximum error, we get

$$2.575 \cdot \frac{6.2}{\sqrt{150}} = 1.30$$

Thus, the efficiency experts can assert with probability 0.99 that their error will be less than 1.30.    ▲

Suppose now that these efficiency experts actually collect the necessary data and get $\bar{x} = 69.5$. Can they still assert with probability 0.99 that the error of their estimate, $\bar{x} = 69.5$, is less than 1.30? After all, $\bar{x} = 69.5$ differs from the true (population) mean by less than 1.30 or it does not, and they have no way of knowing whether it is one or the other. Actually, they can, but it must be understood that the 0.99 probability applies to the method they used to get their estimate and calculate the maximum error (collecting the sample data, determining the value of $\bar{x}$, and using the formula of Theorem 11.1) and not directly to the parameter they are trying to estimate.

To clarify this distinction, it has become the custom to use the word "confidence" here instead of "probability." *In general, we make probability statements about future values of random variables (say, the potential error of an estimate) and confidence statements once the data have been obtained.* Accordingly, we should have said in our example that the efficiency experts can be 99% confident that the error of their estimate, $\bar{x} = 69.5$, is less than $1.30$.

To construct a confidence-interval formula for estimating the mean of a normal population with the known variance $\sigma^2$, we return to the probability

$$P\left(|\bar{X} - \mu| < z_{\alpha/2} \cdot \frac{\sigma}{\sqrt{n}}\right) = 1 - \alpha$$

on page 400, which we now write as

$$P\left(\bar{X} - z_{\alpha/2} \cdot \frac{\sigma}{\sqrt{n}} < \mu < \bar{X} + z_{\alpha/2} \cdot \frac{\sigma}{\sqrt{n}}\right) = 1 - \alpha$$

It follows that

---

**THEOREM 11.2**  If $\bar{x}$ is the value of the mean of a random sample of size $n$ from a normal population with the known variance $\sigma^2$, then

$$\bar{x} - z_{\alpha/2} \cdot \frac{\sigma}{\sqrt{n}} < \mu < \bar{x} + z_{\alpha/2} \cdot \frac{\sigma}{\sqrt{n}}$$

is a $(1 - \alpha)100\%$ confidence interval for the mean of the population.

---

## EXAMPLE 11.2

If a random sample of size $n = 20$ from a normal population with the variance $\sigma^2 = 225$ has the mean $\bar{x} = 64.3$, construct a 95% confidence interval for the population mean $\mu$.

### Solution

Substituting $n = 20$, $\bar{x} = 64.3$, $\sigma = 15$, and $z_{.025} = 1.96$ into the confidence-interval formula of Theorem 11.2, we get

$$64.3 - 1.96 \cdot \frac{15}{\sqrt{20}} < \mu < 64.3 + 1.96 \cdot \frac{15}{\sqrt{20}}$$

which reduces to

$$57.7 < \mu < 70.9 \quad \blacktriangle$$

As we pointed out on page 399, confidence-interval formulas are not unique. This may be seen by changing the confidence-interval formula of Theorem 11.2 to

$$\bar{x} - z_{2\alpha/3} \cdot \frac{\sigma}{\sqrt{n}} < \mu < \bar{x} + z_{\alpha/3} \cdot \frac{\sigma}{\sqrt{n}}$$

or to the **one-sided** $(1 - \alpha)100\%$ **confidence-interval** formula

$$\mu < \bar{x} + z_{\alpha} \cdot \frac{\sigma}{\sqrt{n}}$$

Alternatively, we could base a confidence interval for $\mu$ on the sample median or, say, the midrange.

Strictly speaking, Theorems 11.1 and 11.2 require that we are dealing with a random sample from a normal population with the known variance $\sigma^2$. However, by virtue of the central limit theorem, these results can also be used for random samples from nonnormal populations provided that $n$ is sufficiently large; that is, $n \geq 30$. In that case, we may also substitute for $\sigma$ the value of the sample standard deviation.

**EXAMPLE 11.3**

An industrial designer wants to determine the average amount of time it takes an adult to assemble an "easy-to-assemble" toy. Use the following data (in minutes), a random sample, to construct a 95% confidence interval for the mean of the population sampled:

| | | | | | | | | | | | |
|---|---|---|---|---|---|---|---|---|---|---|---|
| 17 | 13 | 18 | 19 | 17 | 21 | 29 | 22 | 16 | 28 | 21 | 15 |
| 26 | 23 | 24 | 20 | 8 | 17 | 17 | 21 | 32 | 18 | 25 | 22 |
| 16 | 10 | 20 | 22 | 19 | 14 | 30 | 22 | 12 | 24 | 28 | 11 |

*Solution*

Substituting $n = 36$, $\bar{x} = 19.92$, $z_{.025} = 1.96$, and $s = 5.73$ for $\sigma$ into the confidence-interval formula of Theorem 11.2, we get

$$19.92 - 1.96 \cdot \frac{5.73}{\sqrt{36}} < \mu < 19.92 + 1.96 \cdot \frac{5.73}{\sqrt{36}}$$

Thus, the 95% confidence limits are 18.05 and 21.79 minutes.    $\blacktriangle$

When we are dealing with a random sample from a normal population, $n < 30$, and $\sigma$ is unknown, Theorems 11.1 and 11.2 cannot be used. Instead, we make use of the fact that

$$T = \frac{\overline{X} - \mu}{S/\sqrt{n}}$$

is a random variable having the $t$ distribution with $n - 1$ degrees of freedom (see Theorem 8.13). Substituting $\dfrac{\overline{X} - \mu}{S/\sqrt{n}}$ for $T$ in

$$P(-t_{\alpha/2,n-1} < T < t_{\alpha/2,n-1}) = 1 - \alpha$$

where $t_{\alpha/2,n-1}$ is defined as on page 312, we get the following confidence interval for $\mu$.

---

**THEOREM 11.3** If $\bar{x}$ and $s$ are the values of the mean and the standard deviation of a random sample of size $n$ from a normal population, then

$$\bar{x} - t_{\alpha/2,n-1} \cdot \frac{s}{\sqrt{n}} < \mu < \bar{x} + t_{\alpha/2,n-1} \cdot \frac{s}{\sqrt{n}}$$

is a $(1 - \alpha)100\%$ confidence interval for the mean of the population.

---

Since this confidence-interval formula is used mainly when $n$ is small, less than 30, we refer to it as a small-sample confidence interval for $\mu$.

## EXAMPLE 11.4

A paint manufacturer wants to determine the average drying time of a new interior wall paint. If for 12 test areas of equal size he obtained a mean drying time of 66.3 minutes and a standard deviation of 8.4 minutes, construct a 95% confidence interval for the true mean $\mu$.

### Solution

Substituting $\bar{x} = 66.3$, $s = 8.4$, and $t_{.025,11} = 2.201$ (from Table IV), the 95% confidence interval for $\mu$ becomes

$$66.3 - 2.201 \cdot \frac{8.4}{\sqrt{12}} < \mu < 66.3 + 2.201 \cdot \frac{8.4}{\sqrt{12}}$$

or simply

$$61.0 < \mu < 71.6$$

This means that we can assert with 95% confidence that the interval from 61.0 minutes to 71.6 minutes contains the true average drying time of the paint.   ▲

The method by which we constructed confidence intervals in this section consisted essentially of finding a suitable random variable whose values are determined by the sample data as well as the population parameters, yet whose distribution does not involve the parameter we are trying to estimate. This was the case, for example, when we used the random variable

$$Z = \frac{\overline{X} - \mu}{\sigma/\sqrt{n}}$$

whose values cannot be calculated without knowledge of $\mu$, but whose distribution for random samples from normal populations, the standard normal distribution, does not involve $\mu$. This method of confidence-interval construction is called the **pivotal method** and it is widely used, but there exist more general methods, such as the one discussed in the book by Mood, Graybill, and Boes referred to at the end of this chapter.

## 11.3  THE ESTIMATION OF DIFFERENCES BETWEEN MEANS

Using the results of Exercises 8.2 and 8.3, we find that for independent random samples from normal populations

$$Z = \frac{(\overline{X}_1 - \overline{X}_2) - (\mu_1 - \mu_2)}{\sqrt{\dfrac{\sigma_1^2}{n_1} + \dfrac{\sigma_2^2}{n_2}}}$$

has the standard normal distribution. If we substitute this expression for $Z$ into

$$P(-z_{\alpha/2} < Z < z_{\alpha/2}) = 1 - \alpha$$

the pivotal method yields the following confidence-interval formula for $\mu_1 - \mu_2$.

**THEOREM 11.4** If $\bar{x}_1$ and $\bar{x}_2$ are the values of the means of independent random samples of size $n_1$ and $n_2$ from normal populations with the known variances $\sigma_1^2$ and $\sigma_2^2$, then

$$(\bar{x}_1 - \bar{x}_2) - z_{\alpha/2} \cdot \sqrt{\frac{\sigma_1^2}{n_1} + \frac{\sigma_2^2}{n_2}} < \mu_1 - \mu_2 < (\bar{x}_1 - \bar{x}_2) + z_{\alpha/2} \cdot \sqrt{\frac{\sigma_1^2}{n_1} + \frac{\sigma_2^2}{n_2}}$$

is a $(1 - \alpha)100\%$ confidence interval for the difference between the two population means.

By virtue of the central limit theorem, this confidence-interval formula can also be used for independent random samples from nonnormal populations with known variances when $n_1$ and $n_2$ are large—that is, when $n_1 \geqslant 30$ and $n_2 \geqslant 30$.

## EXAMPLE 11.5

Construct a 94% confidence interval for the difference between the mean lifetimes of two kinds of light bulbs, given that a random sample of 40 light bulbs of the first kind lasted on the average 418 hours of continuous use and 50 light bulbs of the second kind lasted on the average 402 hours of continuous use. The population standard deviations are known to be $\sigma_1 = 26$ and $\sigma_2 = 22$.

### Solution

For $\alpha = 0.06$, we find from Table III that $z_{.03} = 1.88$. Therefore, the 94% confidence interval for $\mu_1 - \mu_2$ is

$$(418 - 402) - 1.88 \cdot \sqrt{\frac{26^2}{40} + \frac{22^2}{50}} < \mu_1 - \mu_2$$

$$< (418 - 402) + 1.88 \cdot \sqrt{\frac{26^2}{40} + \frac{22^2}{50}}$$

which reduces to

$$6.3 < \mu_1 - \mu_2 < 25.7$$

Hence, we are 94% confident that the interval from 6.3 to 25.7 hours contains the actual difference between the mean lifetimes of the two kinds of light bulbs. The fact that both confidence limits are positive suggests that on the average the first kind of light bulb is superior to the second kind.    ▲

Chap. 11: Estimation: Applications

To construct a $(1 - \alpha)100\%$ confidence interval for the difference between two means when $n_1 \geqslant 30$, $n_2 \geqslant 30$, but $\sigma_1$ and $\sigma_2$ are unknown, we simply substitute $s_1$ and $s_2$ for $\sigma_1$ and $\sigma_2$ and proceed as before. When $\sigma_1$ and $\sigma_2$ are unknown and either or both of the samples are small, the procedure for estimating the difference between the means of two normal populations is not straightforward unless it can be assumed that $\sigma_1 = \sigma_2$. If $\sigma_1 = \sigma_2 = \sigma$, then

$$Z = \frac{(\overline{X}_1 - \overline{X}_2) - (\mu_1 - \mu_2)}{\sigma \sqrt{\dfrac{1}{n_1} + \dfrac{1}{n_2}}}$$

is a random variable having the standard normal distribution and $\sigma^2$ can be estimated by **pooling** the squared deviations from the means of the two samples. In Exercise 11.8 the reader will be asked to verify that the resulting **pooled estimator**

$$S_p^2 = \frac{(n_1 - 1)S_1^2 + (n_2 - 1)S_2^2}{n_1 + n_2 - 2}$$

is, indeed, an unbiased estimator of $\sigma^2$. Now, by Theorems 8.11 and 8.9, the independent random variables

$$\frac{(n_1 - 1)S_1^2}{\sigma^2} \quad \text{and} \quad \frac{(n_2 - 1)S_2^2}{\sigma^2}$$

have chi-square distributions with $n_1 - 1$ and $n_2 - 1$ degrees of freedom, and their sum

$$Y = \frac{(n_1 - 1)S_1^2}{\sigma^2} + \frac{(n_2 - 1)S_2^2}{\sigma^2} = \frac{(n_1 + n_2 - 2)S_p^2}{\sigma^2}$$

has a chi-square distribution with $n_1 + n_2 - 2$ degrees of freedom. As it can be shown that the above random variables $Z$ and $Y$ are independent (see references at the end of this chapter), it follows from Theorem 8.12 that

$$T = \frac{Z}{\sqrt{\dfrac{Y}{n_1 + n_2 - 2}}}$$

$$= \frac{(\overline{X}_1 - \overline{X}_2) - (\mu_1 - \mu_2)}{S_p \sqrt{\dfrac{1}{n_1} + \dfrac{1}{n_2}}}$$

has a $t$ distribution with $n_1 + n_2 - 2$ degrees of freedom. Substituting this expression for $T$ into

$$P(-t_{\alpha/2,n-1} < T < t_{\alpha/2,n-1}) = 1 - \alpha$$

we arrive at the following $(1 - \alpha)100\%$ confidence interval for $\mu_1 - \mu_2$.

---

**THEOREM 11.5** If $\bar{x}_1$, $\bar{x}_2$, $s_1$, and $s_2$ are the values of the means and the standard deviations of independent random samples of size $n_1$ and $n_2$ from normal populations with equal variances, then

$$(\bar{x}_1 - \bar{x}_2) - t_{\alpha/2,n_1+n_2-2} \cdot s_p \sqrt{\frac{1}{n_1} + \frac{1}{n_2}} < \mu_1 - \mu_2$$

$$< (\bar{x}_1 - \bar{x}_2) + t_{\alpha/2,n_1+n_2-2} \cdot s_p \sqrt{\frac{1}{n_1} + \frac{1}{n_2}}$$

is a $(1 - \alpha)100\%$ confidence interval for the difference between the two population means.

---

Since this confidence-interval formula is used mainly when $n_1$ and/or $n_2$ are small, less than 30, we refer to it as a small-sample confidence interval for $\mu_1 - \mu_2$.

## EXAMPLE 11.6

A study has been made to compare the nicotine contents of two brands of cigarettes. Ten cigarettes of Brand $A$ had an average nicotine content of 3.1 milligrams with a standard deviation of 0.5 milligram, while eight cigarettes of Brand $B$ had an average nicotine content of 2.7 milligrams with a standard deviation of 0.7 milligram. Assuming that the two sets of data are independent random samples from normal populations with equal variances, construct a 95% confidence interval for the difference between the mean nicotine contents of the two brands of cigarettes.

### Solution

First we substitute $n_1 = 10$, $n_2 = 8$, $s_1 = 0.5$, and $s_2 = 0.7$ into the formula for $s_p$, and we get

$$s_p = \sqrt{\frac{9(0.25) + 7(0.49)}{16}} = 0.596$$

Then, substituting this value together with $n_1 = 10$, $n_2 = 8$, $\bar{x}_1 = 3.1$, $\bar{x}_2 = 2.7$, and $t_{.025,16} = 2.120$ (from Table IV) into the confidence-interval formula of Theorem 11.5, we find that the required 95% confidence interval is

$$(3.1 - 2.7) - 2.120(0.596)\sqrt{\tfrac{1}{10} + \tfrac{1}{8}} < \mu_1 - \mu_2$$
$$< (3.1 - 2.7) + 2.120(0.596)\sqrt{\tfrac{1}{10} + \tfrac{1}{8}}$$

which reduces to

$$-0.20 < \mu_1 - \mu_2 < 1.00$$

Thus, the 95% confidence limits are $-0.20$ and $1.00$ milligrams, but observe that since this includes $\mu_1 - \mu_2 = 0$, we cannot conclude that there is a real difference between the average nicotine contents of the two brands of cigarettes. More about that in Chapter 13.    ▲

## EXERCISES

**11.1**  If $x$ is a value of a random variable having an exponential distribution, find $k$ so that the interval from 0 to $kx$ is a $(1 - \alpha)100\%$ confidence interval for the parameter $\theta$.

**11.2**  If $x_1$ and $x_2$ are the values of a random sample of size 2 from a population having a uniform density with $\alpha = 0$ and $\beta = \theta$, find $k$ so that

$$0 < \theta < k(x_1 + x_2)$$

is a $(1 - \alpha)100\%$ confidence interval for $\theta$ when
(a)  $\alpha \leqslant \tfrac{1}{2}$;
(b)  $\alpha > \tfrac{1}{2}$.

**11.3**  Making use of the methods of Section 8.7, it can be shown that for a random sample of size $n = 2$ from the population of Exercise 11.2, the distribution of the sample range is given by

$$f(R) = \begin{cases} \dfrac{2}{\theta^2}(\theta - R) & \text{for } 0 < R < \theta \\ 0 & \text{elsewhere} \end{cases}$$

Use this result to find $c$ so that

$$R < \theta < cR$$

is a $(1 - \alpha)100\%$ confidence interval for $\theta$.

**11.4**  Show that the $(1 - \alpha)100\%$ confidence interval

$$\bar{x} - z_{\alpha/2} \cdot \frac{\sigma}{\sqrt{n}} < \mu < \bar{x} + z_{\alpha/2} \cdot \frac{\sigma}{\sqrt{n}}$$

is shorter than the $(1 - \alpha)100\%$ confidence interval

$$\bar{x} - z_{2\alpha/3} \cdot \frac{\sigma}{\sqrt{n}} < \mu < \bar{x} + z_{\alpha/3} \cdot \frac{\sigma}{\sqrt{n}}$$

**11.5**  Show that if $\bar{x}$ is used as a point estimate of $\mu$ and $\sigma$ is known, the probability is $1 - \alpha$ that $|\bar{x} - \mu|$, the absolute value of our error, will not exceed a specified amount $e$ when

$$n = \left[ z_{\alpha/2} \cdot \frac{\sigma}{e} \right]^2$$

(If it turns out that $n < 30$, this formula cannot be used unless it is reasonable to assume that we are sampling a normal population.)

**11.6**  Modify Theorem 11.1 so that it can be used to appraise the maximum error when $\sigma^2$ is unknown. (Note that this method can be used only after the data have been obtained.)

**11.7**  State a theorem analogous to Theorem 11.1, which enables us to appraise the maximum error in using $\bar{x}_1 - \bar{x}_2$ as an estimate of $\mu_1 - \mu_2$ under the conditions of Theorem 11.4.

**11.8**  Show that $S_p^2$ is an unbiased estimator of $\sigma^2$ and find its variance under the conditions of Theorem 11.5.

**11.9**  Verify the result on page 406, which expresses $T$ in terms of $\bar{X}_1, \bar{X}_2$, and $S_p$.

## APPLICATIONS

**11.10**  A district official intends to use the mean of a random sample of 150 sixth graders from a very large school district to estimate the mean score which all the sixth graders in the district would get if they took a certain arithmetic achievement test. If, based on experience, the official knows that $\sigma = 9.4$ for such data, what can she assert with probability 0.95 about the maximum error?

**11.11**  With reference to Exercise 11.10, suppose that the district official takes her sample and gets $\bar{x} = 61.8$. Use all the given information to construct a 99% confidence interval for the mean score of all the sixth graders in the district.

**11.12** A medical research worker intends to use the mean of a random sample of size $n = 120$ to estimate the mean blood pressure of women in their fifties. If, based on experience, he knows that $\sigma = 10.5$ mm of mercury, what can he assert with probability 0.99 about the maximum error?

**11.13** With reference to Exercise 11.12, suppose that the research worker takes his sample and gets $\bar{x} = 141.8$ mm of mercury. Construct a 98% confidence interval for the mean blood pressure of women in their fifties.

**11.14** A study of the annual growth of certain cacti showed that 64 of them, selected at random in a desert region, grew on the average 52.80 mm with a standard deviation of 4.5 mm. Construct a 99% confidence interval for the true average annual growth of the given kind of cactus.

**11.15** To estimate the average time required for certain repairs, an automobile manufacturer had 40 mechanics, a random sample, timed in the performance of this task. If it took them on the average 24.05 minutes with a standard deviation of 2.68 minutes, what can the manufacturer assert with 95% confidence about the maximum error, if he uses $\bar{x} = 24.05$ minutes as an estimate of the actual mean time required to perform the given repairs?

**11.16** If a sample constitutes an appreciable portion of a population, more than 5 percent of the population according to the rule of thumb on page 302, the formulas of Theorems 11.1 and 11.2 must be modified by using the variance formula of Theorem 8.6 instead of that of Theorem 8.1. For instance, the maximum error in Theorem 11.1 becomes

$$ z_{\alpha/2} \cdot \frac{\sigma}{\sqrt{n}} \cdot \sqrt{\frac{N - n}{N - 1}} $$

Use this modification to rework Exercise 11.10, given that there are 900 sixth graders in the school district.

**11.17** Use the modification suggested in Exercise 11.16 to rework Exercise 11.11, given that there are 900 sixth graders in the school district.

**11.18** An efficiency expert wants to determine the average amount of time it takes a pit crew to change a set of four tires on a race car. Use the formula for $n$ in Exercise 11.5 to determine the sample size that is needed so that the efficiency expert can assert with probability 0.95 that the sample mean will differ from $\mu$, the quantity to be estimated, by less than 2.5 seconds. It is known from previous studies that $\sigma = 12.2$ seconds.

**11.19** In a study of television viewing habits, it is desired to estimate the average number of hours that teenagers spend watching per week. If it is reasonable to assume that $\sigma = 3.2$ hours, how large a sample is needed so that it will be possible to assert with 95% confidence that the sample mean is off by less than 20 minutes. (*Hint*: Refer to Exercise 11.5.)

**11.20** The length of the skulls of 10 fossil skeletons of an extinct species of birds has a mean of 5.68 cm and a standard deviation of 0.29 cm. Assuming that such measurements are normally distributed, find a 95% confidence interval for the mean length of the skulls of this species of birds.

**11.21** A major truck stop has kept extensive records on various transactions with its customers. If a random sample of 18 of these records show average sales of 63.84 gallons of diesel fuel with a standard deviation of 2.75 gallons, construct a 99% confidence interval for the mean of the population sampled.

**11.22** A food inspector, examining 12 jars of a certain brand of peanut butter, obtained the following percentages of impurities: 2.3, 1.9, 2.1, 2.8, 2.3, 3.6, 1.4, 1.8, 2.1, 3.2, 2.0, and 1.9. Based on the modification of Theorem 11.1 of Exercise 11.6, what can she assert with 95% confidence about the maximum error, if she uses the mean of this sample as an estimate of the average percentage of impurities in this brand of peanut butter?

**11.23** Independent random samples of size $n_1 = 16$ and $n_2 = 25$ from normal populations with $\sigma_1 = 4.8$ and $\sigma_2 = 3.5$ have the means $\bar{x}_1 = 18.2$ and $\bar{x}_2 = 23.4$. Find a 90% confidence interval for $\mu_1 - \mu_2$.

**11.24** A study of two kinds of photocopying equipment shows that 61 failures of the first kind of equipment took on the average 80.7 minutes to repair with a standard deviation of 19.4 minutes, whereas 61 failures of the second kind of equipment took on the average 88.1 minutes to repair with a standard deviation of 18.8 minutes. Find a 99% confidence interval for the difference between the true average amounts of time it takes to repair failures of the two kinds of photocopying equipment.

**11.25** Twelve randomly selected mature citrus trees of one variety have a mean height of 13.8 feet with a standard deviation of 1.2 feet, and fifteen randomly selected mature citrus trees of another variety have a mean height of 12.9 feet with a standard deviation of 1.5 feet. Assuming that the random samples were selected from normal populations with equal variances, construct a 95% confidence interval for the difference between the true average heights of the two kinds of citrus trees.

**11.26** The following are the heat-producing capacities of coal from two mines (in millions of calories per ton):

> *Mine A*:    8,500,  8,330,  8,480,  7,960,  8,030
> *Mine B*:    7,710,  7,890,  7,920,  8,270,  7,860

Assuming that the data constitute independent random samples from normal populations with equal variances, construct a 99% confidence interval for the difference between the true average heat-producing capacities of coal from the two mines.

**11.27** To study the effect of alloying on the resistance of electric wires, an engineer plans to measure the resistance of $n_1 = 35$ standard wires and $n_2 = 45$ alloyed wires. If it can be assumed that $\sigma_1 = 0.004$ ohm and $\sigma_2 = 0.005$ ohm for such data, what can she assert with 98% confidence about the maximum error, if she uses $\bar{x}_1 - \bar{x}_2$ as an estimate of $\mu_1 - \mu_2$? (*Hint:* Use the result of Exercise 11.7.)

## 11.4    THE ESTIMATION OF PROPORTIONS

There are many problems in which we must estimate proportions, probabilities, percentages, or rates, such as the proportion of defectives in a large shipment of transistors, the probability that a car stopped at a road block will have faulty lights, the percentage of school children with IQ's over 115, or the mortality rate of a disease. In many of these it is reasonable to assume that we are sampling a binomial population, and, hence, that our problem is to estimate the binomial parameter $\theta$. Thus, we can make use of the fact that for large $n$ the binomial distribution can be approximated with a normal distribution, namely, that

$$Z = \frac{X - n\theta}{\sqrt{n\theta(1 - \theta)}}$$

can be treated as a random variable having approximately the standard normal distribution. Substituting this expression for $Z$ into

$$P(-z_{\alpha/2} < Z < z_{\alpha/2}) = 1 - \alpha$$

we get

$$P\left(-z_{\alpha/2} < \frac{X - n\theta}{\sqrt{n\theta(1 - \theta)}} < z_{\alpha/2}\right) = 1 - \alpha$$

and the two inequalities

$$-z_{\alpha/2} < \frac{x - n\theta}{\sqrt{n\theta(1 - \theta)}} \quad \text{and} \quad \frac{x - n\theta}{\sqrt{n\theta(1 - \theta)}} < z_{\alpha/2}$$

whose solution will yield $(1 - \alpha)100\%$ confidence limits for $\theta$. Leaving the details of this to the reader in Exercise 11.28, let us give here instead a large-sample approximation by rewriting $P(-z_{\alpha/2} < Z < z_{\alpha/2}) = 1 - \alpha$ with $\dfrac{X - n\theta}{\sqrt{n\theta(1 - \theta)}}$ substituted for $Z$ as

$$P\left(\hat{\Theta} - z_{\alpha/2} \cdot \sqrt{\frac{\theta(1 - \theta)}{n}} < \theta < \hat{\Theta} + z_{\alpha/2} \cdot \sqrt{\frac{\theta(1 - \theta)}{n}}\right) = 1 - \alpha$$

where $\hat{\Theta} = \dfrac{X}{n}$. Then, if we substitute $\hat{\theta}$ for $\theta$ inside the radicals, which is a further approximation, we get

---

**THEOREM 11.6**  If $X$ is a binomial random variable with the parameters $n$ and $\theta$, $n$ is large, and $\hat{\theta} = \dfrac{x}{n}$, then

$$\hat{\theta} - z_{\alpha/2} \cdot \sqrt{\frac{\hat{\theta}(1 - \hat{\theta})}{n}} < \theta < \hat{\theta} + z_{\alpha/2} \cdot \sqrt{\frac{\hat{\theta}(1 - \hat{\theta})}{n}}$$

is an approximate $(1 - \alpha)100\%$ confidence interval for $\theta$.

---

## EXAMPLE 11.7

In a random sample, 136 of 400 persons given a flu vaccine experienced some discomfort. Construct a 95% confidence interval for the true proportion of persons who will experience some discomfort from the vaccine.

### Solution

Substituting $n = 400$, $\hat{\theta} = \frac{136}{400} = 0.34$, and $z_{.025} = 1.96$ into the confidence-interval formula of Theorem 11.6, we get

$$0.34 - 1.96\sqrt{\frac{(0.34)(0.66)}{400}} < \theta < 0.34 + 1.96\sqrt{\frac{(0.34)(0.66)}{400}}$$

$$0.294 < \theta < 0.386$$

or, rounding to two decimals, $0.29 < \theta < 0.39$.     ▲

Using the same approximations that led to Theorem 11.6, we can also write

**THEOREM 11.7** If $\hat{\theta} = \dfrac{x}{n}$ is used as an estimate of $\theta$, we can assert with $(1 - \alpha)100\%$ confidence that the error is less than

$$z_{\alpha/2} \cdot \sqrt{\dfrac{\hat{\theta}(1 - \hat{\theta})}{n}}$$

## EXAMPLE 11.8

A study is made to determine the proportion of voters in a sizeable community who favor the construction of a nuclear power plant. If 140 of 400 voters selected at random favor the project and we use $\hat{\theta} = \frac{140}{400} = 0.35$ as an estimate of the actual proportion of all voters in the community who favor the project, what can we say with 99% confidence about the maximum error?

### Solution

Substituting $n = 400$, $\hat{\theta} = 0.35$, and $z_{.005} = 2.575$ into the formula of Theorem 11.7, we get

$$2.575 \cdot \sqrt{\dfrac{(0.35)(0.65)}{400}} = 0.061$$

or 0.06 rounded to two decimals. Thus, if we use $\hat{\theta} = 0.35$ as an estimate of the actual proportion of voters in the community who favor the project, we can assert with 99% confidence that the error is less than 0.06.   ▲

## 11.5 THE ESTIMATION OF DIFFERENCES BETWEEN PROPORTIONS

There are many problems in which we must estimate the difference between the binomial parameters $\theta_1$ and $\theta_2$ on the basis of independent random samples of size $n_1$ and $n_2$ from two binomial populations. This would be the case, for example, if we want to estimate the difference between the proportions of male and female voters who favor a certain candidate for governor of Illinois.

If the respective numbers of successes are $X_1$ and $X_2$ and the corresponding sample proportions are denoted by $\hat{\Theta}_1 = \dfrac{X_1}{n_1}$ and $\hat{\Theta}_2 = \dfrac{X_2}{n_2}$, let us investigate the sampling distribution of $\hat{\Theta}_1 - \hat{\Theta}_2$, which is an obvious estimator of $\theta_1 - \theta_2$. From Exercise 8.5 we have

$$E(\hat{\Theta}_1 - \hat{\Theta}_2) = \theta_1 - \theta_2$$

and

$$\text{var}(\hat{\Theta}_1 - \hat{\Theta}_2) = \frac{\theta_1(1 - \theta_1)}{n_1} + \frac{\theta_2(1 - \theta_2)}{n_2}$$

and since, for large samples, $X_1$ and $X_2$—and hence also their difference—can be approximated with normal distributions, it follows that

$$Z = \frac{(\hat{\Theta}_1 - \hat{\Theta}_2) - (\theta_1 - \theta_2)}{\sqrt{\dfrac{\theta_1(1 - \theta_1)}{n_1} + \dfrac{\theta_2(1 - \theta_2)}{n_2}}}$$

is a random variable having approximately the standard normal distribution. Substituting this expression for $Z$ into $P(-z_{\alpha/2} < Z < z_{\alpha/2}) = 1 - \alpha$, we thus arrive at the following result.

---

**THEOREM 11.8** If $X_1$ is a binomial random variable with the parameters $n_1$ and $\theta_1$, $X_2$ is a binomial random variable with the parameters $n_2$ and $\theta_2$, $n_1$ and $n_2$ are large, and $\hat{\theta}_1 = \dfrac{x_1}{n_1}$ and $\hat{\theta}_2 = \dfrac{x_2}{n_2}$, then

$$(\hat{\theta}_1 - \hat{\theta}_2) - z_{\alpha/2} \cdot \sqrt{\frac{\hat{\theta}_1(1 - \hat{\theta}_1)}{n_1} + \frac{\hat{\theta}_2(1 - \hat{\theta}_2)}{n_2}} < \theta_1 - \theta_2$$

$$< (\hat{\theta}_1 - \hat{\theta}_2) + z_{\alpha/2} \cdot \sqrt{\frac{\hat{\theta}_1(1 - \hat{\theta}_1)}{n_1} + \frac{\hat{\theta}_2(1 - \hat{\theta}_2)}{n_2}}$$

is an approximate $(1 - \alpha)100\%$ confidence interval for $\theta_1 - \theta_2$.

---

## EXAMPLE 11.9

If 132 of 200 male voters and 90 of 159 female voters favor a certain candidate running for governor of Illinois, find a 99% confidence interval for the difference between the actual proportions of male and female voters who favor the candidate.

*Solution*

Substituting $\hat{\theta}_1 = \frac{132}{200} = 0.66$, $\hat{\theta}_2 = \frac{90}{150} = 0.60$, and $z_{.005} = 2.575$ into the confidence-interval formula of Theorem 11.8, we get

$$(0.66 - 0.60) - 2.575 \sqrt{\frac{(0.66)(0.34)}{200} + \frac{(0.60)(0.40)}{150}} < \theta_1 - \theta_2$$

$$< (0.66 - 0.60) + 2.575 \sqrt{\frac{(0.66)(0.34)}{200} + \frac{(0.60)(0.40)}{150}}$$

which reduces to

$$-0.074 < \theta_1 - \theta_2 < 0.194$$

Thus, we are 99% confident that the interval from $-0.074$ to $0.194$ contains the difference between the actual proportions of male and female voters who favor the candidate. Observe that this includes the possibility of a zero difference between the two proportions.    ▲

*EXERCISES*

**11.28** By solving

$$-z_{\alpha/2} = \frac{x - n\theta}{\sqrt{n\theta(1 - \theta)}} \quad \text{and} \quad \frac{x - n\theta}{\sqrt{n\theta(1 - \theta)}} = z_{\alpha/2}$$

for $\theta$, show that

$$\frac{x + \frac{1}{2} \cdot z_{\alpha/2}^2 \pm z_{\alpha/2} \sqrt{\frac{x(n - x)}{n} + \frac{1}{4} \cdot z_{\alpha/2}^2}}{n + z_{\alpha/2}^2}$$

are $(1 - \alpha)100\%$ confidence limits for $\theta$.

**11.29** Use the formula of Theorem 11.7 to show that we can be at least $(1 - \alpha)100\%$ confident that the error we make is less than $e$, when we use a sample proportion $\hat{\theta} = \frac{x}{n}$ with

$$n = \frac{z_{\alpha/2}^2}{4e^2}$$

as an estimate of $\theta$.

**11.30** Find a formula for $n$ analogous to that of the preceding exercise, when it is known that $\theta$ must lie on the interval from $\theta'$ to $\theta''$.

**11.31** Fill in the details that led from the $Z$ statistic on page 415, substituted into $P(-z_{\alpha/2} < Z < z_{\alpha/2}) = 1 - \alpha$, to the confidence-interval formula of Theorem 11.8.

**11.32** Find a formula for the maximum error analogous to that of Theorem 11.7, when we use $\hat{\theta}_1 - \hat{\theta}_2$ as an estimate of $\theta_1 - \theta_2$.

**11.33** Use the result of Exercise 11.32 to show that when $n_1 = n_2 = n$, we can be at least $(1 - \alpha)100\%$ confident that the error we make when using $\hat{\theta}_1 - \hat{\theta}_2$ as an estimate of $\theta_1 - \theta_2$ is less than $e$ when

$$n = \frac{z_{\alpha/2}^2}{2e^2}$$

## APPLICATIONS

**11.34** A sample survey at a supermarket showed that 204 of 300 shoppers regularly use cents-off coupons. Use the large-sample confidence-interval formula of Theorem 11.6 to construct a 95% confidence interval for the corresponding true proportion.

**11.35** With reference to Exercise 11.34, what can we say with 99% confidence about the maximum error, if we use the observed sample proportion as an estimate of the proportion of all shoppers in the population sampled who use cents-off coupons?

**11.36** In a random sample of 250 television viewers in a large city, 190 had seen a certain controversial program. Construct a 99% confidence interval for the corresponding true proportion using

(a) the large-sample confidence-interval formula of Theorem 11.6;

(b) the confidence limits of Exercise 11.28.

**11.37** With reference to Exercise 11.36, what can we say with 95% confidence about the maximum error, if we use the observed sample proportion as an estimate of the corresponding true proportion?

**11.38** Among 100 fish caught in a certain lake, 18 were inedible as a result of the chemical pollution of the environment. Construct a 99% confidence interval for the corresponding true proportion.

**11.39** In a random sample of 120 cheerleaders, 54 had suffered moderate to severe damage to their voices. With 90% confidence, what can we say about the maximum error, if we use the sample proportion, $\frac{54}{120} = 0.45$, as an estimate of the true proportion of cheerleaders who are afflicted in this way?

**11.40** In a random sample of 300 persons eating lunch at a department store cafeteria, only 102 had dessert. If we use $\frac{102}{300} = 0.34$ as an estimate of the corresponding true proportion, with what confidence can we assert that our error is less than 0.05?

**11.41** A private opinion poll is engaged by a politician to estimate what proportion of her constituents favor the decrimilization of certain minor narcotics violations. Use the formula of Exercise 11.29 to determine how large a sample the poll will have to take to be at least 95% confident that the sample proportion is off by less than 0.02.

**11.42** Use the result of Exercise 11.30 to rework Exercise 11.41, given that the poll has reason to believe that the true proportion does not exceed 0.30.

**11.43** Suppose we want to estimate what proportions of all drivers exceed the legal speed limit on a certain stretch of road between Los Angeles and Bakersfield. Use the formula of Exercise 11.29 to determine how large a sample we will need to be at least 99% confident that the resulting estimate, the sample proportion, is off by less than 0.04.

**11.44** Use the result of Exercise 11.30 to rework Exercise 11.43, given that we have good reason to believe that the proportion we are trying to estimate is at least 0.65.

**11.45** In a random sample of visitors to a famous tourist attraction, 84 of 250 men and 156 of 250 women bought souvenirs. Construct a 95% confidence interval for the difference between the true proportions of men and women who buy souvenirs at this tourist attraction.

**11.46** Among 500 marriage license applications chosen at random in a given year, there were 48 in which the woman was at least one year older than the man, and among 400 marriage license applications chosen at random six years later, there were 68 in which the woman was at least a year older than the man. Construct a 99% confidence interval for the difference between the corresponding true proportions of marriage license applications in which the woman was at least one year older than the man.

**11.47** With reference to Exercise 11.46, what can we say with 98% confidence about the maximum error, if we use the difference between the observed sample proportions as an estimate of the difference between the corresponding true proportions? (*Hint*: Use the result of Exercise 11.32.)

**11.48** Suppose we want to determine the difference between the proportions of the customers of a donut chain in North Carolina and Vermont who prefer the chain's donuts to those of all its competitors. Use the formula of Exercise 11.33 to detrmine the size of the samples that are needed to be at least 95% confident that the differrence between the two sample proportions is off by less than 0.05.

## 11.6  THE ESTIMATION OF VARIANCES

Given a random sample of size $n$ from a normal population, we can obtain a $(1 - \alpha)100\%$ confidence interval for $\sigma^2$ by making use of Theorem 8.11, according to which

$$\frac{(n - 1)S^2}{\sigma^2}$$

is a random variable having a chi-square distribution with $n - 1$ degrees of freedom. Thus,

$$P\left[\chi^2_{1-\alpha/2,n-1} < \frac{(n - 1)S^2}{\sigma^2} < \chi^2_{\alpha/2,n-1}\right] = 1 - \alpha$$

$$P\left[\frac{(n - 1)S^2}{\chi^2_{\alpha/2,n-1}} < \sigma^2 < \frac{(n - 1)S^2}{\chi^2_{1-\alpha/2,n-1}}\right] = 1 - \alpha$$

where $\chi^2_{\alpha/2,n-1}$ and $\chi^2_{1-\alpha/2,n-1}$ are as defined on page 312, and we get

---

**THEOREM 11.9**  If $s^2$ is the value of the variance of a random sample of size $n$ from a normal population, then

$$\frac{(n - 1)s^2}{\chi^2_{\alpha/2,n-1}} < \sigma^2 < \frac{(n - 1)s^2}{\chi^2_{1-\alpha/2,n-1}}$$

is a $(1 - \alpha)100\%$ confidence interval for $\sigma^2$.

---

Corresponding $(1 - \alpha)100\%$ confidence limits for $\sigma$ can be obtained by taking the square roots of the confidence limits for $\sigma^2$.

## EXAMPLE 11.10

In 16 test runs the gasoline consumption of an experimental engine had a standard deviation of 2.2 gallons. Construct a 99% confidence interval for $\sigma^2$, which measures the true variability of the gasoline consumption of the engine.

### Solution

Assuming that the observed data can be looked upon as a random sample from a normal population, we substitute $n = 16$ and $s = 2.2$, along with $\chi^2_{.005,15} = 32.801$ and $\chi^2_{.995,15} = 4.601$, obtained from Table V, into the confidence interval formula of Theorem 11.9, and we get

$$\frac{15(2.2)^2}{32.801} < \sigma^2 < \frac{15(2.2)^2}{4.601}$$

or

$$2.21 < \sigma^2 < 15.78 \quad \blacktriangle$$

To get a corresponding 99% confidence interval for $\sigma$, we take square roots and get $1.49 < \sigma < 3.97$.

## 11.7 THE ESTIMATION OF THE RATIO OF TWO VARIANCES

If $S_1^2$ and $S_2^2$ are the variances of independent random samples of size $n_1$ and $n_2$ from normal populations, then, according to Theorem 8.15,

$$F = \frac{\sigma_2^2 S_1^2}{\sigma_1^2 S_2^2}$$

is a random variable having an $F$ distribution with $n_1 - 1$ and $n_2 - 1$ degrees of freedom. Thus, we can write

$$P\left(f_{1-\alpha/2,n_1-1,n_2-1} < \frac{\sigma_2^2 S_1^2}{\sigma_1^2 S_2^2} < f_{\alpha/2,n_1-1,n_2-1}\right) = 1 - \alpha$$

where $f_{\alpha/2,n_1-1,n_2-1}$ and $f_{1-\alpha/2,n_1-1,n_2-1}$ are as defined on page 316. Since

$$f_{1-\alpha/2,n_1-1,n_2-1} = \frac{1}{f_{\alpha/2,n_2-1,n_1-1}}$$

(see Exercise 8.56), it follows that

---

**THEOREM 11.10**  If $s_1^2$ and $s_2^2$ are the values of the variances of independent random samples of size $n_1$ and $n_2$ from normal populations, then

$$\frac{s_1^2}{s_2^2} \cdot \frac{1}{f_{\alpha/2,n_1-1,n_2-1}} < \frac{\sigma_1^2}{\sigma_2^2} < \frac{s_1^2}{s_2^2} \cdot f_{\alpha/2,n_2-1,n_1-1}$$

is a $(1 - \alpha)100\%$ confidence interval for $\dfrac{\sigma_1^2}{\sigma_2^2}$.

---

Corresponding $(1 - \alpha)100\%$ confidence limits for $\dfrac{\sigma_1}{\sigma_2}$ can be obtained by taking the square roots of the confidence limits for $\dfrac{\sigma_1^2}{\sigma_2^2}$.

## EXAMPLE 11.11

With reference to Example 11.6, find a 98% confidence interval for $\dfrac{\sigma_1^2}{\sigma_2^2}$.

*Solution*

Substituting $n_1 = 10$, $n_2 = 8$, $s_1 = 0.5$, $s_2 = 0.7$, and $f_{.01,9,7} = 6.72$ and $f_{.01,7,9} = 5.61$ from Table VI, we get

$$\frac{0.25}{0.49} \cdot \frac{1}{6.72} < \frac{\sigma_1^2}{\sigma_2^2} < \frac{0.25}{0.49} \cdot 5.61$$

or

$$0.076 < \frac{\sigma_1^2}{\sigma_2^2} < 2.862$$

Since the interval obtained here includes the possibility that the ratio is 1, there is no real evidence against the assumption of equal population variances in Example 11.6.    ▲

## EXERCISES

**11.49** Fill in the details that led from the probability on page 420 to the confidence-interval formula of Theorem 11.10.

**11.50** For large $n$, the sampling distribution of $S$ is sometimes approximated with a normal distribution having the mean $\sigma$ and the variance $\dfrac{\sigma^2}{2n}$ (see Exercise 8.42). Show that this approximation leads to the following $(1 - \alpha)100\%$ large-sample confidence interval for $\sigma$:

$$\frac{s}{1 + \dfrac{z_{\alpha/2}}{\sqrt{2n}}} < \sigma < \frac{s}{1 - \dfrac{z_{\alpha/2}}{\sqrt{2n}}}$$

*APPLICATIONS*

**11.51** With reference to Exercise 11.20, construct a 95% confidence interval for the true variance of the skull length of the given species of birds.

**11.52** With reference to Exercise 11.22, construct a 90% confidence interval for the standard deviation of the population sampled; that is, for the percentage of impurities in the given brand of peanut butter.

**11.53** With reference to Exercise 11.14, use the large-sample confidence-interval formula of Exercise 11.50 to construct a 99% confidence interval for the standard deviation of the annual growth of the given kind of cactus.

**11.54** With reference to Exercise 11.15, use the large-sample confidence-interval formula of Exercise 11.50 to construct a 98% confidence interval for the standard deviation of the time it takes a mechanic to perform the given task.

**11.55** With reference to Exercise 11.24, construct a 98% confidence interval for the ratio of the variances of the two populations sampled.

**11.56** With reference to Exercise 11.25, construct a 98% confidence interval for the ratio of the variances of the two populations sampled.

**11.57** With reference to Exercise 11.26, construct a 90% confidence interval for the ratio of the variances of the two populations sampled.

## 11.8  USE OF COMPUTERS

In the examples of this chapter we showed quite some details about substitutions into the various formulas and subsequent calculations. In practice, none of this is really necessary, because there is an abundance of software which requires only that we enter the original **raw** (untreated) **data** into our computer together with the appropriate commands. To illustrate, consider the following example.

EXAMPLE 11.12

To study the durability of a new paint for white center lines, a highway department painted test strips across heavily traveled roads in eight different locations, and electronic counters showed that they deteriorated after having been crossed by (to the nearest hundred) 142,600, 167,800, 136,500, 108,300, 126,400, 133,700, 162,000, and 149,400 cars. Construct a 95% confidence interval for the average amount of traffic (car crossings) this paint can withstand before it deteriorates.

*Solution*

The computer printout of Figure 11.1 shows that the desired confidence interval is

```
MTB  > SET C1
DATA > 142600  167800  136500  108300  126400  133700  162000  149400
MTB  > TINT 95 C1

             N      MEAN    STDEV   SE MEAN    95.0  PERCENT C.I.
     C1      8    140837    19228      6798    (124758,  156917)
```

**Figure 11.1**  Computer printout for Example 11.12.

$$124{,}758 < \mu < 156{,}917$$

car crossings. It also shows the sample size, the mean of the data, their standard deviation, and the estimated standard error of the mean, SE MEAN, which is given by $\dfrac{s}{\sqrt{n}}$.    ▲

As used in this example, computers enable us to do more efficiently—faster, more cheaply, and almost automatically—what was done previously by means of desk calculators, hand-held calculators, or even by hand. However, dealing with a sample of size $n = 8$, the example cannot very well do justice to the power of computers to handle enormous sets of data and perform calculations not even deemed possible until recent years. Also, our example does not show how computers can summarize the output as well the input and the results as well as the original data in various kinds of graphs and charts, which allow for methods of analysis that were not available in the past.

All this is important, but it does not do justice to the phenomenal impact which computers have had on statistics. Among other things, computers can be used to tabulate or graph functions (say, the $t$, $F$, or $\chi^2$ distributions) and thus give the investigator a clear understanding of underlying models and make it possible to study the effects of violations of assumptions. Also important is the use of computers in simulating values of random variables (namely, sampling all kinds of populations), when a formal mathematical approach is not feasible. This provides an important tool when we study the appropriateness of statistical models.

## REFERENCES

A general method for obtaining confidence intervals is given in

Mood, A. M., Graybill, F. A., and Boes, D. C., *Introduction to the Theory of Statistics*, 3rd ed. New York: McGraw-Hill Book Company, 1974, and further criteria for judging the relative merits of confidence intervals may be found in

LEHMANN, E. L., *Testing Statistical Hypotheses*. New York: John Wiley & Sons, Inc., 1959,

and in other advanced texts on mathematical statistics. Special tables for constructing 95% and 99% confidence intervals for proportions are given in the *Biometrika Tables* referred to on page 328. For a proof of the independence of the random variables $Z$ and $Y$ on page 406, see

BRUNK, H. D., *An Introduction to Mathematical Statistics*, 3rd ed. Lexington, Mass.: Xerox Publishing Co., 1975.

# 12
# *Hypothesis Testing: Theory*

## 12.1 INTRODUCTION

If an engineer has to decide on the basis of sample data whether the true average lifetime of a certain kind of tire is at least 22,000 miles, if an agronomist has to decide on the basis of experiments whether one kind of fertilizer produces a higher yield of soybeans than another, and if a manufacturer of pharmaceutical products has to decide on the basis of samples whether 90 percent of all patients given a new medication will recover from a certain disease, these problems can all be translated into the language of **statistical tests of hypotheses**. In the first case we might say that the engineer has to test the hypothesis that $\theta$, the parameter of an exponential population, is at least 22,000; in the second case we might say that the agronomist has to decide whether $\mu_1 > \mu_2$, where $\mu_1$ and $\mu_2$ are the means of two normal populations; and in the third case we might say that the manufacturer has to decide whether $\theta$, the parameter of a binomial population, equals 0.90. In each case it must be assumed, of course, that the chosen distribution correctly describes the experimental conditions, namely, that the distribution provides the correct **statistical model**.

As in the above examples, most tests of statistical hypotheses concern the parameters of distributions, but sometimes they also concern the type, or nature, of the distributions, themselves. For instance, in the first of our three examples the engineer may also have to decide whether he is actually dealing with a sample from an exponential population or whether his data are values of random variables having, say, the Weibull distribution of Exercise 6.23.

---

**DEFINITION 12.1** A **statistical hypothesis** is an assertion or conjecture about the distribution of one or more random variables. If a statistical hypothesis completely specifies the distribution, it is referred to as a **simple hypothesis**; if not, it is referred to as a **composite hypothesis**.

---

A simple hypothesis must therefore specify not only the functional form of the underlying distribution, but also the values of all parameters. Thus, in the third of the above examples, the one dealing with the effectiveness of the new medication, the hypothesis $\theta = 0.90$ is simple, assuming, of course, that we specify the sample size and that the population is binomial. However, in the first of the above examples the hypothesis is composite since $\theta \geqslant 22,000$ does not assign a specific value to the parameter $\theta$.

To be able to construct suitable criteria for testing statistical hypotheses, it is necessary that we also formulate **alternative hypotheses**. For instance, in the example dealing with the lifetimes of the tires, we might formulate the alternative hypothesis that the parameter $\theta$ of the exponential population is less than 22,000; in the example dealing with the two kinds of fertilizer, we might formulate the alternative hypothesis $\mu_1 = \mu_2$; and in the example dealing with the new medication, we might formulate the alternative hypothesis that the parameter $\theta$ of the given binomial population is only 0.60, which is the disease's recovery rate without the new medication.

The concept of simple and composite hypotheses applies also to alternative hypotheses, and in the first example we can now say that we are testing the composite hypothesis $\theta \geqslant 22,000$ against the **composite alternative** $\theta < 22,000$, where $\theta$ is the parameter of an exponential population. Similarly, in the second example we are testing the composite hypothesis $\mu_1 > \mu_2$ against the composite alternative $\mu_1 = \mu_2$, where $\mu_1$ and $\mu_2$ are the means of two normal populations, and in the third example we are testing the simple hypothesis $\theta = 0.90$ against the **simple alternative** $\theta = 0.60$, where $\theta$ is the parameter of a binomial population for which $n$ is given.

Frequently, statisticians formulate as their hypotheses the exact opposite of what they may want to show. For instance, if we want to show that the students in one school have a higher average IQ than those in another school, we might formulate the hypothesis that there is no difference, namely, the hypothesis $\mu_1 = \mu_2$. With this hypothesis we know what to expect, but this would not be the case if we formulated the hypothesis $\mu_1 > \mu_2$; at least, not unless we specify the actual difference between $\mu_1$ and $\mu_2$.

Similarly, if we want to show that one kind of ore has a higher percentage content of uranium than another kind of ore, we might formulate the hypothesis that the two percentages are the same; and if we want to show that there is a greater variability in the quality of one product than there is in the quality of another, we might formulate the hypothesis that there is no difference, namely, that $\sigma_1 = \sigma_2$. In view of the assumptions of "no difference," hypotheses such as these led to the term **null hypothesis**, but nowadays this term is applied to any hypothesis we may want to test.

Symbolically, we shall use the symbol $H_0$ for the null hypothesis we want to test and $H_1$ or $H_A$ for the alternative hypothesis. Problems involving more than two hypotheses—that is, problems involving several alternative hypotheses—tend to be quite complicated, and they will not be studied in this book.

## 12.2 TESTING A STATISTICAL HYPOTHESIS

The testing of a statistical hypothesis is the application of an explicit set of rules for deciding whether to accept the null hypothesis or to reject it in favor of the alternative hypothesis. Suppose, for example, that a statistician wants to test the null hypothesis $\theta = \theta_0$ against the alternative hypothesis $\theta = \theta_1$. In order to make a choice, he will generate sample data by conducting an experiment and then compute the value of a **test statistic**, which will tell him for each possible outcome of the sample space what action to take. The test procedure, therefore, partitions the possible values of the test statistic into two subsets: an **acceptance region** for $H_0$ and a **rejection region** for $H_0$.

The procedure just described can lead to two kinds of errors. For instance, if the true value of the parameter $\theta$ is $\theta_0$ and the statistician incorrectly concludes that $\theta = \theta_1$, he is committing an error referred to as a **type I error**. On the other hand, if the true value of the parameter $\theta$ is $\theta_1$ and the statistician incorrectly concludes that $\theta = \theta_0$, he is committing a second kind of error referred to as a **type II error**.

---

**DEFINITION 12.2**

1.  Rejection of the null hypothesis when it is true is called a **type I error**; the probability of committing a type I error is denoted by $\alpha$.
2.  Acceptance of the null hypothesis when it is false is called a **type II error**; the probability of committing a type II error is denoted by $\beta$.

---

It is customary to refer to the rejection region for $H_0$ as the **critical region** of the test, and to the probability of obtaining a value of the test statistic inside the critical region when $H_0$ is true as the **size** of the critical region. Thus, the size of a critical region is just the probability $\alpha$ of committing a type I error. This probability is also called the **level of significance** of the test (see discussion on page 442).

## EXAMPLE 12.1

With reference to the third illustration on page 425, suppose that the manufacturer of the new medication wants to test the null hypothesis $\theta = 0.90$ against the alternative hypothesis $\theta = 0.60$. His test statistic is $X$, the observed number of successes (recoveries) in 20 trials, and he will accept the null hypothesis if $x > 14$; otherwise, he will reject it. Find $\alpha$ and $\beta$.

### Solution

The acceptance region for the null hypothesis is $x = 15, 16, 17, 18, 19,$ and 20, and, correspondingly, the rejection region (or critical region) is $x = 0, 1, 2, \ldots, 14$. Therefore, from Table I,

$$\alpha = P(X \leqslant 14; \theta = 0.90) = 0.0114$$

and

$$\beta = P(X > 14; \theta = 0.60) = 0.1255 \quad \blacktriangle$$

A good test procedure is one in which both $\alpha$ and $\beta$ are small, thereby giving us a good chance of making the correct decision. The probability of a type II error in Example 12.1 is rather high, but this can be reduced by appropriately changing the critical region. For instance, if we use the acceptance region $x > 15$ in this example, so that the critical region is $x \leqslant 15$, it can easily be checked that this would make $\alpha = 0.0433$ and $\beta = 0.0509$. Thus, although the probability of a type II error is reduced, the probability of a type I error has become larger. The only way in which we can reduce the probabilities of both types of errors is to increase the size of the sample, but so long as $n$ is held fixed, this inverse relationship between the probabilities of type I and type II errors is typical of statistical decision procedures. In other words, if the probability of one type of error is reduced, that of the other type of error is increased.

## EXAMPLE 12.2

Suppose we want to test the null hypothesis that the mean of a normal population with $\sigma^2 = 1$ is $\mu_0$ against the alternative hypothesis that it is $\mu_1$, where $\mu_1 > \mu_0$. Find the value of $K$ such that $\bar{x} > K$ provides a critical region of size $\alpha = 0.05$ for a random sample of size $n$.

*Solution*

Referring to Figure 12.1 and Table III, we find that $z = 1.645$ corresponds to an entry of 0.4500 and, hence, that

$$1.645 = \frac{K - \mu_0}{1/\sqrt{n}}$$

It follows that

$$K = \mu_0 + \frac{1.645}{\sqrt{n}} \qquad \blacktriangle$$

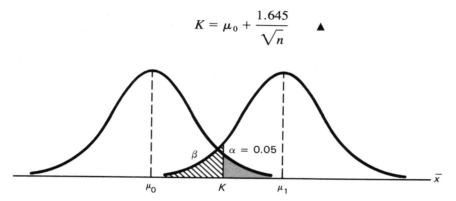

**Figure 12.1**  Diagram for Examples 12.2 and 12.3.

## EXAMPLE 12.3

With reference to Example 12.2, determine the minimum sample size needed to test the null hypothsis $\mu_0 = 10$ against the alternative hypothesis $\mu_1 = 11$ with $\beta \leq 0.06$.

*Solution*

Since $\beta$ is given by the area of the ruled region of Figure 12.1, we get

$$\beta = P\left(\overline{X} < 10 + \frac{1.645}{\sqrt{n}}; \mu = 11\right)$$

$$= P\left[Z < \frac{\left(10 + \dfrac{1.645}{\sqrt{n}}\right) - 11}{1/\sqrt{n}}\right]$$

$$= P(Z < -\sqrt{n} + 1.645)$$

and since $z = 1.555$ corresponds to an entry of $0.5000 - 0.06 = 0.4400$ in Table III, we set $-\sqrt{n} + 1.645$ equal to $-1.555$. It follows that $\sqrt{n} = 1.645 + 1.555 = 3.200$ and $n = 10.24$ or 11 rounded up to the nearest integer.    ▲

## 12.3  LOSSES AND RISKS

The concepts of loss functions and risk functions that were introduced in Chapter 9 also play an important part in the theory of hypothesis testing. In the decision theory approach to testing the null hypothesis that a population parameter $\theta$ equals $\theta_0$ against the alternative that it equals $\theta_1$, the statistician either takes the action $\alpha_0$ and accepts the null hypothesis, or she takes the action $a_1$ and accepts the alternative hypothesis. Depending on the true "state of Nature" and the action which she takes, her losses are shown in the following table:

|  |  | *Statistician* | |
|---|---|---|---|
|  |  | $a_0$ | $a_1$ |
| *Nature* | $\theta_0$ | $L(a_0, \theta_0)$ | $L(a_1, \theta_0)$ |
|  | $\theta_1$ | $L(a_0, \theta_1)$ | $L(a_1, \theta_1)$ |

These losses can be positive or negative (reflecting penalties or rewards), and the only condition which we shall impose is that

$$L(a_0, \theta_0) < L(a_1, \theta_0) \quad \text{and} \quad L(a_1, \theta_1) < L(a_0, \theta_1)$$

namely, that in either case the right decision is more profitable than the wrong one.

As in the statistical games of Section 9.3, the statistician's choice will depend on the outcome of an experiment and the decision function $d$, which tells her for each possible outcome what action to take. If the null hypothesis is true and the statistician accepts the alternative hypothesis, namely, if the value of the parameter is $\theta_0$ and the statistician takes action $a_1$, she commits a type I error; correspondingly, if the value of the parameter is $\theta_1$ and the statistician takes action $a_0$, she commits a type II error. For the decision function $d$, we shall let $\alpha(d)$ denote the probability of committing a type I error and $\beta(d)$ the probability of committing a type II error. The values of the risk function (defined on page 344) are thus

$$R(d, \theta_0) = [1 - \alpha(d)]L(a_0, \theta_0) + \alpha(d)L(a_1, \theta_0)$$

$$= L(a_0, \theta_0) + \alpha(d)[L(a_1, \theta_0) - L(a_0, \theta_0)]$$

and

$$R(d, \theta_1) = \beta(d)L(a_0, \theta_1) + [1 - \beta(d)]L(a_1, \theta_1)$$

$$= L(a_1, \theta_1) + \beta(d)[L(a_0, \theta_1) - L(a_1, \theta_1)]$$

where, by assumption, the quantities in brackets are both positive. It is apparent from this (and should, perhaps, have been obvious from the beginning) that to minimize the risks the statistician must choose a decision function which, in some way, keeps the probabilities of both types of errors as small as possible.

If we could assign prior probabilities to $\theta_0$ and $\theta_1$ and if we knew the exact values of all the losses $L(a_j, \theta_i)$ in the table on page 430, we could calculate the Bayes risk (defined on page 346) and look for the decision function which minimizes this risk. Alternatively, if we looked upon Nature as a malevolent opponent we could use the minimax criterion and choose the decision function which minimizes the maximum risk, but as must have been apparent from the applied exercises on page 352, this is not a very realistic approach in most practical situations.

## 12.4 THE NEYMAN-PEARSON LEMMA

In the theory of hypothesis testing which is nowadays referred to as "classical" or "traditional," namely, the **Neyman-Pearson theory**, we circumvent the dependence between probabilities of type I and type II errors by limiting ourselves to test statistics for which the probability of a type I error is less than or equal to some constant $\alpha$. In other words, we restrict ourselves to critical regions of size less than or equal to $\alpha$. We must allow for the critical region to be of size less than $\alpha$ to take care of discrete random variables, where it may be impossible to find a test statistic for which the size of the critical region is exactly equal to $\alpha$. For all practical purposes, then, we hold the probability of a type I error fixed and look for the test statistic which minimizes the probability of a type II error, or equivalently, which maximizes the quantity $1 - \beta$. When testing the null hypothesis $\theta = \theta_0$ against the alternative hypothesis $\theta = \theta_1$, the quantity $1 - \beta$ is referred to as the **power** of the test at $\theta = \theta_1$.

A critical region for testing a simple null hypothesis $\theta = \theta_0$ against a simple alternative hypothesis $\theta = \theta_1$ is said to be **best** or **most powerful**, if the power of the test at $\theta = \theta_1$ is a maximum. To construct a most powerful critical region in this kind of situation, we refer to the likelihoods (see page 381) of a random sample of size $n$ from the population under consideration when $\theta = \theta_0$ and $\theta = \theta_1$. Denoting these likelihoods by $L_0$ and $L_1$, we thus have

$$L_0 = \prod_{i=1}^{n} f(x_i; \theta_0) \quad \text{and} \quad L_1 = \prod_{i=1}^{n} f(x_i; \theta_1)$$

Intuitively speaking, it stands to reason that $\dfrac{L_0}{L_1}$ should be small for sample points inside the critical region, which lead to type I errors when $\theta = \theta_0$ and to correct decisions when $\theta = \theta_1$; similarly, it stands to reason that $\dfrac{L_0}{L_1}$ should be large for sample points outside the critical region, which lead to correct decisions when $\theta = \theta_0$ and type II errors when $\theta = \theta_1$. The fact that this argument does, indeed, guarantee a most powerful critical region is proved by the following theorem.

---

**THEOREM 12.1**   (*Neyman-Pearson Lemma*) If $C$ is a critical region of size $\alpha$ and $k$ is a constant such that

$$\frac{L_0}{L_1} \leq k \qquad \text{inside } C$$

and

$$\frac{L_0}{L_1} \geq k \qquad \text{outside } C$$

then $C$ is a most powerful critical region of size $\alpha$ for testing $\theta = \theta_0$ against $\theta = \theta_1$.

---

*Proof.*   Suppose that $C$ is a critical region satisfying the conditions of the theorem and that $D$ is some other critical region of size $\alpha$. Thus,

$$\int \cdots \int_C L_0 \, dx = \int \cdots \int_D L_0 \, dx = \alpha$$

where $dx$ stands for $dx_1, dx_2 \ldots dx_n$, and the two multiple integrals are taken over the respective $n$-dimensional regions $C$ and $D$. Now, making use of the fact that $C$ is the union of the disjoint sets $C \cap D$ and $C \cap D'$ while $D$ is the union of the disjoint sets $C \cap D$ and $C' \cap D$, we can write

$$\int \cdots \int_{C \cap D} L_0 \, dx + \int \cdots \int_{C \cap D'} L_0 \, dx = \int \cdots \int_{C \cap D} L_0 \, dx + \int \cdots \int_{C' \cap D} L_0 \, dx = \alpha$$

and, hence,

$$\int \cdots \int_{C \cap D'} L_0 \, dx = \int \cdots \int_{C' \cap D} L_0 \, dx$$

Then, since $L_1 \geq L_0/k$ inside $C$ and $L_1 \leq L_0/k$ outside $C$, it follows that

$$\int \cdots \int_{C \cap D'} L_1 \, dx \geq \int \cdots \int_{C \cap D'} \frac{L_0}{k} \, dx = \int \cdots \int_{C' \cap D} \frac{L_0}{k} \, dx \geq \int \cdots \int_{C' \cap D} L_1 \, dx$$

and, hence, that

$$\int \cdots \int_{C \cap D'} L_1 \, dx \geq \int \cdots \int_{C' \cap D} L_1 \, dx$$

Finally,

$$\int \cdots \int_C L_1 \, dx = \int \cdots \int_{C \cap D} L_1 \, dx + \int \cdots \int_{C \cap D'} L_1 \, dx$$

$$\geq \int \cdots \int_{C \cap D} L_1 \, dx + \int \cdots \int_{C' \cap D} L_1 \, dx = \int \cdots \int_D L_1 \, dx$$

so that

$$\int \cdots \int_C L_1 \, dx \geq \int \cdots \int_D L_1 \, dx$$

and this completes the proof of Theorem 12.1. The final inequality states that for the critical region $C$ the probability of *not* committing a type II error is greater than or equal to the corresponding probability for any other critical region of size $\alpha$. (For the discrete case the proof is the same, with summations taking the place of integrals.)    ▼

## EXAMPLE 12.4

A random sample of size $n$ from a normal population with $\sigma^2 = 1$ is to be used to test the null hypothesis $\mu = \mu_0$ against the alternative hypothesis $\mu = \mu_1$, where $\mu_1 > \mu_0$. Use the Neyman-Pearson lemma to find the most powerful critical region of size $\alpha$.

*Solution*

The two likelihoods are

$$L_0 = \left(\frac{1}{\sqrt{2\pi}}\right)^n \cdot e^{-\frac{1}{2}\Sigma(x_i-\mu_0)^2} \quad \text{and} \quad L_1 = \left(\frac{1}{\sqrt{2\pi}}\right)^n \cdot e^{-\frac{1}{2}\Sigma(x_i-\mu_1)^2}$$

where the summations extend from $i = 1$ to $i = n$, and after some simplifications their ratio becomes

$$\frac{L_0}{L_1} = e^{\frac{n}{2}(\mu_1^2-\mu_0^2)+(\mu_0-\mu_1)\cdot\Sigma x_i}$$

Thus, we must find a constant $k$ and a region $C$ of the sample space such that

$$e^{\frac{n}{2}(\mu_1^2-\mu_0^2)+(\mu_0-\mu_1)\cdot\Sigma x_i} \leq k \qquad \text{inside } C$$

$$e^{\frac{n}{2}(\mu_1^2-\mu_0^2)+(\mu_0-\mu_1)\cdot\Sigma x_i} \geq k \qquad \text{outside } C$$

and after taking logarithms, subtracting $\dfrac{n}{2}(\mu_1^2 - \mu_0^2)$, and dividing by the negative quantity $n(\mu_0 - \mu_1)$, these two inequalities become

$$\bar{x} \geq K \qquad \text{inside } C$$

$$\bar{x} \leq K \qquad \text{outside } C$$

where $K$ is an expression in $k$, $n$, $\mu_0$, and $\mu_1$.

In actual practice, constants like $K$ are determined by making use of the size of the critical region and appropriate statistical theory. In our case (see Example 12.2) we obtain $K = \mu_0 + z_\alpha \cdot \dfrac{1}{\sqrt{n}}$, where $z_\alpha$ is as defined on page 360. Thus, the most powerful critical region of size $\alpha$ for testing the null hypothesis $\mu = \mu_0$ against the alternative $\mu = \mu_1$ (with $\mu_1 > \mu_0$) for the given normal population is

$$\bar{x} \geq \mu_0 + z_\alpha \cdot \frac{1}{\sqrt{n}}$$

and it should be noted that it does not depend on $\mu_1$. This is an important property, to which we shall refer again in Section 12.5.    ▲

Note that we derived the critical region here without first mentioning that the test statistic is to be $\overline{X}$. Since the specification of a critical region thus defines the corresponding test statistic and vice versa, these two terms—"critical region" and "test statistic"—are often used interchangeably in the language of statistics.

## EXERCISES

**12.1**  Decide in each case whether the hypothesis is simple or composite:
(a)  the hypothesis that a random variable has a gamma distribution with $\alpha = 3$ and $\beta = 2$;
(b)  the hypothesis that a random variable has a gamma distribution with $\alpha = 3$ and $\beta \neq 2$;
(c)  the hypothesis that a random variable has an exponential density;
(d)  the hypothesis that a random variable has a beta distribution with the mean $\mu = 0.50$.

**12.2**  Decide in each case whether the hypothesis is simple or composite:
(a)  the hypothesis that a random variable has a Poisson distribution with $\lambda = 1.25$;
(b)  the hypothesis that a random variable has a Poisson distribution with $\lambda > 1.25$;
(c)  the hypothesis that a random variable has a normal distribution with the mean $\mu = 100$;
(d)  the hypothesis that a random variable has a negative binomial distribution with $k = 3$ and $\theta < 0.60$.

**12.3**  A single observation of a random variable having a hypergeometric distribution with $N = 7$ and $n = 2$ is used to test the null hypothesis $k = 2$ against the alternative hypothesis $k = 4$. If the null hypothesis is rejected if and only if the value of the random variable is 2, find the probabilities of type I and type II errors.

**12.4**  With reference to Example 12.1, what would have been the probabilities of type I and type II errors if the acceptance region had been $x > 16$ and the corresponding rejection region had been $x \leq 16$?

**12.5**  A single observation of a random variable having a geometric distribution is used to test the null hypothesis $\theta = \theta_0$ against the alternative hypothesis $\theta = \theta_1 > \theta_0$. If the null hypothesis is rejected if and only if the observed value of the random variable is greater than or equal to the positive integer $k$, find expressions for the probabilities of type I and type II errors.

**12.6**  A single observation of a random variable having an exponential distribution is used to test the null hypothesis that the mean of the distribution is $\theta = 2$ against the alternative that it is $\theta = 5$. If the null hypothesis is accepted if and only if the observed value of the random variable is less than 3, find the probabilities of type I and type II errors.

**12.7** Let $X_1$ and $X_2$ constitute a random sample from a normal population with $\sigma^2 = 1$. If the null hypothesis $\mu = \mu_0$ is to be rejected in favor of the alternative hypothesis $\mu = \mu_1 > \mu_0$ when $\bar{x} > \mu_0 + 1$, what is the size of the critical region?

**12.8** A single observation of a random variable having a uniform density with $\alpha = 0$ is used to test the null hypothesis $\beta = \beta_0$ against the alternative hypothesis $\beta = \beta_0 + 2$. If the null hypothesis is rejected if and only if the random variable takes on a value greater than $\beta_0 + 1$, find the probabilities of type I and type II errors.

**12.9** Let $X_1$ and $X_2$ constitute a random sample of size 2 from the population given by

$$f(x; \theta) = \begin{cases} \theta x^{\theta-1} & \text{for } 0 < x < 1 \\ 0 & \text{elsewhere} \end{cases}$$

If the critical region $x_1 x_2 \geq \frac{3}{4}$ is used to test the null hypothesis $\theta = 1$ against the alternative hypothesis $\theta = 2$, what is the power of this test at $\theta = 2$?

**12.10** Show that if $\mu_1 < \mu_0$ in Example 12.4, the Neyman-Pearson lemma yields the critical region

$$\bar{x} \leq \mu_0 - z_\alpha \cdot \frac{1}{\sqrt{n}}$$

**12.11** A random sample of size $n$ from an exponential population is used to test the null hypothesis $\theta = \theta_0$ against the alternative hypothesis $\theta = \theta_1 > \theta_0$. Use the Neyman-Pearson lemma to find the most powerful critical region of size $\alpha$, and use the result of Example 7.16 to indicate how to evaluate the constant.

**12.12** Use the Neyman-Pearson lemma to indicate how to construct the most powerful critical region of size $\alpha$ to test the null hypothesis $\theta = \theta_0$, where $\theta$ is the parameter of a binomial distribution with a given value of $n$, against the alternative hypothesis $\theta = \theta_1 < \theta_0$.

**12.13** With reference to the preceding exercise, if $n = 100$, $\theta_0 = 0.40$, $\theta_1 = 0.30$, and $\alpha$ is as large as possible without exceeding 0.05, use the normal approximation to the binomial distribution to find the probability of committing a type II error.

**12.14** A single observation of a random variable having a geometric distribution is to be used to test the null hypothesis that its parameter equals $\theta_0$ against the alternative that it equals $\theta_1 > \theta_0$. Use the Neyman-Pearson lemma to find the best critical region of size $\alpha$.

**12.15** Given a random sample of size $n$ from a normal population with $\mu = 0$, use the Neyman-Pearson lemma to construct the most powerful critical region of size $\alpha$ to test the null hypothesis $\sigma = \sigma_0$ against the alternative $\sigma = \sigma_1 > \sigma_0$.

**12.16** Suppose that in Example 12.1 the manufacturer of the new medication feels that the odds are 4 to 1 that with this medication the recovery rate from the disease is 0.90 rather than 0.60. With these odds, what are the probabilities that he will make a wrong decision if he uses the decision function

(a) $d_1(x) = \begin{cases} a_0 & \text{for } x > 14 \\ a_1 & \text{for } x \leqslant 14 \end{cases}$

(b) $d_2(x) = \begin{cases} a_0 & \text{for } x > 15 \\ a_1 & \text{for } x \leqslant 15 \end{cases}$

(c) $d_3(x) = \begin{cases} a_0 & \text{for } x > 16 \\ a_1 & \text{for } x \leqslant 16 \end{cases}$

## APPLICATIONS

**12.17** An airline wants to test the null hypothesis that 60 percent of its passengers object to smoking inside the plane. Explain under what conditions they would be committing a type I error and under what conditions they would be committing a type II error.

**12.18** A doctor is asked to give an executive a thorough physical checkup to test the null hypothesis that he will be able to take on additional responsibilities. Explain under what conditions the doctor would be committing a type I error and under what conditions he would be committing a type II error.

**12.19** The average drying time of a manufacturer's paint is 20 minutes. Investigating the effectiveness of a modification in the chemical composition of her paint, the manufacturer wants to test the null hypothesis $\mu = 20$ minutes against a suitable alternative, where $\mu$ is the average drying time of the modified paint.

(a) What alternative hypothesis should the manufacturer use if she does not want to make the modification in the chemical composition of the paint unless it decreases the drying time?

(b) What alternative hypothesis should the manufacturer use if the new process is actually cheaper and she wants to make the modification unless it increases the drying time of the paint?

**12.20** A biologist wants to test the null hypothesis that the mean wingspan of a certain kind of insect is 12.3 mm against the alternative that it is not 12.3 mm. If she takes a random sample and decides to accept the null hypothesis if and only if the mean of the sample falls between 12.0 mm and 12.6 mm, what decision will she make if she gets $\bar{x} = 12.9$ mm and will it be in error if

(a) $\mu = 12.5$ mm;

(b) $\mu = 12.3$ mm?

**12.21** An employee of a bank wants to test the null hypothesis that on the average the bank cashes 10 bad checks per day against the alternative that this figure is too small. If he takes a random sample and decides to reject the null hypothesis if and only if the mean of the sample exceeds 12.5, what decision will he make if he gets $\bar{x} = 11.2$, and will it be in error if

(a)  $\lambda = 11.5$;

(b)  $\lambda = 10.0$?

Here $\lambda$ is the mean of the Poisson population being sampled.

**12.22** Rework Example 12.3 with

(a)  $\beta = 0.03$;

(b)  $\beta = 0.01$.

**12.23** Suppose we want to test the null hypothesis that a certain kind of tire will last, on the average, 35,000 miles against the alternative hypothesis that it will last, on the average, 45,000 miles. Assuming that we are dealing with a random variable having an exponential distribution, we specify the sample size and the probability of a type I error, and use the Neyman-Pearson lemma to construct a critical region. Would we get the same critical region if we change the alternative hypothesis to

(a)  $\theta_1 = 50,000$ miles;

(b)  $\theta_1 > 35,000$ miles?

## 12.5  THE POWER FUNCTION OF A TEST

In Example 12.1 we were able to give unique values for the probabilities of committing type I and type II errors because we were testing a simple hypothesis against a simple alternative. In actual practice, it is relatively rare, however, that simple hypotheses are tested against simple alternatives; usually one or the other, or both, are composite. For instance, in Example 12.1 it might well have been more realistic to test the null hypothesis that the recovery rate from the disease is $\theta \geq 0.90$ against the alternative hypothesis $\theta < 0.90$, namely, the alternative hypothesis that the new medication is not as effective as claimed.

When we deal with composite hypotheses, the problem of evaluating the merits of a test criterion, or critical region, becomes more involved. In that case we have to consider the probabilities $\alpha(\theta)$ of committing a type I error for all values of $\theta$ within the domain specified under the null hypothesis $H_0$, and the probabiities $\beta(\theta)$ of committing a type II error for all values of $\theta$ within the domain specified under the alternative hypothesis $H_1$. It is customary to combine the two sets of probabilities in the following way.

---

**DEFINITION 12.3**  The **power function** of a test of a statistical hypothesis $H_0$ against an alternative hypothesis $H_1$ is given by

$$\pi(\theta) = \begin{cases} \alpha(\theta) & \text{for values of } \theta \text{ assumed under } H_0 \\ 1 - \beta(\theta) & \text{for values of } \theta \text{ assumed under } H_1 \end{cases}$$

Thus, the values of the power function are the probabilities of rejecting the null hypothesis $H_0$ for various values of the parameter $\theta$. Observe also that for values of $\theta$ assumed under $H_0$, the power function gives the probability of committing a type I error, and for values of $\theta$ assumed under $H_1$, it gives the probability of *not* committing a type II error.

## EXAMPLE 12.5

With reference to Example 12.1, suppose we had wanted to test the null hypothesis $\theta \geq 0.90$ against the alternative hypothesis $\theta < 0.90$. Investigate the power function corresponding to the same test criterion as on page 428, where we accept the null hypothesis if $x > 14$ and reject it if $x \leq 14$. As before, $x$ is the observed number of successes (recoveries) in $n = 20$ trials.

### Solution

Choosing values of $\theta$ for which the respective probabilities, $\alpha(\theta)$ or $\beta(\theta)$, are available from Table I, we find the probabilities $\alpha(\theta)$ of getting at most 14 successes for $\theta = 0.90$ and 0.95 and the probabilities $\beta(\theta)$ of getting more than 14 successes for $\theta = 0.85, 0.80, \ldots, 0.50$. These are shown in the following table, together with the corresponding values of the power function, $\pi(\theta)$:

| $\theta$ | Probability of type I error $\alpha(\theta)$ | Probability of type II error $\beta(\theta)$ | Probability of rejecting $H_0$ $\pi(\theta)$ |
|---|---|---|---|
| 0.95 | 0.0003 | | 0.0003 |
| 0.90 | 0.0114 | | 0.0114 |
| 0.85 | | 0.9326 | 0.0674 |
| 0.80 | | 0.8042 | 0.1958 |
| 0.75 | | 0.6171 | 0.3829 |
| 0.70 | | 0.4163 | 0.5837 |
| 0.65 | | 0.2455 | 0.7545 |
| 0.60 | | 0.1255 | 0.8745 |
| 0.55 | | 0.0553 | 0.9447 |
| 0.50 | | 0.0207 | 0.9793 |

The graph of this power function is shown in Figure 12.2. Of course, it applies only to the decision criterion of Example 12.1, the critical region $x \leq 14$, but it is of interest to note how it compares with the power function of a corresponding ideal (infallible) test criterion, given by the dashed lines of Figure 12.2.    ▲

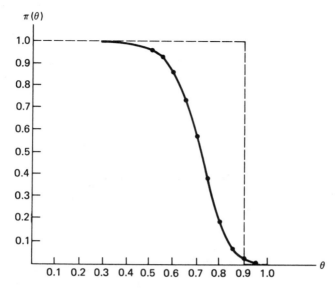

**Figure 12.2** Diagram for Example 12.5.

Power functions play a very important role in the evaluation of statistical tests, particularly in the comparison of several critical regions which might all be used to test a given null hypothesis against a given alternative. Incidentally, if we had plotted in Figure 12.2 the probabilities of accepting $H_0$ (instead of those of rejecting $H_0$), we would have obtained the **operating characteristic curve**, or simply the **OC-curve**, of the given critical region. In other words, the values of the operating characteristic function, used mainly in industrial applications, are given by $1 - \pi(\theta)$.

On page 431 we indicated that in the Neyman-Pearson theory of testing hypotheses we hold $\alpha$, the probability of a type I error, fixed, and this requires that the null hypothesis $H_0$ be a simple hypothesis, say, $\theta = \theta_0$. As a result, the power function of any test of this null hypothesis will pass through the point $(\theta_0, \alpha)$, the only point at which the value of a power function is the probability of making an error. This facilitates the comparison of the power functions of several critical regions, which are all designed to test the simple null hypothesis $\theta = \theta_0$ against a composite alternative, say, the alternative hypothesis $\theta \neq \theta_0$. To illustrate, consider Figure 12.3, giving the power functions of three different critical regions, or test criteria, designed for this purpose. Since for each value of $\theta$ except $\theta_0$ the values of power functions are probabilities of making correct decisions, it is desirable to have them as close to 1 as possible. Thus, it can be seen by inspection that the critical region whose power function is given by the dotted curve of Figure 12.3 is preferable to the critical region whose power function is

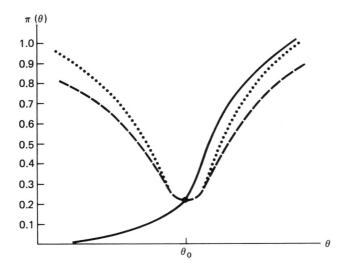

**Figure 12.3** Power functions.

given by the curve which is dashed. The probability of not committing a type II error with the first of these critical regions always exceeds that of the second, and we say that the first critical region is **uniformly more powerful** than the second; also, the second critical region is said to be **inadmissible**.

The same clear-cut distinction is not possible if we attempt to compare the critical regions whose power functions are given by the dotted and solid curves of Figure 12.3; in this case the first one is preferable for $\theta < \theta_0$ while the other is preferable for $\theta > \theta_0$. In situations like this we need further criteria for comparing power functions, for instance that of Exercise 12.34. Note that if the alternative hypothesis had been $\theta > \theta_0$, the critical region whose power function is given by the solid curve would have been uniformly more powerful than the critical region whose power function is given by the dotted curve.

In general, when we test a simple hypothesis against a composite alternative, we specify $\alpha$, the probability of a type I error, and refer to one critical region of size $\alpha$ as uniformly more powerful than another if the values of its power function are always greater than or equal to those of the other, with the strict inequality holding for at least one value of the parameter under consideration. If, for a given problem, a critical region of size $\alpha$ is uniformly more powerful than any other critical region of size $\alpha$, it is said to be **uniformly most powerful**; unfortunately, uniformly most powerful critical regions rarely exist when we test a simple hypothesis against a composite alternative. Of course, when we test a simple hypothesis against a simple alternative, a most powerful critical region of size $\alpha$, as defined on page 431, is, in fact, uniformly most powerful.

Until now we have always assumed that the acceptance of $H_0$ is equivalent to the rejection of $H_1$, and vice versa, but this is not the case, for example, in **multistage** or **sequential tests**, where the alternatives are to accept $H_0$, to accept $H_1$, or to defer the decision until more data have been obtained. It is also not the case in **tests of significance**, where the alternative to rejecting $H_0$ is reserving judgment instead of accepting $H_0$. For instance, if we want to test the null hypothesis that a coin is perfectly balanced against the alternative that this is not the case, and 100 tosses yield 57 heads and 43 tails, this will not enable us to reject the null hypothesis when $\alpha = 0.05$ (see Exercise 12.38). However, since we obtained quite a few more heads than the 50 which we can expect for a balanced coin, we may well be reluctant to accept the null hypothesis as true. To avoid this, we can say that the difference between 50 and 57, the number of heads which we expected and the number of heads which we obtained, may reasonably be attributed to chance—or we can say that this difference is not large enough to reject the null hypothesis. In either case, we do not really commit ourselves one way or the other, and so long as we do not actually accept the null hypothesis, we cannot commit a type II error. It is mainly in connection with tests of this kind that we refer to the probability of a type I error as the **level of significance**.

## 12.6 LIKELIHOOD RATIO TESTS

The Neyman-Pearson lemma provides a means of constructing most powerful critical regions for testing a simple null hypothesis against a simple alternative hypothesis, but it does not always apply to composite hypotheses. We shall now present a general method for constructing critical regions for tests of composite hypotheses which in most cases have very satisfactory properties. The resulting tests, called **likelihood ratio tests**, are based on a generalization of the method of Section 12.4, but they are not necessarily uniformly most powerful. We shall discuss this method here with reference to tests concerning one parameter $\theta$ and continuous populations, but all our arguments can easily be extended to the multiparameter case and to discrete populations.

To illustrate the likelihood ratio technique, let us suppose that $X_1, X_2, \ldots, X_n$ constitute a random sample of size $n$ from a population whose density at $x$ is $f(x; \theta)$, and that $\Omega$ is the set of values that can be taken on by the parameter $\theta$. We often refer to $\Omega$ as the **parameter space** for $\theta$. The null hypothesis we shall want to test is

$$H_0: \quad \theta \in \omega$$

and the alternative hypothesis is

$$H_1: \quad \theta \in \omega'$$

where $\omega$ is a subset of $\Omega$ and $\omega'$ is the complement of $\omega$ with respect to $\Omega$. Thus, the parameter space for $\theta$ is partitioned into the disjoint sets $\omega$ and $\omega'$; according to the null hypothesis $\theta$ is an element of the first set, and according to the alternative hypothesis it is an element of the second set. In most problems $\Omega$ is either the set of all real numbers, the set of all positive real numbers, some interval of real numbers, or a discrete set of real numbers.

When $H_0$ and $H_1$ are both simple hypotheses, $\omega$ and $\omega'$ each have only one element, and in Section 12.4 we constructed tests by comparing the likelihoods $L_0$ and $L_1$. In the general case, where at least one of the two hypotheses is composite, we compare instead the two quantities max $L_0$ and max $L$, where max $L_0$ is the maximum value of the likelihood function (see page 381) for all values of $\theta$ in $\omega$, and max $L$ is the maximum value of the likelihood function for all values of $\theta$ in $\Omega$. In other words, if we have a random sample of size $n$ from a population whose density at $x$ is $f(x; \theta)$, $\hat{\theta}$ is the maximum likelihood estimate of $\theta$ subject to the restriction that $\theta$ must be an element of $\omega$, and $\hat{\hat{\theta}}$ is the maximum likelihood estimate of $\theta$ for all values of $\theta$ in $\Omega$, then

$$\text{max } L_0 = \prod_{i=1}^{n} f(x_i; \hat{\theta})$$

and

$$\text{max } L = \prod_{i=1}^{n} f(x_i; \hat{\hat{\theta}})$$

These quantities are both values of random variables, since they depend on the observed values $x_1, x_2, \ldots, x_n$, and their ratio

$$\lambda = \frac{\text{max } L_0}{\text{max } L}$$

is referred to as a value of the **likelihood ratio statistic** $\Lambda$ (capital Greek *lambda*).

Since max $L_0$ and max $L$ are both values of a likelihood function and therefore are never negative, it follows that $\lambda \geq 0$; also, since $\omega$ is a subset of the parameter space $\Omega$, it follows that $\lambda \leq 1$. When the null hypothesis is false, we would expect max $L_0$ to be small compared to max $L$, in which case $\lambda$ would be close to zero. On the other hand, when the null hypothesis is true and $\theta \in \omega$, we would expect max $L_0$ to be close to max $L$, in which case $\lambda$ would be close to 1. A likelihood ratio test states, therefore, that the null hypothesis $H_0$ is rejected if and only if $\lambda$ falls in a critical region of the form $\lambda \leq k$, where $0 < k < 1$. To summarize,

**DEFINITION 12.4** If $\omega$ and $\omega'$ are complementary subsets of the parameter space $\Omega$, and if

$$\lambda = \frac{\max L_0}{\max L}$$

where $\max L_0$ and $\max L$ are the maximum values of the likelihood function for all values of $\theta$ in $\omega$ and $\Omega$, respectively, then the critical region

$$\lambda \leq k$$

where $0 < k < 1$, defines a **likelihood ratio test** of the null hypothesis $\theta \in \omega$ against the alternative hypothesis $\theta \in \omega'$.

If $H_0$ is a simple hypothesis, $k$ is chosen so that the size of the critical region equals $\alpha$; if $H_0$ is composite, $k$ is chosen so that the probability of a type I error is less than or equal to $\alpha$ for all $\theta$ in $\omega$, and equal to $\alpha$, if possible, for at least one value of $\theta$ in $\omega$. Thus, if $H_0$ is a simple hypothesis and $g(\lambda)$ is the density of $\Lambda$ at $\lambda$ when $H_0$ is true, then $k$ must be such that

$$P(\Lambda \leq k) = \int_0^k g(\lambda) \, d\lambda = \alpha$$

In the discrete case, the integral is replaced by a sum and $k$ is taken to be the largest value for which the sum is less than or equal to $\alpha$.

# EXAMPLE 12.6

Find the critical region of the likelihood ratio test for testing the null hypothesis

$$H_0: \quad \mu = \mu_0$$

against the composite alternative

$$H_1: \quad \mu \neq \mu_0$$

on the basis of a random sample of size $n$ from a normal population with the known variance $\sigma^2$.

## Solution

Since $\omega$ contains only $\mu_0$, it follows that $\hat{\mu} = \mu_0$, and since $\Omega$ is the set of all real numbers, it follows by the method of Section 10.7 that $\hat{\hat{\mu}} = \bar{x}$. Thus,

$$\max L_0 = \left(\frac{1}{\sigma\sqrt{2\pi}}\right)^n \cdot e^{-\frac{1}{2\sigma^2} \cdot \Sigma(x_i - \mu_0)^2}$$

and

$$\max L = \left(\frac{1}{\sigma\sqrt{2\pi}}\right)^n \cdot e^{-\frac{1}{2\sigma^2} \cdot \Sigma(x_i - \bar{x})^2}$$

where the summations extend from $i = 1$ to $i = n$, and the value of the likelihood ratio statistic becomes

$$\lambda = \frac{e^{-\frac{1}{2\sigma^2} \cdot \Sigma(x_i - \mu_0)^2}}{e^{-\frac{1}{2\sigma^2} \cdot \Sigma(x_i - \bar{x})^2}}$$

$$= e^{-\frac{n}{2\sigma^2}(\bar{x} - \mu_0)^2}$$

after suitable simplifications, which the reader will be asked to verify in Exercise 12.26. Hence, the critical region of the likelihood ratio test is

$$e^{-\frac{n}{2\sigma^2}(\bar{x} - \mu_0)^2} \leq k$$

and, after taking logarithms and dividing by $-\dfrac{n}{2\sigma^2}$, it becomes

$$(\bar{x} - \mu_0)^2 \geq -\frac{2\sigma^2}{n} \cdot \ln k$$

or

$$|\bar{x} - \mu_0| \geq K$$

where $K$ will have to be determined so that the size of the critical region is $\alpha$. Note that $\ln k$ is negative in view of the fact that $0 < k < 1$.

Since $\overline{X}$ has a normal distribution with the mean $\mu_0$ and the variance $\dfrac{\sigma^2}{n}$ (see Theorem 8.4), we find that the critical region of this likelihood ratio test is

$$|\bar{x} - \mu_0| \geq z_{\alpha/2} \cdot \frac{\sigma}{\sqrt{n}}$$

or, equivalently,

$$|z| \geq z_{\alpha/2}$$

where

$$z = \frac{\bar{x} - \mu_0}{\sigma/\sqrt{n}}$$

In other words, the null hypothesis must be rejected when $Z$ takes on a value greater than or equal to $z_{\alpha/2}$ or a value less than or equal to $-z_{\alpha/2}$.     ▲

In the preceding example it was easy to find the constant that made the size of the critical region equal to $\alpha$, because we were able to refer to the known distribution of $\overline{X}$ and did not have to derive the distribution of the likelihood ratio statistic $\Lambda$ itself. Since the distribution of $\Lambda$ is usually quite complicated, which makes it difficult to evaluate $k$, it is often preferable to use the following approximation, whose proof is referred to at the end of this chapter.

---

**THEOREM 12.2** For large $n$, the distribution of $-2 \cdot \ln \Lambda$ approaches, under very general conditions, the chi-square distribution with 1 degree of freedom.

---

We should add that this theorem applies only to the one-parameter case; if the population has more than one unknown parameter upon which the null hypothesis imposes $r$ restrictions, the number of degrees of freedom in the chi-square approximation to the distribution of $-2 \cdot \ln \Lambda$ is equal to $r$. For instance, if we want to test the null hypothesis that the unknown mean and variance of a normal population are $\mu_0$ and $\sigma_0^2$ against the alternative hypothesis that $\mu \neq \mu_0$ and $\sigma^2 \neq \sigma_0^2$, the number of degrees of freedom in the chi-square approximation to the distribution of $-2 \cdot \ln \Lambda$ would be 2; the two restrictions are $\mu = \mu_0$ and $\sigma^2 = \sigma_0^2$.

Since small values of $\lambda$ correspond to large values of $-2 \cdot \ln \lambda$, we can use Theorem 12.2 to write the critical region of this approximate likelihood ratio test as

$$-2 \cdot \ln \lambda \geqslant \chi^2_{\alpha,1}$$

where $\chi^2_{\alpha,1}$ is as defined on page 309. In connection with Example 12.6 we find that

$$-2 \cdot \ln \lambda = \frac{n}{\sigma^2} (\bar{x} - \mu_0)^2 = \left( \frac{\bar{x} - \mu_0}{\sigma/\sqrt{n}} \right)^2$$

which actually *is* a value of a random variable having the chi-square distribution with 1 degree of freedom.

As we indicated on page 442, the likelihood ratio technique will generally produce satisfactory results. That this is not always the case is illustrated by the following example, which is somewhat out of the ordinary.

## EXAMPLE 12.7

On the basis of a single observation, we want to test the simple null hypothesis that the probability distribution of $X$ is

| $x$    | 1              | 2              | 3              | 4             | 5             | 6             | 7             |
|--------|----------------|----------------|----------------|---------------|---------------|---------------|---------------|
| $f(x)$ | $\frac{1}{12}$ | $\frac{1}{12}$ | $\frac{1}{12}$ | $\frac{1}{4}$ | $\frac{1}{6}$ | $\frac{1}{6}$ | $\frac{1}{6}$ |

against the composite alternative that the probability distribution is

| $x$    | 1             | 2             | 3             | 4             | 5 | 6 | 7 |
|--------|---------------|---------------|---------------|---------------|---|---|---|
| $g(x)$ | $\frac{a}{3}$ | $\frac{b}{3}$ | $\frac{c}{3}$ | $\frac{2}{3}$ | 0 | 0 | 0 |

where $a + b + c = 1$. Show that the critical region obtained by means of the likelihood ratio technique is inadmissible.

### Solution

The composite alternative hypothesis includes all the probability distributions that we get by assigning different values from 0 to 1 to $a$, $b$, and $c$, subject only to the restriction that $a + b + c = 1$. To determine $\lambda$ for each value of $x$, we first let $x = 1$. For this value we get max $L_0 = \frac{1}{12}$, max $L = \frac{1}{3}$ (corresponding to $a = 1$), and, hence, $\lambda = \frac{1}{4}$. Determining $\lambda$ for the other values of $x$ in the same way, we get the results shown in the following table:

| $x$ | 1 | 2 | 3 | 4 | 5 | 6 | 7 |
|---|---|---|---|---|---|---|---|
| $\lambda$ | $\frac{1}{4}$ | $\frac{1}{4}$ | $\frac{1}{4}$ | $\frac{3}{8}$ | 1 | 1 | 1 |

If the size of the critical region is to be $\alpha = 0.25$, we find that the likelihood ratio technique yields the critical region for which the null hypothesis is rejected when $\lambda = \frac{1}{4}$, namely, when $x = 1$, $x = 2$, or $x = 3$; clearly, $f(1) + f(2) + f(3) = \frac{1}{12} + \frac{1}{12} + \frac{1}{12} = 0.25$. The corresponding probability of a type II error is given by $g(4) + g(5) + g(6) + g(7)$, and hence, it equals $\frac{2}{3}$.

Now let us consider the critical region for which the null hypothesis is rejected only when $x = 4$. Its size is also $\alpha = 0.25$ since $f(4) = \frac{1}{4}$, but the corresponding probability of a type II error is

$$g(1) + g(2) + g(3) + g(5) + g(6) + g(7) = \frac{a}{3} + \frac{b}{3} + \frac{c}{3} + 0 + 0 + 0$$

$$= \frac{1}{3}$$

Since this is less than $\frac{2}{3}$, the critical region obtained by means of the likelihood ratio technique is inadmissible.  ▲

## EXERCISES

**12.24** With reference to Exercise 12.3, suppose that we had wanted to test the null hypothesis $k \leq 2$ against the alternative hypothesis $k > 2$. Find the probabilities of
(a) type I errors for $k = 0$, 1, and 2;
(b) type II errors for $k = 4$, 5, 6, and 7.
Also plot the graph of the corresponding power function.

**12.25** With reference to Example 12.5, suppose we reject the null hypothesis if $x \leq 15$ and accept it if $x > 15$. Calculate $\pi(\theta)$ for the same values of $\theta$ as in the table on page 439 and plot the graph of the power function of this test criterion.

**12.26** In the solution of Example 12.6, verify the step that led to

$$\lambda = e^{-\frac{n}{2\sigma^2}(\bar{x}-\mu_0)^2}$$

**12.27** The number of successes in $n$ trials is to be used to test the null hypothesis that the parameter $\theta$ of a binomial population equals $\frac{1}{2}$ against the alternative that it does not equal $\frac{1}{2}$.

(a)  Find an expression for the likelihood ratio statistic.

(b)  Use the result of part (a) to show that the critical region of the likelihood ratio test can be written as

$$x \cdot \ln x + (n - x) \cdot \ln(n - x) \geqslant K$$

where $x$ is the observed number of successes.

(c)  Study the graph of $f(x) = x \cdot \ln x + (n - x) \cdot \ln(n - x)$, in particular its minimum and its symmetry, to show that the critical region of this likelihood ratio test can also be written as

$$\left| x - \frac{n}{2} \right| \geqslant K$$

where $K$ is a constant which depends on the size of the critical region.

**12.28** A random sample of size $n$ is to be used to test the null hypothesis that the parameter $\theta$ of an exponential population equals $\theta_0$ against the alternative that it does not equal $\theta_0$.

(a)  Find an expression for the likelihood ratio statistic.

(b)  Use the result of part (a) to show that the critical region of the likelihood ratio test can be written as

$$\bar{x} \cdot e^{-\bar{x}/\theta_0} \leqslant K$$

**12.29** A random sample of size $n$ from a normal population with unknown mean and variance is to be used to test the null hypothesis $\mu = \mu_0$ against the alternative $\mu \neq \mu_0$. Using the simultaneous maximum likelihood estimates of $\mu$ and $\sigma^2$ obtained in Example 10.17, show that the values of the likelihood ratio statistic can be written in the form

$$\lambda = \left( 1 + \frac{t^2}{n - 1} \right)^{-n/2}$$

where $t = \dfrac{\bar{x} - \mu_0}{s/\sqrt{n}}$. Note that the likelihood ratio test can, thus, be based on the $t$ distribution.

**12.30** For the likelihood ratio statistic of Exercise 12.29, show that $-2 \cdot \ln \lambda$ approaches $t^2$ as $n \to \infty$. [*Hint*: Use the infinite series for $\ln(1 + x)$ given on page 243.]

**12.31** Given a random sample of size $n$ from a normal population with unknown mean and variance, find an expression for the likelihood ratio statistic for testing the null hypothesis $\sigma = \sigma_0$ against the alternative hypothesis $\sigma \neq \sigma_0$. (*Hint*: See Example 10.17.)

**12.32** Independent random samples of size $n_1, n_2, \ldots ,$ and $n_k$ from $k$ normal populations with unknown means and variances are to be used to test the null hypothesis $\sigma_1^2 = \sigma_2^2 = \cdots = \sigma_k^2$ against the alternative that these variances are not all equal.

(a) Show that under the null hypothesis the maximum likelihood estimates of the means $\mu_i$ and the variances $\sigma_i^2$ are

$$\hat{\mu}_i = \bar{x}_i \quad \text{and} \quad \hat{\sigma}_i^2 = \sum_{i=1}^{k} \frac{(n_i - 1)s_i^2}{n}$$

where $n = \sum_{i=1}^{k} n_i$, while without restrictions the maximum likelihood estimates of the means $\mu_i$ and the variances $\sigma_i^2$ are

$$\hat{\hat{\mu}}_i = \bar{x}_i \quad \text{and} \quad \hat{\hat{\sigma}}_i^2 = \frac{(n_i - 1)s_i^2}{n_i}$$

This follows directly from the results obtained in Section 10.8.

(b) Using the results of part (a), show that the likelihood ratio statistic can be written as

$$\lambda = \frac{\prod_{i=1}^{k} \left[ \frac{(n_i - 1)s_i^2}{n_i} \right]^{n_i/2}}{\left[ \sum_{i=1}^{k} \frac{(n_i - 1)s_i^2}{n} \right]^{n/2}}$$

**12.33** Show that for $k = 2$ the likelihood ratio statistic of Exercise 12.32 can be expressed in terms of the ratio of the two sample variances and that the likelihood ratio test can, therefore, be based on the $F$ distribution.

**12.34** When we test a simple null hypothesis against a composite alternative, a critical region is said to be **unbiased** if the corresponding power function takes on its minimum value at the value of the parameter assumed under the null hypothesis. In other words, a critical region is unbiased if the probability of rejecting the null hypothesis is least when the null hypothesis is true. Given a single observation of the random variable $X$ having the density

$$f(x) = \begin{cases} 1 + \theta^2(\frac{1}{2} - x) & \text{for } 0 < x < 1 \\ 0 & \text{elsewhere} \end{cases}$$

where $-1 \le \theta \le 1$, show that the critical region $x \le \alpha$ provides an unbiased critical region of size $\alpha$ for testing the null hypothesis $\theta = 0$ against the alternative hypothesis $\theta \ne 0$.

## APPLICATIONS

**12.35** A single observation is to be used to test the null hypothesis that the mean waiting time between tremors recorded at a seismological station (the mean of an exponential population) is $\theta = 10$ hours against the alternative that $\theta \ne 10$ hours. If the null hypothesis is to be rejected if and only if the observed value is less than 8 or greater than 12, find

   (a) the probability of a type I error;
   (b) the probabilities of type II errors when $\theta = 2, 4, 6, 8, 12, 16$, and 20.

Also plot the power function of this test criterion.

**12.36** A random sample of size 64 is to be used to test the null hypothesis that for a certain age group the mean score on an achievement test (the mean of a normal population with $\sigma^2 = 256$) is less than or equal to 40.0 against the alternative that it is greater than 40.0. If the null hypothesis is to be rejected if and only if the mean of the random sample exceeds 43.5, find

   (a) the probabilities of type I errors when $\mu = 37.0, 38.0, 39.0$, and 40.0;
   (b) the probabilities of type II errors when $\mu = 41.0, 42.0, 43.0, 44.0, 45.0, 46.0, 47.0$, and 48.0.

Also plot the power function of this test criterion.

**12.37** The sum of the values obtained in a random sample of size $n = 5$ is to be used to test the null hypothesis that on the average there are more than two accidents per week at a certain intersection (that $\lambda > 2$ for this Poisson population) against the alternative hypothesis that on the average the number of accidents is two or less. If the null hypothesis is to be rejected if and only if the sum of the observations is five or less, find

   (a) the probabilities of type I errors when $\lambda = 2.2, 2.4, 2.6, 2.8$, and 3.0;
   (b) the probabilities of type II errors when $\lambda = 2.0, 1.5, 1.0$, and 0.5.

(*Hint*: Use the result of Example 7.15.) Also plot the graph of the power function of this test criterion.

**12.38** Verify the statement on page 442 that 57 heads and 43 tails in 100 flips of a coin does not enable us to reject the null hypothesis that the coin is perfectly balanced (against the alternative that it is not perfectly balanced) at the 0.05 level of significance. (*Hint*: Use the normal approximation to the binomial distribution.)

**12.39** To compare the variations in weight of four breeds of dogs, researchers took independent random samples of size $n_1 = 8$, $n_2 = 10$, $n_3 = 6$, and $n_4 = 8$ and got $s_1^2 = 16$, $s_2^2 = 25$, $s_3^2 = 12$, and $s_4^2 = 24$. Assuming that the populations sampled are normal, use the formula of part (b) of Exercise 12.32 to calculate $-2 \cdot \ln \lambda$ and test the null hypothesis $\sigma_1^2 = \sigma_2^2 = \sigma_3^2 = \sigma_4^2$ at the 0.05 level of significance. Explain why the number of degrees of freedom for this approximate chi-square test is 3.

**12.40** The times to failure of certain electronic components are 15, 28, 3, 12, 42, 19, 20, 2, 25, 30, 62, 12, 18, 16, 44, 65, 33, 51, 4, and 28 minutes. Looking upon these data as a random sample from an exponential population, use the results of Exercise 12.28 and Theorem 12.2 to test the null hypothesis $\theta = 15$ minutes against the alternative hypothesis $\theta \neq 15$ minutes at the 0.05 level of significance. (Use $\ln 1.763 = 0.570$.)

# REFERENCES

Discussions of various properties of likelihood ratio tests, particularly their large-sample properties, and a proof of Theorem 12.2 may be found in most advanced textbooks on the theory of statistics, for example, in,

LEHMANN, E. L., *Testing Statistical Hypotheses*, 2nd ed. New York: John Wiley & Sons, Inc., 1986.

WILKS, S. S., *Mathematical Statistics*. New York: John Wiley & Sons, Inc., 1962.

Much of the original research done in this area is reproduced in

*Selected Papers in Statistics and Probability by Abraham Wald*. Stanford, Calif.: Stanford University Press, 1957.

# 13

# *Hypothesis Testing: Applications*

## 13.1 INTRODUCTION

In Chapter 12 we discussed some of the theory which underlies statistical tests, and in this chapter we shall present some of the standard tests that are most widely used in applications. Most of these tests, at least those based on known population distributions, can be obtained by the likelihood ratio technique.

To explain the terminology we shall use, let us consider a situation in which we want to test the null hypothesis $H_0$: $\theta = \theta_0$ against the **two-sided alternative** hypothesis $H_1$: $\theta \neq \theta_0$. Since it appears reasonable to accept the null hypothesis when our point estimate $\hat{\theta}$ of $\theta$ is close to $\theta_0$ and to reject it when $\hat{\theta}$ is much larger or much smaller than $\theta_0$, it would be logical to let the critical region consist of both tails of the sampling distribution of our test statistic $\hat{\Theta}$. Such a test is referred to as a **two-tailed test**.

On the other hand, if we are testing the null hypothesis $H_0$: $\theta = \theta_0$ against the **one-sided alternative** $H_1$: $\theta < \theta_0$, it would seem reasonable to reject $H_0$ only when $\hat{\theta}$ is much smaller than $\theta_0$. Therefore, in this case it would be logical to let the critical region consist only of the left-hand tail of the sampling distribution of $\hat{\Theta}$. Likewise, in testing $H_0$: $\theta = \theta_0$ against the one-sided alternative $H_1$: $\theta > \theta_0$, we reject $H_0$ only for large values of $\hat{\theta}$ and the critical region consists only of the right tail of the sampling distribution of $\hat{\Theta}$. Any test where the critical region consists only of one tail of the sampling distribution of the test statistic is called a **one-tailed test**.

For instance, for the two-sided alternative $\mu \neq \mu_0$ in Example 12.6, the likelihood ratio technique led to a two-tailed test with the critical region

$$|\bar{x} - \mu_0| \geq z_{\alpha/2} \cdot \frac{\sigma}{\sqrt{n}}$$

or

$$\bar{x} \leq \mu_0 - z_{\alpha/2} \cdot \frac{\sigma}{\sqrt{n}} \quad \text{and} \quad \bar{x} \geq \mu_0 + z_{\alpha/2} \cdot \frac{\sigma}{\sqrt{n}}$$

As is pictured in Figure 13.1, the null hypothesis $\mu = \mu_0$ is rejected if $\bar{X}$ takes on a value falling in either tail of its sampling distribution. Symbolically, this critical region can be written as $z \leq -z_{\alpha/2}$ or $z \geq z_{\alpha/2}$, where

$$z = \frac{\bar{x} - \mu_0}{\sigma/\sqrt{n}}$$

Had we used the one-sided alternative $\mu > \mu_0$, the likelihood ratio technique would have led to the one-tailed test whose critical region is pictured in Figure 13.2, and if we had used the one-sided alternative $\mu < \mu_0$, the likelihood ratio technique would have led to the one-tailed test whose critical region is pictured in Figure 13.3. It stands to reason that in the first case we would reject the null hypothesis only for values of $\bar{X}$ falling into the right-hand tail of its sampling distribution, and in the second case we would reject the null hypothesis only for values of $\bar{X}$ falling into the left-hand tail of its sampling distribution. Symbolically, the corresponding critical regions can be written as $z \geq z_\alpha$ and as $z \leq -z_\alpha$, where $z$ is as defined before. Although there are exceptions to this rule (see Ex-

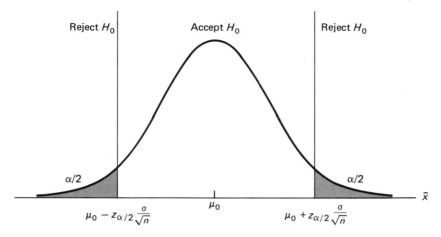

**Figure 13.1**  Critical region for two-tailed test.

ercise 13.1), two-sided alternatives usually lead to two-tailed tests and one-sided alternatives usually lead to one-tailed tests.

Traditionally, it has been the custom to outline tests of hypotheses by means of the following steps:

1.  **Formulate $H_0$ and $H_1$, and specify $\alpha$.**
2.  **Using the sampling distribution of an appropriate test statistic, determine a critical region of size $\alpha$.**

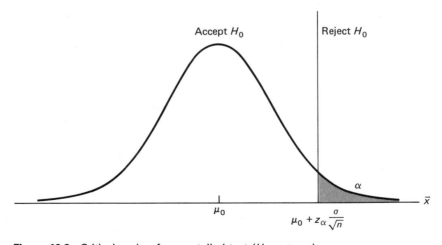

**Figure 13.2**  Critical region for one-tailed test ($H_1$: $\mu > \mu_0$).

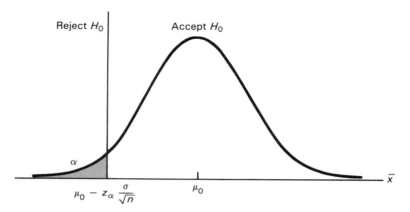

**Figure 13.3**  Critical region for one-tailed test ($H_1$: $\mu < \mu_0$).

3. **Determine the value of the test statistic from the sample data.**
4. **Check whether the value of the test statistic falls into the critical region and, accordingly, reject the null hypothesis, or accept it or reserve judgment.**

In Figures 13.1, 13.2, and 13.3, the dividing lines of the test criteria (that is, the **boundaries** of the critical regions, or the **critical values**) require knowledge of $z_\alpha$ or $z_{\alpha/2}$. These values are readily available from Table III (or more detailed tables of the standard normal distribution) for any level of significance $\alpha$, but the problem is not always this simple. For instance, if the sampling distribution of the test statistic happens to be a $t$ distribution, a chi-square distribution, or an $F$ distribution, the usual tables will provide the necessary values of $t_\alpha$, $t_{\alpha/2}$, $\chi^2_\alpha$, $\chi^2_{\alpha/2}$, $F_\alpha$, or $F_{\alpha/2}$, but only for a few values of $\alpha$. Mainly for this reason, it has been the custom to base tests of statistical hypotheses almost exclusively on the level of significance $\alpha = 0.05$ or $\alpha = 0.01$. This may seem very arbitrary, and of course it is, and this accounts for the current preference for using **P-values** (see Definition 13.1). Alternatively, we could use a decision-theory approach and thus take into account the consequences of all possible actions. However, as we already pointed out in Section 9.1, ". . . there are many problems in which it is difficult, if not impossible, to assign numerical values to the consequences of one's actions and to the probabilities of all eventualities."

With the advent of computers and the general availability of statistical software, the four steps outlined above may be modified to allow for more freedom in the choice of the level of significance $\alpha$. With reference to the test for which the critical region is shown in Figure 13.2, we compare the shaded region of Figure 13.4 with $\alpha$ instead of comparing the observed value of $\overline{X}$ with the boundary of the critical region or the value of

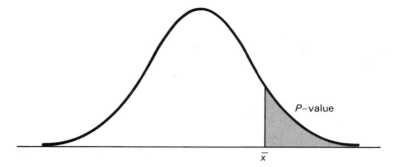

**Figure 13.4**  Diagram for definition of P-values.

$$Z = \frac{\overline{X} - \mu_0}{\sigma/\sqrt{n}}$$

with $z_\alpha$. In other words, we reject the null hypothesis if the shaded region of Figure 13.4 is less than or equal to $\alpha$. This shaded region is referred to as the **P-value**, the **prob-value**, the **tail probability**, or the **observed level of signifi-cance** corresponding to $\bar{x}$, the observed value of $\overline{X}$. In fact, it is the probability $P(\overline{X} \geq \bar{x})$ when the null hypothesis is true.

Correspondingly, when the alternative hypothesis is $\mu < \mu_0$ and the critical region is the one of Figure 13.3, the P-value is the probability $P(\overline{X} \leq \bar{x})$ when the null hypothesis is true; and when the alternative hypothesis is $\mu \neq \mu_0$ and the critical region is the one of Figure 13.1, the P-value is $2P(\overline{X} \geq \bar{x})$ or $2P(\overline{X} \leq \bar{x})$, depending on whether $\bar{x}$ falls into the right-hand tail or the left-hand tail of the sampling distribution of $\overline{X}$. Here it is assumed again that the null hypothesis is true.

More generally, we define P-values as follows.

---

**DEFINITION 13.1**  Corresponding to an observed value of a test statistic, the P-value is the lowest level of significance at which the null hypothesis could have been rejected.

---

With regard to this alternative approach to testing hypotheses, the first of the four steps on pages 455 and 456 remains unchanged, the second step becomes

**2′.   Specify the test statistic**.

the third step becomes

**3′.**   **Determine the value of the test statistic and the corresponding *P*-value from the sample data.**

and the fourth step becomes

**4′.**   **Check whether the *P*-value is less than or equal to $\alpha$ and, accordingly, reject the null hypothesis, or accept it or reserve judgment.**

As we pointed out on page 456, this allows for more freedom in the choice of the level of significance, but it is difficult to conceive of situations in which we could justify using, say, $\alpha = 0.04$ rather than $\alpha = 0.05$ or $\alpha = 0.015$ rather than $\alpha = 0.01$. In practice, it is virtually impossible to avoid some element of arbitrariness, and in most cases we judge subjectively, at least in part, whether $\alpha = 0.05$ or $\alpha = 0.01$ reflects acceptable risks. Of course, when a great deal is at stake and it is practical, we might use a level of significance much smaller than $\alpha = 0.01$.

In any case, it should be understood that the two methods of testing hypotheses, the four steps given on pages 455 and 456 and the four steps described here, are equivalent. This means that no matter which method we use, the ultimate decision—rejecting the null hypothesis, accepting it, or reserving judgment—will be the same. In practice, we use whichever method is most convenient, and this may depend on the sampling distribution of the test statistic, the availability of statistical tables or computer software, and the nature of the problem (see, for instance, Example 13.8 and Exercise 13.48).

There are statisticians who prefer to avoid all problems relating to the choice of the level of significance. Limiting their role to data analysis, they do not specify $\alpha$ and omit step 4′. Of course, it is always desirable to have input from others (research workers or management) in formulating hypotheses and specifying $\alpha$, but it would hardly seem reasonable to dump *P*-values into the laps of persons without adequate training in statistics, and let them take it from there. To compound the difficulties, consider the temptation one might be exposed to when choosing $\alpha$ after having seen the *P*-value with which it is to be compared. Suppose, for instance, that an experiment yields a *P*-value of 0.036. If we are anxious to reject the null hypothesis and thus prove our point, it would be tempting to choose $\alpha = 0.05$; if we are anxious to accept the null hypothesis and thus prove our point, it would be tempting to choose $\alpha = 0.01$.

Nevertheless, in **exploratory data analysis**, where we are not really concerned with making inferences, *P*-values can be used as measures of the strength of evidence. Suppose, for instance, that in cancer research with two drugs, scientists get *P*-values of 0.0735 and 0.0021 for the effectiveness of these drugs in reducing the size of tumors. This suggests that there is more supporting evidence for the effectiveness of the second drug, or that the second drug "looks much more promising."

## 13.2  TESTS CONCERNING MEANS

In this section we shall discuss the most widely used tests concerning the mean of a population, and in Section 13.3 we shall discuss the corresponding tests concerning the means of two populations. Tests concerning the means of more than two populations will be taken up later in Chapter 15. All the tests in this section are based on normal distribution theory, assuming either that the samples come from normal populations or that they are large enough to justify normal approximations; some **nonparametric** alternatives to these tests, which do not require knowledge about the population or populations from which the samples are obtained, will be taken up in Chapter 16.

Suppose that we want to test the null hypothesis $\mu = \mu_0$ against one of the alternatives $\mu \neq \mu_0$, $\mu > \mu_0$, or $\mu < \mu_0$ on the basis of a random sample of size $n$ from a normal population with the known variance $\sigma^2$. This, of course, is the test that was considered in Example 12.6 to illustrate the likelihood ratio technique and the critical regions for the respective alternatives are $|z| \geq z_{\alpha/2}$, $z \geq z_\alpha$, and $z \leq -z_\alpha$, where

$$z = \frac{\bar{x} - \mu_0}{\sigma/\sqrt{n}}$$

As we indicated in Section 13.1, the most commonly used levels of significance are 0.05 and 0.01, and as the reader was asked to show in Exercise 6.60, the corresponding values of $z_\alpha$ and $z_{\alpha/2}$ are $z_{.05} = 1.645$, $z_{.01} = 2.33$, $z_{.025} = 1.96$, and $z_{.005} = 2.575$.

## EXAMPLE 13.1

Suppose that it is known from experience that the standard deviation of the weight of 8-ounce packages of cookies made by a certain bakery is 0.16 ounce. To check whether its production is under control on a given day, namely, to check whether the true average weight of the packages is 8 ounces, employees select a random sample of 25 packages and find that their mean weight is $\bar{x} = 8.091$ ounces. Since the bakery stands to lose money when $\mu > 8$ and the customer loses out when $\mu < 8$, test the null hypothesis $\mu = 8$ against the alternative hypothesis $\mu \neq 8$ at the 0.01 level of significance.

*Solution*

1. $H_0$:  $\mu = 8$
   $H_1$:  $\mu \neq 8$
   $\alpha = 0.01$

2.   Reject the null hypothesis if $z \leqslant -2.575$ or $z \geqslant 2.575$, where

$$z = \frac{\bar{x} - \mu_0}{\sigma/\sqrt{n}}$$

3.   Substituting $\bar{x} = 8.091$, $\mu_0 = 8$, $\sigma = 0.16$, and $n = 25$, we get

$$z = \frac{8.091 - 8}{0.16/\sqrt{25}} = 2.84$$

4.   Since $z = 2.84$ exceeds 2.575, the null hypothesis must be rejected and suitable adjustments should be made in the production process.    ▲

Had we used the alternative approach described on page 457, we would have obtained a P-value of 0.0046 (see Exercise 13.8), and since 0.0046 is less than 0.01, the conclusion would have been the same.

It should be noted that the critical region $z \geqslant z_\alpha$ can also be used to test the null hypothesis $\mu = \mu_0$ against the simple alternative $\mu = \mu_1 > \mu_0$, or the composite null hypothesis $\mu \leqslant \mu_0$ against the composite alternative $\mu > \mu_0$. In the first case we would be testing a simple hypothesis against a simple alternative as in Section 12.4 (see Example 12.4 on page 433, where we studied this test for $\sigma = 1$), and in the second case $\alpha$ would be the maximum probability of committing a type I error for any value of $\mu$ assumed under the null hypothesis. Of course, similar arguments apply to the critical region $z \leqslant -z_\alpha$.

When we are dealing with a large sample of size $n \geqslant 30$ from a population that need not be normal but has a finite variance, we can use the central limit theorem to justify using the test for normal populations, and even when $\sigma^2$ is unknown we can approximate its value with $s^2$ in the computation of the test statistic. To illustrate the use of such an approximate **large-sample test**, consider the following example.

## EXAMPLE 13.2

Suppose that 100 tires made by a certain manufacturer lasted on the average 21,819 miles with a standard deviation of 1,295 miles. Test the null hypothesis $\mu = 22,000$ miles against the alternative hypothesis $\mu < 22,000$ miles at the 0.05 level of significance.

### Solution

1.   $H_0$:   $\mu = 22,000$
     $H_1$:   $\mu < 22,000$
     $\alpha = 0.05$

2.   Reject the null hypothesis if $z \leqslant -1.645$, where

$$z = \frac{\bar{x} - \mu_0}{\sigma/\sqrt{n}}$$

3.   Substituting $\bar{x} = 21{,}819$, $\mu_0 = 22{,}000$, $s = 1{,}295$ for $\sigma$, and $n = 100$, we get

$$z = \frac{21{,}819 - 22{,}000}{1{,}295/\sqrt{100}} = -1.40$$

4.   Since $z = -1.40$ is greater than $-1.645$, the null hypothesis cannot be rejected; there is no real evidence that the tires are not as good as assumed under the null hypothesis.    ▲

Had we used the alternative approach described on page 457, we would have obtained a $P$-value of 0.0808 (see Exercise 13.9), which exceeds 0.05. As should have been expected, the conclusion is the same—the null hypothesis cannot be rejected.

When $n < 30$ and $\sigma^2$ is unknown, the test we have been discussing in this section cannot be used. However, in Exercise 12.29 we saw that for random samples from normal populations, the likelihood ratio technique yields a corresponding test based on

$$t = \frac{\bar{x} - \mu_0}{s/\sqrt{n}}$$

which, according to Theorem 8.13, is a value of a random variable having the $t$ distribution with $n - 1$ degrees of freedom. Thus, critical regions of size $\alpha$ for testing the null hypothesis $\mu = \mu_0$ against the alternatives $\mu \neq \mu_0$, $\mu > \mu_0$, or $\mu < \mu_0$, are, respectively, $|t| \geqslant t_{\alpha/2,n-1}$, $t \geqslant t_{\alpha,n-1}$, and $t \leqslant - t_{\alpha,n-1}$. Note that the comments made on page 460 in connection with the alternative hypothesis $\mu_1 > \mu_0$ and the test of the null hypothesis $\mu \leqslant \mu_0$ against the alternative $\mu > \mu_0$ apply also in this case.

To illustrate this **one-sample $t$ test**, as it is usually called, consider the following example.

## EXAMPLE 13.3

The specifications for a certain kind of ribbon call for a mean breaking strength of 185 pounds. If five pieces randomly selected from different rolls have breaking strengths of 171.6, 191.8, 178.3, 184.9, and 189.1 pounds, test the null hypothesis $\mu = 185$ pounds against the alternative hypothesis $\mu < 185$ pounds at the 0.05 level of significance.

*Solution*

1. $H_0$:  $\mu = 185$
   $H_1$:  $\mu < 185$
   $\alpha = 0.05$

2. Reject the null hypothesis if $t \le -2.132$, where $t$ is determined by means of the formula given above and 2.132 is the value of $t_{.05,4}$.

3. First we calculate the mean and the standard deviation, getting $\bar{x} = 183.1$ and $s = 8.2$. Then, substituting these values together with $\mu_0 = 185$ and $n = 5$ into the formula for $t$, we get

$$t = \frac{183.1 - 185}{8.2/\sqrt{5}} = -0.49$$

4. Since $t = -0.49$ is greater than $-2.132$, the null hypothesis cannot be rejected. If we went beyond this and concluded that the rolls of ribbon from which the sample was selected meet specifications, we would, of course, be exposed to the unknown risk of committing a type II error.    ▲

## 13.3 TESTS CONCERNING DIFFERENCES BETWEEN MEANS

In applied research, there are many problems in which we are interested in hypotheses concerning differences between the means of two populations. For instance, we may want to decide upon the basis of suitable samples whether men can perform a certain task as fast as women, or we may want to decide on the basis of an appropriate sample survey whether the average weekly food expenditures of families in one city exceed those of families in another city by at least $5.00.

Let us suppose that we are dealing with independent random samples of size $n_1$ and $n_2$ from two normal populations having the means $\mu_1$ and $\mu_2$ and the known variances $\sigma_1^2$ and $\sigma_2^2$, and that we want to test the null hypothesis $\mu_1 - \mu_2 = \delta$, where $\delta$ is a given constant, against one of the alternatives $\mu_1 - \mu_2 \ne \delta$, $\mu_1 - \mu_2 > \delta$, or $\mu_1 - \mu_2 < \delta$. Applying the likelihood ratio technique, we will arrive at a test based on $\bar{x}_1 - \bar{x}_2$, and, referring to Exercise 8.3, we find that the respective critical regions can be written as $|z| \ge z_{\alpha/2}$, $z \ge z_{\alpha}$, and $z \le -z_{\alpha}$, where

$$z = \frac{\bar{x}_1 - \bar{x}_2 - \delta}{\sqrt{\dfrac{\sigma_1^2}{n_1} + \dfrac{\sigma_2^2}{n_2}}}$$

When we deal with independent random samples from populations with un-known variances which may not even be normal, we can still use the test which we have just described with $s_1$ substituted for $\sigma_1$ and $s_2$ substituted for $\sigma_2$ so long as both samples are large enough to invoke the central limit theorem.

## EXAMPLE 13.4

An experiment is performed to determine whether the average nicotine content of one kind of cigarette exceeds that of another kind by 0.20 milligram. If $n_1 = 50$ cigarettes of the first kind had an average nicotine content of $\bar{x}_1 = 2.61$ milligrams with a standard deviation of $s_1 = 0.12$ milligram, whereas $n_2 = 40$ cigarettes of the other kind had an average nicotine content of $\bar{x}_2 = 2.38$ milligrams with a standard deviation of $s_2 = 0.14$ milligram, test the null hypothesis $\mu_1 - \mu_2 = 0.20$ against the alternative hypothesis $\mu_1 - \mu_2 \neq 0.20$ at the 0.05 level of significance. Base the decision on the $P$-value corresponding to the value of the appropriate test statistic.

## Solution

1.    $H_0$:   $\mu_1 - \mu_2 = 0.20$
      $H_1$:   $\mu_1 - \mu_2 \neq 0.20$
      $\alpha = 0.05$

2'.   Use the test statistic $Z$, where

$$z = \frac{\bar{x}_1 - \bar{x}_2 - \delta}{\sqrt{\dfrac{\sigma_1^2}{n_1} + \dfrac{\sigma_2^2}{n_2}}}$$

3'.   Substituting $\bar{x}_1 = 2.61$, $\bar{x}_2 = 2.38$, $\delta = 0.20$, $s_1 = 0.12$ for $\sigma_1$, $s_2 = 0.14$ for $\sigma_2$, $n_1 = 50$, and $n_2 = 40$ into this formula, we get

$$z = \frac{2.61 - 2.38 - 0.20}{\sqrt{\dfrac{(0.12)^2}{50} + \dfrac{(0.14)^2}{40}}} = 1.08$$

The corresponding $P$-value is $2(0.5000 - 0.3599) = 0.2802$, where 0.3599 is the entry in Table III for $z = 1.08$.

4'.    Since 0.2802 exceeds 0.05, the null hypothesis cannot be rejected; either
we accept the null hypothesis or we say that the difference between
$2.61 - 2.38 = 0.23$ and 0.20 is not significant. This means that the
difference may well be attributed to chance.    ▲

When $n_1$ and $n_2$ are small and $\sigma_1$ and $\sigma_2$ are unknown, the test we have been
discussing cannot be used. However, for independent random samples from two
normal populations having the same unknown variance $\sigma^2$, the likelihood ratio
technique yields a test based on

$$t = \frac{\bar{x}_1 - \bar{x}_2 - \delta}{s_p \sqrt{\dfrac{1}{n_1} + \dfrac{1}{n_2}}}$$

where

$$s_p^2 = \frac{(n_1 - 1)s_1^2 + (n_2 - 1)s_2^2}{n_1 + n_2 - 2}$$

From Section 11.3, we know that under the given assumptions and the null hypothesis $\mu_1 - \mu_2 = \delta$, this expression for $t$ is a value of a random variable having
the $t$ distribution with $n_1 + n_2 - 2$ degrees of freedom. Thus, the appropriate
critical region of size $\alpha$ for testing the null hypothesis $\mu_1 - \mu_2 = \delta$ against the
alternatives $\mu_1 - \mu_2 \neq \delta$, $\mu_1 - \mu_2 > \delta$, or $\mu_1 - \mu_2 < \delta$ under the given as-
sumptions are, respsectively, $|t| \geq t_{\alpha/2, n_1+n_2-2}$, $t \geq t_{\alpha, n_1+n_2-2}$, and $t \leq -t_{\alpha, n_1+n_2-2}$.
To illustrate this **two-sample $t$ test**, consider the following problem.

## EXAMPLE 13.5

In the comparison of two kinds of paint, a consumer testing service finds that
four 1-gallon cans of one brand cover on the average 546 square feet with a
standard deviation of 31 square feet, whereas four 1-gallon cans of another brand
cover on the average 492 square feet with a standard deviation of 26 square feet.
Assuming that the two populations sampled are normal and have equal variances,
test the null hypothesis $\mu_1 - \mu_2 = 0$ against the alternative hypothesis $\mu_1 - \mu_2$
$> 0$ at the 0.05 level of significance.

### Solution

1.    $H_0$:    $\mu_1 - \mu_2 = 0$
     $H_1$:    $\mu_1 - \mu_2 > 0$
     $\alpha = 0.05$

2.  Reject the null hypothesis if $t \geqslant 1.943$, where $t$ is calculated according to the formula given above and 1.943 is the value of $t_{.05,6}$.
3.  First calculating $s_p$, we get

$$s_p = \sqrt{\frac{3(31)^2 + 3(26)^2}{4 + 4 - 2}} = 28.609$$

and then substituting its value together with $\bar{x}_1 = 546$, $\bar{x}_2 = 492$, $\delta = 0$, and $n_1 = n_2 = 4$ into the formula for $t$, we obtain

$$t = \frac{546 - 492}{28.609\sqrt{\frac{1}{4} + \frac{1}{4}}} = 2.67$$

4.  Since $t = 2.67$ exceeds 1.943, the null hypothesis must be rejected; we conclude that on the average the first kind of paint covers a greater area than the second.    ▲

Note that $n_1 = n_2$ in this example, so that the formula for $s_p^2$ becomes

$$s_p^2 = \tfrac{1}{2}(s_1^2 + s_2^2)$$

Use of this formula would have simplified the calculations.

In Exercise 13.25 the reader will be asked to use suitable computer software to show that the $P$-value would have been 0.0185 in this example, and the conclusion would, of course, have been the same.

If the assumption of equal variances is untenable in a problem of this kind, there are several possibilities. A relatively simple one consists of randomly pairing the values obtained in the two samples and then looking upon their differences as a random sample of size $n_1$ or $n_2$, whichever is smaller, from a normal population which, under the null hypothesis, has the mean $\mu = \delta$. Then we test this null hypothesis against the appropriate alternative by means of the methods of Section 13.2. This is a good reason for having $n_1 = n_2$, but there exist alternative techniques for handling the case where $n_1 \neq n_2$—one of these, the *Smith-Satterthwaite* test, is mentioned among the references at the end of the chapter.

So far we have limited our discussion to random samples that are independent, and the methods we have introduced in this section cannot be used, for example, to decide on the basis of weights "before and after" whether a certain diet is really effective, or whether an observed difference between the average IQ's of husbands and their wives is really significant. In both of these examples the samples are not independent because the data are actually *paired*. A common way of handling this kind of problem is to proceed as in the preceding paragraph, namely, to work with the differences between the paired mesurements or observations. If $n$ is large, we can then use the test described on page 459 to test the null hypothesis $\mu_1 - \mu_2 = \delta$ against the appropriate alternative, and if $n$ is small, we can use the $t$ test described on page 461, provided the differences can be looked upon as a random sample from a normal population.

## EXERCISES

**13.1**  Given a random sample of size $n$ from a normal population with the known variance $\sigma^2$, show that the null hypothesis $\mu = \mu_0$ can be tested against the alternative hypothesis $\mu \neq \mu_0$ with the use of a one-tailed criterion based on the chi-square distribution.

**13.2**  Suppose that a random sample from a normal population with the known variance $\sigma^2$ is to be used to test the null hypothesis $\mu = \mu_0$ against the alternative hypothesis $\mu = \mu_1$, where $\mu_1 > \mu_0$, and that the probabilities of type I and type II errors are to have the preassigned values $\alpha$ and $\beta$. Show that the required size of the sample is given by

$$n = \frac{\sigma^2(z_\alpha + z_\beta)^2}{(\mu_1 - \mu_0)^2}$$

**13.3**  With reference to the preceding exercise, find the required size of the sample when $\sigma = 9$, $\mu_0 = 15$, $\mu_1 = 20$, $\alpha = 0.05$, and $\beta = 0.01$.

**13.4**  Suppose that independent random samples of size $n$ from two normal populations with the known variances $\sigma_1^2$ and $\sigma_2^2$ are to be used to test the null hypothesis $\mu_1 - \mu_2 = \delta$ against the alternative hypothesis $\mu_1 - \mu_2 = \delta'$, and that the probabilities of type I and type II errors are to have the preassigned values $\alpha$ and $\beta$. Show that the required size of the sample is given by

$$n = \frac{(\sigma_1^2 + \sigma_2^2)(z_\alpha + z_\beta)^2}{(\delta - \delta')^2}$$

**13.5**  With reference to the preceding exercise, find the required size of the samples when $\sigma_1 = 9$, $\sigma_2 = 13$, $\delta = 80$, $\delta' = 86$, $\alpha = 0.01$, and $\beta = 0.01$.

## APPLICATIONS

**13.6**  Based on certain data, a null hypothsis is rejected at the 0.05 level of significance. Would it also be rejected at the

(a)   0.01 level of significance;

(b)   0.10 level of significance?

**13.7**  In the test of a certain hypothesis, the $P$-value corresponding to the test statistic is 0.0316. Can the null hypothesis be rejected at the

(a)   0.01 level of significance;
(b)   0.05 level of significance;
(c)   0.10 level of significance?

**13.8**   With reference to Example 13.1, verify that the $P$-value corresponding to the observed value of the test statistic is 0.0046.

**13.9**   With reference to Example 13.2, verify that the $P$-value corresponding to the observed value of the test statistic is 0.0808.

**13.10**   With reference to Example 13.3, use suitable statistical software to find the $P$-value which corresponds to $t = -0.49$, where $t$ is a value of a random variable having the $t$ distribution with 4 degrees of freedom. Use this $P$-value to rework the example.

**13.11**   According to the norms established for a reading comprehension test, eighth graders should average 84.3 with a standard deviation of 8.6. If 45 randomly selected eighth graders from a certain school district averaged 87.8, use the four steps on pages 455 and 456 to test the null hypothesis $\mu = 84.3$ against the alternative $\mu > 84.3$ at the 0.01 level of significance.

**13.12**   Rework Exercise 13.11, basing the decision on the $P$-value corresponding to the observed value of the test statistic.

**13.13**   The security department of a factory wants to know whether or not the true average time required by the night guard to walk his round is 30 minutes. If, in a random sample of 32 rounds, the night guard averaged 30.8 minutes with a standard deviation of 1.5 minutes, determine whether this is sufficient evidence to reject the null hypothesis $\mu = 30$ minutes in favor of the alternative hypothesis $\mu \neq 30$ minutes. Use the four steps on pages 455 and 456 and the 0.01 level of significance.

**13.14**   Rework Exercise 13.13, basing the decision on the $P$-value corresponding to the observed value of the test statistic.

**13.15**   In 12 test runs over a marked course, a newly designed motorboat averaged 33.6 seconds with a standard deviation of 2.3 seconds. Assuming that it is reasonable to treat the data as a random sample from a normal population, use the four steps on pages 455 and 456 to test the null hypothesis $\mu = 35$ against the alternative $\mu < 35$ at the 0.05 level of significance.

**13.16**   Five measurements of the tar content of a certain kind of cigarette yielded 14.5, 14.2, 14.4, 14.3, and 14.6 mg/cigarette. Assuming that the data are a random sample from a normal population, use the four steps on pages 455 and 456 to show that at the 0.05 level of significance the null hypothesis $\mu = 14.0$ must be rejected in favor of the alternative $\mu \neq 14.0$.

**13.17**   With reference to Exercise 13.16, show that if the first measurement is recorded incorrectly as 16.0 instead of 14.5, this will reverse the result. Explain the apparent paradox that even though the difference between the sample mean and $\mu_0$ has increased, it is no longer significant.

**13.18**   With reference to Exercise 13.16, use suitable statistical software to find the $P$-value which corresponds to the observed value of the test statistic. Use this $P$-value to rework the exercise.

**13.19** With reference to Example 13.4, for what values of $\bar{x}_1 - \bar{x}_2$ would the null hypothesis have been rejected? Also find the probabilities of type II errors with the given criterion if

(a)  $\mu_1 - \mu_2 = 0.12$;
(b)  $\mu_1 - \mu_2 = 0.16$;
(c)  $\mu_1 - \mu_2 = 0.24$;
(d)  $\mu_1 - \mu_2 = 0.28$.

**13.20** A study of the number of business lunches that executives in the insurance and banking industries claim as deductible expenses per month was based on random samples and yielded the following results:

$$n_1 = 40 \qquad \bar{x}_1 = 9.1 \qquad s_1 = 1.9$$

$$n_2 = 50 \qquad \bar{x}_2 = 8.0 \qquad s_2 = 2.1$$

Use the four steps on pages 455 and 456 and the 0.05 level of significance to test the null hypothesis $\mu_1 - \mu_2 = 0$ against the alternative hypothesis $\mu_1 - \mu_2 \neq 0$.

**13.21** Rework Exercise 13.20, basing the decision on the $P$-value corresponding to the observed value of the test statistic.

**13.22** Sample surveys conducted in a large county in a certain year and again 20 years later showed that originally the average height of 400 ten-year-old boys was 53.8 inches with a standard deviation of 2.4 inches, whereas 20 years later the average height of 500 ten-year-old boys was 54.5 inches with a standard deviation of 2.5 inches. Use the four steps on pages 455 and 456 and the 0.05 level of significance to test the null hypothesis $\mu_1 - \mu_2 = -0.5$ against the alternative hypothesis $\mu_1 - \mu_2 < -0.5$.

**13.23** Rework Exercise 13.22, basing the decision on the $P$-value corresponding to the observed value of the test statistic.

**13.24** To find out whether the inhabitants of two South Pacific islands may be regarded as having the same racial ancestry, an anthropologist determines the cephalic indices of six adult males from each island, getting $\bar{x}_1 = 77.4$, $\bar{x}_2 = 72.2$, and the corresponding standard deviations $s_1 = 3.3$ and $s_2 = 2.1$. Use the four steps on pages 455 and 456 and the 0.01 level of significance to see whether the difference between the two sample means can reasonably be attributed to chance. Assume that the populations sampled are normal and have equal variances.

**13.25** With reference to Example 13.5, use suitable statistical software to show that the $P$-value corresponding to $t = 2.67$ is 0.0185.

**13.26** To compare two kinds of bumper guards, six of each kind were mounted on a certain make compact car. Then each car was run into a concrete wall at 5 miles per hour, and the following are the costs of the repairs (in dollars):

*Bumper guard 1*:    127  168  143  165  122  139
*Bumper guard 2*:    154  135  132  171  153  149

Use the four steps on pages 455 and 456 to test at the 0.01 level of significance whether the difference between the means of these two samples is significant.

**13.27** With reference to Exercise 13.26, use suitable statistical software to find the $P$-value corresponding to the observed value of the test statistic. Use this $P$-value to rework the exercise.

**13.28** In a study of the effectiveness of certain exercises in weight reduction, a group of 16 persons engaged in these exercises for one month and showed the following results:

| Weight before | Weight after | Weight before | Weight after |
|---|---|---|---|
| 211 | 198 | 172 | 166 |
| 180 | 173 | 155 | 154 |
| 171 | 172 | 185 | 181 |
| 214 | 209 | 167 | 164 |
| 182 | 179 | 203 | 201 |
| 194 | 192 | 181 | 175 |
| 160 | 161 | 245 | 233 |
| 182 | 182 | 146 | 142 |

Use the 0.05 level of significance to test the null hypothesis $\mu_1 - \mu_2 = 0$ against the alternative hypothesis $\mu_1 - \mu_2 > 0$, and thus judge whether the exercises are effective in weight reduction.

**13.29** The following are the average weekly losses of man-hours due to accidents in 10 industrial plants before and after a certain safety program was put into operation:

45 and 36, 73 and 60, 46 and 44, 124 and 119, 33 and 35,
57 and 51, 83 and 77, 34 and 29, 26 and 24, and 17 and 11

Use the four steps on pages 455 and 456 and the 0.05 level of significance to test whether the safety program is effective.

**13.30** With reference to Exercise 13.29, use suitable statistical software to find the $P$-value which corresponds to the observed value of the test statistic. Use this $P$-value to rework the exercise.

## 13.4 TESTS CONCERNING VARIANCES

There are several reasons why it is important to test hypotheses concerning the variances of populations. So far as direct applications are concerned, a manufacturer who has to meet rigid specifications will have to perform tests about the variability of his product, a teacher may want to know whether certain statements are true about the variability that he or she can expect in the performance of a student, and a pharmacist may have to check whether the variation in the potency of a medicine is within permissible limits. So far as indirect applications are concerned, tests about variances are often prerequisites for tests concerning other parameters. For instance, the two-sample $t$ test described on page 464 requires that the two population variances are equal, and in practice this means that we may have to check on the reasonableness of this assumption before we perform the test concerning the means.

The tests which we shall study in this section include a test of the null hypothesis that the variance of a normal population equals a given constant, and the likelihood ratio test of the equality of the variances of two normal populations (which was referred to in Exercise 12.33).

The first of these tests is essentially that of Exercise 12.31. Given a random sample of size $n$ from a normal population, we shall want to test the null hypothesis $\sigma^2 = \sigma_0^2$ against one of the alternatives $\sigma^2 \neq \sigma_0^2$, $\sigma^2 > \sigma_0^2$, or $\sigma^2 < \sigma_0^2$, and, as the reader should have discovered in Exercise 12.31, the likelihood ratio technique leads to a test based on $s^2$, the value of the sample variance. Based on Theorem 8.10, we can thus write the critical regions for testing the null hypothesis against the two one-sided alternatives as $\chi^2 \geq \chi_{\alpha,n-1}^2$ and $\chi^2 \leq \chi_{1-\alpha,n-1}^2$, where

$$\chi^2 = \frac{(n-1)s^2}{\sigma_0^2}$$

So far as the two-sided alternative is concerned, we reject the null hypothesis if $\chi^2 \geq \chi_{\alpha/2,n-1}^2$ or $\chi^2 \leq \chi_{1-\alpha/2,n-1}^2$, and the size of all these critical regions is, of course, equal to $\alpha$.

## EXAMPLE 13.6

Suppose that the thickness of a part used in a semiconductor is its critical dimension and that measurements of the thickness of a random sample of 18 such parts have the variance $s^2 = 0.68$, where the measurements are in thousandths of an inch. The process is considered to be under control if the variation of the thicknesses is given by a variance not greater than 0.36. Assuming that the measurements constitute a random sample from a normal population, test the null hypothesis $\sigma^2 = 0.36$ against the alternative hypothesis $\sigma^2 > 0.36$ at the 0.05 level of significance.

### Solution

1. $H_0$: $\sigma^2 = 0.36$
   $H_1$: $\sigma^2 > 0.36$
   $\alpha = 0.05$

2. Reject the null hypothesis if $\chi^2 \geq 27.587$, where

$$\chi^2 = \frac{(n-1)s^2}{\sigma_0^2}$$

   and 27.587 is the value of $\chi^2_{.05,17}$.

3. Substituting $s^2 = 0.68$, $\sigma_0^2 = 0.36$, and $n = 18$, we get

$$\chi^2 = \frac{17(0.68)}{0.36} = 32.11$$

4. Since $\chi^2 = 32.11$ exceeds 27.587, the null hypothesis must be rejected and the process used in the manufacture of the parts must be adjusted.   ▲

Note that if $\alpha$ had been 0.01 in the preceding example, the null hypothesis could not have been rejected, since $\chi^2 = 32.11$ does not exceed $\chi^2_{.01,17} = 33.409$. This serves to indicate again that the choice of the level of significance is something which must always be specified in advance, so we will be spared the temptation of choosing a value that happens to suit our purpose (see also page 458).

In Exercise 12.33 the reader was asked to show that the likelihood ratio statistic for testing the equality of the variances of two normal populations can be expressed in terms of the ratio of the two sample variances. Given independent random samples of size $n_1$ and $n_2$ from two normal populations with the variances $\sigma_1^2$ and $\sigma_2^2$, we thus find from Theorem 8.15 that corresponding critical regions of size $\alpha$ for testing the null hypothesis $\sigma_1^2 = \sigma_2^2$ against the one-sided alternatives $\sigma_1^2 > \sigma_2^2$ or $\sigma_1^2 < \sigma_2^2$ are, respectively,

$$\frac{s_1^2}{s_2^2} \geq f_{\alpha, n_1-1, n_2-1} \quad \text{and} \quad \frac{s_2^2}{s_1^2} \geq f_{\alpha, n_2-1, n_1-1}$$

where $f_{\alpha, n_1-1, n_2-1}$ and $f_{\alpha, n_2-1, n_1-1}$ are as defined on page 316. The appropriate critical region for testing the null hypothesis against the two-sided alternative $\sigma_1^2 \neq \sigma_2^2$ is

$$\frac{s_1^2}{s_2^2} \geq f_{\alpha/2, n_1-1, n_2-1} \qquad \text{if } s_1^2 \geq s_2^2$$

and

$$\frac{s_2^2}{s_1^2} \geq f_{\alpha/2, n_2-1, n_1-1} \qquad \text{if } s_1^2 < s_2^2$$

Note that this test is based entirely on the right-hand tail of the $F$ distribution, which is made possible by the result of Exercise 8.55—namely, by the fact that if the random variable $X$ has an $F$ distribution with $\nu_1$ and $\nu_2$ degrees of freedom, then $\dfrac{1}{X}$ has an $F$ distribution with $\nu_2$ and $\nu_1$ degrees of freedom.

## EXAMPLE 13.7

In comparing the variability of the tensile strength of two kinds of structural steel, an experiment yielded the following results: $n_1 = 13$, $s_1^2 = 19.2$, $n_2 = 16$, and $s_2^2 = 3.5$, where the units of measurement are 1,000 pounds per square inch. Assuming that the measurements constitute independent random samples from two normal populations, test the null hypothesis $\sigma_1^2 = \sigma_2^2$ against the alternative $\sigma_1^2 \neq \sigma_2^2$ at the 0.02 level of significance.

### Solution

1. $H_0$: $\sigma_1^2 = \sigma_2^2$
   $H_1$: $\sigma_1^2 \neq \sigma_2^2$
   $\alpha = 0.02$

2. Since $s_1^2 \geq s_2^2$, reject the null hypothesis if $\dfrac{s_1^2}{s_2^2} \geq 3.67$, where 3.67 is the value of $f_{.01, 12, 15}$.

3. Substituting $s_1^2 = 19.2$ and $s_2^2 = 3.5$, we get

$$\frac{s_1^2}{s_2^2} = \frac{19.2}{3.5} = 5.49$$

4. Since $f = 5.49$ exceeds 3.67, the null hypothesis must be rejected; we conclude that the variability of the tensile strength of the two kinds of steel is not the same.    ▲

*EXERCISES*

**13.31** Making use of the fact that the chi-square distribution can be approximated with a normal distribution when $\nu$, the number of degrees of freedom, is large, show that for large samples from normal populations

$$s^2 \geq \sigma_0^2 \left[ 1 + z_\alpha \sqrt{\frac{2}{n-1}} \right]$$

is an approximate critical region of size $\alpha$ for testing the null hypothesis $\sigma^2 = \sigma_0^2$ against the alternative $\sigma^2 > \sigma_0^2$. Also construct corresponding critical regions for testing this null hypothesis against the alternatives $\sigma^2 < \sigma_0^2$ and $\sigma^2 \neq \sigma_0^2$ (see Exercise 8.37).

**13.32** Making use of the result of Exercise 8.42, show that for large random samples from normal populations, tests of the null hypothesis $\sigma^2 = \sigma_0^2$ can be based on the statistic

$$\left( \frac{s}{\sigma_0} - 1 \right) \sqrt{2(n-1)}$$

which has approximately the standard normal distribution.

*APPLICATIONS*

**13.33** Nine determinations of the specific heat of iron had a standard deviation of 0.0086. Assuming that these determinations constitute a random sample from a normal population, test the null hypothesis $\sigma = 0.0100$ against the alternative hypothesis $\sigma < 0.0100$ at the 0.05 level of significance.

**13.34** In a random sample, the weights of 24 Black Angus steers of a certain age have a standard deviation of 238 pounds. Assuming that the weights constitute a random sample from a normal population, test the null hypothesis $\sigma = 250$ pounds against the two-sided alternative $\sigma \neq 250$ pounds at the 0.01 level of significance.

**13.35** In a random sample, $s = 2.53$ minutes for the amount of time that 30 women took to complete the written test for their driver's licenses. At the 0.05 level of significance, test the null hypothesis $\sigma = 2.85$ minutes against the alternative hypothesis $\sigma < 2.85$ minutes. (Use the method described in the text).

**13.36** Use the method of Exercise 13.32 to rework Exercise 13.35.

**13.37** Past data indicate that the standard deviation of measurements made on sheet metal stampings by experienced inspectors is 0.41 square inch. If a new inspector measures 50 stampings with a standard deviation of 0.49 square inch, use the method of Exercise 13.32 to test the null hypothesis

$\sigma = 0.41$ square inch against the alternative hypothesis $\sigma > 0.41$ square inch at the 0.05 level of significance.

**13.38** With reference to Exercise 13.37, find the *P*-value corresponding to the observed value of the test statistic and use it to decide whether the null hypothesis could have been rejected at the 0.015 level of significance.

**13.39** With reference to Example 13.5, test the null hypothesis $\sigma_1 - \sigma_2 = 0$ against the alternative hypothesis $\sigma_1 - \sigma_2 > 0$ at the 0.05 level of significance.

**13.40** With reference to Exercise 13.24, test at the 0.10 level of significance whether it is reasonable to assume that the two populations sampled have equal variances.

**13.41** With reference to Exercise 13.26, test at the 0.02 level of significance whether it is reasonable to assume that the two populations sampled have equal variances.

## 13.5 TESTS CONCERNING PROPORTIONS

If an outcome of an experiment is the number of votes which a candidate receives in a poll, the number of imperfections found in a piece of cloth, the number of children who are absent from school on a given day, ... , we refer to these data as **count data**. Appropriate models for the analysis of count data are the binomial distribution, the Poisson distribution, the multinomial distribution, and some of the other discrete distributions which we studied in Chapter 5. In this section we shall present one of the most common tests based on count data, namely, a test concerning the parameter $\theta$ of the binomial distribution. Thus, we might test on the basis of a sample whether the true proportion of cures from a certain disease is 0.90 or whether the true proportion of defectives coming off an assembly line is 0.02.

In Exercise 12.12 the reader was asked to show that the most powerful critical region for testing the null hypothesis $\theta = \theta_0$ against the alternative hypothesis $\theta = \theta_1 < \theta_0$, where $\theta$ is the parameter of a binomial population, is based on the value of $X$, the number of "successes" obtained in $n$ trials. When it comes to composite alternatives, the likelihood ratio technique also yields tests based on the observed number of successes (as we saw in Exercise 12.27 for the special case where $\theta_0 = \frac{1}{2}$). In fact, if we want to test the null hypothesis $\theta = \theta_0$ against the one-sided alternative $\theta > \theta_0$, the critical region of size $\alpha$ of the likelihood ratio criterion is

$$x \geq k_\alpha$$

where $k_\alpha$ is the smallest integer for which

$$\sum_{y=k_\alpha}^{n} b(y; n, \theta_0) \leq \alpha$$

and $b(y; n, \theta_0)$ is the probability of getting $y$ successes in $n$ binomial trials when $\theta = \theta_0$. The size of this critical region, as well as the ones which follow, is thus as close as possible to $\alpha$ without exceeding it.

The corresponding critical region for testing the null hypothesis $\theta = \theta_0$ against the one-sided alternative $\theta < \theta_0$ is

$$x \leq k'_\alpha$$

where $k'_\alpha$ is the largest integer for which

$$\sum_{y=0}^{k'_\alpha} b(y; n, \theta_0) \leq \alpha$$

and, finally, the critical region for testing the null hypothesis $\theta = \theta_0$ against the two-sided alternative $\theta \neq \theta_0$ is

$$x \geq k_{\alpha/2} \quad \text{or} \quad x \leq k'_{\alpha/2}$$

We shall not illustrate this method of determining critical regions for tests concerning the binomial parameter $\theta$ because, in actual practice, it is much less tedious to base the decisions on $P$-values.

## EXAMPLE 13.8

If $x = 4$ of $n = 20$ patients suffered serious side effects from a new medication, test the null hypothesis $\theta = 0.50$ against the alternative hypothesis $\theta \neq 0.50$ at the 0.05 level of significance. Here $\theta$ is the true proportion of patients suffering serious side effects from the new medication.

### Solution

1. $H_0$:  $\theta = 0.50$
   $H_1$:  $\theta \neq 0.50$
   $\alpha = 0.05$
2'. Use the test statistic $X$, the observed number of successes.
3'. $x = 4$, and since $P(X \leq 4) = 0.0059$, the $P$-value is $2(0.0059) = 0.0118$.
4'. Since the $P$-value, 0.0118, is less than 0.05, the null hypothesis must be rejected; we conclude that $\theta \neq 0.50$.    ▲

The tests we have described require the use of a table of binomial probabilities, regardless of whether we use the four steps on pages 455 and 456 or those on pages 457 and 458. For $n \leq 20$ we can use Table I at the end of the book, and for values of $n$ up to 100 we can use the tables referred to at the end of Chapter 5. Alternatively, for large values of $n$ we can use the normal approximation to the binomial distribution and treat

$$z = \frac{x - n\theta}{\sqrt{n\theta(1 - \theta)}}$$

as a value of a random variable having the standard normal distribution. For large $n$, we can thus test the null hypothesis $\theta = \theta_0$ against the alternatives $\theta \neq \theta_0$, $\theta > \theta_0$, or $\theta < \theta_0$ using, respectively, the critical regions $|z| \geq z_{\alpha/2}$, $z \geq z_\alpha$, and $z \leq -z_\alpha$, where

$$z = \frac{x - n\theta_0}{\sqrt{n\theta_0(1 - \theta_0)}}$$

or

$$z = \frac{\left(x \pm \dfrac{1}{2}\right) - n\theta_0}{\sqrt{n\theta_0(1 - \theta_0)}}$$

if we use the continuity correction introduced in Example 6.5. We use the minus sign when $x$ exceeds $n\theta_0$ and the plus sign when $x$ is less than $n\theta_0$.

## EXAMPLE 13.9

An oil company claims that less than 20 percent of all car owners have not tried its gasoline. Test this claim at the 0.01 level of significance, if a random check reveals that 22 of 200 car owners have not tried the oil company's gasoline.

### Solution

1. $H_0$: $\theta = 0.20$
   $H_1$: $\theta < 0.20$
   $\alpha = 0.01$

2. Reject the null hypothesis of $z \leq -2.33$, where (without the continuity correction)

$$z = \frac{x - n\theta_0}{\sqrt{n\theta_0(1 - \theta_0)}}$$

3. Substituting $x = 22$, $n = 200$, and $\theta_0 = 0.20$, we get

$$z = \frac{22 - 200(0.20)}{\sqrt{200(0.20)(0.80)}} = -3.18$$

4. Since $z = -3.18$ is less than $-2.33$, the null hypothesis must be rejected; we conclude that, as claimed, less than 20 percent of all car owners have not tried the oil company's gasoline.    ▲

Note that if we had used the continuity correction in the preceding example, we would have obtained $z = -3.09$ and the conclusion would have been the same.

## 13.6 TESTS CONCERNING DIFFERENCES AMONG $k$ PROPORTIONS

In applied research there are many problems in which we must decide whether observed differences among sample proportions, or percentages, are significant or whether they can be attributed to chance. For instance, if 6 percent of the frozen chickens in a sample from one supplier fail to meet certain standards and only 4 percent in a sample from another supplier fail to meet the standards, we may want to investigate whether the difference between these two percentages is significant. Similarly, we may want to judge on the basis of sample data whether equal proportions of voters in four different cities favor a certain candidate for governor.

To indicate a general method for handling problems of this kind, suppose that $x_1, x_2, \ldots, x_k$ are observed values of $k$ independent random variables $X_1$, $X_2, \ldots, X_k$ having binomial distributions with the parameters $n_1$ and $\theta_1$, $n_2$ and $\theta_2, \ldots, n_k$ and $\theta_k$. If the $n$'s are sufficiently large, we can approximate the distributions of the independent random variables

$$Z_i = \frac{X_i - n_i\theta_i}{\sqrt{n_i\theta_i(1 - \theta_i)}} \qquad \text{for } i = 1, 2, \ldots, k$$

with standard normal distributions, and, according to Theorem 8.8, we can then look upon

$$\chi^2 = \sum_{i=1}^{k} \frac{(x_i - n_i\theta_i)^2}{n_i\theta_i(1 - \theta_i)}$$

as a value of a random variable having the chi-square distribution with $k$ degrees of freedom. To test the null hypothesis, $\theta_1 = \theta_2 = \cdots = \theta_k = \theta_0$ (against the alternative that at least one of the $\theta$'s does not equal $\theta_0$) we can thus use the critical region $\chi^2 \geqslant \chi^2_{\alpha,k}$, where

$$\chi^2 = \sum_{i=1}^{k} \frac{(x_i - n_i\theta_0)^2}{n_i\theta_0(1 - \theta_0)}$$

When $\theta_0$ is not specified, that is, when we are interested only in the null hypothesis $\theta_1 = \theta_2 = \cdots = \theta_k$, we substitute for $\theta$ the pooled estimate

$$\hat{\theta} = \frac{x_1 + x_2 + \cdots + x_k}{n_1 + n_2 + \cdots + n_k}$$

and the critical region becomes $\chi^2 \geqslant \chi^2_{\alpha,k-1}$, where

$$\chi^2 = \sum_{i=1}^{k} \frac{(x_i - n_i\hat{\theta})^2}{n_i\hat{\theta}(1 - \hat{\theta})}$$

The loss of 1 degree of freedom, namely, the change in the critical region from $\chi^2_{\alpha,k}$ to $\chi^2_{\alpha,k-1}$, is due to the fact that an estimate is substituted for the unknown parameter $\theta$; a formal discussion of this is referred to on page 493.

Let us now present an alternative formula for the chi-square statistic immediately above, which, as we shall see in Section 13.7, lends itself more rapidly to other applications. Arranging the data as in the following table,

|  | *Successes* | *Failures* |
|---|:---:|:---:|
| *Sample 1* | $x_1$ | $n_1 - x_1$ |
| *Sample 2* | $x_2$ | $n_2 - x_2$ |
| | $\cdots$ | $\cdots$ |
| *Sample k* | $x_k$ | $n_k - x_k$ |

let us refer to its entries as the **observed cell frequencies** $f_{ij}$, where the first subscript indicates the row and the second subscript indicates the column of this $k \times 2$ table.

Under the null hypothesis $\theta_1 = \theta_2 = \cdots = \theta_k = \theta_0$ the **expected cell frequencies** for the first column are $n_i\theta_0$ for $i = 1, 2, \ldots, k$, and those for the second column are $n_i(1 - \theta_0)$. When $\theta_0$ is not known, we substitute for it, as before, the pooled estimate $\hat{\theta}$, and estimate the expected cell frequencies as

$$e_{i1} = n_i \hat{\theta} \quad \text{and} \quad e_{i2} = n_i(1 - \hat{\theta})$$

for $i = 1, 2, \ldots, k$. It will be left to the reader to show in Exercise 13.42 that the chi-square statistic

$$\chi^2 = \sum_{i=1}^{k} \frac{(x_i - n_i\hat{\theta})^2}{n_i\hat{\theta}(1 - \hat{\theta})}$$

can also be written as

$$\chi^2 = \sum_{i=1}^{k} \sum_{j=1}^{2} \frac{(f_{ij} - e_{ij})^2}{e_{ij}}$$

## EXAMPLE 13.10

Determine, on the basis of the sample data shown in the following table, whether the true proportion of shoppers favoring detergent $A$ over detergent $B$ is the same in all three cities:

| | Number favoring detergent A | Number favoring detergent B | |
|---|---|---|---|
| Los Angeles | 232 | 168 | 400 |
| San Diego | 260 | 240 | 500 |
| Fresno | 197 | 203 | 400 |

Use the 0.05 level of significance.

## Solution

1. $H_0$: $\theta_1 = \theta_2 = \theta_3$
   $H_1$: $\theta_1$, $\theta_2$, and $\theta_3$ are not all equal.
   $\alpha = 0.05$
2. Reject the null hypothesis if $\chi^2 \geq 5.991$, where

$$\chi^2 = \sum_{i=1}^{3} \sum_{j=1}^{2} \frac{(f_{ij} - e_{ij})^2}{e_{ij}}$$

and 5.991 is the value of $\chi^2_{.05,2}$.

3. Since the pooled estimate of $\theta$ is

$$\hat{\theta} = \frac{232 + 260 + 197}{400 + 500 + 400} = \frac{689}{1,300} = 0.53$$

the expected cell frequencies are

$$e_{11} = 400(0.53) = 212 \quad \text{and} \quad e_{12} = 400(0.47) = 188$$
$$e_{21} = 500(0.53) = 265 \quad \text{and} \quad e_{22} = 500(0.47) = 235$$
$$e_{31} = 400(0.53) = 212 \quad \text{and} \quad e_{32} = 400(0.47) = 188$$

and substitution into the formula for $\chi^2$ given above yields

$$\chi^2 = \frac{(232 - 212)^2}{212} + \frac{(260 - 265)^2}{265} + \frac{(197 - 212)^2}{212}$$
$$+ \frac{(168 - 188)^2}{188} + \frac{(240 - 235)^2}{235} + \frac{(203 - 188)^2}{188}$$
$$= 6.48$$

4. Since $\chi^2 = 6.48$ exceeds 5.991, the null hypothesis must be rejected; in other words, the true proportions of shoppers favoring detergent $A$ over detergent $B$ in the three cities are not the same.    ▲

## EXERCISES

**13.42** Show that the two formulas for $\chi^2$ on page 479 are equivalent.

**13.43** Modify the critical regions on pages 474 and 475 so that they can be used to test the null hypothesis $\lambda = \lambda_0$ against the alternative hypotheses $\lambda > \lambda_0$, $\lambda < \lambda_0$, and $\lambda \neq \lambda_0$ on the basis of $n$ observations. Here $\lambda$ is the parameter of the Poisson distribution. (*Hint*: Use the result of Example 7.15.)

**13.44** With reference to Exercise 13.43, use Table II to find values corresponding to $k_{.025}$ and $k'_{.025}$ to test the null hypothesis $\lambda = 3.6$ against the alternative hypothesis $\lambda \neq 3.6$ on the basis of five observations. Use the 0.05 level of significance.

**13.45** For $k = 2$, show that the $\chi^2$ formula on page 479 can be written as

$$\chi^2 = \frac{(n_1 + n_2)(n_2 x_1 - n_1 x_2)^2}{n_1 n_2 (x_1 + x_2)[(n_1 + n_2) - (x_1 + x_2)]}$$

**13.46** Given large random samples from two binomial populations, show that the null hypothesis $\theta_1 = \theta_2$ can be tested on the basis of the statistic

$$z = \frac{\dfrac{x_1}{n_1} - \dfrac{x_2}{n_2}}{\sqrt{\hat{\theta}(1 - \hat{\theta})\left(\dfrac{1}{n_1} + \dfrac{1}{n_2}\right)}}$$

where $\hat{\theta} = \dfrac{x_1 + x_2}{n_1 + n_2}$. (*Hint*: Refer to Exercise 8.5.)

**13.47** Show that the square of the expression for $z$ in Exercise 13.46 equals

$$\chi^2 = \sum_{i=1}^{2} \frac{(x_i - n_i\hat{\theta})^2}{n_i\hat{\theta}(1 - \hat{\theta})}$$

so that the two tests are actually equivalent when the alternative hypothesis is $\theta_1 \neq \theta_2$. Note that the test described in Exercise 13.46, but not the one based on the $\chi^2$ statistic, can be used when the alternative hypothesis is $\theta_1 < \theta_2$ or $\theta_1 > \theta_2$.

## APPLICATIONS

**13.48** With reference to Example 13.8, show that the critical region is $x \leqslant 5$ or $x \geqslant 15$ and that, corresponding to this critical region, the level of significance is actually 0.0414.

**13.49** It has been claimed that more than 40 percent of all shoppers can identify a highly advertised trademark. If, in a random sample, 10 of 18 shoppers were able to identify the trademark, test at the 0.05 level of significance whether the null hypothesis $\theta = 0.40$ can be rejected against the alternative hypothesis $\theta > 0.40$.

**13.50** With reference to the preceding exercise, find the critical region and the actual level of significance corresponding to this critical region.

**13.51** A doctor claims that less than 30 percent of all persons exposed to a certain amount of radiation will feel any ill effects. If, in a random sample, only 1 of 19 persons exposed to such radiation felt any ill effects, test the null hypothesis $\theta = 0.30$ against the alternative hypothesis $\theta < 0.30$ at the 0.05 level of significance.

**13.52** With reference to the preceding exercise, find the critical region and the actual level of significance corresponding to this critical region.

**13.53** In a random sample, 12 of 14 industrial accidents were due to unsafe working conditions. Use the 0.01 level of significance to test the null hypothesis $\theta = 0.40$ against the alternative hypothesis $\theta \neq 0.40$.

**13.54** With reference to the preceding exercise, find the critical region and the actual level of significance corresponding to this critical region.

**13.55** In a random sample of 12 undergraduate business students, six said that they will take advanced work in accounting. Use the 0.01 level of significance to test the null hypothesis $\theta = 0.20$, namely, that 20 percent of all undergraduate business students will take advanced work in accounting, against the alternative hypothesis $\theta > 0.20$.

**13.56** A food processor wants to know whether the probability is really 0.60 that a customer will prefer a new kind of packaging to the old kind. If, in a random sample, seven of 18 customers prefer the new kind of packaging to the old kind, test the null hypothesis $p = 0.60$ against the alternative hypothesis $p \neq 0.60$ at the 0.05 level of significance.

**13.57** In a random sample of 600 cars making a right turn at a certain intersection, 157 pulled into the wrong lane. Use the 0.05 level of significance to test the null hypothesis that the actual proportion of drivers who make this mistake at the given intersection is $\theta = 0.30$ against the alternative hypothesis $\theta \neq 0.30$.

**13.58** The manufacturer of a spot remover claims that his product removes 90 percent of all spots. If, in a random sample, only 174 of 200 spots were removed with the manufacturer's product, test the null hypothesis $\theta = 0.90$ against the alternative hypothesis $\theta < 0.90$ at the 0.05 level of significance.

**13.59** In random samples, 74 of 250 persons who watched a certain television program in black and white and 92 of 250 persons who watched the same program in color remembered 2 hours later what products were advertised. Use the $\chi^2$ statistic to test the null hypothesis $\theta_1 = \theta_2$ against the alternative hypothesis $\theta_1 \neq \theta_2$ at the 0.01 level of significance.

**13.60** Use the statistic of Exercise 13.46 to rework the preceding exercise.

**13.61** In random samples, 46 of 400 tulip bulbs from one nursery failed to bloom and 18 of 200 tulip bulbs from another nursery failed to bloom. Use the $\chi^2$ statistic to test the null hypothesis $\theta_1 = \theta_2$ against the alternative hypothesis $\theta_1 \neq \theta_2$ at the 0.05 level of significance.

**13.62** Use the statistic of Exercise 13.46 to rework Exercise 13.61, and verify that the square of the value obtained for $z$ equals the value obtained for $\chi^2$.

**13.63** In a random sample of 200 persons who skipped breakfast, 82 reported that they experienced midmorning fatigue, and in a random sample of 300 persons who ate breakfast, 87 reported that they experienced midmorning fatigue. Use the method of Exercise 13.46 and the 0.05 level of significance to test the null hypothesis that there is no difference between the corresponding population proportions against the alternative hypothesis that midmorning fatigue is more prevalent among persons who skip breakfast.

**13.64** If 26 of 200 tires of brand A failed to last 30,000 miles, whereas the corresponding figures for 200 tires of brands, B, C, and D were 23, 15, and 32, test the null hypothesis that there is no difference in the quality of the four kinds of tires at the 0.05 level of significance.

**13.65** In random samples of 250 persons with low incomes, 200 persons with average incomes, and 150 persons with high incomes, there were, respectively, 155, 118, and 87 who favor a certain piece of legislation. Use the 0.05 level of significance to test the null hypothesis $\theta_1 = \theta_2 = \theta_3$ (that the proportion of persons favoring the legislation is the same for all three income groups) against the alternative hypothesis that the three $\theta$'s are not all equal.

## 13.7  THE ANALYSIS OF AN $r \times c$ TABLE

The method we shall describe in this section applies to two kinds of problems, which differ conceptually but are analyzed in the same way. In the first kind of problem we deal with samples from $r$ multinomial populations, with each trial permitting $c$ possible outcomes. This would be the case, for instance, when persons interviewed in five different precincts are asked whether they are for a candidate, against her, or undecided. Here $r = 5$ and $c = 3$.

It would also have been the case in Example 13.10 if each shopper had been asked whether he or she favors detergent $A$, detergent $B$, or does not care one way or the other. We might thus have obtained the results shown in the following $3 \times 3$ table:

|  | Number favoring detergent A | Number favoring detergent B | Number indifferent |  |
|---|---|---|---|---|
| Los Angeles | 174 | 93 | 133 | 400 |
| San Diego | 196 | 124 | 180 | 500 |
| Fresno | 148 | 105 | 147 | 400 |

The null hypothesis we would want to test in a problem like this is that we are sampling $r$ identical multinomial populations. Symbolically, if $\theta_{ij}$ is the probability of the $j$th outcome for the $i$th population, we would want to test the null hypothesis

$$\theta_{1j} = \theta_{2j} = \cdots = \theta_{rj}$$

for $j = 1, 2, \ldots, c$. The alternative hypothesis would be that $\theta_{1j}, \theta_{2j}, \ldots$, and $\theta_{rj}$ are not all equal for at least one value of $j$.

In the preceding example we dealt with three samples, whose fixed sizes were given by the row totals, 400, 500, and 400; on the other hand, the column totals were left to chance. In the other kind of problem where the method of this section applies, we are dealing with one sample and the row totals as well as the column totals are left to chance.

To give an example, let us consider the following table obtained in a study of the relationship, if any, of the IQ's of persons who have gone through a large company's job-training program and their subsequent performance on the job:

|  |  | *Poor* | *Fair* | *Good* |  |
|---|---|---|---|---|---|
|  |  | spanning *Performance* |  |  |  |
|  | *Below average* | 67 | 64 | 25 | *156* |
| *IQ* | *Average* | 42 | 76 | 56 | *174* |
|  | *Above average* | 10 | 23 | 37 | *70* |
|  |  | *119* | *163* | *118·* | *400* |

Here there is one sample of size 400, and the row totals as well as the column totals are left to chance. It is mainly in connection with problems like this that $r \times c$ tables are referred to as **contingency tables**.

The null hypothesis we shall want to test by means of the preceding table is that the on the job performance of persons who have gone through the training program is independent of their I.Q. In general, if $\theta_{ij}$ is the probability that an item will fall into the cell belonging to the $i$th row and the $j$th column, $\theta_{i\cdot}$ is the probability that an item will fall into the $i$th row, and $\theta_{\cdot j}$ is the probability that an item will fall into the $j$th column, the null hypothesis we want to test is

$$\theta_{ij} = \theta_{i\cdot} \cdot \theta_{\cdot j}$$

for $i = 1, 2, \ldots, r$ and $j = 1, 2, \ldots, c$. Correspondingly, the alternative hypothesis is $\theta_{ij} \neq \theta_{i\cdot} \cdot \theta_{\cdot j}$ for at least one pair of values of $i$ and $j$.

Since the method by which we analyze an $r \times c$ table is the same regardless of whether we are dealing with $r$ samples from multinomial populations with $c$ different outcomes or one sample from a multinomial population with $rc$ different outcomes, let us discuss it here with regard to the latter. In Exercise 13.67 the reader will be asked to parallel the work for the first kind of problem.

In what follows, we shall denote the observed frequency for the cell in the $i$th row and the $j$th column by $f_{ij}$, the row totals by $f_{i\cdot}$, the column totals by $f_{\cdot j}$, and the grand total, the sum of all the cell frequencies, by $f$. With this notation, we estimate the probabilities $\theta_{i\cdot}$ and $\theta_{\cdot j}$ as

$$\hat{\theta}_{i \cdot} = \frac{f_{i \cdot}}{f} \quad \text{and} \quad \hat{\theta}_{\cdot j} = \frac{f_{\cdot j}}{f}$$

and under the null hypothesis of independence we get

$$e_{ij} = \hat{\theta}_{i \cdot} \cdot \hat{\theta}_{\cdot j} \cdot f = \frac{f_{i \cdot}}{f} \cdot \frac{f_{\cdot j}}{f} \cdot f = \frac{f_{i \cdot} \cdot f_{\cdot j}}{f}$$

for the expected frequency for the cell in the $i$th row and the $j$th column. Note that $e_{ij}$ is thus obtained by *multiplying the total of the row to which the cell belongs by the total of the column to which it belongs, and then dividing by the grand total.*

Once we have calculated the $e_{ij}$, we base our decision on the value of

$$\chi^2 = \sum_{i=1}^{r} \sum_{j=1}^{c} \frac{(f_{ij} - e_{ij})^2}{e_{ij}}$$

and reject the null hypothesis if it exceeds $\chi^2_{\alpha,(r-1)(c-1)}$.

The number of degrees of freedom is $(r - 1)(c - 1)$, and in connection with this let us make the following observation: whenever expected cell frequencies in chi-square formulas are estimated on the basis of sample count data, the number of degrees of freedom is $s - t - 1$, where $s$ is the number of terms in the summation and $t$ is the number of independent parameters replaced by estimates. When testing for differences among $k$ proportions with the chi-square statistic of Section 13.6, we had $s = 2k$ and $t = k$, since we had to estimate the $k$ parameters $\theta_1, \theta_2, \ldots, \theta_k$, and the number of degrees of freedom was $2k - k - 1 = k - 1$. When testing for independence with an $r \times c$ contingency table, we have $s = rc$ and $t = r + c - 2$, since the $r$ parameters $\theta_{i \cdot}$ and the $c$ parameters $\theta_{\cdot j}$ are not all independent—their respective sums must equal 1. Thus, we get $s - t - 1 = rc - (r + c - 2) - 1 = (r - 1)(c - 1)$.

Since the test statistic which we have described has only approximately a chi-square distribution with $(r - 1)(c - 1)$ degrees of freedom, it is customary to use this test only when none of the $e_{ij}$ is less than 5; sometimes this requires that we combine some of the cells with a corresponding loss in the number of degrees of freedom.

## EXAMPLE 13.11

Use the data shown in the following table to test at the 0.01 level of significance whether a person's ability in mathematics is independent of his or her interest in statistics.

|  | | Ability in mathematics | | |
| --- | --- | --- | --- | --- |
|  | | Low | Average | High |
| *Interest in statistics* | *Low* | 63 | 42 | 15 |
|  | *Average* | 58 | 61 | 31 |
|  | *High* | 14 | 47 | 29 |

## Solution

1. $H_0$: Ability in mathematics and interest in statistics are independent.
   $H_1$: Ability in mathematics and interest in statistics are not independent.
   $\alpha = 0.01$

2. Reject the null hypothesis if $\chi^2 \geq 13.277$, where

$$\chi^2 = \sum_{i=1}^{r} \sum_{j=1}^{c} \frac{(f_{ij} - e_{ij})^2}{e_{ij}}$$

and 13.277 is the value of $\chi^2_{.01,4}$.

3. The expected frequencies for the first row are $\dfrac{120 \cdot 135}{360} = 45.0$, $\dfrac{120 \cdot 150}{360} = 50.0$, and $120 - 45.0 - 50.0 = 25.0$, where we made use of the fact that for each row or column the sum of the expected cell frequencies equals the sum of the corresponding observed frequencies (see Exercise 13.66). Similarly, the expected frequencies for the second row are 56.25, 62.5, and 31.25, and those for the third row (all obtained by subtraction from the column totals) are 33.75, 37.5, and 18.75. Then, substituting into the formula for $\chi^2$ yields

$$\chi^2 = \frac{(63 - 45.0)^2}{45.0} + \frac{(42 - 50.0)^2}{50.0} + \cdots + \frac{(29 - 18.75)^2}{18.75}$$

$$= 32.14$$

4. Since $\chi^2 = 32.14$ exceeds 13.277, the null hypothesis must be rejected; we conclude that there is a relationship between a person's ability in mathematics and his or her interest in statistics. ▲

A shortcoming of the chi-square analysis of an $r \times c$ table is that it does not take into account a possible ordering of the rows and/or columns. For instance, in Example 13.11 ability in mathematics, as well as interest in statistics, is ordered from low to average to high, and the value we get for $\chi^2$ would remain the same if the rows and/or columns were interchanged among themselves. Also, the columns of the table on page 483 reflect a definite ordering from favoring $B$ (not favoring $A$) to being indifferent to favoring $A$, but in this case there is no specific ordering of the rows. How such orderings can be taken into account is explained in Exercises 14.59 and 15.12.

## 13.8  GOODNESS OF FIT

The goodness-of-fit test considered here applies to situations in which we want to determine whether a set of data may be looked upon as a random sample from a population having a given distribution. A second kind of "goodness of fit" which applies to the fitting of a curve to a set of paired data will be discussed in Chapter 14. To illustrate, suppose that we want to decide on the basis of the data (observed frequencies) shown in the following table whether the number of errors a compositor makes in setting a galley of type is a random variable having a Poisson distribution:

| Number of errors | Observed frequencies $f_i$ | Poisson probabilities with $\lambda = 3$ | Expected frequencies $e_i$ |
|---|---|---|---|
| 0 | 18 | 0.0498 | 21.9 |
| 1 | 53 | 0.1494 | 65.7 |
| 2 | 103 | 0.2240 | 98.6 |
| 3 | 107 | 0.2240 | 98.6 |
| 4 | 82 | 0.1680 | 73.9 |
| 5 | 46 | 0.1008 | 44.4 |
| 6 | 18 | 0.0504 | 22.2 |
| 7 | 10 | 0.0216 | 9.5 |
| 8 | 2 ⎫3 | 0.0081 | 3.6 ⎫5.3 |
| 9 | 1 ⎭ | 0.0038 | 1.7 ⎭ |

To determine a corresponding set of expected frequencies for a random sample from a Poisson population, we first use the mean of the observed distribution to estimate the Poisson parameter $\lambda$, getting $\hat{\lambda} = \dfrac{1{,}341\ \Sigma\ x_i/n}{440\ \Sigma\ f_i} = 3.05$ or, approximately, $\hat{\lambda} = 3$. Then, copying the Poisson probabilities for $\lambda = 3$ from Table II (with the probability of 9 *or more* used instead of the probability of 9) and multiplying by 440, the total frequency, we get the expected frequencies shown in the right-hand column of the table. To test the null hypothesis that the observed frequencies constitute a random sample from a Poisson population, we must judge how good a fit, or how close an agreement, we have between the two sets of frequencies. In general, to test the null hypothesis $H_0$ that a set of observed data comes from a population having a specified distribution against the alternative that the population has some other distribution, we compute

$$\chi^2 = \sum_{i=1}^{m} \frac{(f_i - e_i)^2}{e_i}$$

and reject $H_0$ at the level of significance $\alpha$ if $\chi^2 \geq \chi^2_{\alpha, m-t-1}$, where $m$ is the number of terms in the summation and $t$ is the number of independent parameters estimated on the basis of the sample data (see discussion page 485). In the above illustration, $t = 1$ since only one parameter is estimated on the basis of the data, and the number of degrees of freedom is $m - 2$.

## EXAMPLE 13.12

For the data in the table on page 487, test at the 0.05 level of significance whether the number of errors the compositor makes in setting a galley of type is a random variable having a Poisson distribution.

### Solution

(Since the expected frequencies corresponding to 8 and 9 errors are less than 5, the two classes are combined.)

1. $H_0$:   Number of errors is a Poisson random variable.
   $H_1$:   Number of errors is not a Poisson random variable.
   $\alpha = 0.05$
2. Reject the null hypothesis if $\chi^2 \geq 14.067$, where

$$\chi^2 = \sum_{i=1}^{m} \frac{(f_i - e_i)^2}{e_i}$$

and 14.067 is the value of $\chi^2_{.05,7}$.
3. Substituting into the formula for $\chi^2$, we get

$$\chi^2 = \frac{(18 - 21.9)^2}{21.9} + \frac{(53 - 65.7)^2}{65.7} + \cdots + \frac{(3 - 5.3)^2}{5.3}$$

$$= 6.83$$

4. Since $\chi^2 = 6.83$ is less than 14.067, the null hypothesis cannot be

rejected; indeed, the close agreement between the observed and expected frequencies suggests that the Poisson distribution provides a "good fit."    ▲

## EXERCISES

**13.66** Verify that if the expected cell frequencies are calculated in accordance with the rule on page 485, their sum for any row or column equals the sum of the corresponding observed frequencies.

**13.67** Show that the rule on page 485 for calculating the expected cell frequencies applies also when we test the null hypothesis that we are sampling $r$ populations with identical multinomial distributions.

**13.68** Show that the following computing formula for $\chi^2$ is equivalent to the formula on page 485:

$$\chi^2 = \sum_{i=1}^{r} \sum_{j=1}^{c} \frac{f_{ij}^2}{e_{ij}} - f$$

**13.69** Use the formula of the preceding exercise to recalculate $\chi^2$ for Example 13.10.

**13.70** If the analysis of a contingency table shows that there is a relationship between the two variables under consideration, the strength of this relationship may be measured by means of the **contingency coefficient**

$$C = \sqrt{\frac{\chi^2}{\chi^2 + f}}$$

where $\chi^2$ is the value obtained for the test statistic and $f$ is the grand total as defined on page 484. Show that

(a)  for a $2 \times 2$ contingency table the maximum value of $C$ is $\frac{1}{2}\sqrt{2}$;

(b)  for a $3 \times 3$ contingency table the maximum value of $C$ is $\frac{1}{3}\sqrt{6}$.

## APPLICATIONS

**13.71** In a study of parents' feelings about a required course in sex education, 360 parents, a random sample, are classified according to whether they have one, two, or three or more children in the school system and also whether they feel that the course is poor, adequate, or good. Based on the results shown in the following table, test at the 0.05 level of significance whether there is a relationship between parents' reaction to the course and the number of children they have in the school system:

|  | Number of children | | |
|---|---|---|---|
|  | *1* | *2* | *3 or more* |
| *Poor* | 48 | 40 | 12 |
| *Adequate* | 55 | 53 | 29 |
| *Good* | 57 | 46 | 20 |

**13.72** Tests of the fidelity and the selectivity of 190 radios produced the results shown in the following table:

|  |  | Fidelity | | |
|---|---|---|---|---|
|  |  | *Low* | *Average* | *High* |
| | *Low* | 7 | 12 | 31 |
| *Selectivity* | *Average* | 35 | 59 | 18 |
| | *High* | 15 | 13 | 0 |

Use the 0.01 level of significance to test the null hypothesis that fidelity is independent of selectivity.

**13.73** The following sample data pertain to the shipments received by a large firm from three different vendors:

|  | *Number rejected* | *Number imperfect but acceptable* | *Number perfect* |
|---|---|---|---|
| *Vendor A* | 12 | 23 | 89 |
| *Vendor B* | 8 | 12 | 62 |
| *Vendor C* | 21 | 30 | 119 |

Test at the 0.01 level of significance whether the three vendors ship products of equal quality.

**13.74** Analyze the 3 × 3 table on page 483, which pertains to the responses of shoppers in three different cities with regard to two detergents. Use the 0.05 level of significance.

**13.75** Four coins are tossed 160 times and 0, 1, 2, 3, or 4 heads showed, respectively, 19, 54, 58, 23, and 6 times. Use the 0.05 level of significance to test whether it is reasonable to suppose that the coins are balanced and randomly tossed.

**13.76** It is desired to test whether the number of gamma rays emitted per second by a certain radioactive substance is a random variable having the Poisson distribution with $\lambda = 2.4$. Use the following data obtained for 300 one-second intervals to test this null hypothesis at the 0.05 level of significance:

| Number of gamma rays | Frequency |
|:---:|:---:|
| 0 | 19 |
| 1 | 48 |
| 2 | 66 |
| 3 | 74 |
| 4 | 44 |
| 5 | 35 |
| 6 | 10 |
| 7 or more | 4 |

**13.77** Each day, Monday through Saturday, a baker bakes three large chocolate cakes, and those not sold on the same day are given away to a food bank. Use the data shown in the following table to test at the 0.05 level of significance whether they may be looked upon as values of a binomial random variable:

| Number of cakes sold | Number of days |
|:---:|:---:|
| 0 | 1 |
| 1 | 16 |
| 2 | 55 |
| 3 | 228 |

**13.78** The following is the distribution of the readings obtained with a Geiger counter of the number of particles emitted by a radioactive substance in 100 successive 40-second intervals:

| Number of particles | Frequency |
|:---:|:---:|
| 5–9 | 1 |
| 10–14 | 10 |
| 15–19 | 37 |
| 20–24 | 36 |
| 25–29 | 13 |
| 30–34 | 2 |
| 35–39 | 1 |

(a)  Verify that the mean and the standard deviation of this distribution are $\bar{x} = 20$ and $s = 5$.
(b)  Find the probabilities that a random variable having a normal distribution with $\mu = 20$ and $\sigma = 5$ will take on a value less than 9.5, between 9.5 and 14.5, between 14.5 and 19.5, between 19.5 and 24.5, between 24.5 and 29.5, between 29.5 and 34.5, and greater than 34.5.
(c)  Find the expected normal curve frequencies for the various classes by multiplying the probabilities obtained in part (b) by the total frequency, and then test at the 0.05 level of significance whether the data may be looked upon as a random sample from a normal population.

## 13.9  USE OF COMPUTERS

As in Chapter 11, there exists computer software for all the tests we have discussed. Again, we have only to enter the original raw (untreated) data into our computer together with the appropriate command. To illustrate, consider the following example.

## EXAMPLE 13.13

The following random samples are measurements of the heat-producing capacity (in millions of calories per ton) of specimens of coal from two mines:

| | | | | | |
|---|---|---|---|---|---|
| Mine 1: | 8,400 | 8,230 | 8,380 | 7,860 | 7,930 |
| Mine 2: | 7,510 | 7,690 | 7,720 | 8,070 | 7,660 |

Use the 0.05 level of significance to test whether the difference between the means of these two samples is significant.

### Solution

The computer printout of Figure 13.5 shows that the value of the test statistic is $t = 2.95$, the number of degrees of freedom is 8, and the $P$-value is 0.018. Since 0.018 is less than 0.05, we conclude that the difference between the means of the two samples is significant.    ▲

As we indicated in Section 11.8, the impact of computers on statistics goes far beyond what we did in Example 11.12. This also goes for Example 13.13, but we wanted to make the point that there exists software for all the standard testing procedures we have discussed.

```
MTB  > SET C1
DATA > 8400  8230  8380  7860  7930
MTB  > SET C2
DATA > 7510  7690  7720  8070  7660
MTB     > POOL C1 C2

TWOSAMPLE T FOR C1 VS C2
      N       MEAN      STDEV    SE MEAN
C1   5        8160        252        113
C2   5        7730        207         92

95 PCT CI FOR MU C1 - MU C2: (94, 766)
TTEST MU C1 = MU C2 (VS NE): T=2.95 P=0.018 DF=8.0
```

**Figure 13.5**  Computer printout for Example 13.13.

## REFERENCES

The problem of determining the appropriate number of degrees of freedom for various uses of the chi-square statistic is discussed in

CRAMÉR, H., *Mathematical Methods of Statistics*. Princeton, N.J.: Princeton University Press, 1946.

The *Smith-Satterthwaite* test of the null hypothesis that two normal populations with unequal variances have the same mean is given in

MILLER, I., FREUND, J. E., and JOHNSON, R. A., *Probability and Statistics for Engineers*, 4th ed. Englewood Cliffs, N.J.: Prentice Hall, Inc., 1990.

Additional comments relative to using the continuity correction for testing hypotheses concerning binomial parameters can be found in

BROWNLEE, K. A., *Statistical Theory and Methodology in Science and Engineering*, 2nd ed. New York: John Wiley & Sons, Inc., 1965.

Details about the analysis of contingency tables may be found in

EVERITT, B. S., *The Analysis of Contingency Tables*. New York: John Wiley & Sons, Inc., 1977.

In recent years, research has been done on the analysis of $r \times c$ tables, where the categories represented by the rows and/or columns are ordered. This work is beyond the level of this text, but some introductory material may be found in

AGRESTI, A., *Analysis of Ordinal Categorical Data*. New York: John Wiley & Sons, Inc., 1984.

AGRESTI, A., *Categorical Data Analysis*. New York: John Wiley & Sons, Inc., 1990.

GOODMAN, L. A., *The Analysis of Cross-Classified Data Having Ordered Categories*. Cambridge, Mass.: Harvard University Press, 1984.

# 14

# *Regression and Correlation*

## 14.1 INTRODUCTION

A major objective of many statistical investigations is to establish relationships which make it possible to predict one or more variables in terms of others. Thus, studies are made to predict the potential sales of a new product in terms of its price, a patient's weight in terms of the number of weeks he or she has been on a diet, family expenditures on entertainment in terms of family income, the per capita consumption of certain foods in terms of their nutritional values and the amount of money spent advertising them on television, and so forth.

Although it is, of course, desirable to be able to predict one quantity exactly in terms of others, this is seldom possible, and in most instances we have to be satisfied with predicting averages or expected values. Thus, we may not be able to predict exactly how much money Mr. Brown will make ten years after graduating from college, but, given suitable data, we can predict the average income of a college graduate in terms of the number of years he has been out of college. Similarly, we can at best predict the average yield of a given variety of wheat in terms of data on the rainfall in July, and we can at best predict the average performance of students starting college in terms of their IQ's.

Formally, if we are given the joint distribution of two random variables $X$ and $Y$, and $X$ is known to take on the value $x$, the basic problem of **bivariate regression** is that of determining the conditional mean $\mu_{Y|x}$, namely, the "average" value of $Y$ for the given value of $X$. The term "regression," as it is used here, dates back to Francis Galton, who employed it to indicate certain relationships in the theory of heredity. In problems involving more than two random variables—that is, in **multiple regression**—we are concerned with quantities such as $\mu_{Z|x,y}$, the mean of $Z$ for given values of $X$ and $Y$, $\mu_{X_4|x_1,x_2,x_3}$, the mean of $X_4$ for given values of $X_1$, $X_2$, and $X_3$, and so on.

If $f(x, y)$ is the value of the joint density of two random variables $X$ and $Y$ at $(x, y)$, the problem of bivariate regression is simply that of determining the conditional density of $Y$ given $X = x$ and then evaluating the integral

$$\mu_{Y|x} = E(Y|x) = \int_{-\infty}^{\infty} y \cdot w(y|x) \, dy$$

as outlined in Section 4.8. The resulting equation is called the **regression equation** of $Y$ on $X$. Alternatively, we might be interested in the regression equation

$$\mu_{X|y} = E(X|y) = \int_{-\infty}^{\infty} x \cdot f(x|y) \, dx$$

In the discrete case, when we are dealing with probability distributions instead of probability densities, the integrals in the two regression equations given above are simply replaced by sums.

When we do not know the joint probability density or distribution of the two random variables, or at least not all its parameters, the determination of $\mu_{Y|x}$ or $\mu_{X|y}$ becomes a problem of estimation based on sample data; this is an entirely different problem, which we shall discuss in Sections 14.3 and 14.4.

## EXAMPLE 14.1

Given the two random variables $X$ and $Y$ which have the joint density

$$f(x, y) = \begin{cases} x \cdot e^{-x(1 + y)} & \text{for } x > 0 \text{ and } y > 0 \\ 0 & \text{elsewhere} \end{cases}$$

find the regression equation of $Y$ on $X$ and sketch the regression curve.

*Solution*

Integrating out $y$ we find that the marginal density of $X$ is given by

$$g(x) = \begin{cases} e^{-x} & \text{for } x > 0 \\ 0 & \text{elsewhere} \end{cases}$$

and, hence, the conditional density of $Y$ given $X = x$ is given by

$$w(y|x) = \frac{f(x, y)}{g(x)} = \frac{x \cdot e^{-x(1+y)}}{e^{-x}} = x \cdot e^{-xy}$$

for $y > 0$ and $w(y|x) = 0$ elsewhere, which we recognize as an exponential density with $\theta = \dfrac{1}{x}$. Hence, by evaluating

$$\mu_{Y|x} = \int_0^\infty y \cdot x \cdot e^{-xy} \, dy$$

or by referring to Corollary 1 of Theorem 6.3, we find that the regression equation of $Y$ on $X$ is given by

$$\mu_{Y|x} = \frac{1}{x}$$

The corresponding regression curve is shown in Figure 14.1.    ▲

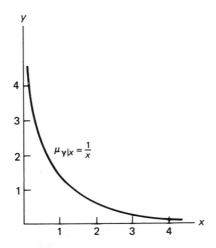

**Figure 14.1** Regression curve of Example 14.1.

## EXAMPLE 14.2

If $X$ and $Y$ have the multinomial distribution

$$f(x, y) = \binom{n}{x, y, n - x - y} \cdot \theta_1^x \theta_2^y (1 - \theta_1 - \theta_2)^{n-x-y}$$

for $x = 0, 1, 2, \ldots, n$, and $y = 0, 1, 2, \ldots, n$, with $x + y \leq n$, find the regression equation of $Y$ on $X$.

### Solution

The marginal distribution of $X$ is given by

$$g(x) = \sum_{y=0}^{n-x} \binom{n}{x, y, n - x - y} \cdot \theta_1^x \theta_2^y (1 - \theta_1 - \theta_2)^{n-x-y}$$

$$= \binom{n}{x} \theta_1^x (1 - \theta_1)^{n-x}$$

for $x = 0, 1, 2, \ldots, n$, which we recognize as a binomial distribution with the parameters $n$ and $\theta_1$. Hence,

$$w(y|x) = \frac{f(x, y)}{g(x)} = \frac{\binom{n - x}{y} \theta_2^y (1 - \theta_1 - \theta_2)^{n-x-y}}{(1 - \theta_1)^{n-x}}$$

for $y = 0, 1, 2, \ldots, n - x$, and, rewriting this formula as

$$w(y|x) = \binom{n - x}{y} \left(\frac{\theta_2}{1 - \theta_1}\right)^y \left(\frac{1 - \theta_1 - \theta_2}{1 - \theta_1}\right)^{n-x-y}$$

we find by inspection that the conditional distribution of $Y$ given $X = x$ is a binomial distribution with the parameters $n - x$ and $\dfrac{\theta_2}{1 - \theta_1}$, so that the regression equation of $Y$ on $X$ is

$$\mu_{Y|x} = \frac{(n - x)\theta_2}{1 - \theta_1}$$

according to Theorem 5.2.    ▲

With reference to the preceding example, if we let $X$ be the number of times an even number comes up in 30 rolls of a balanced die and $Y$ be the number of times the result is a five, then the regression equation becomes

$$\mu_{Y|x} = \frac{(30 - x)\frac{1}{6}}{1 - \frac{1}{2}} = \frac{1}{3}(30 - x)$$

This stands to reason, because there are three equally likely possibilities, 1, 3, or 5, for each of the $30 - x$ outcomes that are not even.

## EXAMPLE 14.3

If the joint density of $X_1$, $X_2$, and $X_3$ is given by

$$f(x_1, x_2, x_3) = \begin{cases} (x_1 + x_2)e^{-x_3} & \text{for } 0 < x_1 < 1, 0 < x_2 < 1, x_3 > 0 \\ 0 & \text{elsewhere} \end{cases}$$

find the regression equation of $X_2$ on $X_1$ and $X_3$.

### Solution

Referring to Example 3.22, we find that the joint marginal density of $X_1$ and $X_3$ is given by

$$m(x_1, x_3) = \begin{cases} (x_1 + \frac{1}{2})e^{-x_3} & \text{for } 0 < x_1 < 1, x_3 > 0 \\ 0 & \text{elsewhere} \end{cases}$$

Therefore,

$$\mu_{X_2|x_1, x_3} = \int_{-\infty}^{\infty} x_2 \cdot \frac{f(x_1, x_2, x_3)}{m(x_1, x_3)} \, dx_2 = \int_0^1 \frac{x_2(x_1 + x_2)}{(x_1 + \frac{1}{2})} \, dx_2$$

$$= \frac{x_1 + \frac{2}{3}}{2x_1 + 1} \qquad \blacktriangle$$

Note that the conditional expectation obtained in the preceding example depends on $x_1$ but not on $x_3$. This could have been expected, since we indicated on page 134 that there is a pairwise independence between $X_2$ and $X_3$.

## 14.2  LINEAR REGRESSION

An important feature of Example 14.2 is that the regression equaton is **linear**, namely, that it is of the form

$$\mu_{Y|x} = \alpha + \beta x$$

where $\alpha$ and $\beta$ are constants, called the **regression coefficients**. There are several reasons why linear regression equations are of special interest: First, they lend themselves readily to further mathematical treatment; then, they often provide good approximations to otherwise complicated regression equations; and, finally, in the case of the bivariate normal distribution, which we studied in Section 6.7, the regression equations are, in fact, linear.

To simplify the study of linear regression equations, let us express the regression coefficients $\alpha$ and $\beta$ in terms of some of the lower moments of the joint distribution of $X$ and $Y$, namely, in terms of $E(X) = \mu_1$, $E(Y) = \mu_2$, var$(X) = \sigma_1^2$, vary$(Y) = \sigma_2^2$, and cov$(X, Y) = \sigma_{12}$. Then, also using the correlation coefficient

$$\rho = \frac{\sigma_{12}}{\sigma_1 \sigma_2}$$

defined in Section 6.7, we can prove the following results.

---

**THEOREM 14.1**  If the regression of $Y$ on $X$ is linear, then

$$\mu_{Y|x} = \mu_2 + \rho \frac{\sigma_2}{\sigma_1} (x - \mu_1)$$

and if the regression of $X$ on $Y$ is linear, then

$$\mu_{X|y} = \mu_1 + \rho \frac{\sigma_1}{\sigma_2} (y - \mu_2)$$

---

***Proof.***  Since $\mu_{Y|x} = \alpha + \beta x$, it follows that

$$\int y \cdot w(y|x) \, dy = \alpha + \beta x$$

and if we multiply the expression on both sides of this equation by $g(x)$, the corresponding value of the marginal density of $X$, and integrate on $x$, we obtain

$$\int\int y \cdot w(y|x)g(x) \, dy \, dx = \alpha \int g(x) \, dx + \beta \int x \cdot g(x) \, dx$$

or

$$\mu_2 = \alpha + \beta\mu_1$$

since $w(y|x)g(x) = f(x, y)$. If we had multiplied the equation for $\mu_{Y|x}$ on both sides by $x \cdot g(x)$ before integrating on $x$, we would have obtained

$$\int\int xy \cdot f(x, y) \, dy \, dx = \alpha \int x \cdot g(x) \, dx + \beta \int x^2 \cdot g(x) \, dx$$

or

$$E(XY) = \alpha\mu_1 + \beta E(X^2)$$

Solving $\mu_2 = \alpha + \beta\mu_1$ and $E(XY) = \alpha\mu_1 + \beta E(X^2)$ for $\alpha$ and $\beta$ and making use of the fact that $E(XY) = \sigma_{12} + \mu_1\mu_2$ and $E(X^2) = \sigma_1^2 + \mu_1^2$, we find that

$$\alpha = \mu_2 - \frac{\sigma_{12}}{\sigma_1^2} \cdot \mu_1 = \mu_2 - \rho\frac{\sigma_2}{\sigma_1} \cdot \mu_1$$

and

$$\beta = \frac{\sigma_{12}}{\sigma_1^2} = \rho\frac{\sigma_2}{\sigma_1}$$

This enables us to write the linear regression equation of $Y$ on $X$ as

$$\mu_{Y|x} = \mu_2 + \rho\frac{\sigma_2}{\sigma_1}(x - \mu_1)$$

When the regression of $X$ on $Y$ is linear, similar steps lead to the equation

$$\mu_{X|y} = \mu_1 + \rho\frac{\sigma_1}{\sigma_2}(y - \mu_2) \qquad \blacktriangledown$$

It follows from Theorem 14.1 that if the regression equation is linear and $\rho = 0$, then $\mu_{Y|x}$ does not depend on $x$ (or $\mu_{X|y}$ does not depend on $y$). When $\rho = 0$ and, hence, $\sigma_{12} = 0$, the two random variables $X$ and $Y$ are **uncorrelated**, and we can paraphrase the assertion which we made on page 172 by saying that if two random variables are independent they are also uncorrelated, but if two random variables are uncorrelated they are not necessarily independent; the latter is again illustrated in Exercise 14.9.

The correlation coefficient and its estimates are of importance in many statistical investigations, and they will be discussed in some detail in Section 14.5. At this time, let us again point out that $-1 \leqslant \rho \leqslant +1$, as the reader will be asked to prove in Exercise 14.11, and the sign of $\rho$ tells us directly whether the slope of a regression line is upward or downward.

## 14.3 THE METHOD OF LEAST SQUARES

In the preceding sections we have discussed the problem of regression only in connection with random variables having known joint distributions. In actual practice, there are many problems where a set of **paired data** gives the indication that the regression is linear, where we do not know the joint distribution of the random variables under consideration but, nevertheless, want to estimate the regression coefficients $\alpha$ and $\beta$. Problems of this kind are usually handled by the **method of least squares**, a method of curve fitting suggested early in the nineteenth century by the French mathematician Adrien Legendre.

To illustrate this technique, let us consider the following data on the number of hours which 10 persons studied for a French test and their scores on the test:

| Hours studied $x$ | Test score $y$ |
|:---:|:---:|
| 4 | 31 |
| 9 | 58 |
| 10 | 65 |
| 14 | 73 |
| 4 | 37 |
| 7 | 44 |
| 12 | 60 |
| 22 | 91 |
| 1 | 21 |
| 17 | 84 |

Plotting these data as in Figure 14.2, we get the impression that a straight line provides a reasonably good fit. Although the points do not all fall on a straight

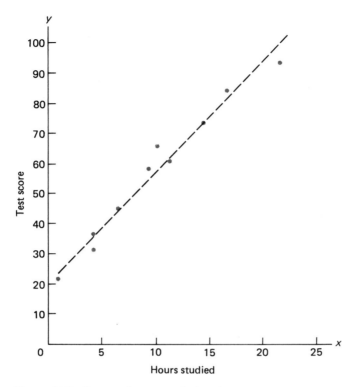

**Figure 14.2**  Data on hours studied and test scores.

line, the overall pattern suggests that the average test score for a given number of hours studied may well be related to the number of hours studied by means of an equation of the form $\mu_{Y|x} = \alpha + \beta x$.

Once we have decided in a given problem that the regression is approximately linear, we face the problem of estimating the coefficients $\alpha$ and $\beta$ from the sample data. In other words, we face the problem of obtaining estimates $\hat{\alpha}$ and $\hat{\beta}$ such that the estimated regression line $\hat{y} = \hat{\alpha} + \hat{\beta}x$ in some sense provides the best possible fit to the given data.

Denoting the vertical deviation from a point to the line by $e_i$, as indicated in Figure 14.3, the least squares criterion on which we shall base this "goodness of fit" requires that we minimize the sum of the squares of these deviations. Thus, if we are given a set of paired data $\{(x_i, y_i); i = 1, 2, \ldots, n\}$, the **least squares estimates** of the regression coefficients are the values $\hat{\alpha}$ and $\hat{\beta}$ for which the quantity

$$q = \sum_{i=1}^{n} e_i^2 = \sum_{i=1}^{n} [y_i - (\hat{\alpha} + \hat{\beta}x_i)]^2$$

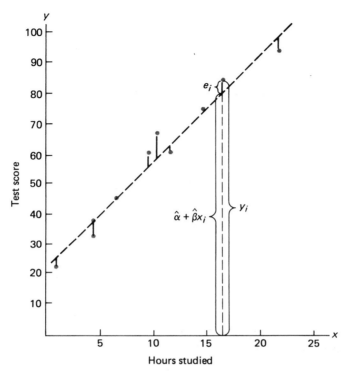

**Figure 14.3** Least squares criterion.

is a minimum. Differentiating partially with respect to $\hat{\alpha}$ and $\hat{\beta}$, and equating these partial derivatives to zero, we obtain

$$\frac{\partial q}{\partial \hat{\alpha}} = \sum_{i=1}^{n} (-2)[y_i - (\hat{\alpha} + \hat{\beta}x_i)] = 0$$

and

$$\frac{\partial q}{\partial \hat{\beta}} = \sum_{i=1}^{n} (-2)x_i[y_i - (\hat{\alpha} + \hat{\beta}x_i)] = 0$$

which yield the so-called system of **normal equations**

$$\sum_{i=1}^{n} y_i = \hat{\alpha}n + \hat{\beta} \cdot \sum_{i=1}^{n} x_i$$

$$\sum_{i=1}^{n} x_i y_i = \hat{\alpha} \cdot \sum_{i=1}^{n} x_i + \hat{\beta} \cdot \sum_{i=1}^{n} x_i^2$$

Solving this system of equations by using determinants or the method of elimination, we find that the least squares estimate of $\beta$ is

$$\hat{\beta} = \frac{n\left(\sum_{i=1}^{n} x_i y_i\right) - \left(\sum_{i=1}^{n} x_i\right)\left(\sum_{i=1}^{n} y_i\right)}{n\left(\sum_{i=1}^{n} x_i^2\right) - \left(\sum_{i=1}^{n} x_i\right)^2}$$

Then we can write the least squares estimate of $\alpha$ as

$$\hat{\alpha} = \frac{\sum_{i=1}^{n} y_i - \hat{\beta} \cdot \sum_{i=1}^{n} x_i}{n}$$

by solving the first of the two normal equations for $\hat{\alpha}$. This formula for $\hat{\alpha}$ can also be written as

$$\hat{\alpha} = \bar{y} - \hat{\beta} \cdot \bar{x}$$

To simplify the formula for $\hat{\beta}$ as well as some of the formulas we shall meet in Sections 14.4 and 14.5, let us introduce the following notation:

$$S_{xx} = \sum_{i=1}^{n} (x_i - \bar{x})^2 = \sum_{i=1}^{n} x_i^2 - \frac{1}{n}\left(\sum_{i=1}^{n} x_i\right)^2$$

$$S_{yy} = \sum_{i=1}^{n} (y_i - \bar{y})^2 = \sum_{i=1}^{n} y_i^2 - \frac{1}{n}\left(\sum_{i=1}^{n} y_i\right)^2$$

and

$$S_{xy} = \sum_{i=1}^{n} (x_i - \bar{x})(y_i - \bar{y}) = \sum_{i=1}^{n} x_i y_i - \frac{1}{n}\left(\sum_{i=1}^{n} x_i\right)\left(\sum_{i=1}^{n} y_i\right)$$

We can thus write

---

**THEOREM 14.2** Given the sample data $\{(x_i, y_i); i = 1, 2, \ldots, n\}$, the coefficients of the least squares line $\hat{y} = \hat{\alpha} + \hat{\beta}x$ are

$$\hat{\beta} = \frac{S_{xy}}{S_{xx}}$$

and

$$\hat{\alpha} = \bar{y} - \hat{\beta} \cdot \bar{x}$$

---

## EXAMPLE 14.4

With reference to the data on page 501,

(a)  find the equation of the least squares line that approximates the regression of the test scores on the number of hours studied;

(b)  predict the average test score of a person who studied 14 hours for the test.

*Solution*

(a)  Getting $n = 10$, $\Sigma x = 100$, $\Sigma x^2 = 1{,}376$, $\Sigma y = 564$, and $\Sigma xy = 6{,}945$ from the data, we find that

$$S_{xx} = 1{,}376 - \tfrac{1}{10} (100)^2 = 376$$

and

$$S_{xy} = 6{,}945 - \tfrac{1}{10} (100)(564) = 1{,}305$$

Thus, $\hat{\beta} = \dfrac{1{,}305}{376} = 3.471$ and $\hat{\alpha} = \dfrac{564}{10} - 3.471 \cdot \dfrac{100}{10} = 21.69$, and the equation of the least squares line is

$$\hat{y} = 21.69 + 3.471x$$

(b)  Substituting $x = 14$ into the equation obtained in part (a), we get

$$\hat{y} = 21.69 + 3.471(14) = 70.284$$

or $\hat{y} = 70$, rounded to the nearest unit.    ▲

Since we did not make any assumptions about the joint distribution of the random variables with which we were concerned in the preceding example, we cannot judge the "goodness" of the prediction obtained in part (b); also, we cannot judge the "goodness" of the estimates $\hat{\alpha} = 21.69$ and $\hat{\beta} = 3.471$ obtained in part (a). Problems like this will be discussed in Section 14.4.

The least squares criterion, or in other words, the method of least squares, is used in many problems of curve fitting which are more general than the one treated in this section. Above all, it will be used in Sections 14.6 and 14.7 to estimate the coefficients of **multiple regression equations** of the form

$$\mu_{Y|x_1,\ldots,x_k} = \beta_0 + \beta_1 x_1 + \cdots + \beta_k x_k$$

## EXERCISES

**14.1** With reference to Example 14.1, show that the regression equation of $X$ on $Y$ is

$$\mu_{X|y} = \frac{2}{1 + y}$$

Also sketch the regression curve.

**14.2** Given the joint density

$$f(x, y) = \begin{cases} \frac{2}{5} (2x + 3y) & \text{for } 0 < x < 1 \text{ and } 0 < y < 1 \\ 0 & \text{elsewhere} \end{cases}$$

find $\mu_{Y|x}$ and $\mu_{X|y}$.

**14.3** Given the joint density

$$f(x, y) = \begin{cases} 6x & \text{for } 0 < x < y < 1 \\ 0 & \text{elsewhere} \end{cases}$$

find $\mu_{Y|x}$ and $\mu_{X|y}$.

**14.4** Given the joint density

$$f(x, y) = \begin{cases} \dfrac{2x}{(1 + x + xy)^3} & \text{for } x > 0 \text{ and } y > 0 \\ 0 & \text{elsewhere} \end{cases}$$

show that $\mu_{Y|x} = 1 + \dfrac{1}{x}$ and that $\text{var}(Y|x)$ does not exist.

**14.5**  With reference to Exercise 3.86, use the results of parts (c) and (d) to find $\mu_{X|1}$ and $\mu_{Y|0}$.

**14.6**  With reference to Exercise 3.87, find an expression for $\mu_{Y|x}$.

**14.7**  Given the joint density

$$f(x, y) = \begin{cases} 2 & \text{for } 0 < y < x < 1 \\ 0 & \text{elsewhere} \end{cases}$$

show that

(a)   $\mu_{Y|x} = \dfrac{x}{2}$ and $\mu_{X|y} = \dfrac{1 + y}{2}$;

(b)   $E(X^m Y^n) = \dfrac{2}{(n + 1)(m + n + 2)}.$

Also,

(c)   verify the results of part (a) by substituting the values of $\mu_1$, $\mu_2$, $\sigma_1$, $\sigma_2$, and $\rho$, obtained with the formula of part (b), into the formulas of Theorem 14.1.

**14.8**  Given the joint density

$$f(x, y) = \begin{cases} 24xy & \text{for } x > 0, y > 0, \text{ and } x + y < 1 \\ 0 & \text{elsewhere} \end{cases}$$

show that $\mu_{Y|x} = \frac{2}{3}(1 - x)$ and verify this result by determining the values of $\mu_1$, $\mu_2$, $\sigma_1$, $\sigma_2$, and $\rho$, and substituting them into the first formula of Theorem 14.1.

**14.9**  Given the joint density

$$f(x, y) = \begin{cases} 1 & \text{for } -y < x < y \text{ and } 0 < y < 1 \\ 0 & \text{elsewhere} \end{cases}$$

show that the random variables $X$ and $Y$ are uncorrelated but not independent.

**14.10**  Show that if $\mu_{Y|x}$ is linear in $x$ and $\text{var}(Y|x)$ is constant, then $\text{var}(Y|x) = \sigma_2^2(1 - \rho^2)$.

**14.11**  Given a pair of random variables $X$ and $Y$ having the variances $\sigma_1^2$ and $\sigma_2^2$ and the correlation coefficient $\rho$, use Theorem 4.14 to express $\text{var}\left(\dfrac{X}{\sigma_1} + \dfrac{Y}{\sigma_2}\right)$ and $\text{var}\left(\dfrac{X}{\sigma_1} - \dfrac{Y}{\sigma_2}\right)$ in terms of $\sigma_1$, $\sigma_2$, and $\rho$. Then, making use of the fact that variances cannot be negative, show that $-1 \leq \rho \leq +1$.

**14.12** Given the random variables $X_1$, $X_2$, and $X_3$ having the joint density $f(x_1, x_2, x_3)$, show that if the regression of $X_3$ on $X_1$ and $X_2$ is linear and written as

$$\mu_{X_3|x_1,x_2} = \alpha + \beta_1(x_1 - \mu_1) + \beta_2(x_2 - \mu_2)$$

then

$$\alpha = \mu_3$$

$$\beta_1 = \frac{\sigma_{13}\sigma_2^2 - \sigma_{12}\sigma_{23}}{\sigma_1^2\sigma_2^2 - \sigma_{12}^2}$$

$$\beta_2 = \frac{\sigma_{23}\sigma_1^2 - \sigma_{12}\sigma_{13}}{\sigma_1^2\sigma_2^2 - \sigma_{12}^2}$$

where $\mu_i = E(X_i)$, $\sigma_i^2 = \mathrm{var}(X_i)$, and $\sigma_{ij} = \mathrm{cov}(X_i, X_j)$. [*Hint*: Proceed as on page 500, multiplying by $(x_1 - \mu_1)$ and $(x_2 - \mu_2)$, respectively, to obtain the second and third equations.]

**14.13** Find the least squares estimate of the parameter $\beta$ in the regression equation $\mu_{Y|x} = \beta x$.

**14.14** Solve the normal equations on pages 503 and 504 simultaneously to show that

$$\hat{\alpha} = \frac{\left(\sum_{i=1}^{n} x_i^2\right)\left(\sum_{i=1}^{n} y_i\right) - \left(\sum_{i=1}^{n} x_i\right)\left(\sum_{i=1}^{n} x_i y_i\right)}{n\left(\sum_{i=1}^{n} x_i^2\right) - \left(\sum_{i=1}^{n} x_i\right)^2}$$

**14.15** When the $x$'s are equally spaced, the calculation of $\hat{\alpha}$ and $\hat{\beta}$ can be simplified by coding the $x$'s by assigning them the values ... , $-3, -2, -1,$ $0, 1, 2, 3, \ldots$ when $n$ is odd, or the values ... , $-5, -3, -1, 1, 3, 5,$ ... when $n$ is even. Show that with this coding the formulas for $\hat{\alpha}$ and $\hat{\beta}$ become

$$\hat{\alpha} = \frac{\sum_{i=1}^{n} y_i}{n} \quad \text{and} \quad \hat{\beta} = \frac{\sum_{i=1}^{n} x_i y_i}{\sum_{i=1}^{n} x_i^2}$$

## APPLICATIONS

**14.16** Various doses of a poisonous substance were given to groups of 25 mice and the following results were observed:

| Dose (mg) | Number of deaths |
|-----------|------------------|
| x | y |
| 4 | 1 |
| 6 | 3 |
| 8 | 6 |
| 10 | 8 |
| 12 | 14 |
| 14 | 16 |
| 16 | 20 |

(a)  Find the equation of the least squares line fit to these data.

(b)  Estimate the number of deaths in a group of 25 mice who receive a 7-milligram dose of this poison.

**14.17**  The following are the scores which 12 students obtained in the midterm and final examinations in a course in statistics:

| Midterm examination | Final examination |
|---------------------|-------------------|
| x | y |
| 71 | 83    5893 |
| 49 | 62    3058 |
| 80 | 76    6080 |
| 73 | 77    5621 |
| 93 | 89    8277 |
| 85 | 74    6290 |
| 58 | 48    2784 |
| 82 | 78    6396 |
| 64 | 76    4864 |
| 32 | 51    1632 |
| 87 | 73    6351 |
| 80 | 89    7120 |

(a)  Find the equation of the least squares line which will enable us to predict a student's final examination score in this course on the basis of his or her score on the midterm examination.

(b)  Predict the final examination score of a student who received an 84 on the midterm examination.

**14.18**  Raw material used in the production of a synthetic fiber is stored in a place which has no humidity control. Measurements of the relative humidity and the moisture content of samples of the raw material (both in percentages) on 12 days yielded the following results:

| Humidity | Moisture content |
|----------|------------------|
| 46 | 12 |
| 53 | 14 |
| 37 | 11 |
| 42 | 13 |
| 34 | 10 |
| 29 | 8 |
| 60 | 17 |
| 44 | 12 |
| 41 | 10 |
| 48 | 15 |
| 33 | 9 |
| 40 | 13 |

(a)  Fit a least squares line which will enable us to predict the moisture content in terms of the relative humidity.

(b)  Use the result of part (a) to estimate (predict) the moisture content when the relative humidity is 38 percent.

**14.19**  The following data pertain to the chlorine residual in a swimming pool at various times after it has been treated with chemicals:

| Number of hours | Chlorine residual (parts per million) |
|-----------------|----------------------------------------|
| 2 | 1.8 |
| 4 | 1.5 |
| 6 | 1.4 |
| 8 | 1.1 |
| 10 | 1.1 |
| 12 | 0.9 |

(a)  Fit a least squares line from which we can predict the chlorine residual in terms of the number of hours since the pool has been treated with chemicals.

(b)  Use the equation of the least squares line to estimate the chlorine residual in the pool 5 hours after it has been treated with chemicals.

**14.20**  Use the coding of Exercise 14.15 to rework both parts of Exercise 14.16.

**14.21**  Use the coding of Exercise 14.15 to rework both parts of Exercise 14.19.

**14.22**  During its first five years of operation, a company's gross income from sales was 1.4, 2.1, 2.6, 3.5, and 3.7 million dollars. Use the coding of Exercise 14.15 to fit a least squares line and, assuming that the trend continues, predict the company's gross income from sales during its sixth year of operation.

**14.23** If a set of paired data gives the indication that the regression equation is of the form $\mu_{Y|x} = \alpha \cdot \beta^x$, it is customary to estimate $\alpha$ and $\beta$ by fitting the line

$$\log \hat{y} = \log \hat{\alpha} + x \cdot \log \hat{\beta}$$

to the points $\{(x_i, \log y_i); i = 1, 2, \ldots, n\}$ by the method of least squares. Use this technique to fit an exponential curve of the form $\hat{y} = \hat{\alpha} \cdot \hat{\beta}^x$ to the following data on the growth of cactus grafts under controlled environmental conditions:

| Weeks after grafting $x$ | Height (inches) $y$ |
|---|---|
| 1 | 2.0 |
| 2 | 2.4 |
| 4 | 5.1 |
| 5 | 7.3 |
| 6 | 9.4 |
| 8 | 18.3 |

**14.24** If a set of paired data gives the indication that the regression equation is of the form $\mu_{Y|x} = \alpha \cdot x^\beta$, it is customary to estimate $\alpha$ and $\beta$ by fitting the line

$$\log \hat{y} = \log \hat{\alpha} + \hat{\beta} \cdot \log x$$

to the points $\{(\log x_i, \log y_i); i = 1, 2, \ldots, n\}$ by the method of least squares.

(a)  Use this technique to fit a power function of the form $\hat{y} = \hat{\alpha} \cdot x^{\hat{\beta}}$ to the following data on the unit cost of producing certain electronic components and the number of units produced:

| Lot size $x$ | Unit cost $y$ |
|---|---|
| 50 | $108 |
| 100 | $53 |
| 250 | $24 |
| 500 | $9 |
| 1,000 | $5 |

(b)    Use the result of part (a) to estimate the unit cost for a lot of 300 components.

## 14.4  NORMAL REGRESSION ANALYSIS

When we analyze a set of paired data $\{(x_i, y_i); 1, 2, \ldots, n\}$ by **regression analysis**, we look upon the $x_i$ as constants and the $y_i$ as values of corresponding independent random variables $Y_i$. This clearly differs from **correlation analysis**, which we shall take up in Section 14.5, where we look upon the $x_i$ and the $y_i$ as values of corresponding random variables $X_i$ and $Y_i$. For example, if we want to analyze data on the ages and prices of used cars, treating the ages as known constants and the prices as values of random variables, this is a problem of regression analysis. On the other hand, if we want to analyze data on the height and weight of certain animals, and height and weight are both looked upon as random variables, this is a problem of correlation analysis.

This section will be devoted to some of the basic problems of **normal regression analysis**, where it is assumed that for each fixed $x_i$ the conditional density of the corresponding random variable $Y_i$ is the normal density

$$w(y_i|x_i) = \frac{1}{\sigma\sqrt{2\pi}} \cdot e^{-\frac{1}{2}\left[\frac{y_i-(\alpha+\beta x_i)}{\sigma}\right]^2} \qquad -\infty < y_i < \infty$$

where $\alpha$, $\beta$, and $\sigma$ are the same for each $i$. Given a random sample of such paired data, normal regression analysis concerns itself mainly with the estimation of $\sigma$ and the regression coefficients $\alpha$ and $\beta$, with tests of hypotheses concerning these three parameters, and with predictions based on the estimated regression equation $\hat{y} = \hat{\alpha} + \hat{\beta}x$, where $\hat{\alpha}$ and $\hat{\beta}$ are estimates of $\alpha$ and $\beta$.

To obtain maximum likelihood estimates of the parameters $\alpha$, $\beta$, and $\sigma$, we partially differentiate the likelihood function (or its logarithm, which is easier) with respect to $\alpha$, $\beta$, and $\sigma$, equate the expressions to zero, and then solve the resulting system of equations. Thus, differentiating

$$\ln L = -n \cdot \ln \sigma - \frac{n}{2} \cdot \ln 2\pi - \frac{1}{2\sigma^2} \cdot \sum_{i=1}^{n} [y_i - (\alpha + \beta x_i)]^2$$

partially with respect to $\alpha$, $\beta$, and $\sigma$, and equating the expressions which we obtain to zero, we get

$$\frac{\partial \ln L}{\partial \alpha} = \frac{1}{\sigma^2} \cdot \sum_{i=1}^{n} [y_i - (\alpha + \beta x_i)] = 0$$

$$\frac{\partial \ln L}{\partial \beta} = \frac{1}{\sigma^2} \cdot \sum_{i=1}^{n} x_i [y_i - (\alpha + \beta x_i)] = 0$$

$$\frac{\partial \ln L}{\partial \sigma} = -\frac{n}{\sigma} + \frac{1}{\sigma^3} \cdot \sum_{i=1}^{n} [y_i - (\alpha + \beta x_i)]^2 = 0$$

Since the first two equations are equivalent to the two normal equations on pages 503 and 504, the maximum likelihood estimates of $\alpha$ and $\beta$ are identical with the least squares estimate of Theorem 14.2. Also, if we substitute these estimates of $\alpha$ and $\beta$ into the equation obtained by equating $\dfrac{\partial \ln L}{\partial \sigma}$ to zero, it follows immediately that the maximum likelihood estimate of $\sigma$ is given by

$$\hat{\sigma} = \sqrt{\frac{1}{n} \cdot \sum_{i=1}^{n} [y_i - (\hat{\alpha} + \hat{\beta} x_i)]^2}$$

This can also be written as

$$\hat{\sigma} = \sqrt{\frac{1}{n} (S_{yy} - \hat{\beta} \cdot S_{xy})}$$

as the reader will be asked to verify in Exercise 14.25.

Having obtained maximum likelihood estimators of the regression coefficients, let us now investigate their use in testing hypotheses concerning $\alpha$ and $\beta$, and in constructing confidence intervals for these two parameters. Since problems concerning $\beta$ are usually of more immediate interest than problems concerning $\alpha$ ($\beta$ is the slope of the regression line, whereas $\alpha$ is merely the $y$-intercept; also, the null hypothesis $\beta = 0$ is equivalent to the null hypothesis $\rho = 0$), we shall discuss here some of the sampling theory relating to $\hat{B}$, where B is the capital Greek letter *beta*. Corresponding theory relating to $\hat{A}$, where A is the capital Greek letter *alpha*, will be treated in Exercises 14.28 and 14.30.

To study the sampling distribution of $\hat{B}$, let us write

$$\hat{B} = \frac{S_{xY}}{S_{xx}} = \frac{\sum_{i=1}^{n} (x_i - \bar{x})(Y_i - \bar{Y})}{S_{xx}}$$

$$= \sum_{i=1}^{n} \left( \frac{x_i - \bar{x}}{S_{xx}} \right) Y_i$$

which is seen to be a linear combination of the $n$ independent normal random variables $Y_i$. It follows from Exercise 7.58 that $\hat{B}$ itself has a normal distribution with the mean

$$E(\hat{B}) = \sum_{i=1}^{n} \left[ \frac{x_i - \bar{x}}{S_{xx}} \right] \cdot E(Y_i|x_i)$$

$$= \sum_{i=1}^{n} \left[ \frac{x_i - \bar{x}}{S_{xx}} \right] (\alpha + \beta x_i) = \beta$$

and the variance

$$\text{var}(\hat{B}) = \sum_{i=1}^{n} \left[ \frac{x_i - \bar{x}}{S_{xx}} \right]^2 \cdot \text{var}(Y_i|x_i)$$

$$= \sum_{i=1}^{n} \left[ \frac{x_i - \bar{x}}{S_{xx}} \right]^2 \cdot \sigma^2 = \frac{\sigma^2}{S_{xx}}$$

In order to apply this theory to test hypotheses about $\beta$ or construct confidence intervals for $\beta$, we shall have to use the following theorem:

---

**THEOREM 14.3** Under the assumptions of normal regression analysis, $\dfrac{n\hat{\sigma}^2}{\sigma^2}$ is a value of a random variable having the chi-square distribution with $n - 2$ degrees of freedom. Furthermore, this random variable and $\hat{B}$ are independent.

---

A proof of this theorem is referred to at the end of this chapter.

Making use of this theorem as well as the result proved earlier that $\hat{B}$ has a normal distribution with the mean $\beta$ and the variance $\dfrac{\sigma^2}{S_{xx}}$, we find that the definition of the $t$ distribution in Section 8.5 leads to

---

**THEOREM 14.4** Under the assumptions of normal regression analysis,

$$t = \frac{\dfrac{\hat{\beta} - \beta}{\sigma/\sqrt{S_{xx}}}}{\sqrt{\dfrac{n\hat{\sigma}^2}{\sigma^2} \Big/ (n - 2)}} = \frac{\hat{\beta} - \beta}{\hat{\sigma}} \sqrt{\frac{(n - 2)S_{xx}}{n}}$$

is a value of a random variable having the $t$ distribution with $n - 2$ degrees of freedom.

---

Based on this statistic, let us now test a hypothesis about the regression coefficient $\beta$.

## EXAMPLE 14.5

With reference to the data on page 501 pertaining to the amount of time that 10 persons studied for a certain test and the scores which they obtained, test the null hypothesis $\beta = 3$ against the alternative hypothesis $\beta > 3$ at the 0.01 level of significance.

### Solution

1. $H_0$:  $\beta = 3$
   $H_1$:  $\beta > 3$
   $\alpha = 0.01$

2. Reject the null hypothesis if $t \geqslant 2.896$, where $t$ is determined in accordance with Theorem 14.4 and 2.896 is the value of $t_{.01,8}$ obtained from Table IV.

3. Calculating $\Sigma\, y^2 = 36{,}562$ from the original data and copying the other quantities from page 505, we get

$$S_{yy} = 36{,}562 - \tfrac{1}{10}\,(564)^2 = 4{,}752.4$$

and

$$\hat{\sigma} = \sqrt{\tfrac{1}{10}\,[4{,}752.4 - (3.471)(1{,}305)]} = 4.720$$

so that

$$t = \frac{3.471 - 3}{4.720}\sqrt{\frac{8 \cdot 376}{10}} = 1.73$$

4. Since $t = 1.73$ is less than 2.896, the null hypothesis cannot be rejected; we cannot conclude that on the average an extra hour of study will increase the score by more than 3 points.    ▲

Letting $\hat{\Sigma}$ be the random variable whose values are $\hat{\sigma}$, we have

$$P\left(-t_{\alpha/2,n-2} < \frac{\hat{B} - \beta}{\hat{\Sigma}}\sqrt{\frac{(n-2)S_{xx}}{n}} < t_{\alpha/2,n-2}\right) = 1 - \alpha$$

according to Theorem 14.4. Writing this as

$$P\left[\hat{B} - t_{\alpha/2,n-2} \cdot \hat{\Sigma}\sqrt{\frac{n}{(n-2)S_{xx}}} < \beta < \hat{B} + t_{\alpha/2,n-2} \cdot \hat{\Sigma}\sqrt{\frac{n}{(n-2)S_{xx}}}\right] = 1 - \alpha$$

we arrive at the following confidence interval formula.

---

**THEOREM 14.5**  Under the assumptions of normal regression analysis,

$$\hat{\beta} - t_{\alpha/2,n-2} \cdot \hat{\sigma}\sqrt{\frac{n}{(n-2)S_{xx}}} < \beta < \hat{\beta} + t_{\alpha/2,n-2} \cdot \hat{\sigma}\sqrt{\frac{n}{(n-2)S_{xx}}}$$

is a $(1 - \alpha)100\%$ confidence interval for the parameter $\beta$.

---

## EXAMPLE 14.6

With reference to the same data as in Example 14.5, construct a 95% confidence interval for $\beta$.

### Solution

Copying the various quantities from pages 505 and 515 and substituting them together with $t_{.025,8} = 2.306$ into the confidence interval formula of Theorem 14.5, we get

$$3.471 - (2.306)(4.720)\sqrt{\frac{10}{8(376)}} < \beta < 3.471 + (2.306)(4.720)\sqrt{\frac{10}{8(376)}}$$

or

$$2.84 < \beta < 4.10 \quad \blacktriangle$$

Since most realistically complex regression problems require fairly extensive calculations, they are virtually always done nowadays by using appropriate computer software. A printout thus obtained for our illustration is shown in Figure 14.4; as can be seen, it provides not only the values of $\hat{\alpha}$ and $\hat{\beta}$ in the column headed "COEFFICIENT," but also estimates of the standard deviations of the sampling distributions of $\hat{A}$ and $\hat{B}$ in the column headed "ST. DEV. of COEF." Had we used this printout in Example 14.5, we could have written the value of the $t$ statistic directly as

```
MTB  > NAME C1 = 'X'
MTB  > NAME C2 = 'Y'
MTB  > SET C1
DATA > 4   9   10   14   4   7   12   22   1   17
MTB  > SET C2
DATA > 31  58  65  73  37  44  60  91  21  84
MTB  > REGR C2 1 C1

THE REGRESSION EQUATION IS
Y = 21.7 + 3.47 X

                                     ST. DEV.    T-RATIO =
COLUMN          COEFFICIENT          OF COEF.    COEF/S.D.
                21.693               3.194       6.79
X               3.4707               0.2723      12.74
```

**Figure 14.4**  Computer printout for Examples 14.4, 14.5, and 14.6.

$$t = \frac{3.471 - 3}{0.2723} = 1.73$$

and in Example 14.6 we could have written the confidence limits directly as 3.471 ± (2.306)(0.2723).

## EXERCISES

**14.25** Making use of the fact that $\hat{\alpha} = \bar{y} - \hat{\beta}\bar{x}$ and $\hat{\beta} = \dfrac{S_{xy}}{S_{xx}}$, show that

$$\sum_{i=1}^{n} [y_i - (\hat{\alpha} + \hat{\beta}x_i)]^2 = S_{yy} - \hat{\beta}S_{xy}$$

**14.26** Show that
  (a) $\hat{\Sigma}^2$, the random variable corresponding to $\hat{\sigma}^2$, is not an unbiased estimator of $\sigma^2$;
  (b) $S_e^2 = \dfrac{n \cdot \hat{\Sigma}^2}{n - 2}$ is an unbiased estimator of $\sigma^2$.
  The quantity $s_e$ is often referred to as the **standard error of estimate**.

**14.27** Using $s_e$ (see Exercise 14.26) instead of $\hat{\sigma}$, rewrite
  (a) the expression for $t$ in Theorem 14.4;
  (b) the confidence interval formula of Theorem 14.5.

**14.28** Under the assumptions of normal regression analysis, show that

(a)  the least squares estimate of $\alpha$ in Theorem 14.2 can be written in the form

$$\hat{\alpha} = \sum_{i=1}^{n} \left[ \frac{S_{xx} + n\bar{x}^2 - n\bar{x}x_i}{nS_{xx}} \right] y_i$$

(b)  $\hat{A}$ has a normal distribution with

$$E(\hat{A}) = \alpha \quad \text{and} \quad \text{var}(\hat{A}) = \frac{(S_{xx} + n\bar{x}^2)\sigma^2}{nS_{xx}}$$

**14.29** Use Theorem 4.15 to show that

$$\text{cov}(\hat{A}, \hat{B}) = -\frac{\bar{x}}{S_{xx}} \cdot \sigma^2$$

**14.30** Use the result of part (b) of Exercise 14.28 to show that

$$z = \frac{(\hat{\alpha} - \alpha)\sqrt{nS_{xx}}}{\sigma\sqrt{S_{xx} + n\bar{x}^2}}$$

is a value of a random variable having the standard normal distribution. Also, use the first part of Theorem 14.3 and the fact that $\hat{A}$ and $\dfrac{n\hat{\Sigma}^2}{\sigma^2}$ are independent to show that

$$t = \frac{(\hat{\alpha} - \alpha)\sqrt{(n-2)S_{xx}}}{\hat{\sigma}\sqrt{S_{xx} + n\bar{x}^2}}$$

is a value of a random variable having the $t$ distribution with $n - 2$ degrees of freedom.

**14.31** Use the results of Exercises 14.28 and 14.29, and the fact that $E(\hat{B}) = \beta$ and $\text{var}(\hat{B}) = \dfrac{\sigma^2}{S_{xx}}$, to show that $\hat{Y}_0 = \hat{A} + \hat{B}x_0$ is a random variable having a normal distribution with the mean

$$\alpha + \beta x_0 = \mu_{Y|x_0}$$

and the variance

$$\sigma^2 \left[ \frac{1}{n} + \frac{(x_0 - \bar{x})^2}{S_{xx}} \right]$$

Also, use the first part of Theorem 14.3 as well as the fact that $\hat{Y}_0$ and $\dfrac{n\hat{\Sigma}^2}{\sigma^2}$ are independent to show that

$$t = \frac{(\hat{y}_0 - \mu_{Y|x_0})\sqrt{n-2}}{\hat{\sigma}\sqrt{1 + \dfrac{n(x_0 - \bar{x})^2}{S_{xx}}}}$$

is a value of a random variable having the $t$ distribution with $n-2$ degrees of freedom.

**14.32** Derive a $(1 - \alpha)100\%$ confidence interval for $\mu_{Y|x_0}$, the mean of $Y$ at $x = x_0$, by solving the double inequality $-t_{\alpha/2,n-2} < t < t_{\alpha/2,n-2}$ with $t$ given by the formula of Exercise 14.31.

**14.33** Use the results of Exercises 14.28 and 14.29 and the fact that $E(\hat{B}) = \beta$ and $\mathrm{var}(\hat{B}) = \dfrac{\sigma^2}{S_{xx}}$ to show that $Y_0 - (\hat{A} + \hat{B}x_0)$ is a random variable having a normal distribution with zero mean and the variance

$$\sigma^2\left[1 + \frac{1}{n} + \frac{(x_0 - \bar{x})^2}{S_{xx}}\right]$$

Here $Y_0$ has a normal distribution with the mean $\alpha + \beta x_0$ and the variance $\sigma^2$; that is, $Y_0$ is a future observation of $Y$ corresponding to $x = x_0$. Also, use the first part of Theorem 14.3 as well as the fact that $Y_0 - (\hat{A} + \hat{B}x_0)$ and $\dfrac{n\hat{\Sigma}^2}{\sigma^2}$ are independent to show that

$$t = \frac{[y_0 - (\hat{\alpha} + \hat{\beta}x_0)]\sqrt{n-2}}{\hat{\sigma}\sqrt{1 + n + \dfrac{n(x_0 - \bar{x})^2}{S_{xx}}}}$$

is a value of a random variable having the $t$ distribution with $n-2$ degrees of freedom.

**14.34** Solve the double inequality $-t_{\alpha/2,n-2} < t < t_{\alpha/2,n-2}$ with $t$ given by the formula of Exercise 14.33, so that the middle term is $y_0$ and the two limits can be calculated without knowledge of $y_0$. Note that although the resulting double inequality may be interpreted like a confidence interval, it is not designed to estimate a parameter; instead, it provides **limits of prediction** for a future observation of $Y$ which corresponds to the (given or observed) value $x_0$.

## APPLICATIONS

**14.35** With reference to Exercise 14.16, test the null hypothesis $\beta = 1.25$ against the alternative hypothesis $\beta > 1.25$ at the 0.01 level of significance.

**14.36** With reference to Exercise 14.18, test the null hypothesis $\beta = 0.350$ against the alternative hypothesis $\beta < 0.350$ at the 0.05 level of significance.

**14.37** The following table shows the assessed values and the selling prices of eight houses, constituting a random sample of all the houses sold recently in a metropolitan area:

| Assessed value (thousands of dollars) | Selling price (thousands of dollars) |
| --- | --- |
| 70.3 | 114.4 |
| 102.0 | 169.3 |
| 62.5 | 106.2 |
| 74.8 | 125.0 |
| 57.9 | 99.8 |
| 81.6 | 132.1 |
| 110.4 | 174.2 |
| 88.0 | 143.5 |

(a) Fit a least squares line which will enable us to predict the selling price of a house in that metropolitan area in terms of its assessed value.

(b) Test the null hypothesis $\beta = 1.30$ against the alternative hypothesis $\beta > 1.30$ at the 0.05 level of significance.

**14.38** With reference to Exercise 14.17, construct a 99% confidence interval for the regression coefficient $\beta$.

**14.39** With reference to Exercise 14.19, construct a 98% confidence interval for the regression coefficient $\beta$.

**14.40** With reference to Example 14.4, use the theory of Exercise 14.30 to test the null hypothesis $\alpha = 21.50$ against the alternative hypothesis $\alpha \neq 21.50$ at the 0.01 level of significance.

**14.41** The following data show the advertising expenses (expressed as a percentage of total expenses) and the net operating profits (expressed as a percentage of total sales) in a random sample of six drugstores:

| Advertising expenses | Net operating profits |
| --- | --- |
| 1.5 | 3.6 |
| 1.0 | 2.8 |
| 2.8 | 5.4 |
| 0.4 | 1.9 |
| 1.3 | 2.9 |
| 2.0 | 4.3 |

(a)   Fit a least squares line which will enable us to predict net operating profits in terms of advertising expenses.

(b)   Test the null hypothesis $\alpha = 0.8$ against the alternative hypothesis $\alpha > 0.8$ at the 0.01 level of significance.

**14.42** With reference to Exercise 14.16, use the theory of Exercise 14.30 to construct a 95% confidence interval for $\alpha$.

**14.43** With reference to Exercise 14.17, use the theory of Exercise 14.30 to construct a 99% confidence interval for $\alpha$.

**14.44** Use the theory of Exercises 14.32 and 14.34, as well as the quantities already calculated in Examples 14.4 and 14.5, to construct

(a)   a 95% confidence interval for the mean test score of persons who have studied 14 hours for the test;

(b)   95% limits of prediction for the test score of a person who has studied 14 hours for the test.

**14.45** Use the theory of Exercises 14.32 and 14.34, as well as the quantities already calculated in Exercise 14.35 for the data of Exercise 14.16, to find

(a)   a 99% confidence interval for the expected number of deaths in a group of 25 mice when the dosage is 9 milligrams;

(b)   99% limits of prediction of the number of deaths in a group of 25 mice when the dosage is 9 milligrams.

## 14.5  NORMAL CORRELATION ANALYSIS

In normal correlation analysis we analyze a set of paired data $\{(x_i, y_i); i = 1, 2, \ldots, n\}$, where the $x_i$'s and $y_i$'s are values of a random sample from a bivariate normal population with the parameters $\mu_1$, $\mu_2$, $\sigma_1$, $\sigma_2$, and $\rho$. To estimate these parameters by the method of maximum likelihood, we shall have to maximize the likelihood

$$L = \prod_{i=1}^{n} f(x_i, y_i)$$

where $f(x_i, y_i)$ is given by Definition 6.8, and to this end we shall have to differentiate $L$, or $\ln L$, partially with respect to $\mu_1$, $\mu_2$, $\sigma_1$, $\sigma_2$, and $\rho$, equate the resulting expressions to zero, and then solve the resulting system of equations for the five parameters. Leaving the details to the reader, let us merely state that when $\dfrac{\partial \ln L}{\partial \mu_1}$ and $\dfrac{\partial \ln L}{\partial \mu_2}$ are equated to zero, we get

$$-\frac{\sum_{i=1}^{n}(x_i - \mu_1)}{\sigma_1^2} + \frac{\rho \sum_{i=1}^{n}(y_i - \mu_2)}{\sigma_1 \sigma_2} = 0$$

and

$$-\frac{\rho \sum\limits_{i=1}^{n} (x_i - \mu_1)}{\sigma_1 \sigma_2} + \frac{\sum\limits_{i=1}^{n} (y_i - \mu_2)}{\sigma_2^2} = 0$$

Solving these two equations for $\mu_1$ and $\mu_2$, we find that the maximum likelihood estimates of these two parameters are

$$\hat{\mu}_1 = \bar{x} \quad \text{and} \quad \hat{\mu}_2 = \bar{y}$$

namely, the respective sample means. Subsequently, equating $\dfrac{\partial \ln L}{\partial \sigma_1}$, $\dfrac{\partial \ln L}{\partial \sigma_2}$, and $\dfrac{\partial \ln L}{\partial \rho}$ to zero, and substituting $\bar{x}$ and $\bar{y}$ for $\mu_1$ and $\mu_2$, we obtain a system of equations whose solution is

$$\hat{\sigma}_1 = \sqrt{\frac{\sum\limits_{i=1}^{n} (x_i - \bar{x})^2}{n}}, \quad \hat{\sigma}_2 = \sqrt{\frac{\sum\limits_{i=1}^{n} (y_i - \bar{y})^2}{n}}$$

$$\hat{\rho} = \frac{\sum\limits_{i=1}^{n} (x_i - \bar{x})(y_i - \bar{y})}{\sqrt{\sum\limits_{i=1}^{n} (x_i - \bar{x})^2}\sqrt{\sum\limits_{i=1}^{n} (y_i - \bar{y})^2}}$$

(A detailed derivation of these maximum likelihood estimates is referred to at the end of this chapter.) It is of interest to note that the maximum likelihood estimates of $\sigma_1$ and $\sigma_2$ are identical with the one obtained on page 385 for the standard deviation of the univariate normal distribution; they differ from the respective sample standard deviations $s_1$ and $s_2$ only by the factor $\sqrt{\dfrac{n-1}{n}}$.

The estimate $\hat{\rho}$, called the **sample correlation coefficient**, is usually denoted by the letter $r$, and its calculation is facilitated by using the following alternative, but equivalent, computing formula.

---

**THEOREM 14.6**  If $\{(x_i, y_i); i = 1, 2, \ldots, n\}$ are the values of a random sample from a bivariate population, then

$$r = \frac{S_{xy}}{\sqrt{S_{xx} \cdot S_{yy}}}$$

---

Since $\rho$ measures the strength of the linear relationship between $X$ and $Y$, there are many problems in which the estimation of $\rho$ and tests concerning $\rho$ are of special interest. When $\rho = 0$, the two random variables are uncorrelated, and as we have already seen, in the case of the bivariate normal distribution this means that they are also independent. When $\rho$ equals $+1$ or $-1$, it follows from the relationship

$$\sigma^2_{Y|x} = \sigma^2 = \sigma^2_2(1 - \rho^2)$$

established in Theorem 6.9, that $\sigma = 0$, and this means that there is a perfect linear relationship between $X$ and $Y$. Using the invariance property of maximum likelihood estimators, we can write

$$\hat{\sigma}^2 = \hat{\sigma}^2_2(1 - r^2)$$

which not only provides an alternative computing formula for finding $\hat{\sigma}^2$, but also serves to tie together the concepts of regression and correlation. From this formula for $\hat{\sigma}^2$ it is clear that when $\hat{\sigma}^2 = 0$, namely, when the set of data points $\{(x_i, y_i);$ $i = 1, 2, \ldots , n\}$ fall on a straight line, then $r$ will equal $+1$ or $-1$, depending on whether the line has an upward or downward slope. In order to interpret values of $r$ between 0 and $+1$ or 0 and $-1$, we solve the preceding equation for $r^2$ and multiply by 100, getting

$$100r^2 = \frac{\hat{\sigma}^2_2 - \hat{\sigma}^2}{\hat{\sigma}^2_2} \cdot 100$$

where $\hat{\sigma}^2_2$ measures the total variation of the $y$'s, $\hat{\sigma}^2$ measures the conditional variation of the $y$'s for fixed values of $x$, and, hence, $\hat{\sigma}^2_2 - \hat{\sigma}^2$ measures that part of the total variation of the $y$'s which is accounted for by the relationship with $x$. *Thus, $100r^2$ is the percentage of the total variation of the $y$'s which is accounted for by the relationship with $x$.* For instance, when $r = 0.5$ then 25 percent of the variation of the $y$'s is accounted for by the relationship with $x$, when $r = 0.7$ then 49 percent of the variation of the $y$'s is accounted for by the relationship with $x$, and we might thus say that a correlation of $r = 0.7$ is almost "twice as strong" as a correlation of $r = 0.5$. Similarly, we might say that a correlation of $r = 0.6$ is "nine times as strong" as a correlation of $r = 0.2$.

# EXAMPLE 14.7

Suppose that we want to determine on the basis of the following data whether there is a relationshp between the time, in minutes, it takes a secretary to complete a certain form in the morning and in the late afternoon:

| Morning $x$ | Afternoon $y$ |
|---|---|
| 8.2 | 8.7 |
| 9.6 | 9.6 |
| 7.0 | 6.9 |
| 9.4 | 8.5 |
| 10.9 | 11.3 |
| 7.1 | 7.6 |
| 9.0 | 9.2 |
| 6.6 | 6.3 |
| 8.4 | 8.4 |
| 10.5 | 12.3 |

Compute and interpret the sample correlation coefficient.

*Solution*

From the data we get $n = 10$, $\Sigma x = 86.7$, $\Sigma x^2 = 771.35$, $\Sigma y = 88.8$, $\Sigma y^2 = 819.34$, and $\Sigma xy = 792.92$, so that

$$S_{xx} = 771.35 - \tfrac{1}{10}(86.7)^2 = 19.661$$
$$S_{yy} = 819.34 - \tfrac{1}{10}(88.8)^2 = 30.796$$
$$S_{xy} = 792.92 - \tfrac{1}{10}(86.7)(88.8) = 23.024$$

and

$$r = \frac{23.024}{\sqrt{(19.661)(30.796)}} = 0.936$$

This is indicative of a positive association between the time it takes a secretary to perform the given task in the morning and in the late afternoon, and this is also apparent from the **scattergram** of Figure 14.5. Since $100r^2 = 100(0.936)^2 = 87.6$, we can say that almost 88 percent of the variation of the $y$'s is accounted for by a linear relationship with $x$. ▲

Since the sampling distribution of $R$ for random samples from bivariate normal populations is rather complicated, it is common practice to base confidence intervals for $\rho$ and tests concerning $\rho$ on the statistic

$$\frac{1}{2} \cdot \ln \frac{1+R}{1-R}$$

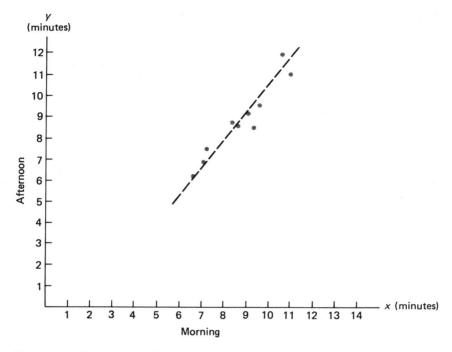

**Figure 14.5**  Scattergram of data of Example 14.7.

whose distribution is approximately normal with the mean $\frac{1}{2} \cdot \ln \frac{1 + \rho}{1 - \rho}$ and the variance $\frac{1}{n - 3}$. Thus,

$$
z = \frac{\frac{1}{2} \cdot \ln \dfrac{1 + r}{1 - r} - \frac{1}{2} \cdot \ln \dfrac{1 + \rho}{1 - \rho}}{\dfrac{1}{\sqrt{n - 3}}}
$$

$$
= \frac{\sqrt{n - 3}}{2} \cdot \ln \frac{(1 + r)(1 - \rho)}{(1 - r)(1 + \rho)}
$$

can be looked upon as a value of a random variable having approximately the standard normal distribution. Using this approximation, we can test the null hypothesis $\rho = \rho_0$ against an appropriate alternative, as illustrated in Example 14.8, or calculate confidence intervals for $\rho$ by the method suggested in Exercise 14.49.

## EXAMPLE 14.8

With reference to Example 14.7, test the null hypothesis $\rho = 0$ against the alternative hypothesis $\rho \neq 0$ at the 0.01 level of significance.

### Solution

1.  $H_0$:  $\rho = 0$
    $H_1$:  $\rho \neq 0$
    $\alpha = 0.01$

2.  Reject the null hypothesis if $z \leq -2.575$ or $z \geq 2.575$, where

$$z = \frac{\sqrt{n-3}}{2} \cdot \ln \frac{1+r}{1-r}$$

3.  Substituting $n = 10$ and $r = 0.936$. we get

$$z = \frac{\sqrt{7}}{2} \cdot \ln \frac{1.936}{0.064} = 4.5$$

4.  Since $z = 4.5$ exceeds 2.575, the null hypothesis must be rejected; we conclude that there is a relationship between the time it takes a secretary to complete the form in the morning and in the late afternoon.   ▲

### EXERCISES

**14.46** Verify that the formula for $t$ of Theorem 14.4 can be written as

$$t = \left(1 - \frac{\beta}{\hat{\beta}}\right) \frac{r\sqrt{n-2}}{\sqrt{1-r^2}}$$

**14.47** Use the formula for $t$ of the preceding exercise to derive the following $(1 - \alpha)100\%$ confidence limits for $\beta$:

$$\hat{\beta}\left[1 \pm t_{\alpha/2,n-2} \cdot \frac{\sqrt{1-r^2}}{r\sqrt{n-2}}\right]$$

**14.48** Use the formula for $t$ of Exercise 14.46 to show that if the assumptions underlying normal regression analysis are met and $\beta = 0$, then $R^2$ has a beta distribution with the mean $\dfrac{1}{n-1}$.

**14.49** By solving the double inequality $-z_{\alpha/2} \leq z \leq z_{\alpha/2}$ (with $z$ given by the formula on page 525) for $\rho$, derive a $(1 - \alpha)100\%$ confidence interval formula for $\rho$.

**14.50** In a random sample of $n$ pairs of values of $X$ and $Y$, $(x_i, y_j)$ occurs $f_{ij}$ times for $i = 1, 2, \ldots, r$ and $j = 1, 2, \ldots, c$. Letting $f_{i\cdot}$ denote the number of pairs where $X$ takes on the value $x_i$ and $f_{\cdot j}$ the number of pairs where $Y$ takes on the value $y_j$, write a formula for the coefficient of correlation.

## APPLICATIONS

**14.51** An achievement test is said to be reliable if a student who takes the test several times will consistently get high (or low) scores. One way of checking the reliability of a test is to divide it into two parts, usually the even-numbered problems and the odd-numbered problems, and observe the correlation between the scores which students get in both halves of the test. Thus, the following data represent the grades, $x$ and $y$, which 20 students obtained for the even-numbered problems and the odd-numbered problems of a new objective test designed to test eighth grade achievement in general science:

| $x$ | $y$ | $x$ | $y$ |
|---|---|---|---|
| 27 | 29 | 33 | 42 |
| 36 | 44 | 39 | 31 |
| 44 | 49 | 38 | 38 |
| 32 | 27 | 24 | 22 |
| 27 | 35 | 33 | 34 |
| 41 | 33 | 32 | 37 |
| 38 | 29 | 37 | 38 |
| 44 | 40 | 33 | 35 |
| 30 | 27 | 34 | 32 |
| 27 | 38 | 39 | 43 |

Calculate $r$ for these data and test its significance, namely, the null hypothesis $\rho = 0$ against the alternative hypothesis $\rho \neq 0$ at the 0.05 level of significance.

**14.52** With reference to Exercise 14.51, use the formula obtained in Exercise 14.49 to construct a 95% confidence interval for $\rho$.

**14.53** The following data pertain to $x$, the amount of fertilizer (in pounds) which a farmer applies to his soil, and $y$, his yield of wheat (in bushels per acre):

| x | y | x | y | x | y |
|---|---|---|---|---|---|
| *3696* 112 | 33 | *2112* 88 | 24 | *999* 37 | 27 |
| *2576* 92 | 28 | *748* 44 | 17 | *207* 23 | 9 |
| *2736* 72 | 38 | *4757* 132 | 36 | *2464* 77 | 32 |
| *1122* 66 | 17 | *322* 23 | 14 | *5396* 142 | 38 |
| *3920* 112 | 35 | *1425* 57 | 25 | *481* 37 | 13 |
| *2728* 88 | 31 | *4440* 111 | 40 | *2921* 127 | 23 |
| *336* 42 | 8 | *2001* 69 | 29 | *2728* 88 | 31 |
| *4662* 126 | 37 | *228* 19 | 12 | *1776* 48 | 37 |
| *2304* 72 | 32 | *2781* 103 | 27 | *1525* 61 | 25 |
| *1040* 52 | 20 | *5640* 141 | 40 | *994* 71 | 14 |
| *476* 28 | 17 | *2002* 77 | 26 | *2938* 113 | 26 |

Assuming that the data can be looked upon as a random sample from a bivariate normal population, calculate $r$ and test its significance at the 0.01 level of significance. Also, draw a scattergram of these paired data and judge whether the assumption seems reasonable.

**14.54** With reference to Exercise 14.53, use the formula obtained in Exercise 14.49 to construct a 99% confidence interval for $\rho$.

**14.55** Use the formula of Exercise 14.46 to calculate a 95% confidence interval for $\beta$ for the numbers of hours studied and the test scores on page 501, and compare this interval with the one obtained in Example 14.6.

**14.56** The calculation of $r$ can often be simplified by adding the same constant to each $x$, adding the same constant to each $y$, or by multiplying each $x$ and/or $y$ by the same positive constants. Recalculate $r$ for the data of Example 14.7 by first multiplying each $x$ and each $y$ by 10, and then subtracting 70 from each $x$ and 60 from each $y$.

**14.57** The following table shows how the history and economics scores of 25 students are distributed:

| | *History scores* | | | | |
| | *21–25* | *26–30* | *31–35* | *36–40* | *41–45* |
|---|---|---|---|---|---|
| *21–25* | 1 | | | | |
| *26–30* | | 3 | 1 | | |
| *31–35* | | 2 | 5 | 2 | |
| *36–40* | | | 1 | 4 | 1 |
| *41–45* | | | 1 | 3 | |
| *46–50* | | | | | 1 |

*Economics scores* (row label, vertical)

Use the method of Exercise 14.50 to determine the value of $r$, replacing the row headings by the corresponding **class marks** (midpoints) 23, 28, 33, 38, 43, and 48 and the column headings by the corresponding class marks 23, 28, 33, 38, and 43. Use this value of $r$ to test at the 0.05 level of significance whether there is a relationship between scores in the two subjects.

**14.58** Rework Exercise 14.57, coding the class marks of the history scores $-2$, $-1$, 0, 1, and 2, and the class marks of the economics scores $-2$, $-1$, 0, 1, 2, and 3. (It follows from Exercise 14.56 that this kind of coding will not affect the value of $r$.)

**14.59** If the row categories as well as the column categories of an $r \times c$ table are ordered, we can replace the row headings and also the column headings by consecutive integers and then calculate $r$ with the formula obtained in Exercise 14.50. Use this method to rework Example 13.11, replacing "Low," "Average," and "High" in each case by $-1$, 0, and 1.

**14.60** With reference to the $r \times c$ table on page 484, use the method suggested in the preceding exercise to test at the 0.05 level of significance whether there is a relationship between IQ and on-the-job performance. Replace the row headings as well as the column headings by $-1$, 0, and 1.

## 14.6 MULTIPLE LINEAR REGRESSION

Although there are many problems in which one variable can be predicted quite accurately in terms of another, it stands to reason that predictions should improve if one considers additional relevant information. For instance, we should be able to make better predictions of the performance of newly hired teachers if we consider not only their education, but also their years of experience and their personality. Also, we should be able to make better predictions of a new textbook's success if we consider not only the quality of the work, but also the potential demand and the competition.

Although there are many different formulas that can be used to express regression relationships among more than two variables (see, for instance, Example 14.3), most widely used are linear equations of the form

$$\mu_{Y|x_1, x_2, \ldots, x_k} = \beta_0 + \beta_1 x_1 + \beta_2 x_2 + \cdots + \beta_k x_k$$

This is partly a matter of mathematical convenience and partly due to the fact that many relationships are actually of this form or can be approximated closely by linear equations.

In the equation above, $Y$ is the random variable whose values we want to predict in terms of given values of $x_1$, $x_2$, $\ldots$, and $x_k$, and $\beta_0$, $\beta_1$, $\beta_2$, $\ldots$, and $\beta_k$, the **multiple regression coefficients**, are numerical constants that must be determined from observed data.

To illustrate, consider the following equation, which was obtained in a study of the demand for different meats:

$$\hat{y} = 3.489 - 0.090x_1 + 0.064x_2 + 0.019x_3$$

Here $\hat{y}$ denotes the estimated consumption of federally inspected beef and veal in millions of pounds, $x_1$ denotes a composite retail price of beef in cents per pound, $x_2$ denotes a composite retail price of pork in cents per pound, and $x_3$ denotes income as measured by a certain payroll index.

As in Section 14.3, where there was only one independent variable $x$, multiple regression coefficients are usually estimated by the method of least squares. For $n$ data points

$$\{(x_{i1}, x_{i2}, \ldots, x_{ik}, y_i); \; i = 1, 2, \ldots, n\}$$

the least squares estimates of the $\beta$'s are the values $\hat{\beta}_0$, $\hat{\beta}_1$, $\hat{\beta}_2$, $\ldots$ , and $\hat{\beta}_k$ for which the quantity

$$q = \sum_{i=1}^{n} [y_i - (\hat{\beta}_0 + \hat{\beta}_1 x_{i1} + \hat{\beta}_2 x_{i2} + \cdots + \hat{\beta}_k x_{ik})]^2$$

is a minimum. In this notation, $x_{i1}$ is the $i$th value of the variable $x_1$, $x_{i2}$ is the $i$th value of the variable $x_2$, and so on.

So, we differentiate partially with respect to the $\hat{\beta}$'s, and equating these partial derivatives to zero, we get

$$\frac{\partial q}{\partial \hat{\beta}_0} = \sum_{i=1}^{n} (-2)[y_i - (\hat{\beta}_0 + \hat{\beta}_1 x_{i1} + \hat{\beta}_2 x_{i2} + \cdots + \hat{\beta}_k x_{ik})] = 0$$

$$\frac{\partial q}{\partial \hat{\beta}_1} = \sum_{i=1}^{n} (-2)x_{i1}[y_i - (\hat{\beta}_0 + \hat{\beta}_1 x_{i1} + \hat{\beta}_2 x_{i2} + \cdots + \hat{\beta}_k x_{ik})] = 0$$

$$\frac{\partial q}{\partial \hat{\beta}_2} = \sum_{i=1}^{n} (-2)x_{i2}[y_i - (\hat{\beta}_0 + \hat{\beta}_1 x_{i1} + \hat{\beta}_2 x_{i2} + \cdots + \hat{\beta}_k x_{ik})] = 0$$

$$\cdot \quad \cdot \quad \cdot$$

$$\frac{\partial q}{\partial \hat{\beta}_k} = \sum_{i=1}^{n} (-2)x_{ik}[y_i - (\hat{\beta}_0 + \hat{\beta}_1 x_{i1} + \hat{\beta}_2 x_{i2} + \cdots + \hat{\beta}_k x_{ik})] = 0$$

and finally the $k + 1$ normal equations

$$\Sigma y = \hat{\beta}_0 \cdot n \quad + \hat{\beta}_1 \cdot \Sigma x_1 \quad + \hat{\beta}_2 \cdot \Sigma x_2 \quad + \cdots + \hat{\beta}_k \cdot \Sigma x_k$$

$$\Sigma x_1 y = \hat{\beta}_0 \cdot \Sigma x_1 + \hat{\beta}_1 \cdot \Sigma x_1^2 \quad + \hat{\beta}_2 \cdot \Sigma x_1 x_2 + \cdots + \hat{\beta}_k \cdot \Sigma x_1 x_k$$

$$\Sigma x_2 y = \hat{\beta}_0 \cdot \Sigma x_2 + \hat{\beta}_1 \cdot \Sigma x_2 x_1 + \hat{\beta}_2 \cdot \Sigma x_2^2 \quad + \cdots + \hat{\beta}_k \cdot \Sigma x_2 x_k$$

$$\cdot \quad \cdot \quad \cdot$$

$$\Sigma x_k y = \hat{\beta}_0 \cdot \Sigma x_k + \hat{\beta}_1 \cdot \Sigma x_k x_1 + \hat{\beta}_2 \cdot \Sigma x_k x_2 + \cdots + \hat{\beta}_k \cdot \Sigma x_k^2$$

Here we abbreviated our notation by writing $\sum x_{i1}$ as $\Sigma x_1$, $\sum_{i=1}^{n} x_{i1} x_{i2}$ as $\Sigma x_1 x_2$, and so on.

## EXAMPLE 14.9

The following data show the number of bedrooms, the number of baths, and the prices at which a random sample of eight one-family houses sold recently in a certain large housing development:

| Number of bedrooms $x_1$ | Number of baths $x_2$ | Price (dollars) $y$ |
|---|---|---|
| 3 | 2 | 78,800 |
| 2 | 1 | 74,300 |
| 4 | 3 | 83,800 |
| 2 | 1 | 74,200 |
| 3 | 2 | 79,700 |
| 2 | 2 | 74,900 |
| 5 | 3 | 88,400 |
| 4 | 2 | 82,900 |

Use the method of least squares to find a linear equation which will enable us to predict the average sales price of a one-family house in the given housing development in terms of the number of bedrooms and the number of baths.

## Solution

The quantities we need for substitution into the three normal equations are $n = 8$, $\Sigma x_1 = 25$, $\Sigma x_2 = 16$, $\Sigma y = 637,000$, $\Sigma x_1^2 = 87$, $\Sigma x_1 x_2 = 55$, $\Sigma x_2^2 = 36$, $\Sigma x_1 y = 2,031,100$, and $\Sigma x_2 y = 1,297,700$, and we get

$$637{,}000 = 8\hat{\beta}_0 + 25\hat{\beta}_1 + 16\hat{\beta}_2$$

$$2{,}031{,}100 = 25\hat{\beta}_0 + 87\hat{\beta}_1 + 55\hat{\beta}_2$$

$$1{,}297{,}700 = 16\hat{\beta}_0 + 55\hat{\beta}_1 + 36\hat{\beta}_2$$

We could solve these equations by the method of elimination or by using determinants, but in view of the rather tedious calculations, such work is usually left to computers. Thus, let us refer to the printout of Figure 14.6, which shows in the column headed "COEFFICIENT" that $\hat{\beta}_0 = 65{,}191.7$, $\hat{\beta}_1 = 4{,}133.3$, and $\hat{\beta}_2 = 758.3$. After rounding, the least squares equation becomes

$$\hat{y} = 65{,}192 + 4{,}133x_1 + 758x_2$$

and this tells us that (in the given housing development and at the time the study was made) each extra bedroom adds on the average $4,133, and each bath $758, to the sales price of a house.    ▲

**EXAMPLE 14.10**

Based on the result obtained in Example 14.9, predict the sales price of a three-bedroom house with two baths in the large housing development.

```
MTB  > SET C1
DATA > 3  2  4  2  3  2  5  4
MTB  > SET C2
DATA > 2  1  3  1  2  2  3  2
MTB  > SET C3
DATA > 78800  74300  83800  74200  79700  74900  88400  82900
MTB  > REGR C3 2 C1 C2

THE REGRESSION EQUATION IS
C3 = 65192 + 4133 C1 + 758 C2

                                   ST. DEV.      T-RATIO =
     COLUMN      COEFFICIENT       OF COEF.      COEF/S.D.
                   65191.7           418.0         155.96
     C1             4133.3           228.6          18.08
     C2              758.3           340.5           2.23

     S = 370.4
```

**Figure 14.6**  Computer printout for Example 14.9.

*Solution*

Substituting $x_1 = 3$ and $x_2 = 2$ into the equation obtained above, we get

$$\hat{y} = 65{,}192 + 4{,}133(3) + 758(2)$$
$$= \$79{,}107$$

or approximately $\$79{,}100$.  ▲

Printouts like those of Figure 14.6 also provide information that is needed to make inferences about the multiple regression coefficients and to judge the merits of estimates or predictions based on the least squares equations. This corresponds to the work of Section 14.4, but we shall defer it until Section 14.7, where we shall study the whole problem of multiple linear regression in a much more compact notation.

## 14.7 MULTIPLE LINEAR REGRESSION (MATRIX NOTATION)[†]

The model we are using in multiple linear regression lends itself uniquely to a unified treatment in matrix notation. This notation makes it possible to state general results in compact form and to utilize many results of matrix theory to great advantage. As is customary, we shall denote matrices by capital letters in boldface type.

We could introduce the matrix approach by expressing the sum of squares $q$ (which we minimized in the preceding section by differentiating partially with respect to the $\hat{\beta}$'s) in matrix notation and take it from there, but leaving this to the reader in Exercise 14.61, let us begin here with the normal equations on page 531.

To express the normal equations in matrix notation, let us define the following three matrices:

$$\mathbf{X} = \begin{pmatrix} 1 & x_{11} & x_{12} & \cdots & x_{1k} \\ 1 & x_{21} & x_{22} & \cdots & x_{2k} \\ \cdot & \cdot & \cdot & \cdot & \cdot \\ 1 & x_{n1} & x_{n2} & \cdots & x_{nk} \end{pmatrix},$$

---

[†]It is assumed for this section that the reader is familiar with the material ordinarily covered in a first course on matrix algebra. Since matrix notation is not used elsewhere in this book, this section may be omitted without loss of continuity.

$$\mathbf{Y} = \begin{pmatrix} y_1 \\ y_2 \\ \vdots \\ y_n \end{pmatrix}, \quad \text{and} \quad \mathbf{B} = \begin{pmatrix} \hat{\beta}_0 \\ \hat{\beta}_1 \\ \vdots \\ \hat{\beta}_k \end{pmatrix}$$

The first one, $\mathbf{X}$, is an $n \times (k + 1)$ matrix consisting essentially of the given values of the $x$'s, with the column of 1's appended to accommodate the constant terms. $\mathbf{Y}$ is an $n \times 1$ matrix (or column vector) consisting of the observed values of $Y$, and $\mathbf{B}$ is a $(k + 1) \times 1$ matrix (or column vector) consisting of the least squares estimates of the regression coefficients.

Using these matrices, we can now write the following symbolic solution of the normal equations on page 531.

---

**THEOREM 14.7** The least squares estimates of the multiple regression coefficients are given by

$$\mathbf{B} = (\mathbf{X}'\mathbf{X})^{-1}\mathbf{X}'\mathbf{Y}$$

where $\mathbf{X}'$ is the transpose of $\mathbf{X}$ and $(\mathbf{X}'\mathbf{X})^{-1}$ is the inverse of $\mathbf{X}'\mathbf{X}$.

---

**Proof.** First we determine $\mathbf{X}'\mathbf{X}$, $\mathbf{X}'\mathbf{XB}$, and $\mathbf{X}'\mathbf{Y}$, getting

$$\mathbf{X}'\mathbf{X} = \begin{pmatrix} n & \Sigma x_1 & \Sigma x_2 & \cdots & \Sigma x_k \\ \Sigma x_1 & \Sigma x_1^2 & \Sigma x_1 x_2 & \cdots & \Sigma x_1 x_k \\ \Sigma x_2 & \Sigma x_2 x_1 & \Sigma x_2^2 & \cdots & \Sigma x_2 x_k \\ & & \cdot \quad \cdot \quad \cdot & \\ \Sigma x_k & \Sigma x_k x_1 & \Sigma x_k x_2 & \cdots & \Sigma x_k^2 \end{pmatrix}$$

$$\mathbf{X}'\mathbf{XB} = \begin{pmatrix} \hat{\beta}_0 \cdot n & + \hat{\beta}_1 \cdot \Sigma x_1 & + \hat{\beta}_2 \cdot \Sigma x_2 & + \cdots + \hat{\beta}_k \cdot \Sigma x_k \\ \hat{\beta}_0 \cdot \Sigma x_1 & + \hat{\beta}_1 \cdot \Sigma x_1^2 & + \hat{\beta}_2 \cdot \Sigma x_1 x_2 & + \cdots + \hat{\beta}_k \cdot \Sigma x_1 x_k \\ \hat{\beta}_0 \cdot \Sigma x_2 & + \hat{\beta}_1 \cdot \Sigma x_2 x_1 & + \hat{\beta}_2 \cdot \Sigma x_2^2 & + \cdots + \hat{\beta}_k \cdot \Sigma x_2 x_k \\ & & \cdot \quad \cdot \quad \cdot & \\ \hat{\beta}_0 \cdot \Sigma x_k & + \hat{\beta}_1 \cdot \Sigma x_k x_1 & + \hat{\beta}_2 \cdot \Sigma x_k x_2 & + \cdots + \hat{\beta}_k \cdot \Sigma x_k^2 \end{pmatrix}$$

$$\mathbf{X}'\mathbf{Y} = \begin{pmatrix} \Sigma y \\ \Sigma x_1 y \\ \Sigma x_2 y \\ \cdots \\ \Sigma x_k y \end{pmatrix}$$

Identifying the elements of $\mathbf{X'XB}$ as the expressions on the right-hand side of the normal equations on page 531 and those of $\mathbf{X'Y}$ as the expressions on the left-hand side, we can write

$$\mathbf{X'XB} = \mathbf{X'Y}$$

Multiplying on the left by $(\mathbf{X'X})^{-1}$, we get

$$(\mathbf{X'X})^{-1}\mathbf{X'XB} = (\mathbf{X'X})^{-1}\mathbf{X'Y}$$

and finally

$$\mathbf{B} = (\mathbf{X'X})^{-1}\mathbf{X'Y}$$

since $(\mathbf{X'X})^{-1}\mathbf{X'X}$ equals the $(k + 1) \times (k + 1)$ identity matrix $\mathbf{I}$ and by definition $\mathbf{IB} = \mathbf{B}$. We have assumed here that $\mathbf{X'X}$ is nonsingular, so that its inverse exists.    ▼

## EXAMPLE 14.11

With reference to Example 14.9, use Theorem 14.7 to determine the least squares estimates of the multiple regression coefficients.

### Solution

Substituting $\Sigma x_1 = 25$, $\Sigma x_2 = 16$, $\Sigma x_1^2 = 87$, $\Sigma x_1 x_2 = 55$, $\Sigma x_2^2 = 36$, and $n = 8$ from page 531 into the expression for $\mathbf{X'X}$ on page 534, we get

$$\mathbf{X'X} = \begin{pmatrix} 8 & 25 & 16 \\ 25 & 87 & 55 \\ 16 & 55 & 36 \end{pmatrix}$$

Then, the inverse of this matrix can be obtained by any one of a number of different techniques; using the one based on cofactors, we find that

$$(\mathbf{X'X})^{-1} = \frac{1}{84} \cdot \begin{pmatrix} 107 & -20 & -17 \\ -20 & 32 & -40 \\ -17 & -40 & 71 \end{pmatrix}$$

where 84 is the value of $|\mathbf{X'X}|$, the determinant of $\mathbf{X'X}$.

Substituting $\Sigma y = 637{,}000$, $\Sigma x_1 y = 2{,}031{,}100$, and $\Sigma x_2 y = 1{,}297{,}700$ from page 531 into the expression for $\mathbf{X'Y}$ on page 534, we then get

$$\mathbf{X'Y} = \begin{pmatrix} 637{,}000 \\ 2{,}031{,}100 \\ 1{,}297{,}700 \end{pmatrix}$$

and finally

$$(\mathbf{X'X})^{-1}\mathbf{X'Y} = \frac{1}{84} \cdot \begin{pmatrix} 107 & -20 & -17 \\ -20 & 32 & -40 \\ -17 & -40 & 71 \end{pmatrix} \begin{pmatrix} 637{,}000 \\ 2{,}031{,}100 \\ 1{,}297{,}700 \end{pmatrix}$$

$$= \frac{1}{84} \cdot \begin{pmatrix} 5{,}476{,}100 \\ 347{,}200 \\ 63{,}700 \end{pmatrix}$$

$$= \begin{pmatrix} 65{,}191.7 \\ 4{,}133.3 \\ 758.3 \end{pmatrix}$$

where the $\hat{\beta}$'s are rounded to one decimal. Note that the results obtained here are identical with those shown in the computer printout of Figure 14.6. ▲

Next, to generalize the work of Section 14.4, we make assumptions that are very similar to those on page 512—we assume that for $i = 1, 2, \ldots$, and $n$, the $Y_i$ are independent random variables having normal distributions with the means $\beta_0 + \beta_1 x_{i1} + \beta_2 x_{i2} + \cdots + \beta_k x_{ik}$ and the common standard deviation $\sigma$. Based on $n$ data points

$$(x_{i1}, x_{i2}, \ldots, x_{ik}, y_i)$$

we can then make all sorts of inferences about the parameters of our model, the $\beta$'s and $\sigma$, and judge the merits of estimates and predictions based on the estimated multiple regression equation.

Finding maximum likelihood estimates of the $\beta$'s and $\sigma$ is straightforward, as on pages 512 and 513, and it will be left to the reader in Exercise 14.61. The results are as follows: The maximum likelihood estimates of the $\beta$'s equal the corresponding least squares estimates, so they are given by the elements of the $(k + 1) \times 1$ column matrix

$$\mathbf{B} = (\mathbf{X'X})^{-1}\mathbf{X'Y}$$

The maximum likelihood estimate of $\sigma$ is given by

$$\hat{\sigma} = \sqrt{\frac{1}{n} \cdot \sum_{i=1}^{n} [y_i - (\hat{\beta}_0 + \hat{\beta}_1 x_{i1} + \hat{\beta}_2 x_{i2} + \cdots + \hat{\beta}_k x_{ik})]^2}$$

where the $\hat{\beta}$'s are the maximum likelihood estimates of the $\beta$'s, and as the reader will be asked to verify in Exercise 14.63, it can also be written as

$$\hat{\sigma} = \sqrt{\frac{\mathbf{Y'Y} - \mathbf{B'X'Y}}{n}}$$

in matrix notation.

## EXAMPLE 14.12

Use the results of Example 14.11 to determine the value of $\hat{\sigma}$ for the data of Example 14.9.

### Solution

First let us calculate $\mathbf{Y'Y}$, which is simply $\sum_{i=1}^{n} y_i^2$, so we get

$$\mathbf{Y'Y} = 78{,}800^2 + 74{,}300^2 + \cdots + 82{,}900^2$$

$$= 50{,}907{,}080{,}000$$

Then, copying $\mathbf{B}$ and $\mathbf{X'Y}$ from page 536, we get

$$\mathbf{B'X'Y} = \frac{1}{84} \cdot (5{,}476{,}100 \quad 347{,}200 \quad 63{,}700) \begin{pmatrix} 637{,}000 \\ 2{,}031{,}100 \\ 1{,}297{,}700 \end{pmatrix}$$

$$= 50{,}906{,}394{,}166$$

and it follows that

$$\hat{\sigma} = \sqrt{\frac{50{,}907{,}080{,}000 - 50{,}906{,}394{,}166}{8}}$$

$$= 292.8 \qquad \qquad \blacktriangle$$

It is of interest to note that the estimate which we obtained here does not equal the one shown in the computer printout of Figure 14.6. The estimate shown there, $S = 370.4$, is such that $S^2$ is an unbiased estimator of $\sigma^2$, analogous to the standard error of estimate which we defined on page 517. It differs from $\hat{\sigma}$ in that we divide by $n - k - 1$ instead of $n$, and if we had done so in our example, we would have obtained

$$s_e = \sqrt{\frac{50{,}907{,}080{,}000 - 50{,}906{,}394{,}166}{8 - 2 - 1}}$$

$$= 370.4$$

Proceeding as in Section 14.4, we investigate next the sampling distribution of the $\hat{B}_i$ for $i = 0, 1, \ldots, k$, and $\hat{\Sigma}$. Leaving the details to the reader, let us merely point out that arguments similar to those on pages 513 and 514 lead to the results that the $\hat{B}_i$ are linear combinations of the $n$ independent random variables $Y_i$, so that the $\hat{B}_i$ themselves have normal distributions. Furthermore, they are unbiased estimators, that is,

$$E(\hat{B}_i) = \beta_i \qquad \text{for } i = 0, 1, \ldots, k$$

and their variances are given by

$$\text{var}(\hat{B}_i) = c_{ii}\sigma^2 \qquad \text{for } i = 0, 1, \ldots, k$$

Here $c_{ij}$ is the element in the $i$th row and the $j$th column of the matrix $(\mathbf{X'X})^{-1}$, with $i$ and $j$ taking on the values $0, 1, \ldots, k$.

Let us also state the result that, analogous to Theorem 14.3, the sampling distribution of $\dfrac{n\hat{\Sigma}^2}{\sigma^2}$, the random variable corresponding to $\dfrac{n\hat{\sigma}^2}{\sigma^2}$, is the chi-square distribution with $n - k - 1$ degrees of freedom, and that $\dfrac{n\hat{\Sigma}^2}{\sigma^2}$ and $\hat{B}_i$ are independent for $i = 0, 1, \ldots, k$. Combining all these results, we find that the definition of the $t$ distribution in Section 8.5 leads to

---

**THEOREM 14.8** Under the assumptions of normal multiple regression analysis,

$$t = \frac{\hat{\beta}_i - \beta_i}{\hat{\sigma} \cdot \sqrt{\dfrac{n|c_{ii}|}{n - k - 1}}} \qquad \text{for } i = 0, 1, \ldots, k$$

are values of random variables having the $t$ distribution with $n - k - 1$ degrees of freedom.

---

Based on this theorem, let us now test a hypothesis about one of the multiple regression coefficients.

## EXAMPLE 14.13

With reference to Example 14.9, test the null hypothesis $\beta_1 = \$3,500$ against the alternative hypothesis $\beta_1 > \$3,500$ at the 0.05 level of significance.

*Solution*

1.  $H_0$:  $\beta_1 = 3,500$
    $H_1$:  $\beta_1 > 3,500$
    $\alpha = 0.05$

2.  Reject the null hypothesis if $t \geq 2.015$, where $t$ is determined in accordance with Theorem 14.8 and 2.015 is the value of $t_{.05,5}$ according to Table IV.

3.  Substituting $n = 8$, $\hat{\beta}_1 = 4,133.3$, and $c_{11} = \frac{32}{84}$ from Example 14.11, and $\hat{\sigma} = 292.8$ from Example 14.12 into the formula for $t$, we get

$$t = \frac{4,133.3 - 3,500}{292.8 \cdot \sqrt{\dfrac{8 \cdot \left|\frac{32}{84}\right|}{5}}}$$

$$= \frac{4,133.3 - 3,500}{228.6}$$

$$= 2.77$$

4.  Since $t = 2.77$ exceeds 2.015, the null hypothsis must be rejected; we conclude that on the average each additional bedroom adds more than $3,500 to the sales price of such a house. (Note that the value in the denominator of the $t$ statistic, 228.6, equals the second value in the column headed "ST. DEV. OF COEF." in the computer printout of Figure 14.6.)     ▲

Analogous to Theorem 14.5, we can also use the $t$ statistic of Theorem 14.8 to construct confidence intervals for regression coefficients (see Exercise 14.66).

## *EXERCISES*

**14.61** If **b** is the column vector of the $\beta$'s, verify in matrix notation that $\mathbf{q} = (\mathbf{Y} - \mathbf{Xb})'(\mathbf{Y} - \mathbf{Xb})$ is a minimum when $\mathbf{b} = \mathbf{B} = (\mathbf{X'X})^{-1}(\mathbf{X'Y})$.

**14.62** Verify that under the assumptions of normal multiple regression analysis,
  (a) the maximum likelihood estimates of the $\beta$'s equal the corresponding least squares estimates;
  (b) the maximum likelihood estimate of $\sigma$ is

$$\hat{\sigma} = \sqrt{\frac{(\mathbf{Y} - \mathbf{XB})'(\mathbf{Y} - \mathbf{XB})}{n}}$$

**14.63** Verify that the estimate of part (b) of Exercise 14.62 can also be written as

$$\hat{\sigma} = \sqrt{\frac{\mathbf{Y'Y} - \mathbf{B'X'Y}}{n}}$$

**14.64** Show that under the assumptions of normal multiple regression analysis
  (a) $E(\hat{B}_i) = \beta_i$ for $i = 0, 1, \ldots, k$;
  (b) $\text{var}(\hat{B}_i) = c_{ii}\sigma^2$ for $i = 0, 1, \ldots, k$;
  (c) $\text{cov}(\hat{B}_i, \hat{B}_j) = c_{ij}\sigma^2$ for $i \neq j = 0, 1, \ldots, k$.

**14.65** Show that for $k = 1$, the formulas of Exercise 14.64 are equivalent to those given on page 514 and in Exercises 14.28 and 14.29.

**14.66** Use the $t$ statistic of Theorem 14.8 to construct a $(1 - \alpha)100\%$ confidence interval formula for $\beta_i$ for $i = 0, 1, \ldots, k$.

**14.67** If $x_{01}, x_{02}, \ldots, x_{0k}$ are given values of $x_1, x_2, \ldots, x_k$ and $\mathbf{X}_0$ is the column vector

$$\mathbf{X}_0 = \begin{pmatrix} 1 \\ x_{01} \\ x_{02} \\ \ldots \\ x_{0k} \end{pmatrix}$$

it can be shown that

$$t = \frac{\mathbf{B'X}_0 - \mu_{Y|x_{01}, x_{02}, \ldots, x_{0k}}}{\hat{\sigma} \cdot \sqrt{\dfrac{n[\mathbf{X}_0'(\mathbf{X'X})^{-1}\mathbf{X}_0]}{n - k - 1}}}$$

is a value of a random variable having the $t$ distribution with $n - k - 1$ degrees of freedom.
  (a) Show that for $k = 1$ this statistic is equivalent to the one of Exercise 14.31.

(b)   Derive a $(1 - \alpha)100\%$ confidence interval formula for

$$\mu_{Y|x_{01}, x_{02}, \ldots, x_{0k}}$$

**14.68** With $x_{01}, x_{02}, \ldots, x_{0k}$ and $\mathbf{X}_0$ as defined in Exercise 14.67 and $Y_0$ being a random variable having a normal distribution with the mean $\beta_0 + \beta_1 x_{01} + \cdots + \beta_k x_{0k}$ and the variance $\sigma^2$, it can be shown that

$$t = \frac{y_0 - \mathbf{B}'\mathbf{X}_0}{\hat{\sigma} \cdot \sqrt{\dfrac{n[1 + \mathbf{X}_0'(\mathbf{X}'\mathbf{X})^{-1}\mathbf{X}_0]}{n - k - 1}}}$$

is a value of a random variable having the $t$ distribution with $n - k - 1$ degrees of freedom.
(a)   Show that for $k = 1$ this statistic is equivalent to the one of Exercise 14.33.
(b)   Derive a formula for $(1 - \alpha)100\%$ limits of prediction for a future observation of $Y_0$.

## APPLICATIONS

**14.69** The following are sample data provided by a moving company on the weights of six shipments, the distances they were moved, and the damage that was incurred:

| Weight (1,000 lb) $x_1$ | Distance (1,000 miles) $x_2$ | Damage (dollars) $y$ |
|---|---|---|
| 4.0 | 1.5 | 160 |
| 3.0 | 2.2 | 112 |
| 1.6 | 1.0 | 69 |
| 1.2 | 2.0 | 90 |
| 3.4 | 0.8 | 123 |
| 4.8 | 1.6 | 186 |

(a)   Assuming that the regression is linear, estimate $\beta_0$, $\beta_1$, and $\beta_2$.
(b)   Use the results of part (a) to estimate the damage when a shipment weighing 2,400 pounds is moved 1,200 miles.

**14.70** The following are data on the average weekly profits (in $1,000) of five restaurants, their seating capacities, and the average daily traffic (in thousands of cars) which passes their locations:

| Seating capacity $x_1$ | Traffic count $x_2$ | Weekly net profit $y$ |
|---|---|---|
| 120 | 19 | 23.8 |
| 200 | 8 | 24.2 |
| 150 | 12 | 22.0 |
| 180 | 15 | 26.2 |
| 240 | 16 | 33.5 |

(a) Assumming that the regression is linear, estimate $\beta_0$, $\beta_1$, and $\beta_2$.

(b) Use the results of part (a) to predict the average weekly net profit of a restaurant with a seating capacity of 210 at a location where the daily traffic count averages 14,000 cars.

**14.71** The following data consist of the scores which ten students obtained in an examination, their IQ's, and the numbers of hours they spent studying for the examination:

| IQ $x_1$ | Number of hours studied $x_2$ | Score $y$ |
|---|---|---|
| 112 | 5 | 79 |
| 126 | 13 | 97 |
| 100 | 3 | 51 |
| 114 | 7 | 65 |
| 112 | 11 | 82 |
| 121 | 9 | 93 |
| 110 | 8 | 81 |
| 103 | 4 | 38 |
| 111 | 6 | 60 |
| 124 | 2 | 86 |

(a) Assuming that the regression is linear, estimate $\beta_0$, $\beta_1$, and $\beta_2$.

(b) Predict the score of a student with an IQ of 108 who studied 6 hours for the examination.

**14.72** The following data were collected to determine the relationship between two processing variables and the hardness of a certain kind of steel:

| Hardness (Rockwell 30-T) y | Copper content (percent) $x_1$ | Annealing temperature (degrees F) $x_2$ |
|---|---|---|
| 78.9 | 0.02 | 1,000 |
| 55.2 | 0.02 | 1,200 |
| 80.9 | 0.10 | 1,000 |
| 57.4 | 0.10 | 1,200 |
| 85.3 | 0.18 | 1,000 |
| 60.7 | 0.18 | 1,200 |

Fit a plane by the method of least squares, and use it to estimate the average hardness of this kind of steel when the copper content is 0.14 percent and the annealing temperature is 1,100 degrees F.

**14.73** When the $x_1$'s, $x_2$'s, ... , and/or the $x_k$'s are equally spaced, the calculation of the $\hat{\beta}$'s can be simplified by using the coding suggested in Exercise 14.15. Rework Exercise 14.72 coding the $x_1$-values $-1$, 0, and 1, and the $x_2$-values $-1$ and 1. (Note that for the coded $x_1$'s and $x_2$'s, call them $z_1$'s and $z_2$'s, we have not only $\Sigma z_1 = 0$ and $\Sigma z_2 = 0$, but also $\Sigma z_1 z_2 = 0$.)

**14.74** The following are data on the percent effectiveness of a pain reliever and the amounts of three different medications (in milligrams) present in each capsule:

| Medication A $x_1$ | Medication B $x_2$ | Medication C $x_3$ | Percent effective y |
|---|---|---|---|
| 15 | 20 | 10 | 47 |
| 15 | 20 | 20 | 54 |
| 15 | 30 | 10 | 58 |
| 15 | 30 | 20 | 66 |
| 30 | 20 | 10 | 59 |
| 30 | 20 | 20 | 67 |
| 30 | 30 | 10 | 71 |
| 30 | 30 | 20 | 83 |
| 45 | 20 | 10 | 72 |
| 45 | 20 | 20 | 82 |
| 45 | 30 | 10 | 85 |
| 45 | 30 | 20 | 94 |

Assuming that the regression is linear, estimate the regression coefficients after suitably coding each of the $x$'s, and express the estimated regression equation in terms of the original variables.

**14.75** The regression models we introduced in Sections 14.2 and 14.6 are linear in the $x$'s, but more importantly, they are also linear in the $\beta$'s. Indeed, they can be used in some problems where the relationship between the $x$'s and $y$ is not linear. For instance, when the regression is parabolic and of the form

$$\mu_{Y|x} = \beta_0 + \beta_1 x + \beta_2 x^2$$

we simply use the regression equation $\mu_{Y|x} = \beta_0 + \beta_1 x_1 + \beta_2 x_2$ with $x_1 = x$ and $x_2 = x^2$. Use this method to fit a parabola to the following data on the drying time of a varnish and the amount of a certain chemical that has been added:

| Amount of additive (grams) $x$ | Drying time (hours) $y$ |
|---|---|
| 1 | 8.5 |
| 2 | 8.0 |
| 3 | 6.0 |
| 4 | 5.0 |
| 5 | 6.0 |
| 6 | 5.5 |
| 7 | 6.5 |
| 8 | 7.0 |

Also, predict the drying time when 6.5 grams of the chemical are added.

**14.76** The following data pertain to the demand for a product (in thousands of units) and its price (in cents) charged in five different market areas:

| Price $x$ | Demand $y$ |
|---|---|
| 20 | 22 |
| 16 | 41 |
| 10 | 120 |
| 11 | 89 |
| 14 | 56 |

Fit a parabola to these data by the method suggested in the preceding exercise.

**14.77** To judge whether it was worthwhile to fit a parabola in the preceding exercise and not just a straight line, test the null hypothesis $\beta_2 = 0$ against the alternative hypothesis $\beta_2 \neq 0$ at the 0.05 level of significance.

**14.78**  Use the results obtained for the data of Example 14.9 in Section 14.7 to construct a 90% confidence interval for the regression coefficient $\beta_2$ (see Exercise 14.66).

**14.79**  With reference to Exercise 14.69, test the null hypothsis $\beta_2 = 10.0$ against the alternative hypothesis $\beta_2 \neq 10.0$ at the 0.05 level of significance.

**14.80**  With reference to Exercise 14.69, construct a 95% confidence interval for the regression coefficient $\beta_1$.

**14.81**  With reference to Exercise 14.70, test the null hypothesis $\beta_1 = 0.12$ against the alternative hypothesis $\beta_1 < 0.12$ at the 0.05 level of significance.

**14.82**  With reference to Exercise 14.70, construct a 98% confidence interval for the regression coefficient $\beta_2$.

**14.83**  Use the results obtained for the data of Example 14.9 in Section 14.7 and the result of part (b) of Exercise 14.67 to construct a 95% confidence interval for the mean sales price of a three-bedroom house with two baths in the given housing development.

**14.84**  Use the results obtained for the data of Example 14.9 in Section 14.7 and the result of part (b) of Exercise 14.68 to construct 99% limits of prediction for the sales price of a three-bedroom house with two baths in the given housing development.

**14.85**  With reference to Exercise 14.69, use the result of part (b) of Exercise 14.67 to construct a 98% confidence interval for the mean damage of 2,400-pound shipments that are moved 1,200 miles.

**14.86**  With reference to Exercise 14.69, use the result of part (b) of Exercise 14.68 to construct 95% limits of prediction for the damage that will be incurred by a 2,400-pound shipment that is moved 1,200 miles.

**14.87**  With reference to Exercise 14.70, use the result of part (b) of Exercise 14.67 to construct a 99% confidence interval for the mean weekly net profit of restaurants with a seating capacity of 210 at a location where the daily traffic count averages 14,000 cars.

**14.88**  With reference to Exercise 14.70, use the result of part (b) of Exercise 14.68 to construct 98% limits of predition for the average weekly net profit of a restaurant with a seating capacity of 210 at a location where the daily traffic count averages 14,000 cars.

## REFERENCES

A proof of Theorem 14.3 and other mathematical details left out in the text may be found in the book by S. S. Wilks referred to at the end of Chapter 7, and information about the distribution of $\frac{1}{2} \cdot \ln \frac{1+R}{1-R}$ may be found in the book by Kendall and Stuart referred to at the end of Chapter 3. A derivation of the maximum likelihood estimates of $\sigma_1$, $\sigma_2$, and $\rho$ is given in the third edition (but not in the fourth edition) of

HOEL, P., *Introduction to Mathematical Statistics*, 3rd ed. New York: John Wiley & Sons, Inc., 1962.

More detailed treatments of multiple regression may be found in numerous more advanced books; for instance, in

MORRISON, D. F., *Applied Linear Statistical Methods*. Englewood Cliffs, N.J.: Prentice-Hall, Inc., 1983,

WEISBERG, S., *Applied Linear Regression*, 2nd ed. New York: John Wiley & Sons, Inc., 1985,

WONNACOTT, T. H., and WONNACOTT, R. J., *Regression: A Second Course in Statistics*. New York: John Wiley & Sons, Inc., 1981.

# 15

# *Analysis of Variance*

## 15.1 INTRODUCTION

In this chapter we shall generalize the work of Section 13.3 and consider the problem of deciding whether observed differences among more than two sample means can be attributed to chance, or whether there are real differences among the means of the populations sampled. For instance, we may want to decide on the basis of sample data whether there really is a difference in the effectiveness of three methods of teaching a foreign language, we may want to compare the average yields per acre of six varieties of wheat, or we may want to see whether there really is a difference in the average mileage obtained with four kinds of gasoline.

Since observed differences can always be due to causes other than those postulated—for instance, differences in the performance of students taught a foreign language by three different methods may be due to differences in intelligence, and differences in the average mileage obtained with four kinds of gasoline may be due to differences in road conditions—we shall also discuss some questions of **experimental design**, so that, with reasonable assurance, statistically significant results can be attributed to particular causes.

## 15.2 ONE-WAY ANALYSIS OF VARIANCE

To give an example of a typical situation where we would perform a one-way analysis of variance, suppose that we want to compare the cleansing action of three detergents on the basis of the following whiteness readings made on 15 swatches of white cloth, which were first soiled with India ink and then washed in an agitator-type machine with the respective detergents:

$$Detergent\ A: \quad 77, 81, 71, 76, 80$$

$$Detergent\ B: \quad 72, 58, 74, 66, 70$$

$$Detergent\ C: \quad 76, 85, 82, 80, 77$$

The means of these three samples are 77, 68, and 80, and we want to know whether the differences among them are significant or whether they can be attributed to chance.

In general, in a problem like this, we have independent random samples of size $n$ from $k$ populations. The $j$th value from the $i$th population is denoted $x_{ij}$, that is,

$$Population\ 1: \quad x_{11}, x_{12}, \ldots, x_{1n}$$

$$Population\ 2: \quad x_{21}, x_{22}, \ldots, x_{2n}$$

. . .

$$Population\ k: \quad x_{k1}, x_{k2}, \ldots, x_{kn}$$

and we shall assume that the corresponding random variables $X_{ij}$, which are all independent, have normal distributions with the respective means $\mu_i$ and the common variance $\sigma^2$. Stating these assumptions somewhat differently, we could say that the model for the observations is given by

$$x_{ij} = \mu_i + e_{ij}$$

for $i = 1, 2, \ldots, k$, and $j = 1, 2, \ldots, n$, where the $e_{ij}$ are values of $nk$ independent random variables having normal distributions with zero means and the common variance $\sigma^2$. To permit the generalization of this model to more complicated kinds of situations (see page 559), it is usually written in the form

$$x_{ij} = \mu + \alpha_i + e_{ij}$$

for $i = 1, 2, \ldots, k$, and $j = 1, 2, \ldots, n$. Here $\mu$ is referred to as the **grand mean**, and the $\alpha_i$, called the **treatment effects**, are such that $\sum_{i=1}^{k} \alpha_i = 0$. Note that we have merely written the mean of the $i$th population as $\mu_i = \mu + \alpha_i$ and imposed the condition $\sum_{i=1}^{k} \alpha_i = 0$ so that the mean of the $\mu_i$ equals the grand mean $\mu$. The practice of referring to the different populations as different **treaments** is due to the fact that many analysis-of-variance techniques were originally developed in connection with agricultural experiments where different fertilizers, for example, were regarded as different treatments applied to the soil. Thus, we shall refer to the three detergents of our example on page 548 as three different treatments, and in other problems we may refer to four nationalities as four different treatments, five kinds of advertising campaigns as five different treatments, and so on. "Levels" is another term often used instead of "treatments."

The null hypothesis we shall want to test is that the population means are all equal, namely, that $\mu_1 = \mu_2 = \cdots = \mu_k$ or equivalently that

$$H_0: \quad \alpha_i = 0 \qquad \text{for } i = 1, 2, \ldots, k$$

Correspondingly, the alternative hypothesis is that the population means are not all equal, namely, that

$$H_1: \quad \alpha_i \neq 0 \qquad \text{for at least one value of } i$$

The test, itself, is based on an analysis of the total variability of the combined data ($nk - 1$ times their variance), which is given by

$$\sum_{i=1}^{k} \sum_{j=1}^{n} (x_{ij} - \bar{x}_{..})^2 \quad \text{where} \quad \bar{x}_{..} = \frac{1}{nk} \cdot \sum_{i=1}^{k} \sum_{j=1}^{n} x_{ij}$$

If the null hypothesis is true, all this variability is due to chance, but if it is not true, then part of the above sum of squares is due to the differences among the population means. To isolate, or separate, these two contributions to the total variability of the data, we refer to the following theorem.

---

**THEOREM 15.1**

$$\sum_{i=1}^{k} \sum_{j=1}^{n} (x_{ij} - \bar{x}_{..})^2 = n \cdot \sum_{i=1}^{k} (\bar{x}_{i.} - \bar{x}_{..})^2 + \sum_{i=1}^{k} \sum_{j=1}^{n} (x_{ij} - \bar{x}_{i.})^2$$

where $\bar{x}_{i.}$ is the mean of the observations from the $i$th population and $\bar{x}_{..}$ is the mean of all $nk$ observations.

---

**_Proof._**

$$\sum_{i=1}^{k}\sum_{j=1}^{n}(x_{ij}-\bar{x}_{..})^2 = \sum_{i=1}^{k}\sum_{j=1}^{n}[(\bar{x}_{i.}-\bar{x}_{..})+(x_{ij}-\bar{x}_{i.})]^2$$

$$= \sum_{i=1}^{k}\sum_{j=1}^{n}[(\bar{x}_{i.}-\bar{x}_{..})^2 + 2(\bar{x}_{i.}-\bar{x}_{..})(x_{ij}-\bar{x}_{i.})$$

$$+ (x_{ij}-\bar{x}_{i.})^2]$$

$$= \sum_{i=1}^{k}\sum_{j=1}^{n}(\bar{x}_{i.}-\bar{x}_{..})^2 + 2\sum_{i=1}^{k}\sum_{j=1}^{n}(\bar{x}_{i.}-\bar{x}_{..})(x_{ij}-\bar{x}_{i.})$$

$$+ \sum_{i=1}^{k}\sum_{j=1}^{n}(x_{ij}-\bar{x}_{i.})^2$$

$$= n \cdot \sum_{i=1}^{k}(\bar{x}_{i.}-\bar{x}_{..})^2 + \sum_{i=1}^{k}\sum_{j=1}^{n}(x_{ij}-\bar{x}_{i.})^2$$

since $\sum_{j=1}^{n}(x_{ij}-\bar{x}_{i.}) = 0$ for each value of $i$.    ▼

It is customary to refer to the expression on the left-hand side of the identity of Theorem 15.1 as the **total sum of squares**, to the first term of the expression on the right-hand side as the **treatment sum of squares**, and to the second term as the **error sum of squares**, where "error" denotes the **experimental error**, or chance. Correspondingly, we denote these three sums of squares by SST, SS(Tr), and SSE, and we can write

$$SST = SS(Tr) + SSE$$

Now we have accomplished what we set out to do: We have partitioned SST, a measure of the total variation of the combined data into two components—_the second component, SSE, measures chance variation (namely, the variation within the samples); the first component, SS(Tr), also measures chance variation when the null hypothesis is true, but it also reflects the variation among the population means when the null hypothesis is false._

Since, for each value of $i$, the $x_{ij}$ are values of a random sample of size $n$ from a normal population with the variance $\sigma^2$, it follows from Theorem 8.11 that for each value of $i$

$$\frac{1}{\sigma^2} \cdot \sum_{j=1}^{n}(X_{ij}-\bar{X}_{i.})^2$$

is a random variable having the chi-square distribution with $n - 1$ degrees of freedom. Furthermore, since the $k$ random samples are independent, it follows from Theorem 8.9 that

$$\frac{1}{\sigma^2} \cdot \sum_{i=1}^{k} \sum_{j=1}^{n} (X_{ij} - \overline{X}_{i.})^2$$

is a random variable having the chi-square distribution with $k(n - 1)$ degrees of freedom. Since the mean of a chi-square distribution equals its degrees of freedom, we find that $\frac{1}{\sigma^2} \cdot SSE$ is a value of a random variable having the mean $k(n - 1)$, and, hence, that $\dfrac{SSE}{k(n - 1)}$ can serve as an estimate of $\sigma^2$. This quantity, $\dfrac{SSE}{k(n - 1)}$, is called the **error mean square**, and it is denoted by MSE.

Also, since under the null hypothesis the $\bar{x}_{i.}$ are values of independent random variables having identical normal distributions with the mean $\mu$ and the variance $\dfrac{\sigma^2}{n}$, it follows from Theorem 8.11 that

$$\frac{n}{\sigma^2} \cdot \sum_{i=1}^{k} (\overline{X}_{i.} - \overline{X}_{..})^2$$

is a random variable having the chi-square distribution with $k - 1$ degrees of freedom. Since the mean of this distribution is $k - 1$, it follows that $\dfrac{SS(Tr)}{k - 1}$ provides a second estimate of $\sigma^2$. This quantity, $\dfrac{SS(Tr)}{k - 1}$, is called the **treatment mean square** and it is denoted by MS(Tr).

Of course, if the null hypothesis is false, then, according to Exercise 15.1, MS(Tr) provides an estimate of $\sigma^2$ plus whatever variation there may be among the population means. This suggests that we reject the null hypothesis that the population means are all equal when MS(Tr) is appreciably greater than MSE. To put this decision on a precise basis, we shall have to assume without proof that the corresponding estimators are independent, for with this assumption we can utilize Theorem 8.14, according to which

$$f = \frac{\dfrac{SS(Tr)}{(k - 1)\sigma^2}}{\dfrac{SSE}{k(n - 1)\sigma^2}} = \frac{MS(Tr)}{MSE}$$

is a value of a random variable having the $F$ distribution with $k - 1$ and $k(n - 1)$ degrees of freedom.[†] Thus, we reject the null hypothesis that the population means are all equal if the value we obtain for $f$ exceeds $f_{\alpha, k-1, k(n-1)}$, where $\alpha$ is the level of significance.

The procedure we have described in this section is called a **one-way analysis of variance**, and the necessary details are usually presented in the following kind of **analysis-of-variance table**:

| Source of variation | Degrees of freedom | Sum of squares | Mean square | $f$ |
|---|---|---|---|---|
| Treatments | $k - 1$ | SS(Tr) | MS(Tr) | $\dfrac{\text{MS(Tr)}}{\text{MSE}}$ |
| Error | $k(n - 1)$ | SSE | MSE | |
| Total | $kn - 1$ | SST | | |

To simplify the calculation of the various sums of squares, we usually use the following computing formulas, which the reader will be asked to derive in Exercise 15.2.

---

**THEOREM 15.2**

$$\text{SST} = \sum_{i=1}^{k} \sum_{j=1}^{n} x_{ij}^2 - \frac{1}{kn} \cdot T_{..}^2$$

and

$$\text{SS(Tr)} = \frac{1}{n} \cdot \sum_{i=1}^{k} T_{i.}^2 - \frac{1}{kn} \cdot T_{..}^2$$

where $T_{i.}$ is the total of the values obtained for the $i$th treatment and $T_{..}$ is the grand total of all $nk$ observations.

---

Then, the value of SSE can be obtained by subtracting SS(Tr) from SST.

---

[†] A proof of this independence may be found in the book by H. Scheffé referred to at the end of this chapter.

## EXAMPLE 15.1

With reference to the illustration on page 548, test at the 0.01 level of significance whether the differences among the means of the whiteness readings are significant.

### Solution

1. $H_0$: $\alpha_i = 0$ for $i = 1, 2, 3$
   $H_1$: $\alpha_i \neq 0$ for at least one value of $i$
   $\alpha = 0.01$

2. Reject the null hypothesis if $f \geq 6.93$, where $f$ is obtained by a one-way analysis of variance and 6.93 is the value of $f_{.01,2,12}$.

3. The required sums and sums of squares are $T_{1.} = 385$, $T_{2.} = 340$, $T_{3.} = 400$, $T_{..} = 1,125$, and $\Sigma\Sigma\, x^2 = 85,041$, and substitution of these values together with $k = 3$ and $n = 5$ into the formulas of Theorem 15.2 yields

$$\text{SST} = 85,041 - \tfrac{1}{15}\,(1,125)^2$$

$$= 666$$

and

$$\text{SS(Tr)} = \tfrac{1}{5}\,(385^2 + 340^2 + 400^2) - \tfrac{1}{15}\,(1,125)^2$$

$$= 390$$

Then, by subtraction, SSE $= 666 - 390 = 276$, and the remaining calculations are shown in the following analysis-of-variance table:

| Source of variation | Degrees of freedom | Sum of squares | Mean square | $f$ |
|---|---|---|---|---|
| Treatments | 2 | 390 | $\dfrac{390}{2} = 195$ | $\dfrac{195}{23} = 8.48$ |
| Error | 12 | 276 | $\dfrac{276}{12} = 23$ | |
| Total | 14 | 666 | | |

Note that the mean squares are simply the sums of squares divided by the corresponding degrees of freedom.

4. Since $f = 8.48$ exceeds 6.93, the null hypothesis must be rejected, and we conclude that the three detergents are not all equally effective.  ▲

The parameters of the model given on page 548, namely, $\mu$ and the $\alpha_i$, are usually estimated by the method of least squares. That is, their estimates are the values which minimize

$$\sum_{i=1}^{k} \sum_{j=1}^{n} [x_{ij} - (\mu + \alpha_i)]^2$$

subject to the restriction that $\sum_{i=1}^{k} \alpha_i = 0$; as the reader will be asked to verify in Exercise 15.6, these least squares estimates are $\hat{\mu} = \bar{x}_{..}$ and $\hat{\alpha}_i = \bar{x}_{i.} - \bar{x}_{...}$

### EXERCISES

**15.1** For the one-way analysis of variance with $k$ independent samples of size $n$, show that

$$E\left[\frac{n \cdot \sum_{i=1}^{k} (\bar{X}_{i.} - \bar{X}_{..})^2}{k - 1}\right] = \sigma^2 + \frac{n \cdot \sum_{i=1}^{k} \alpha_i^2}{k - 1}$$

**15.2** Prove Theorem 15.2.

**15.3** If, in a one-way analysis of variance, the sample sizes are unequal and there are $n_i$ observations for the $i$th treatment, show that

$$\sum_{i=1}^{k} \sum_{j=1}^{n_i} (x_{ij} - \bar{x}_{..})^2 = \sum_{i=1}^{k} n_i(\bar{x}_{i.} - \bar{x}_{..})^2 + \sum_{i=1}^{k} \sum_{j=1}^{n_i} (x_{ij} - \bar{x}_{i.})^2$$

is analogous to the identity of Theorem 15.1. Also show that the degrees of freedom for SST, SS(Tr), and SSE are, respectively, $N - 1$, $k - 1$, and $N - k$, where $N = \sum_{i=1}^{k} n_i$.

**15.4** With reference to Exercise 15.3, show that the computing formulas for the sums of squares are

$$\text{SST} = \sum_{i=1}^{k} \sum_{j=1}^{n_i} x_{ij}^2 - \frac{1}{N} \cdot T_{..}^2$$

$$\text{SS(Tr)} = \sum_{i=1}^{k} \frac{T_{i.}^2}{n_i} - \frac{1}{N} \cdot T_{..}^2$$

and

$$\text{SSE} = \text{SST} - \text{SS(Tr)}$$

**15.5** Show that for $k = 2$ the $F$ test of a one-way analysis of variance is equivalent to the $t$ test of Section 13.3 with $\delta = 0$ and the alternative hypothesis $\mu_1 - \mu_2 \neq 0$.

**15.6** Use Lagrange multipliers to show that the least squares estimates of the parameters of the model on page 548 are $\hat{\mu} = \bar{x}_{..}$ and $\hat{\alpha}_i = \bar{x}_{i.} - \bar{x}_{...}$.

## APPLICATIONS

**15.7** To compare the effectiveness of three different types of phosphorescent coatings of airplane instrument dials, eight dials each are coated with the three types. Then the dials are illuminated by an ultraviolet light, and the following are the number of minutes each glowed after the light source was shut off:

*Type 1*:   52.9, 62.1, 57.4, 50.0, 59.3, 61.2, 60.8, 53.1
*Type 2*:   58.4, 55.0, 59.8, 62.5, 64.7, 59.9, 54.7, 58.4
*Type 3*:   71.3, 66.6, 63.4, 64.7, 75.8, 65.6, 72.9, 67.3

Test the null hypothesis that there is no difference in the effectiveness of the three coatings at the 0.01 level of significance.

**15.8** The following are the numbers of mistakes made in five successive weeks by four technicians working for a medical laboratory:

*Technician I*:     13, 16, 12, 14, 15
*Technician II*:    14, 16, 11, 19, 15
*Technician III*:   13, 18, 16, 14, 18
*Technician IV*:    18, 10, 14, 15, 12

Test at the 0.05 level of significance whether the differences among the four sample means can be attributed to chance.

**15.9** Three groups of six guinea pigs each were injected, respectively, with 0.5 milligram, 1.0 milligram, and 1.5 milligrams of a new tranquilizer, and the following are the numbers of minutes it took them to fall asleep:

*0.5 mg*:   21, 23, 19, 24, 25, 23
*1.0 mg*:   19, 21, 20, 18, 22, 20
*1.5 mg*:   15, 10, 13, 14, 11, 15

Test at the 0.05 level of significance whether the null hypothesis that differences in dosage have no effect can be rejected. Also estimate the parameters $\mu$, $\alpha_1$, $\alpha_2$, and $\alpha_3$ of the model used in the analysis.

15.10 The following are the numbers of words per minute which a secretary typed on several occasions on four different typewriters:

> *Typewriter C*:    71, 75, 69, 77, 61, 72, 71, 78
> *Typewriter D*:    68, 71, 74, 66, 69, 67, 70, 62
> *Typewriter E*:    75, 70, 81, 73, 78, 72
> *Typewriter F*:    62, 59, 71, 68, 63, 65, 72, 60, 64

Use the computing formulas of Exercise 15.4 to calculate the sums of squares required to test at the 0.05 level of significance whether the differences among the four sample means can be attributed to chance.

15.11 A consumer testing service, wishing to test the accuracy of the thermostats of three different kinds of electric irons, set them at 480° F and obtained the following actual temperature readings by means of a thermocouple:

> *Iron X*:    474, 496, 467, 471
> *Iron Y*:    492, 498
> *Iron Z*:    460, 495, 490

Use the computing formulas of Exercise 15.4 to calculate the sums of squares required to test at the 0.05 level of significance whether the differences among the three sample means can be attributed to chance.

15.12 In Section 13.7 we pointed out that in the chi-square analysis of an $r \times c$ table we do not take into account a possible ordering of the rows and/or columns. When the rows and the columns are both ordered, we indicated an alternative to the chi-square analysis in Exercises 14.59 and 14.60. When only the rows or only the columns are ordered, we look upon the categories that are not ordered as treatments, and we replace the ones that are ordered by consecutive integers. For instance, in the $3 \times 3$ table on page 483 we look upon the three cities as three different treatments, and we replace the column headings by 1, $-1$, and 0, reflecting an ordering from favoring $B$ (not favoring $A$) to being indifferent to favoring $A$. Thus, the sample of size $n_1 = 400$ from Los Angeles consists of 174 ones, 93 minus ones, and 133 zeros; the sample of size $n_2 = 500$ from San Diego consists of 196 ones, 124 minus ones, and 180 zeros; and so on. Looking at the $r \times c$ table in this way, we then perform a one-way analysis of variance. Use this method to analyze the $3 \times 3$ table on page 483, testing the null hypothesis that the treatment effects are all equal to zero at the 0.05 level of significance, and compare the result with that obtained in Exercise 13.74.

15.13 Use the method of Exercise 15.12 to analyze the $3 \times 3$ table of Exercise 13.73, and compare the result with the result obtained in that exercise.

## 15.3  EXPERIMENTAL DESIGN

In Example 15.1 it may have seemed reasonable to conclude that the three detergents are not equally effective; yet, a moment's reflection will show that this conclusion is not so "reasonable" at all. For all we know, the swatches cleaned with detergent $B$ may have been more soiled than the others, the washing times may have been longer for detergent $C$, there may have been differences in water hardness or water temperature, and even the instruments used to make the whiteness readings may have gone out of adjustment after the readings for detergents $A$ and $C$ were made.

It is entirely possible, of course, that the differences among the three sample means are due largely to differences in the effectiveness of the detergents, but we have just listed several other factors which could be held responsible. It is important to remember that *a significance test may show that differences among sample means are too large to be attributed to chance, but such a test cannot say why the differences occurred.*

In general, if we want to show that one factor (among various others) can be considered the cause of an observed phenomenon, we must somehow make sure that none of the other factors can reasonably be held responsible. There are various ways in which this can be done; for instance, we can conduct a rigorously **controlled experiment** in which all variables except the one of concern are held fixed. To do this in the example dealing with the three detergents, we might soil the swatches with exactly equal amounts of India ink, always use the same washing time, water of exactly the same hardness and temperature, and inspect (and, if necessary, adjust) the measuring instruments after each use. Under such rigidly controlled conditions, significant differences among the sample means cannot be due to differently soiled swatches, or differences in washing time, water temperature, water hardness, or measuring instruments. On the positive side, the differences among the means show that the detergents are not all equally effective *if they are used in this narrowly restricted way.* Of course, we cannot say whether the same differences would exist if the washing time were longer or shorter, if the water had a different temperature or hardness, and so on.

In most cases, "overcontrolled" experiments like the one just described do not really provide us with the kind of information we want. So, we look for alternatives, and at the other extreme we can conduct experiments in which none of the extraneous factors is controlled, but in which we protect ourselves against their effects by **randomization**. That is, we design, or plan, the experiments in such a way that the variations caused by extraneous factors can all be combined under the general heading of "chance." For instance, in our example we could accomplish this by randomly assigning five of the soiled swatches to each detergent, and randomly specifying the order in which they are to be washed and measured. When all the variations due to uncontrolled extraneous factors can thus be included under the heading of chance variation, we refer to the design of the experiment as a **completely randomized design**.

It should be apparent, however, that randomization protects against the effects of the extraneous factors only in a probabilistic sort of way. For instance, in our example it is possible, though very unlikely, that detergent $A$ will be randomly assigned to the five swatches which happen to be the least soiled or that the water happens to be coldest when we wash the five swatches with detergent $B$. It is partly for this reason that we often try to control some of the factors and randomize the others, and thus use designs that are somewhere between the two extremes which we have described.

To introduce another important concept in the design of experiments, let us consider the following data on the amount of time (in minutes) it took a certain person to drive to work, Monday through Friday, along four different routes:

| | |
|---|---|
| *Route 1*: | 22, 26, 25, 25, 31 |
| *Route 2*: | 25, 27, 28, 26, 29 |
| *Route 3*: | 26, 29, 33, 30, 33 |
| *Route 4*: | 26, 28, 27, 30, 30 |

The means of these four samples are 25.8, 27.0, 30.2, and 28.2, and since the differences among them are fairly large, it would seem reasonable to conclude that there are some real differences in the true average time it takes the person to drive to work along the four different routes. This does not follow, however, from a one-way analysis of variance. We get $f = 2.80$, and since this does not exceed $f_{.05,3,16} = 3.24$, the null hypothesis cannot be rejected.

Of course, the null hypothesis may be true, but observe that there are not only considerable differences among the four means, but also large differences among the values within the samples. In the first sample they range from 22 to 31, in the second sample from 25 to 29, in the third sample from 26 to 33, and in the fourth sample from 26 to 30. Not only that, but in each sample the first value is the smallest and the last value is the largest. The latter suggests that the variation within the samples may well be due to differences in driving conditions on the different days of the week. If this is the case, variations due to driving conditions were included in the error sum of squares of the one-way analysis of variance, the denominator of the $f$ statistic was "inflated," and this may be why the results were not significant.

To avoid this kind of situation, we could hold the extraneous factor fixed, but this will seldom give us the information we want. In our example, we could limit the study to driving conditions on Monday, but then we would have no assurance that the results would apply also to driving conditions on Tuesday or on any other day of the week. Another possibility is to vary the extraneous factor deliberately over as wide a range as necessary, so that the variation it causes can be measured and, hence, eliminated from the error sum of squares. This means that we must plan the experiment in such a way that we can perform a **two-way analysis of variance**, in which the total variation of the data is partitioned into three components attributed, respectively, to treatments (in our example, the four routes), the extraneous factor (in our example, driving conditions on the different days of the week), and experimental error, or chance.

What we have suggested here is called **blocking** and the different days of the week are referred to as **blocks**. In general, blocks are the levels at which we hold an extraneous factor fixed, so that we can measure its contribution to the total variation of the data. If each treatment appears the same number of times in each block (in our example, each route is used once each day of the week), we say that the design of the experiment is a **complete block design**. Furthermore, if the treatments are distributed at random within each block (in our example, we would randomly distribute the four routes among the four Mondays, the four Tuesdays, etc.), we say that the design of the experiment is a **randomized block design**.

## 15.4 TWO-WAY ANALYSIS OF VARIANCE

There are essentially two different ways of analyzing two-variable experiments, and they depend on whether the two variables are independent or whether they **interact**. To illustrate what we mean here by "interact," suppose that a tire manufacturer is experimenting with different treads, and that she finds that one kind is especially good for use on dirt roads while another kind is especially good for use on hard pavement. If this is the case, we say that there is an **interaction** between road conditions and tread design. In this book we shall study only the no-interaction case.

To present the theory of a two-way analysis of variance, we shall use the terminology introduced in the preceding sections and refer to the two variables as treatments and blocks; alternatively, we could refer to them also as **factor A** and **factor B**, or as **rows** and **columns**. Thus, if $x_{ij}$ for $i = 1, 2, \ldots , k$ and $j = 1, 2, \ldots , n$ are values of independent random variables having normal distributions with the respective means $\mu_{ij}$ and the common variance $\sigma^2$, we shall consider the array

|  | Block 1 | Block 2 | ... | Block n |
|---|---|---|---|---|
| Treatment 1 | $x_{11}$ | $x_{12}$ | ... | $x_{1n}$ |
| Treatment 2 | $x_{21}$ | $x_{22}$ | ... | $x_{2n}$ |
| ... | ... | ... | ... | ... |
| Treatment k | $x_{k1}$ | $x_{k2}$ | ... | $x_{kn}$ |

and write the model for a two-way analysis of variance (without interaction) as

$$x_{ij} = \mu + \alpha_i + \beta_j + e_{ij}$$

for $i = 1, 2, \ldots, k$ and $j = 1, 2, \ldots, n$. Here $\mu$ is the **grand mean**, the **treatment effects** $\alpha_i$ are such that $\sum\limits_{i=1}^{k} \alpha_i = 0$, the **block effects** $\beta_j$ are such that $\sum\limits_{j=1}^{n} \beta_j = 0$, and the $e_{ij}$ are values of independent random variables having normal distributions with zero means and the common variance $\sigma^2$. Note that

$$\mu_{ij} = \mu + \alpha_i + \beta_j$$

and, as the reader is asked to verify in Exercise 15.15,

$$\frac{\sum\limits_{i=1}^{k} \sum\limits_{j=1}^{n} \mu_{ij}}{nk} = \mu$$

The two null hypotheses we shall want to test are that the treatment effects are all equal to zero and that the block effects are all equal to zero, namely,

$$H_0: \quad \alpha_i = 0 \qquad \text{for } i = 1, 2, \ldots, k$$

and

$$H_0': \quad \beta_j = 0 \qquad \text{for } j = 1, 2, \ldots, n$$

The alternative to $H_0$ is that the treatment effects are not all equal to zero, and the alternative to $H_0'$ is that the block effects are not all equal to zero. Symbolically,

$$H_1: \quad \alpha_i \neq 0 \qquad \text{for at least one value of } i$$

and

$$H_1': \quad \beta_j \neq 0 \qquad \text{for at least one value of } j$$

The two-way analysis, itself, is based on the following generalization of Theorem 15.1, which the reader will be asked to prove in Exercise 15.14.

**THEOREM 15.3**

$$\sum_{i=1}^{k} \sum_{j=1}^{n} (x_{ij} - \bar{x}_{..})^2 = n \cdot \sum_{i=1}^{k} (\bar{x}_{i.} - \bar{x}_{..})^2 + k \cdot \sum_{j=1}^{n} (\bar{x}_{.j} - \bar{x}_{..})^2$$
$$+ \sum_{i=1}^{k} \sum_{j=1}^{n} (x_{ij} - \bar{x}_{i.} - \bar{x}_{.j} + \bar{x}_{..})^2$$

where $\bar{x}_{i.}$ is the mean of the observations for the $i$th treatment, $\bar{x}_{.j}$ is the mean of the observations for the $j$th block, and $\bar{x}_{..}$ is the mean of all $nk$ observations.

The expression on the left-hand side of the identity of Theorem 15.3 is the total sum of squares SST as defined on page 550 and the first term on the right-hand side is the treatment sum of squares SS(Tr). Measuring the variation among the $\bar{x}_{.j}$, the second term on the right-hand side is the **block sum of squares** SSB, and the third term on the right-hand side is the *new* error sum of squares SSE. Thus, we have

$$SST = SS(Tr) + SSB + SSE$$

and it can be shown that if $H_0$ is true, then $\dfrac{SS(Tr)}{\sigma^2}$ and $\dfrac{SSE}{\sigma^2}$ are values of independent random variables having chi-square distributions with $k-1$ and $(n-1)(k-1)$ degrees of freedom. If $H_0$ is not true, then SS(Tr) will also reflect the variation among the $\alpha_i$, and according to Theorem 8.14 we reject $H_0$ if $f_{Tr} \geq f_{\alpha,k-1,(n-1)(k-1)}$, where

$$f_{Tr} = \frac{\dfrac{SS(Tr)}{(k-1)\sigma^2}}{\dfrac{SSE}{(n-1)(k-1)\sigma^2}} = \frac{MS(Tr)}{MSE}$$

Here and below, the mean squares are again the respective sums of squares divided by their degrees of freedom.

Similarly, if $H_0'$ is true, then $\dfrac{SSB}{\sigma^2}$ and $\dfrac{SSE}{\sigma^2}$ are values of independent random variables having chi-square distributions with $n-1$ and $(n-1)(k-1)$ degrees of freedom. If $H_0'$ is not true, then SSB will also reflect the variation among the $\beta_j$, and according to Theorem 8.14 we reject $H_0'$ if $f_B \geq f_{\alpha,n-1,(n-1)(k-1)}$, where

$$f_B = \frac{\dfrac{SSB}{(n-1)\sigma^2}}{\dfrac{SSE}{(n-1)(k-1)\sigma^2}} = \frac{MSB}{MSE}$$

This kind of analysis is called a **two-way analysis of variance**, and the necessary details are usually presented in the following kind of analysis-of-variance table:

| Source of variation | Degrees of freedom | Sum of squares | Mean square | $f$ |
|---|---|---|---|---|
| Treatments | $k - 1$ | SS(Tr) | MS(Tr) | $f_{\text{Tr}} = \dfrac{\text{MS(Tr)}}{\text{MSE}}$ |
| Blocks | $n - 1$ | SSB | MSB | $f_B = \dfrac{\text{MSB}}{\text{MSE}}$ |
| Error | $(n - 1)(k - 1)$ | SSE | MSE | |
| Total | $nk - 1$ | SST | | |

To simplify the calculations, SST and SS(Tr) are usually determined by means of the formulas of Theorem 15.2, and SSB can be determined by means of the following formula, which the reader will be asked to derive in Exercise 15.17.

---

**THEOREM 15.4**

$$\text{SSB} = \frac{1}{k} \cdot \sum_{j=1}^{n} T_{\cdot j}^2 - \frac{1}{kn} \cdot T_{\cdot\cdot}^2$$

where $T_{\cdot j}$ is the total of the values obtained for the $j$th block and $T_{\cdot\cdot}$ is the grand total of all $nk$ observations.

---

Then, the value of SSE can be obtained by subtracting SS(Tr) and SSB from SST.

## EXAMPLE 15.2

With reference to the illustration on page 558, where we had

| | Monday | Tuesday | Wednesday | Thursday | Friday |
|---|---|---|---|---|---|
| Route 1 | 22 | 26 | 25 | 25 | 31 |
| Route 2 | 25 | 27 | 28 | 26 | 29 |
| Route 3 | 26 | 29 | 33 | 30 | 33 |
| Route 4 | 26 | 28 | 27 | 30 | 30 |

test at the 0.05 level of significance whether the differences among the means obtained for the different routes (treatments) are significant, and also whether the differences among the means obtained for the different days of the week (blocks) are significant.

## Solution

1. $H_0$:   $\alpha_i = 0$ for $i = 1, 2, 3, 4$
   $H_0'$:   $\beta_j = 0$ for $j = 1, 2, 3, 4, 5$
   $H_1$:   $\alpha_i \neq 0$ for at least one value of $i$
   $H_1'$:   $\beta_j \neq 0$ for at least one value of $j$
   $\alpha = 0.05$ for both tests.

2. Reject the null hypothesis for treatments if $f_{\text{Tr}} \geqslant 3.49$ and reject the null hypothesis for blocks if $f_B \geqslant 3.26$, where $f_{\text{Tr}}$ and $f_B$ are obtained by means of a two-way analysis of variance and 3.49 and 3.26 are, respectively, the values of $f_{.05,3,12}$ and $f_{.05,4,12}$.

3. The required sums and sums of squares are $T_1. = 129$, $T_2. = 135$, $T_3. = 151$, $T_4. = 141$, $T_{.1} = 99$, $T_{.2} = 110$, $T_{.3} = 113$, $T_{.4} = 111$, $T_{.5} = 123$, $T_{..} = 556$, and $\Sigma\Sigma\, x^2 = 15{,}610$, and substitution of these values together with $k = 4$ and $n = 5$ into the formulas of Theorems 15.2 and 15.4 yields

$$\text{SST} = 15{,}610 - \tfrac{1}{20}\,(556)^2$$

$$= 153.2$$

$$\text{SS(Tr)} = \tfrac{1}{5}\,(129^2 + 135^2 + 151^2 + 141^2) - \tfrac{1}{20}\,(556)^2$$

$$= 52.8$$

$$\text{SSB} = \tfrac{1}{4}\,(99^2 + 110^2 + 113^2 + 111^2 + 123^2) - \tfrac{1}{20}\,(556)^2$$

$$= 73.2$$

and, hence,

$$\text{SSE} = 153.2 - 52.8 - 73.2$$

$$= 27.2$$

The remaining calculations are shown in the following analysis-of-variance table:

| Source of variation | Degrees of freedom | Sum of squares | Mean square | $f$ |
|---|---|---|---|---|
| Treatments | 3 | 52.8 | $\frac{52.8}{3} = 17.6$ | $\frac{17.6}{2.27} = 7.75$ |
| Blocks | 4 | 73.2 | $\frac{73.2}{4} = 18.3$ | $\frac{18.3}{2.27} = 8.06$ |
| Error | 12 | 27.2 | $\frac{27.2}{12} = 2.27$ | |
| Total | 19 | 153.2 | | |

4.  Since $f_{\text{Tr}} = 7.75$ exceeds 3.49 and $f_B = 8.06$ exceeds 3.26, both null hypotheses must be rejected. In other words, the differences among the means obtained for the four routes are significant, and so are the differences among the means obtained for the different days of the week. Note, however, that we cannot conclude that Route 1 is necessarily fastest and that on Fridays traffic conditions are always the worst. All we have shown by means of the analysis is that differences exist, and if we want to go one step further and pinpoint the nature of the differences, we will have to use one of the **multiple comparisons tests** referred to on page 569.    ▲

## EXERCISES

**15.14** Make use of the identity

$$x_{ij} - \bar{x}.. = (\bar{x}_{i.} - \bar{x}..) + (\bar{x}_{.j} - \bar{x}..) + (x_{ij} - \bar{x}_{i.} - \bar{x}_{.j} + \bar{x}..)$$

to prove Theorem 15.3.

**15.15** With reference to the notation on page 560, show that

$$\frac{\sum_{i=1}^{k} \sum_{j=1}^{n} \mu_{ij}}{nk} = \mu$$

**15.16** For the two-way analysis of variance with $k$ treatments and $n$ blocks, show that

$$E\left[\frac{k \cdot \sum_{j=1}^{n} (\bar{X}_{.j} - \bar{X}..)^2}{n-1}\right] = \sigma^2 + \frac{k \cdot \sum_{j=1}^{n} \beta_j^2}{n-1}$$

**15.17** Prove Theorem 15.4.

**15.18** A **Latin square** is a square array in which each letter (or some other kind of symbol) appears exactly once in each row and once in each column. For instance,

| A | B | C | D |
|---|---|---|---|
| B | C | D | A |
| C | D | A | B |
| D | A | B | C |

is a $4 \times 4$ Latin square. If we look upon the $m$ rows of a Latin square as the levels of one variable, the $m$ columns as the levels of a second variable, and $A$, $B$, $C$, $\ldots$ , as $m$ "treatments," namely, as the levels of a third variable, it is possible to test hypotheses concerning all three of these variables on the basis of as few as $m^2$ observations (provided there are no interactions). Letting $x_{ij(k)}$ denote the observation in the $i$th row and the $j$th column of a Latin square (so that $k$, denoting the treatment, is determined when we give $i$ and $j$), we write the model equation as

$$x_{ij(k)} = \mu + \alpha_i + \beta_j + \tau_k + e_{ij}$$

for $i = 1, 2, \ldots , m$, $j = 1, 2, \ldots , m$, and $k = 1, 2, \ldots , m$, where $\mu$ is the grand mean, the **row effects** $\alpha_i$ are such that $\sum_{i=1}^{m} \alpha_i = 0$, the **column effects** $\beta_j$ are such that $\sum_{j=1}^{m} \beta_j = 0$, the treatment effects $\tau_k$ are such that $\sum_{k=1}^{m} \tau_k = 0$, and the $e_{ij}$ are values of independent random variables having normal distributions with zero means and the common variance $\sigma^2$. The null hypotheses we shall want to test (against appropriate alternatives) are that the row effects are all zero, that the column effects are all zero, and that the treatment effects are all zero.

(a)    Show that

$$\sum_{i=1}^{m} \sum_{j=1}^{m} (x_{ij(k)} - \bar{x}_{..})^2 = m \cdot \sum_{i=1}^{m} (\bar{x}_{i.} - \bar{x}_{..})^2 + m \cdot \sum_{j=1}^{m} (\bar{x}_{.j} - \bar{x}_{..})^2$$

$$+ m \cdot \sum_{k=1}^{m} (\bar{x}_{(k)} - \bar{x}_{..})^2 + \sum_{i=1}^{m} \sum_{j=1}^{m} (x_{ij(k)} - \bar{x}_{i.} - \bar{x}_{.j} - \bar{x}_{(k)} + 2\bar{x}_{..})^2$$

where $\bar{x}_{(k)}$ is the mean of all the observations for the $k$th treatment and the other means are as defined in Theorem 15.3. The expression on the left-hand side of the above identity is the total sum of squares SST, while those on the right-hand side are, respectively, the **row sum of squares** SSR, the **column sum of squares** SSC, the treatment sum of squares SS(Tr), and the error sum of squares SSE.

(b)   Construct an analysis-of-variance table for this kind of experiment, determining the degrees of freedom for SSE by subtracting those for SSR, SSC, and SS(Tr) from $m^2 - 1$, the degrees of freedom for SST.

## APPLICATIONS

**15.19**   An experiment was performed to judge the effect of four different fuels and three different types of launchers on the range of a certain rocket. Test, on the basis of the following ranges, in miles, whether there is a significant effect due to differences in fuels and whether there is a significant effect due to differences in launchers:

|  | Fuel 1 | Fuel 2 | Fuel 3 | Fuel 4 |
|---|---|---|---|---|
| Launcher X | 45.9 | 57.6 | 52.2 | 41.7 |
| Launcher Y | 46.0 | 51.0 | 50.1 | 38.8 |
| Launcher Z | 45.7 | 56.9 | 55.3 | 48.1 |

Use the 0.01 level of significance.

**15.20**   The following are the cholesterol contents, in milligrams per package, which four laboratories obtained for 6-ounce packages of three very similar diet foods:

|  | Diet food A | Diet food B | Diet food C |
|---|---|---|---|
| Laboratory 1 | 3.4 | 2.6 | 2.8 |
| Laboratory 2 | 3.0 | 2.7 | 3.1 |
| Laboratory 3 | 3.3 | 3.0 | 3.4 |
| Laboratory 4 | 3.5 | 3.1 | 3.7 |

Perform a two-way analysis of variance and test the null hypotheses concerning the diet foods and the laboratories at the 0.05 level of significance.

**15.21**   A laboratory technician measures the breaking strength of each of five kinds of linen threads by using four different measuring instruments, $I_1$, $I_2$, $I_3$, and $I_4$, and obtains the following results, in ounces:

|  | $I_1$ | $I_2$ | $I_3$ | $I_4$ |
|---|---|---|---|---|
| *Thread 1* | 20.9 | 20.4 | 19.9 | 21.9 |
| *Thread 2* | 25.0 | 26.2 | 27.0 | 24.8 |
| *Thread 3* | 25.5 | 23.1 | 21.5 | 24.4 |
| *Thread 4* | 24.8 | 21.2 | 23.5 | 25.7 |
| *Thread 5* | 19.6 | 21.2 | 22.1 | 22.1 |

Perform a two-way analysis of variance, using the 0.05 level of significance for both tests.

**15.22** The sample data in the following Latin square (see Exercise 15.18) are the scores obtained by nine college students of various ethnic backgrounds and various professional interests in an American history test:

|  | *Ethnic background* | | |
|---|---|---|---|
|  | *Mexican* | *German* | *Polish* |
| *Law* | A 75 | B 86 | C 69 |
| *Medicine* | B 95 | C 79 | A 86 |
| *Engineering* | C 70 | A 83 | B 93 |

In this table, $A$, $B$, and $C$ are the three instructors by whom the nine college students were taught the course in American history. Analyze these data by the method of Exercise 15.18 and test the following hypotheses at the 0.05 level of significance:

(a)  Having a different instructor has no effect on the scores.
(b)  Differences in ethnic background have no effect on the scores.
(c)  Differences in professional interest have no effect on the scores.

**15.23** Among the nine persons interviewed in a poll, three are Easterners, three are Southerners, and three are Westerners. By profession, three of them are teachers, three are lawyers, and three are doctors, and no two of the same profession come from the same part of the United States. Also, three are Democrats, three are Republicans, and three are Independents, and no two of the same political affiliation are of the same profession or come from the same part of the United States. If one of the teachers is an Easterner and an Independent, another teacher is a Southerner and a Republican, and one of the lawyers is a Southerner and a Democrat, what is the political affiliation of the doctor who is a Westerner? [*Hint*: Construct a Latin square (see Exercise 15.18) with $m = 3$.] This exercise is a simplified version of a famous problem posed by R. A. Fisher in his classical work, *The Design of Experiments*.

## 15.5  SOME FURTHER CONSIDERATIONS

In this chapter we have presented a brief introduction to some of the basic methods and ideas of analysis of variance and experimental design. The scope of these subjects, which are closely interrelated, is vast, and new methods are constantly being developed as their need arises in experimentation.

The designs we have discussed all had the special feature that there were observations corresponding to all possible combinations of the values (levels) of the variables under consideration. To show that this can be very impractical or even physically impossible, we have only to consider an experiment in which we want to compare the yield of 25 varieties of wheat and, at the same time, the effect of 12 different fertilizers. To perform an experiment in which each of the 25 varieties of wheat is used in conjunction with each of the 12 fertilizers, we would have to plant 300 plots, and it does not require much imagination to see how difficult it would be to find that many test plots for which soil composition, irrigation, slope, . . . , are constant or otherwise controllable. Consequently, there is a need for designs which make it possible to test hypotheses concerning the most relevant (though not all) parameters of the model on the basis of experiments which are feasible from a practical point of view. This leads to so-called **incomplete block designs**, which are discussed in the general references on experimental design listed at the end of the chapter.

Further complications arise when there are extraneous variables which can be measured but not controlled. For example, in a comparison of various kinds of "teaching machines" it may be impossible to use persons who all have the same IQ, but at least their IQ's can be determined. In a situation like that we might use an **analysis-of-covariance** model such as

$$x_{ij} = \mu + \alpha_i + \beta y_{ij} + e_{ij}$$

which differs from the one-way analysis of variance model in that we added the term $\beta y_{ij}$, where the $y_{ij}$ are the given IQ's. Note that in this model the estimation of $\beta$ is essentially a problem of regression.

Other difficulties arise when the parameters $\alpha_i$ and $\beta_j$ in an analysis-of-variance model are not constants, but values of random variables. This kind of situation would arise, for example, if there are 25 varieties of wheat and 12 kinds of fertilizers and we randomly select, say, six of the varieties of wheat and three of the fertilizers to be included in an experiment.

These are just some of the generalizations of the methods we have presented in this chapter; they are treated in detail in the general texts on analysis of variance and experimental design which are listed below.

## REFERENCES

A proof of the independence of the chi-square random variables whose values constitute the various sums of squares in an analysis of variance, for instance SS(Tr) and SSE in a one-way analysis, may be found in

SCHEFFÉ, H., *The Analysis of Variance*. New York: John Wiley & Sons, Inc., 1959,

and a discusson of various multiple comparisons tests is given in

FEDERER, W. T., *Experimental Design, Theory and Application*. New York: Macmillan Publishing Co., Inc., 1955.

The following are some general texts on analysis of variance and experimental design:

ANDERSON, V. L., and MCLEAN, R. A., *Design of Experiments: A Realistic Approach*. New York: Marcel Dekker, Inc., 1974,

COCHRAN, W. G., and COX, G. M., *Experimental Design*, 2nd ed. New York: John Wiley & Sons, Inc., 1957,

FINNEY, D. J., *An Introduction to the Theory of Experimental Design*. Chicago: University of Chicago Press, 1960,

GUENTHER, W. C., *Analysis of Variance*. Englewood Cliffs, N.J.: Prentice-Hall, Inc., 1964,

HICKS, C. R., *Fundamental Concepts in the Design of Experiments*, 2nd ed. New York: Holt, Rinehart and Winston, Inc., 1973,

MONTGOMERY, D. C., *Design and Analysis of Experiments*, 3rd ed. New York: John Wiley & Sons, Inc., 1991,

SNEDECOR, G. W., and COCHRAN, W. G., *Statistical Methods*, 8th ed. Ames, Iowa: Iowa University Press, 1989.

# 16

# *Nonparametric Tests*

## 16.1 INTRODUCTION

In Chapter 10 we introduced the concept of **robustness** in connection with problems of estimation. Let us now extend this concept to tests of hypotheses, which are said to be robust if the sampling distributions of the test statistics are not seriously affected by violations of underlying assumptions.

In connection with tests of hypotheses, it is especially important to know whether violations of underlying assumptions might affect the level of significance. As we saw in Section 12.5, any comparison of the power functions of two or more tests requires that the levels of significance are equal; and if this is not the case, the comparison is invalid. For instance, the one-sample $t$ test of Section 13.3 requires that our sample comes from a normal population. So, what happens when the population is "not quite normal"—say, if it is bell-shaped but not perfectly symmetrical? Computer simulations have shown that even though a population may depart somewhat from normality, most of the time the level of significance will still be close to the prescribed value $\alpha$.

The following examples show how violating underlying assumptions about a population may affect the level of significance. Suppose that we want to test the null hypothesis $\mu = \mu_0$ at the 0.05 level of significance, where $\mu$ is the mean of a normal population with the *known* standard deviation $\sigma$, but there is an appreciable probability (say, one in 50) that one of the values will be recorded incorrectly. In connection with the test illustrated by Example 13.1, we are thus violating the assumption that we are dealing with a random sample from a normal population. If one of the values in Example 13.1 had been recorded incorrectly, say, as 7.452 ounces instead of 7.952 ounces, the mean weight of the 25 packages of cookies would have been reduced by

$$\frac{7.952 - 7.452}{25} = 0.020$$

ounce, $z$ would have been reduced from 2.48 to 2.22, and the corresponding $P$-value would have increased from 0.0046 to 0.0264. Since the new $P$-value exceeds 0.025, the null hypothesis can no longer be rejected; this shows how $P$-values and, hence, the level of significance can be affected when we allow for the possibility of incorrectly recording the data.

Now suppose that in a problem like the one given above, $\sigma$ is *unknown*, so that the standard procedure would be the one-sample $t$ test illustrated by Example 13.3. In that case, an error in recording a value will affect the sample standard deviation as well as the sample mean, which appear, respectively, in the denominator and the numerator of the test statistic. As is illustrated for a special case by Exercise 16.1, this will often yield values of $t$ closer to $+1$ or $-1$ and, hence, make it more difficult to reject the null hypothesis. In other words, with the risk of such an error, the level of significance may well be less than the prescribed value $\alpha$. This is also illustrated by Exercises 13.16 and 13.17 on page 467.

Since there are many situations in which we face serious questions about the robustness of tests of hypotheses, especially with regard to the assumption of normality, statisticians have developed alternative techniques which require fewer assumptions, if any. These tests are generally referred to as **nonparametric**; they include tests that are **distribution-free** (where we make no assumptions about the populations, except, perhaps, that they are continuous) and also tests that are nonparametric only in the sense that we are not concerned with specific parameters of given populations.

Aside from the fact that nonparametric tests can be used under more general conditions than the standard tests which they replace, they have considerable intuitive appeal—for the most part, they are easy to explain and easy to understand.

Also, in many nonparametric tests the computational burden is so light that they come under the heading of "quick-and-easy," or "shortcut," techniques. Partly for these reasons, nonparametric tests have become very popular, and extensive literature is devoted to their theory and application.

The main disadvantage of nonparametric tests is that they are often wasteful of information and thus less efficient than the standard techniques which they replace. It should be observed, however, that efficiency comparisons usually assume that the conditions underlying the standard tests are met, and hence they tend to understate the real worth of nonparametric methods when it comes to questions of robustness. In general, it is true that *the less one assumes, the less one can infer from a set of data*, but it is also true that *the less one assumes, the more one broadens the applicability of one's method*.

## 16.2 THE SIGN TEST

The **sign test** is often used as a nonparametric alternative to the one-sample $t$ test, where we test the null hypothesis $\mu = \mu_0$ against a suitable alternative. For the sign test, we assume merely that the population sampled is continuous and symmetrical. We assume that the population is continuous so that there is zero probability of getting a value equal to $\mu_0$, and we do not even need the assumption of symmetry if we change the null hypothesis to $\tilde{\mu} = \tilde{\mu}_0$, where $\tilde{\mu}$ is the population median.

In the sign test we replace each sample value exceeding $\mu_0$ with a plus sign and each value less than $\mu_0$ with a minus sign, and then we test the null hypothesis that the number of plus signs is a value of a random variable having the binomial distribution with the parameters $n$ (the total number of plus or minus signs) and $\theta = \frac{1}{2}$. The two-sided alternative $\mu \neq \mu_0$ thus becomes $\theta \neq \frac{1}{2}$, and the one-sided alternatives $\mu < \mu_0$ and $\mu > \mu_0$ become $\theta < \frac{1}{2}$ and $\theta > \frac{1}{2}$, respectively. If a sample value equals $\mu_0$, which can happen when we deal with rounded data even though the population is continuous, we simply discard it.

To perform a sign test when the sample size is very small, we refer directly to a table of binomial probabilities such as Table I; when the sample size is large, we use the normal approximation to the binomial distribution.

## EXAMPLE 16.1

The following are measurements of the breaking strength of a certain kind of 2-inch cotton ribbon in pounds:

| 163 | 165 | 160 | 189 | 161 | 171 | 158 | 151 | 169 | 162 |
|-----|-----|-----|-----|-----|-----|-----|-----|-----|-----|
| 163 | 139 | 172 | 165 | 148 | 166 | 172 | 163 | 187 | 173 |

Use the sign test to test the null hypothesis $\mu = 160$ against the alternative hypothesis $\mu > 160$ at the 0.05 level of significance.

*Solution*

1.  $H_0$:  $\mu = 160$
    $H_1$:  $\mu > 160$
    $\alpha = 0.05$

2′.  Use the test statistic $X$, the observed number of plus signs.

3′.  Replacing each value exceeding 160 with a plus sign, each value less than 160 with a minus sign, and discarding the one value which equals 160, we get

$$+ + + + + - - + + + - + + - + + + + +$$

so that $n = 19$ and $x = 15$. From Table I we find that $P(X \geq 15) = 0.0095$ for $\theta = \frac{1}{2}$.

4′.  Since the $P$-value, 0.0095, is less than 0.05, the null hypothesis must be rejected and we conclude that the mean breaking strength of the given kind of ribbon exceeds 160 pounds.    ▲

## EXAMPLE 16.2

The following data, in tons, are the amounts of sulfur oxides emitted by a large industrial plant in 40 days:

| | | | | | | | | | |
|--|--|--|--|--|--|--|--|--|--|
| 17 | 15 | 20 | 29 | 19 | 18 | 22 | 25 | 27 | 9 |
| 24 | 20 | 17 | 6 | 24 | 14 | 15 | 23 | 24 | 26 |
| 19 | 23 | 28 | 19 | 16 | 22 | 24 | 17 | 20 | 13 |
| 19 | 10 | 23 | 18 | 31 | 13 | 20 | 17 | 24 | 14 |

Use the sign test to test the null hypothesis $\mu = 21.5$ against the alternative hypothesis $\mu < 21.5$ at the 0.01 level of significance.

*Solution*

1.  $H_0$:  $\mu = 21.5$
    $H_1$:  $\mu < 21.5$
    $\alpha = 0.01$

2.  Reject the null hypothesis if $z \leq -z_{.01} = -2.33$, where

$$z = \frac{x - n\theta}{\sqrt{n\theta(1 - \theta)}}$$

with $\theta = \frac{1}{2}$, and $x$ is the number of plus signs (values exceeding 21.5).

3. Since $n = 40$ and $x = 16$, we get $n\theta = 40 \cdot \frac{1}{2} = 20$, $\sqrt{n\theta(1 - \theta)} = \sqrt{40(0.5)(0.5)} = 3.16$, and, hence,

$$z = \frac{16 - 20}{3.16} = -1.26$$

4. Since $z = -1.26$ exceeds $-2.33$, the null hypothesis cannot be rejected.   ▲

The sign test can also be used when we deal with paired data, as in Exercises 13.28 and 13.29. In such problems, each pair of sample values is replaced by a plus sign if the difference between the paired observations is positive (that is, if the first value exceeds the second value) and by a minus sign if the difference between the paired observations is negative (that is, if the first value is less than the second value), and it is discarded if the difference is zero. To test the null hypothesis that two continuous symmetrical populations have equal means (or that two continuous populations have equal medians), we can thus use the sign test, which, in connection with this kind of problem, is referred to as the **paired-sample sign test**. When the sign test is used as in Examples 16.1 and 16.2, we refer to it as the **one-sample sign test**.

## EXAMPLE 16.3

To determine the effectiveness of a new traffic-control system, the number of accidents that occurred at 12 dangerous intersections during four weeks before and four weeks after the installation of the new system was observed, and the following data were obtained:

3 and 1,   5 and 2,   2 and 0,   3 and 2,   3 and 2,   3 and 0
0 and 2,   4 and 3,   1 and 3,   6 and 4,   4 and 1,   1 and 0

Use the paired-sample sign test at the 0.05 level of significance to test the null hypothesis that the new traffic-control system is only as effective as the old system. (The populations sampled are not continuous, but this does not matter so long as zero differences are discarded.)

*Solution*

1. $H_0$: $\mu_1 = \mu_2$
   $H_1$: $\mu_1 > \mu_2$
   $\alpha = 0.05$

2'. Use the test statistic $X$, the observed number of plus signs.

3'. Replacing each positive difference by a plus sign and each negative difference by a minus sign, we get

$$+ + + + + + - + - + + +$$

so that $n = 12$ and $x = 10$. From Table I we find that $P(X \geq 10) = 0.0192$ for $\theta = \frac{1}{2}$.

4'. Since the $P$-value, 0.0192, is less than 0.05, the null hypothesis must be rejected, and we conclude that the new traffic-control system is effective in reducing the number of accidents at dangerous intersections.  ▲

## 16.3 THE SIGNED-RANK TEST

As we saw in Section 16.2, the sign test is easy to perform, but since we utilize only the signs of the differences between the observations and $\mu_0$ in the one-sample case, or the signs of the differences between the pairs of observations in the paired-sample case, it tends to be wasteful of information. An alternative non-parametric test, the **Wilcoxon signed-rank test**, is less wasteful in that it takes into account also the magnitudes of the differences. In this test, we rank the differences without regard to their signs, assigning rank 1 to the smallest difference in absolute value, rank 2 to the second smallest difference in absolute value, ... , and rank $n$ to the largest difference in absolute value. Zero differences are again discarded, and if the absolute values of two or more differences are the same, we assign each one the mean of the ranks which they jointly occupy. Then, the signed-rank test is based on $T^+$, the sum of the ranks assigned to the positive differences, $T^-$, the sum of the ranks assigned to the negative differences, $T^+ - T^-$, or $T = \min(T^+, T^-)$. Since

$$T^+ + T^- = \frac{n(n + 1)}{2}$$

the resulting tests are all equivalent. (Note that we are using this traditional notation even though it conflicts with the practice of using capital letters for random variables and the corresponding lowercase letters for their values. This avoids confusion between the statistics used here and the $t$ statistics of Chapter 13.)

Since the sum of $T^+$ and $T^-$ is always $\dfrac{n(n+1)}{2}$ and they are both values of random variables which take on values on the interval from 0 to $\dfrac{n(n+1)}{2}$ with distributions that are symmetrical about $\dfrac{n(n+1)}{4}$, we can picture the relationship among the distributions of the random variables corresponding to $T^+$, $T^-$, and $T$ as in Figure 16.1 for $n = 5$.

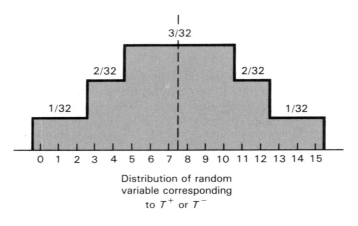

Distribution of random
variable corresponding
to $T^+$ or $T^-$

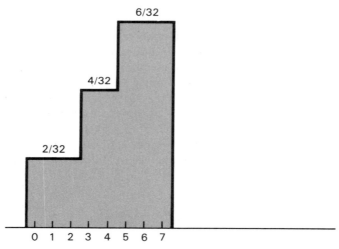

Distribution of random variable
corresponding to $T$

**Figure 16.1**   Distributions of random variables corresponding to $T^+$, $T^-$, and $T$ for $n = 5$.

Depending on the alternative hypothesis, we base the signed-rank test on $T$, $T^+$, or $T^-$, with the assumptions and the null hypotheses being the same as in Sections 16.1 and 16.2. We have to be careful, though, to use the right statistic and the right critical value, as summarized in the following table, where in each case the level of significance is $\alpha$:

| Alternative hypothesis | Reject the null hypothesis if: |
|---|---|
| $\mu \neq \mu_0$ | $T \leq T_\alpha$ |
| $\mu > \mu_0$ | $T^- \leq T_{2\alpha}$ |
| $\mu < \mu_0$ | $T^+ \leq T_{2\alpha}$ |

The critical values in the right-hand column of this table, $T_\alpha$ or $T_{2\alpha}$, are the largest values for which the corresponding $P$-values do not exceed $\alpha$ or $2\alpha$, respectively. They may be obtained from Table IX for values of $n$ not exceeding 25. Note that the same critical values can serve for tests at different levels of significance depending on whether the alternative hypothesis is one-sided or two-sided. For instance, $T_{.02}$ can serve as a critical value at the 0.02 level of significance when the alternative hypothesis is two-sided and at the 0.01 level of significance when the alternative hypothesis is one-sided. This may seem confusing, but it is how these critical values are tabulated in some texts.

## EXAMPLE 16.4

The following are 15 measurements of the octane rating of a certain kind of gasoline: 97.5, 95.2, 97.3, 96.0, 96.8, 100.3, 97.4, 95.3, 93.2, 99.1, 96.1, 97.6, 98.2, 98.5, and 94.9. Use the signed-rank test at the 0.05 level of significance to test whether or not the mean octane rating of the given kind of gasoline is 98.5.

### Solution

1. $H_0$:   $\mu = 98.5$
   $H_1$:   $\mu \neq 98.5$
   $\alpha = 0.05$

2. Reject the null hypothesis if $T \leq T_{.05}$, where $T_{.05}$ must be read from Table IX for the appropriate value of $n$.

3. Subtracting 98.5 from each value and ranking the differences without regard to their sign, we get

| Measurement | Difference | Rank |
|:---:|:---:|:---:|
| 97.5 | −1.0 | 4 |
| 95.2 | −3.3 | 12 |
| 97.3 | −1.2 | 6 |
| 96.0 | −2.5 | 10 |
| 96.8 | −1.7 | 7 |
| 100.3 | 1.8 | 8 |
| 97.4 | −1.1 | 5 |
| 95.3 | −3.2 | 11 |
| 93.2 | −5.3 | 14 |
| 99.1 | 0.6 | 2 |
| 96.1 | −2.4 | 9 |
| 97.6 | −0.9 | 3 |
| 98.2 | −0.3 | 1 |
| 98.5 | 0.0 | |
| 94.9 | −3.6 | 13 |

so that $T^- = 4 + 12 + 6 + 10 + 7 + 5 + 11 + 14 + 9 + 3 + 1 + 13 = 95$, $T^+ = 8 + 2 = 10$, and $T = 10$. From Table IX we find that $T_{.05} = 21$ for $n = 14$.

4.   Since $T = 10$ is less than $T_{.05} = 21$, the null hypothesis must be rejected; the mean octane rating of the given kind of gasoline is not 98.5.   ▲

When we deal with paired data, the signed-rank test can also be used in place of the paired-sample sign test. In that case, we test the null hypothesis $\mu_1 = \mu_2$ using the test criteria given in the table on page 577, except that the alternative hypotheses are now $\mu_1 \neq \mu_2$, $\mu_1 > \mu_2$, or $\mu_1 < \mu_2$ instead of $\mu \neq \mu_0$, $\mu > \mu_0$, or $\mu < \mu_0$.

For $n \geq 15$ it is considered reasonable to assume that $T^+$ is a value of a random variable having approximately a normal distribution. To perform the signed-rank test based on this assumption, we need the following results, which apply regardless of whether the null hypothesis is $\mu = \mu_0$ or $\mu_1 = \mu_2$.

---

**THEOREM 16.1**   Under the assumptions required by the signed-rank test, $T^+$ is a value of a random variable with the mean

$$\mu = \frac{n(n + 1)}{4}$$

and the variance

$$\sigma^2 = \frac{n(n + 1)(2n + 1)}{24}$$

---

***Proof.***   Expressed in terms of ranks and signed differences, the null hypotheses for the one-sample and paired-sample signed-rank tests may be stated as follows: For each rank, the probabilities that it will be assigned to a positive difference or to a negative difference are both $\frac{1}{2}$. Thus, we can write

$$T^+ = 1 \cdot x_1 + 2 \cdot x_2 + \cdots + n \cdot x_n$$

where $x_1, x_2, \ldots, x_n$ are values of independent random variables having the Bernoulli distribution with $\theta = \frac{1}{2}$. Since $E(X_i) = \theta = \frac{1}{2}$ and $\text{var}(X_i) = \theta(1 - \theta) = \frac{1}{4}$ for $i = 1, 2, \ldots, n$ by Theorem 5.2 with $n = 1$, it follows that

$$\begin{aligned}
\mu &= 1 \cdot \tfrac{1}{2} + 2 \cdot \tfrac{1}{2} + \cdots + n \cdot \tfrac{1}{2} \\
&= \frac{1 + 2 + \cdots + n}{2} \\
&= \frac{n(n + 1)}{4}
\end{aligned}$$

Also, according to the corollary to Theorem 14.4, we find that

$$\begin{aligned}
\sigma^2 &= 1^2 \cdot \tfrac{1}{4} + 2^2 \cdot \tfrac{1}{4} + \cdots + n^2 \cdot \tfrac{1}{4} \\
&= \frac{1^2 + 2^2 + \cdots + n^2}{4} \\
&= \frac{n(n + 1)(2n + 1)}{24}
\end{aligned}$$

We made use here of the familiar formulas for the sum and the sum of the squares of the first $n$ positive integers, which are proved in the appendix at the end of the book.   ▼

Note also that, by symmetry, the results of Theorem 16.1 apply if we substitute $T^-$ for $T^+$.

## EXAMPLE 16.5

The following are the weights in pounds, before and after, of 16 persons who stayed on a certain reducing diet for four weeks:

| Before | After |
|--------|-------|
| 147.0 | 137.9 |
| 183.5 | 176.2 |
| 232.1 | 219.0 |
| 161.6 | 163.8 |
| 197.5 | 193.5 |
| 206.3 | 201.4 |
| 177.0 | 180.6 |
| 215.4 | 203.2 |
| 147.7 | 149.0 |
| 208.1 | 195.4 |
| 166.8 | 158.5 |
| 131.9 | 134.4 |
| 150.3 | 149.3 |
| 197.2 | 189.1 |
| 159.8 | 159.1 |
| 171.7 | 173.2 |

Use the signed-rank test to test at the 0.05 level of significance whether the weight-reducing diet is effective.

### Solution

1.  $H_0$:  $\mu_1 = \mu_2$
    $H_1$:  $\mu_1 > \mu_2$
    $\alpha = 0.05$

2.  Reject the null hypothesis if $z \geq z_{.05} = 1.645$, where

    $$z = \frac{T^+ - \mu}{\sigma}$$

    and $\mu$ and $\sigma^2$ are given by the formulas of Theorem 16.1.

3.  The differences between the respective pairs are 9.1, 7.3, 13.1, −2.2, 4.0, 4.9, −3.6, 12.2, −1.3, 12.7, 8.3, −2.5, 1.0, 8.1, 0.7, −1.5, and if their absolute values are ranked, we find that the positive differences occupy ranks 13, 10, 16, 8, 9, 14, 15, 12, 2, 11, and 1. Thus,

    $$T^+ = 13 + 10 + 16 + 8 + 9 + 14 + 15 + 12 + 2 + 11 + 1$$
    $$= 111$$

    Since $\mu = \dfrac{16 \cdot 17}{4} = 68$ and $\sigma^2 = \dfrac{16 \cdot 17 \cdot 33}{24} = 374$, we get

$$z = \frac{111 - 68}{\sqrt{374}} = 2.22$$

4.  Since $z = 2.22$ exceeds $z_{.05} = 1.645$, the null hypothesis must be rejected; we conclude that the diet is, indeed, effective in reducing weight.    ▲

## EXERCISES

**16.1**  A random sample of size $n = 2$ is taken to test whether a normal population has the mean $\mu = 0$.
   (a)  If the observed sample values are $x_1$ and $x_2$ with $x_1 > x_2 > 0$, show that the statistic for the one-sample $t$ test can be written as

$$t = \frac{x_1 + x_2}{x_1 - x_2}$$

   (b)  If the decimal point is erroneously moved one place to the right when recording $x_1$, find an expression for $t'$, the corresponding value of the $t$ statistic, and verify that

$$1 < t' < t$$

**16.2**  Show that under the null hypotheses of Section 16.3, $T^+$ is a value of a random variable whose distribution is symmetrical about $\dfrac{n(n + 1)}{4}$.

**16.3**  With reference to the signed-rank test, find the mean and the variance of the random variable whose values are given by $T^+ - T^-$.

**16.4**  Explain why, among others, there is a blank in Table IX for $n = 5$ in the column for $T_{.02}$.

## APPLICATIONS

**16.5**  The following are the amounts of time, in minutes, which it took a random sample of 20 technicians to perform a certain task: 18.1, 20.3, 18.3, 15.6, 22.5, 16.8, 17.6, 16.9, 18.2, 17.0, 19.3, 16.5, 19.5, 18.6, 20.0, 18.8, 19.1, 17.5, 18.5, and 18.0. Assuming that this sample came from a symmetrical continuous population, use the sign test at the 0.05 level of significance to test the null hypothesis that the mean of this population is 19.4 minutes against the alternative hypothesis that it is not 19.4 minutes. Perform the test using
   (a)  Table I;
   (b)  the normal approximation to the binomial distribution.

**16.6**   Rework Exercise 16.5 using the signed-rank test based on Table IX.

**16.7**   The following are the amounts of money (in dollars) spent by 16 persons at an amusement park: 20.15, 19.85, 23.75, 18.63, 21.09, 25.63, 16.65, 19.27, 18.80, 21.45, 20.29, 19.51, 23.80, 20.00, 17.48, and 19.11. Assuming that this is a random sample from a symmetrical population and that the probability that a person will spend exactly $19.00 is extremely small, use the sign test at the 0.05 level of significance to test the null hypothesis that on the average a person spends $19.00 at the park against the alternative hypothesis that this figure is too low. Base the test on Table I.

**16.8**   Rework Exercise 16.7 using the signed-rank test based on Table IX.

**16.9**   The following are the miles per gallon obtained with 40 tankfuls of a certain kind of gasoline:

| | | | | |
|---|---|---|---|---|
| 24.1 | 25.0 | 24.8 | 24.3 | 24.2 |
| 25.3 | 24.2 | 23.6 | 24.5 | 24.4 |
| 24.5 | 23.2 | 24.0 | 23.8 | 23.8 |
| 25.3 | 24.5 | 24.6 | 24.0 | 25.2 |
| 25.2 | 24.4 | 24.7 | 24.1 | 24.6 |
| 24.9 | 24.1 | 25.8 | 24.2 | 24.2 |
| 24.8 | 24.1 | 25.6 | 24.5 | 25.1 |
| 24.6 | 24.3 | 25.2 | 24.7 | 23.3 |

Assuming that the underlying conditions are met, use the sign test at the 0.01 level of significance to test the null hypothesis $\tilde{\mu} = 24.2$ against the alternative hypothesis $\tilde{\mu} > 24.2$.

**16.10**   Rework the preceding exercise using the signed-rank test.

**16.11**   The following are the numbers of passengers carried on flights 136 and 137 between Chicago and Phoenix on 12 days:

| | | | |
|---|---|---|---|
| 232 and 189 | 265 and 230 | 249 and 236 | 250 and 261 |
| 255 and 249 | 236 and 218 | 270 and 258 | 247 and 253 |
| 249 and 251 | 240 and 233 | 257 and 254 | 239 and 249 |

Use the sign test at the 0.01 level of significance to test the null hypothesis $\mu_1 = \mu_2$ (that on the average the two flights carry equally many passengers) against the alternative hypothesis $\mu_1 > \mu_2$. Base the test on Table I.

**16.12**   Rework Exercise 16.11 using the signed-rank test based on Table IX.

**16.13**   The following are the numbers of employees absent from two government agencies on 25 days: 24 and 29, 32 and 45, 36 and 36, 33 and 39, 41 and 48, 45 and 36, 33 and 41, 38 and 39, 46 and 40, 32 and 39, 37 and 30, 34 and 45, 41 and 42, 32 and 40, 30 and 33, 46 and 42, 38 and 50, 34 and 37, 45 and 39, 32 and 37, 44 and 32, 25 and 33, 45 and 48, 35 and 33, and 30 and 35. Use the sign test at the 0.05 level of significance to test the null hypothesis $\mu_1 = \mu_2$ (that on the average equally many employees are absent from the two agencies) against the alternative hypothesis $\mu_1 < \mu_2$.

**16.14**  Rework Exercise 16.13 using the signed-rank test based on Table IX.

**16.15**  On what statistic do we base our decision and for what values of the statistic do we reject the null hypothesis, if we have a random sample of size $n = 10$ and are using the signed-rank test at the 0.05 level of significance to test the null hypothesis $\mu = \mu_0$ against the alternative hypothesis

(a)  $\mu \neq \mu_0$;
(b)  $\mu > \mu_0$;
(c)  $\mu < \mu_0$?

**16.16**  Rework Exercise 16.15 with the level of significance changed to 0.01.

**16.17**  In a random sample taken at a public playground, it took 38, 43, 36, 29, 44, 28, 40, 50, 39, 47, and 33 minutes to play a set of tennis. Use the signed-rank test at the 0.05 level of significance to test whether or not it takes on the average 35 minutes to play a set of tennis at that public playground.

**16.18**  A sample of 24 suitcases carried by an airline on transoceanic flights weighed 32.0, 46.4, 48.1, 27.7, 35.5, 52.6, 66.0, 41.3, 49.9, 36.1, 50.0, 44.7, 48.2, 36.9, 40.8, 35.1, 63.3, 42.5, 52.4, 40.9, 38.6, 43.2, 41.7, and 35.6 pounds. Test at the 0.05 level of significance whether or not the mean weight of suitcases carried by the airline on such flights is 37.0 pounds, using the signed-rank test based on

(a)  Table IX;
(b)  the results of Theorem 16.1.

**16.19**  The following is a random sample of the IQ's of husbands and wives: 108 and 103, 104 and 116, 103 and 106, 112 and 104, 99 and 99, 105 and 94, 102 and 110, 112 and 128, 119 and 106, 106 and 103, 125 and 120, 96 and 98, 107 and 117, 115 and 130, 101 and 100, 110 and 101, 103 and 96, 105 and 99, 124 and 120, and 113 and 116. Test at the 0.05 level of significance whether or not husbands and wives are on the average equally intelligent in the population sampled, using the signed-rank test based on

(a)  Table IX;
(b)  the results of Theorem 16.1.

## 16.4  RANK-SUM TESTS: THE *U* TEST

In this section we shall present a nonparametric alternative to the two-sample *t* test, which is called the ***U* test**, the **Wilcoxon test**, or the **Mann-Whitney test**, named after the statisticians who contributed to its development. Without having to assume that the two populations sampled have normal distributions, we will be able to test the null hypothesis that we are sampling identical continuous populations against the alternative that the two populations have unequal means.

To illustrate the procedure, suppose that we want to compare two kinds of emergency flares on the basis of the following burning times (rounded to the nearest tenth of a minute):

*Brand A*:   14.9, 11.3, 13.2, 16.6, 17.0, 14.1, 15.4, 13.0, 16.9
*Brand B*:   15.2, 19.8, 14.7, 18.3, 16.2, 21.2, 18.9, 12.2, 15.3, 19.4

Arranging these values jointly (as if they were one sample) in an increasing order of magnitude and assigning them in this order the ranks 1, 2, 3, ... , and 19, we find that the values of the first sample (Brand A) occupy ranks 1, 3, 4, 5, 7, 10, 12, 13, and 14, while those of the second sample (Brand B) occupy ranks 2, 6, 8, 9, 11, 15, 16, 17, 18, and 19. Had there been ties, we would have assigned to each of the tied observations the mean of the ranks which they jointly occupy.

If there is an appreciable difference between the means of the two populations, most of the lower ranks are likely to go to the values of one sample, while most of the higher ranks are likely to go to the values of the other sample. As originally proposed by Wilcoxon, the test is thus based on $W_1$, the sum of the ranks of the values of the first sample, or on $W_2$, the sum of the ranks of the values of the second sample. It does not matter whether we choose $W_1$ or $W_2$, for if there are $n_1$ values in the first sample and $n_2$ values in the second sample, $W_1 + W_2$ is always the sum of the first $n_1 + n_2$ positive integers, namely,

$$\frac{(n_1 + n_2)(n_1 + n_2 + 1)}{2}$$

In actual practice, we seldom base tests directly on $W_1$ or $W_2$; instead, we use the related statistics

$$U_1 = W_1 - \frac{n_1(n_1 + 1)}{2}$$

or

$$U_2 = W_2 - \frac{n_2(n_2 + 1)}{2}$$

or the smaller of the two, which we denote by $U$. Note that we are again departing from the practice of using capital letters for random variables and the corresponding lowercase letters for their values. (Traditionally, $U_1$, $U_2$, and $U$ have been used in connection with this test for the values of corresponding random variables, and $u$ has been used in connection with another nonparametric test, which we shall discuss in Section 16.6.)

The tests based on $U_1$, $U_2$, or $U$ are all equivalent to those based on $W_1$ or $W_2$, but they have the advantage that they lend themselves more readily to the construction of tables of critical values. As the reader will be asked to verify in Exercise 16.20, the sum of $U_1$ and $U_2$ is always equal to $n_1 n_2$, and the corresponding random variables both take on values from 0 to $n_1 n_2$. Indeed, these random variables have identical distributions, which are symmetrical about $\dfrac{n_1 n_2}{2}$.

Regardless of the alternative hypothesis, we can thus base all tests of the null hypothesis on the distribution of the random variable corresponding to $U = \min(U_1, U_2)$, but as on page 577 we have to be careful to use the right statistic and the right critical value, as summarized in the following table, where in each case the level of significance is $\alpha$:

| Alternative hypothesis | Reject the null hypothesis if: |
|---|---|
| $\mu_1 \neq \mu_2$ | $U \leq U_\alpha$ |
| $\mu_1 > \mu_2$ | $U_2 \leq U_{2\alpha}$ |
| $\mu_1 < \mu_2$ | $U_1 \leq U_{2\alpha}$ |

The critical values in the right-hand column of this table, $U_\alpha$ or $U_{2\alpha}$, are the largest values for which the corresponding *P*-values do not exceed $\alpha$ or $2\alpha$, respectively. They may be obtained from Table X for values of $n_1$ and $n_2$ not exceeding 15. Note that as in Table IX, the same critical values serve for tests at different levels of significance, depending on whether the alternative hypothesis is one-sided or two-sided. For instance, $U_{.10}$ can serve as a critical value at the 0.10 level of significance when the alternative hypothesis is two-sided and at the 0.05 level of significance when the alternative hypothesis is one-sided. As on page 577, we mention this primarily because this is how these critical values are tabulated in some texts.

## EXAMPLE 16.6

With reference to the data on page 584, test at the 0.05 level of significance whether the two samples come from identical continuous populations or whether the average burning time of Brand *A* flares is less than that of Brand *B* flares.

### Solution

1.  $H_0$:  $\mu_1 = \mu_2$
    $H_1$:  $\mu_1 < \mu_2$
    $\alpha = 0.05$

2.   Since $n_1 = 9$ and $n_2 = 10$, reject the null hypothesis if $U_1 \leq 24$, where 24 is the corresponding value of $U_{.10}$.
3.   Using the ranks obtained on page 584, we get

$$W_1 = 1 + 3 + 4 + 5 + 7 + 10 + 12 + 13 + 14$$
$$= 69$$

so that $U_1 = 69 - \dfrac{9 \cdot 10}{2} = 24.$

4.   Since $U_1 = 24$ equals $U_{.10} = 24$, the null hypothesis must be rejected; we conclude that on the average Brand $A$ flares have a shorter burning time than Brand $B$ flares.   ▲

When $n_1$ and $n_2$ are both greater than 8, it is considered reasonable to assume that $U_1$ and $U_2$ are values of random variables having approximately normal distributions. To perform the $U$ test based on this assumption, we need the following results:

---

**THEOREM 16.2**   Under the assumptions required by the $U$ test, $U_1$ and $U_2$ are values of random variables having the mean

$$\mu = \frac{n_1 n_2}{2}$$

and the variance

$$\sigma^2 = \frac{n_1 n_2 (n_1 + n_2 + 1)}{12}$$

---

***Proof.***   Under the null hypothesis that the two samples come from identical populations which are continuous (so that the probability is zero that there will be any ties), $W_1$ is the sum of $n_1$ positive integers selected at random from among the first $n_1 + n_2$ positive integers. Making use of the results of part (c) of Exercise 8.13 with $n = n_1$ and $N = n_1 + n_2$, we find that $W_1$ is the value of a random variable with the mean

$$\frac{n_1(n_1 + n_2 + 1)}{2}$$

and the variance

$$\frac{n_1 n_2 (n_1 + n_2 + 1)}{12}$$

Since $U_1 = W_1 - \dfrac{n_1(n_1 + 1)}{2}$, it follows that the mean and the variance of the random variable corresponding to $U_1$ are

$$\mu = \frac{n_1(n_1 + n_2 + 1)}{2} - \frac{n_1(n_1 + 1)}{2} = \frac{n_1 n_2}{2}$$

and

$$\sigma^2 = \frac{n_1 n_2 (n_1 + n_2 + 1)}{12}$$

Also, since $U_1 + U_2$ is always equal to $n_1 n_2$, the mean and the variance of the random variable corresponding to $U_2$ are equal to those of the random variable corresponding to $U_1$ [see part (a) of Exercise 16.20]. ▼

## EXAMPLE 16.7

The following are the weight gains (in pounds) of two random samples of young turkeys fed two different diets but otherwise kept under identical conditions:

*Diet 1*:   16.3, 10.1, 10.7, 13.5, 14.9, 11.8, 14.3, 10.2,
            12.0, 14.7, 23.6, 15.1, 14.5, 18.4, 13.2, 14.0

*Diet 2*:   21.3, 23.8, 15.4, 19.6, 12.0, 13.9, 18.8, 19.2,
            15.3, 20.1, 14.8, 18.9, 20.7, 21.1, 15.8, 16.2

Use the *U* test at the 0.01 level of significance to test the null hypothesis that the two populations sampled are identical against the alternative hypothesis that on the average the second diet produces a greater gain in weight.

### Solution

(It does not matter here whether we base the test on $U_1$ or on $U_2$.)

1. $H_0$:   $\mu_1 = \mu_2$
   $H_1$:   $\mu_1 < \mu_2$
   $\alpha = 0.01$

2. Reject the null hypothesis if $z \leq -2.33$, where

$$z = \frac{U_1 - \mu}{\sigma}$$

and $\mu$ and $\sigma^2$ are given by the formulas of Theorem 16.2.

3. Ranking the data jointly according to size, we find that the values of the first sample occupy ranks 21, 1, 3, 8, 15, 4, 11, 2, 5.5, 13, 31, 16, 12, 22, 7, and 10. (The fifth and sixth values are both 12.0, so we assigned each the rank 5.5.) Thus,

$$W_1 = 1 + 2 + 3 + 4 + 5.5 + 7 + 8 + 10 + 11 + 12 + 13$$

$$+ 15 + 16 + 21 + 22 + 31$$

$$= 181.5$$

and

$$U_1 = 181.5 - \frac{16 \cdot 17}{2}$$

$$= 45.5$$

Since $\mu = \dfrac{16 \cdot 16}{2} = 128$ and $\sigma^2 = \dfrac{16 \cdot 16 \cdot 33}{12} = 704$, we get

$$z = \frac{45.5 - 128}{\sqrt{704}} = -3.11$$

4. Since $z = -3.11$ is less than $-2.33$, the null hypothesis must be rejected; we conclude that on the average the second diet produces a greater gain in weight.     ▲

## 16.5 RANK-SUM TESTS: THE *H* TEST

The *H* **test**, also called the **Kruskal-Wallis test**, is a generalization of the rank-sum test of the preceding section to the case where we test the null hypothesis that $k$ samples come from identical continuous populations. In other words, it is a nonparametric alternative to the one-way analysis of variance.

As in the *U* test, the data are ranked jointly from low to high, as though they constitute one sample. Then, letting $R_i$ be the sum of the ranks of the values of the *i*th sample, we base the test on the statistic

$$H = \frac{12}{n(n + 1)} \cdot \sum_{i=1}^{k} \frac{R_i^2}{n_i} - 3(n + 1)$$

where $n = n_1 + n_2 + \cdots + n_k$, and $k$ is the number of populations sampled. As it can be shown (see Exercise 16.24) that $H$ is proportional to a weighted mean of the squared differences $\left( \dfrac{R_i}{n_i} - \dfrac{n + 1}{2} \right)^2$, where $\dfrac{R_i}{n_i}$ is the mean rank of the values of the *i*th sample and $\dfrac{n + 1}{2}$ is the mean rank of all the data, it follows that the null hypothesis must be rejected for large values of $H$.

For very small values of $k$, $n_1$, ... , and $n_k$, the test of the null hypothesis may be based on special tables (see references on page 606), but since the sampling distribution of the random variable corresponding to $H$ depends on the values of the $n_i$, it is impossible to tabulate it in a compact form. Hence, the test is usually based on the large-sample theory that the sampling distribution of the random variable corresponding to $H$ can be approximated closely with a chi-square distribution with $k - 1$ degrees of freedom. Proofs of this result may be found in some of the books on nonparametric statistics listed on page 606, and they are based on the form of the $H$ statistic as it is given in Exercise 16.24.

## EXAMPLE 16.8

The following are the final examination grades of samples from three groups of students who were taught German by three different methods (classroom instruction and language laboratory, only classroom instruction, and only self-study in language laboratory):

> *First method*:   94, 88, 91, 74, 87, 97

> *Second method*:   85, 82, 79, 84, 61, 72, 80

> *Third method*:   89, 67, 72, 76, 69

Use the $H$ test at the 0.05 level of significance to test the null hypothesis that the three methods are equally effective.

## Solution

1.   $H_0$:   $\mu_1 = \mu_2 = \mu_3$
     $H_1$:   $\mu_1$, $\mu_2$, and $\mu_3$ are not all equal
     $\alpha = 0.05$

2.  Reject the null hypothesis if $H \geqslant 5.991$, where $5.991$ is the value of $\chi^2_{.05,2}$.

3.  Ranking the grades from 1 to 18, we find that $R_1 = 6 + 13 + 14 + 16 + 17 + 18 = 84$, $R_2 = 1 + 4.5 + 8 + 9 + 10 + 11 + 12 = 55.5$, and $R_3 = 2 + 3 + 4.5 + 7 + 15 = 31.5$, where there is one tie and the tied grades are each assigned the rank 4.5. Substituting the values of $R_1$, $R_2$, and $R_3$ together with $n_1 = 6$, $n_2 = 7$, $n_3 = 5$, and $n = 18$ into the formula for $H$, we get

$$H = \frac{12}{18 \cdot 19} \left( \frac{84^2}{6} + \frac{55.5^2}{7} + \frac{31.5^2}{5} \right) - 3 \cdot 19$$

$$= 6.67$$

4.  Since $H = 6.67$ exceeds $\chi^2_{.05,2} = 5.991$, the null hypothesis must be rejected; we conclude that the three methods are not all equally effective.    ▲

### EXERCISES

**16.20** Show that

(a)  $U_1 + U_2 = n_1 n_2$ for any pair of values of the corresponding random variables;

(b)  the random variables corresponding to $U_1$ and $U_2$ both take on values on the range from 0 to $n_1 n_2$.

**16.21** Show that the distribution of the random variable corresponding to $W_1$ is symmetrical about

$$\frac{n_1(n_1 + n_2 + 1)}{2}$$

and, hence, that the distribution of the random variable corresponding to $U_1$ is symmetrical about $\frac{n_1 n_2}{2}$. (*Hint*: Rank the combined data in an increasing as well as a decreasing order of magnitude.)

**16.22** Verify that $U_1$ and $U_2$ are also given by

$$U_1 = n_1 n_2 + \frac{n_2(n_2 + 1)}{2} - W_2$$

and

$$U_2 = n_1 n_2 + \frac{n_1(n_1 + 1)}{2} - W_1$$

**16.23** If $X_1, X_2, \ldots, X_{n_1}$ and $Y_1, Y_2, \ldots, Y_{n_2}$ are independent random samples, we can test the null hypothesis that they come from identical continuous populations on the basis of the Mann-Whitney statistic $U$, which is simply the number of pairs $(x_i, y_j)$ for which $x_i > y_j$. Symbolically,

$$U = \sum_{i=1}^{n_1} \sum_{j=1}^{n_2} d_{ij}$$

where

$$d_{ij} = \begin{cases} 1 & \text{if } x_i > y_j \\ 0 & \text{if } x_i < y_j \end{cases}$$

for $i = 1, 2, \ldots, n_1$, and $j = 1, 2, \ldots, n_2$. Show that this Mann-Whitney $U$ statistic is the same as the $U_1$ statistic of Section 16.4.

**16.24** Verify that the Kruskal-Wallis statistic on page 589 is equivalent to

$$H = \frac{12}{n(n+1)} \cdot \sum_{i=1}^{k} n_i \left( \frac{R_i}{n_i} - \frac{n+1}{2} \right)^2$$

**16.25** Show that if a one-way analysis of variance is performed on the ranks of the observations instead of the observations themselves, it becomes equivalent to a test based on the *H* statistic.

## *APPLICATIONS*

**16.26** The following are figures on the numbers of burglaries committed in a city in random samples of six days in the spring and six days in the fall:

> *Spring:*    36, 25, 32, 38, 28, 35
> *Fall:*    27, 20, 15, 29, 18, 22

Use the *U* test at the 0.01 level of significance to test the claim that on the average there are equally many burglaries per day in the spring as in the fall against the alternative that there are fewer in the fall.

**16.27** The following are the Rockwell hardness numbers obtained for six aluminum die castings randomly selected from production lot *A* and eight from production lot *B*:

> *Production lot A:*    75, 56, 63, 70, 58, 74
> *Production lot B:*    63, 85, 77, 80, 86, 76, 72, 82

Use the *U* test at the 0.05 level of significance to test whether the castings of production lot *B* are on the average equally hard or whether they are harder than those of production lot *A*.

**16.28** The following are the numbers of minutes it took random samples of 15 men and 12 women to complete a written test given for the renewal of their driver's licenses:

*Men*: 9.9, 7.4, 8.9, 9.1, 7.7, 9.7, 11.8, 9.2, 10.0, 10.2, 9.5, 10.8, 8.0, 11.0, 7.5

*Women*: 8.6, 10.9, 9.8, 10.7, 9.4, 10.3, 7.3, 11.5, 7.6, 9.3, 8.8, 9.6

Use the $U$ test based on Table X at the 0.05 level of significance to decide whether to accept the null hypothesis $\mu_1 = \mu_2$ or the alternative hypothesis $\mu_1 \neq \mu_2$, where $\mu_1$ and $\mu_2$ are the average amounts of time it takes men and women to complete the test.

**16.29** Rework Exercise 16.28 using the normal approximation to the distribution of the test statistic.

**16.30** An examination designed to measure basic knowledge of American history was given to random samples of freshmen at two major universities, and their grades were

*University A*: 77, 72, 58, 92, 87, 93, 97, 91, 70, 98, 76, 90, 62, 69, 90, 78, 96, 84, 73, 80

*University B*: 89, 74, 45, 56, 71, 74, 94, 88, 66, 62, 88, 63, 88, 37, 63, 75, 78, 34, 75, 68

Use the $U$ test at the 0.05 level of significance to test the null hypothesis that there is no difference in the average knowledge of American history between freshmen entering the two universities.

**16.31** The following are data on the breaking strength (in pounds) of random samples of two kinds of 2-inch cotton ribbons:

*Type I ribbon*: 144, 181, 200, 187, 169, 171, 186, 194, 176, 182, 133, 183, 197, 165, 180, 198

*Type II ribbon*: 175, 164, 172, 194, 176, 198, 154, 134, 169, 164, 185, 159, 161, 189, 170, 164

Use the $U$ test at the 0.05 level of significance to test the claim that Type I ribbon is, on the average, stronger than Type II ribbon.

**16.32** With reference to the data on page 584 and Example 16.6, calculate $U$ as defined in Exercise 16.23 and verify that it equals the value obtained for $U_1$.

**16.33** With reference to Exercise 16.26, calculate $U$ as defined in Exercise 16.23 and verify that it equals the value obtained for $U_1$.

**16.34** To compare four bowling balls, a professional bowler bowls five games with each ball and gets the following results:

> *Ball D*:   208, 220, 247, 192, 229
> *Ball E*:   216, 196, 189, 205, 210
> *Ball F*:   226, 218, 252, 225, 202
> *Ball G*:   212, 198, 207, 232, 221

Use the Kruskal-Wallis test at the 0.05 level of significance to test whether or not the bowler can expect to score equally well with the four bowling balls.

**16.35** The following are the miles per gallon which a test driver got for 10 tankfuls of each of three kinds of gasoline:

> *Gasoline A*:   20, 31, 24, 33, 23, 24, 28, 16, 19, 26
> *Gasoline B*:   29, 18, 29, 19, 20, 21, 34, 33, 30, 23
> *Gasoline C*:   19, 31, 16, 26, 31, 33, 28, 28, 25, 30

Use the Kruskal-Wallis test at the 0.05 level of significance to test whether or not there is a difference in the actual average mileage yield of the three kinds of gasoline.

**16.36** Three groups of guinea pigs were injected, respectively, with 0.5, 1.0, and 1.5 milligrams of a tranquilizer, and the following are the numbers of seconds it took them to fall asleep:

| *0.5-mg dose*: | 8.2 | 10.0 | 10.2 | 13.7 | 14.0 | 7.8 |
| | 12.7 | 10.9 | | | | |
| *1.0-mg dose*: | 9.7 | 13.1 | 11.0 | 7.5 | 13.3 | 12.5 |
| | 8.8 | 12.9 | 7.9 | 10.5 | | |
| *1.5-mg dose*: | 12.0 | 7.2 | 8.0 | 9.4 | 11.3 | 9.0 |
| | 11.5 | 8.5 | | | | |

Use the $H$ test at the 0.01 level of significance to test the null hypothesis that the differences in dosage have no effect on the length of time it takes guinea pigs to fall asleep.

## 16.6 TESTS BASED ON RUNS

There are several nonparametric methods for testing the randomness of observed data on the basis of the order in which they were obtained. The technique we shall describe here is based on the **theory of runs**, where a **run** is a succession of identical letters (or other kinds of symbols) which is preceded and followed by different letters or no letters at all. To illustrate, consider the following arrangement of defective, $d$, and nondefective, $n$, pieces produced in the given order by a certain machine:

$$\underbrace{n\,n\,n\,n\,n}\,\underbrace{d\,d\,d\,d}\,\underbrace{n\,n\,n\,n\,n\,n\,n\,n\,n\,n}\,\underbrace{d\,d}\,\underbrace{n\,n}\,\underbrace{d\,d\,d\,d}\,\underbrace{n}\,\underbrace{d\,d}\,\underbrace{n\,n}$$

Using braces to combine the letters which constitute a run, we find that there is first a run of five $n$'s, then a run of four $d$'s, then a run of ten $n$'s, ... , and finally a run of two $n$'s; in all, there are nine runs of varying lengths.

The total number of runs appearing in an arrangement of this kind is often a good indication of a possible lack of randomness. If there are too few runs, we might suspect a definite grouping or clustering, or perhaps a trend; if there are too many runs, we might suspect some sort of repeated alternating pattern. In our illustration there seems to be a definite clustering, the defective pieces seem to come in groups, but it remains to be seen whether this is significant or whether it can be attributed to chance.

To find the probability that $n_1$ letters of one kind and $n_2$ letters of another kind will form $u$ runs when each of the $\binom{n_1 + n_2}{n_1}$ possible arrangements of these letters is regarded as equally likely, let us first investigate the case where $u$ is even, namely, where $u = 2k$ and $k$ is a positive integer. In that case there will have to be $k$ runs of each kind alternating with one another. To find the number of ways in which $n_1$ letters can form $k$ runs, let us first consider the very simple case where we have five letters $c$ which are to be divided up into three runs. Using vertical bars to separate the five letters into three runs, we find that there are the *six* possibilities

$$c|c|ccc \quad c|cc|cc \quad c|ccc|c$$
$$cc|c|cc \quad cc|cc|c \quad ccc|c|c$$

corresponding to the $\binom{4}{2}$ ways in which we can put two vertical bars into two of the four spaces between the five $c$'s. By the same token there are $\binom{n_1 - 1}{k - 1}$ ways in which the $n_1$ letters of the first kind can form $k$ runs, $\binom{n_2 - 1}{k - 1}$ ways in which the $n_2$ letters of the second kind can form $k$ runs, and it follows that there are altogether $2\binom{n_1 - 1}{k - 1}\binom{n_2 - 1}{k - 1}$ ways in which these $n_1 + n_2$ letters can form $2k$ runs. The factor 2 is accounted for by the fact that when we combine the two kinds of runs so that they alternate, we can begin either with a run of the first kind of letter or with a run of the second kind. Thus, when $u = 2k$ (where $k$ is a positive integer), the probability of getting that many runs is

$$f(u) = \frac{2\binom{n_1 - 1}{k - 1}\binom{n_2 - 1}{k - 1}}{\binom{n_1 + n_2}{n_1}}$$

and it is left to the reader to show in Exercise 16.37 that similar arguments lead to

$$f(u) = \frac{\binom{n_1 - 1}{k}\binom{n_2 - 1}{k - 1} + \binom{n_1 - 1}{k - 1}\binom{n_2 - 1}{k}}{\binom{n_1 + n_2}{n_1}}$$

when $u = 2k + 1$ (where $k$ is a positive integer).

When $n_1$ and $n_2$ are small, tests of randomness based on $u$ are usually performed with the use of special tables such as Table XI at the end of the book. We reject the null hypothesis of randomness at the level of significance $\alpha$ if

$$u \leq u'_{\alpha/2} \quad \text{or} \quad u \geq u_{\alpha/2}$$

where $u'_{\alpha/2}$ is the largest value for which the probability of getting a value less than or equal to it does not exceed $\alpha/2$ and $u_{\alpha/2}$ is the smallest value for which the probability of getting a value greater than or equal to it does not exceed $\alpha/2$.

## EXAMPLE 16.9

Checking on elm trees that were planted many years ago along a country road, a county official obtained the following arrangement of healthy, $H$, and diseased, $D$, trees:

$$HHHHDDDHHHHHHHHDDHHDDDD$$

Test at the 0.05 level of significance whether this arrangement may be regarded as random.

### Solution

1.  $H_0$: Arrangement is random.
    $H_1$: Arrangement is not random.
    $\alpha = 0.05$

2.  Since $n_1 = 13$ and $n_2 = 9$, reject the null hypothesis if $u \leqslant 6$ or $u \geqslant 17$, where 6 and 17 are the corresponding values of $u'_{.025}$ and $u_{.025}$.

3.  $u = 6$ by inspection of the data.

4.  Since $u = 6$ is equal to $u'_{.025} = 6$, the null hypothesis must be rejected; the arrangement of healthy and diseased elm trees is not random. It appears that the diseased trees come in clusters.   ▲

When $n_1$ and $n_2$ are both greater than or equal to 10, it is considered reasonable to assume that the distribution of the random variable corresponding to $u$ can be approximated closely with a normal curve. To perform the runs test on the basis of this assumption, we need the following results.

---

**THEOREM 16.3** Under the null hypothesis of randomness, the mean and the variance of the random variable corresponding to $u$ are

$$\mu = \frac{2n_1 n_2}{n_1 + n_2} + 1$$

and

$$\sigma^2 = \frac{2n_1 n_2 (2n_1 n_2 - n_1 - n_2)}{(n_1 + n_2)^2 (n_1 + n_2 - 1)}$$

---

These results can be obtained directly with the use of the formulas given on page 595. The details of such a proof, as well as an alternative approach which is easier, may be found in the book by J. D. Gibbons listed among the references at the end of the chapter.

## EXAMPLE 16.10

The following is an arrangement of men, $M$, and women, $W$, lined up to purchase tickets for a rock concert:

$$M \; W \; M \; W \; M \; M \; M \; W \; M \; W \; M \; M \; M \; W \; W \; M \; M \; M \; W \; W \; M \; W \; M$$

$$M \; M \; W \; M \; M \; M \; W \; W \; W \; M \; W \; M \; M \; M \; W \; M \; W \; M \; M \; M \; M \; W \; W \; M$$

Test for randomness at the 0.05 level of significance.

*Solution*

1.  $H_0$:  Arrangement is random.
    $H_1$:  Arrangement is not random.
    $\alpha = 0.05$
2.  Reject the null hypothesis if $z \leqslant -1.96$ or $z \geqslant 1.96$, where

$$z = \frac{u - \mu}{\sigma}$$

and $\mu$ and $\sigma^2$ are given by the formulas of Theorem 16.3.

3.  Since $n_1 = 30$, $n_2 = 18$, and $u = 27$, we get

$$\mu = \frac{2 \cdot 30 \cdot 18}{30 + 18} + 1 = 23.5$$

$$\sigma = \sqrt{\frac{2 \cdot 30 \cdot 18(2 \cdot 30 \cdot 18 - 30 - 18)}{(30 + 18)^2(30 + 18 - 1)}} = 3.21$$

and, hence,

$$z = \frac{27 - 23.5}{3.21} = 1.09$$

4.  Since $z = 1.09$ falls between $-1.96$ and $1.96$, the null hypothesis cannot be rejected; in other words, there is no real evidence to indicate that the arrangement is not random.    ▲

The method we have discussed in this section is not limited to tests of the randomness of series of attributes (such as the $d$'s and $n$'s of the example on page 594). Any sample which consists of numerical measurements or observations can be treated similarly by using the letters $a$ and $b$ to denote, respectively, values falling above and below the median of the sample. (Numbers equaling the median are omitted.) The resulting series of $a$'s and $b$'s can then be tested for randomness on the basis of the total number of runs of $a$'s and $b$'s, namely, the total number of **runs above and below the median.**

## EXAMPLE 16.11

The following are the speeds (in miles per hour) at which every fifth passenger car was timed at a certain checkpoint: 46, 58, 60, 56, 70, 66, 48, 54, 62, 41, 39, 52, 45, 62, 53, 69, 65, 65, 67, 76, 52, 52, 59, 59, 67, 51, 46, 61, 40, 43,

42, 77, 67, 63, 59, 63, 63, 72, 57, 59, 42, 56, 47, 62, 67, 70, 63, 66, 69, and 73. Test the null hypothesis of randomness at the 0.05 level of significance.

*Solution*

1.  $H_0$:   The sample is random.
    $H_1$:   The sample is not random.
    $\alpha = 0.05$

2.  Reject the null hypothesis if $z \leqslant -1.96$ or $z \geqslant 1.96$, where

$$z = \frac{u - \mu}{\sigma}$$

$u$ is the number of runs above and below the median, and $\mu$ and $\sigma^2$ are given by the formulas of Theorem 16.3.

3.  Since the median of the speeds is 59.5, we get the following arrangement of $a$'s and $b$'s:

$$b\,b\,a\,b\,a\,a\,b\,b\,a\,b\,b\,b\,b\,a\,b\,a\,a\,a\,a\,a\,b\,b\,b\,b\,a$$

$$b\,b\,a\,b\,b\,b\,a\,a\,a\,b\,a\,a\,a\,b\,b\,b\,b\,b\,a\,a\,a\,a\,a\,a\,a$$

Then, since $n_1 = 25$, $n_2 = 25$, and $u = 20$, we get

$$\mu = \frac{2 \cdot 25 \cdot 25}{25 + 25} + 1 = 26$$

$$\sigma^2 = \frac{2 \cdot 25 \cdot 25(2 \cdot 25 \cdot 25 - 25 - 25)}{(25 + 25)^2(25 + 25 - 1)} = 12.2$$

and

$$z = \frac{20 - 26}{\sqrt{12.2}} = -1.72$$

4.  Since $z = -1.72$ falls between $-1.96$ and $1.96$, the null hypothesis cannot be rejected; there is no real evidence that the sample should not be regarded as random.   ▲

## EXERCISES

**16.37** Verify the formula given on page 595 for the probability of getting $u$ runs when $u = 2k + 1$, where $k$ is a positive integer.

**16.38** If a person gets seven heads and three tails in 10 tosses of a balanced coin, find the probabilities for 2, 3, 4, 5, 6, and 7 runs.

**16.39** Find the probability that $n_1 = 6$ letters of one kind and $n_2 = 5$ letters of another kind will form at least 8 runs.

**16.40** If there are $n_1 = 8$ letters of one kind and $n_2 = 8$ letters of another kind, for how many runs would we reject the null hypothesis of randomness at the 0.01 level of significance?

## APPLICATIONS

**16.41** The following is the order in which a broker received buy, $B$, and sell, $S$, orders for a certain stock:

$$B\,B\,B\,B\,B\,B\,B\,B\,S\,S\,B\,S\,S\,S\,S\,S\,B\,B\,B\,B\,B$$

Test for randomness at the 0.05 level of significance.

**16.42** A driver buys gasoline either at a Texaco station, $T$, or at a Mobil station, $M$, and the following arrangement shows the order of the stations from which she bought gasoline over a certain period of time:

$$T\,T\,T\,M\,T\,M\,T\,M\,M\,T\,T\,M\,T\,M\,T\,M\,T\,M\,M\,T\,M\,T$$

Test for randomness at the 0.05 level of significance.

**16.43** The following is the order in which red, $R$, and black, $B$, cards were dealt to a bridge player:

$$B\,B\,B\,R\,R\,R\,R\,R\,B\,B\,R\,R\,R$$

Test for randomness at the 0.05 level of significance.

**16.44** The following arrangement indicates whether 60 consecutive cars that went by the toll booth of a bridge had local plates, $L$, or out-of-state plates, $O$:

$$L\,L\,O\,L\,L\,L\,L\,O\,O\,L\,L\,L\,L\,O\,L\,O\,O\,L\,L\,L\,L\,O\,L\,O\,O\,L\,L\,L\,L\,L$$
$$O\,L\,L\,O\,L\,O\,L\,L\,L\,L\,O\,O\,L\,O\,O\,O\,O\,L\,L\,L\,L\,O\,L\,O\,O\,L\,L\,L\,O$$

Test at the 0.05 level of significance whether this arrangement of $L$'s and $O$'s may be regarded as random.

**16.45** To test whether a radio signal contains a message or constitutes random noise, an interval of time is subdivided into a number of very short intervals and for each of these it is determined whether the signal strength exceeds, $E$, or does not exceed, $N$, a certain level of background noise. Test at the 0.01 level of significance whether the following arrangement, thus obtained, may be regarded as random, and hence that the signal does not contain a message:

$$N N N E N E N E N E E E N E E E N E E E N E E N E E$$

$$N E E N N E N E E E N E N N N E N N E N N N N E$$

**16.46** Write a sequence of 100 $H$'s and $T$'s, supposedly representing a random sequence of heads and tails, and test for randomness at the 0.05 level of significance.

**16.47** The following are the numbers of students absent from school on 24 consecutive school days: 29, 25, 31, 28, 30, 28, 33, 31, 35, 29, 31, 33, 35, 28, 36, 30, 33, 26, 30, 28, 32, 31, 38, and 27. Test for randomness at the 0.01 level of significance.

**16.48** The following are the numbers of defective pieces produced by a machine on fifty consecutive days: 7, 14, 17, 10, 18, 19, 23, 19, 14, 10, 12, 18, 19, 13, 24, 26, 9, 16, 19, 14, 19, 10, 15, 22, 25, 24, 20, 9, 17, 28, 29, 19, 25, 23, 24, 28, 31, 19, 24, 30, 27, 24, 39, 35, 23, 26, 28, 31, 37, and 40. Test at the 0.025 level of significance whether there might be a trend.

**16.49** The following are the numbers of lunches that an insurance agent claimed as business deductions in 30 consecutive months : 6, 7, 5, 6, 8, 6, 8, 6, 6, 4, 3, 2, 4, 4, 3, 4, 7, 5, 6, 8, 6, 6, 3, 4, 2, 5, 4, 4, 3, and 7. Use the runs test based on Table XI to test for randomness at the 0.01 level of significance.

**16.50** The numbers of retail stores that opened for business and also quit business in the same year were 108, 103, 109, 107, 125, 142, 147, 122, 116, 153, 144, 162, 143, 126, 145, 129, 134, 137, 143, 150, 148, 152, 125, 106, 112, 139, 132, 122, 138, 148, 155, 146, and 158 during a period of 33 years. Making use of the fact that the median is 138, test at the 0.05 level of significance whether there is a real trend.

**16.51** The following are six year's quarterly sales (in millions of dollars) of a manufacturer of heavy machinery: 83.8, 102.5, 121.0, 90.5, 106.6, 104.8, 114.7, 93.6, 98.9, 96.9, 122.6, 85.6, 103.2, 96.9, 118.0, 92.1, 100.5, 92.9, 125.6, 79.2, 110.8, 95.1, 125.6, and 86.7. At the 0.05 level of significance, is there a real cyclical pattern?

**16.52** The theory of runs may also be used as an alternative to the rank-sum test of Section 16.4, namely, the test of the null hypothesis that two independent random samples come from identical continuous populations. We simply rank the data jointly, write a 1 below each value belonging to the first sample, a 2 below each value belonging to the second sample, and then test the randomness of the resulting arrangement of 1's and 2's. If there are too few runs, this may well be accounted for by the fact that the two samples come from populations with unequal means. With reference to the data on page 584, use this technique to test at the 0.05 level of significance whether the two samples came from identical continuous populations or whether the two populations have unequal means.

## 16.7 THE RANK CORRELATION COEFFICIENT

Since the assumptions underlying the significance test for correlation coefficients of Section 14.5 are rather stringent, it is sometimes preferable to use a nonparametric alternative. Most popular among such nonparametric measures of association is the **rank correlation coefficient**, also called **Spearman's rank correlation coefficient**, $r_S$. For a given set of paired data $\{(x_i, y_i); i = 1, 2, \ldots , n\}$, it is obtained by ranking the $x$'s among themselves and also the $y$'s, both from low to high or from high to low, and then substituting into the following formula.

---

**DEFINITION 16.1**   The rank correlation coefficient is given by

$$r_S = 1 - \frac{6 \cdot \sum\limits_{i=1}^{n} d_i^2}{n(n^2 - 1)}$$

where $d_i$ is the difference between the ranks assigned to $x_i$ and $y_i$.

---

When there are ties in rank, we proceed as before and assign the tied observations the mean of the ranks which they jointly occupy.

When there are no ties in rank, $r_S$ actually equals the correlation coefficient $r$ calculated for the ranks. To verify this, let $r_i$ and $s_i$ be the ranks of $x_i$ and $y_i$. Making use of the fact that the sum and the sum of the squares of the first $n$ positive integers are $\dfrac{n(n + 1)}{2}$ and $\dfrac{n(n + 1)(2n + 1)}{6}$, respectively, we find that

$$\sum_{i=1}^{n} r_i = \sum_{i=1}^{n} s_i = \frac{n(n + 1)}{2}$$

$$\sum_{i=1}^{n} r_i^2 = \sum_{i=1}^{n} s_i^2 = \frac{n(n + 1)(2n + 1)}{6}$$

$$\sum_{i=1}^{n} r_i s_i = \frac{n(n + 1)(2n + 1)}{6} - \frac{1}{2} \cdot \sum_{i=1}^{n} d_i^2$$

and if we substitute these expressions into the formula for $r$, we get the above formula for $r_S$.

## EXAMPLE 16.12

The following are the numbers of hours which 10 students studied for an examination and the scores they obtained:

| Number of hours studied $x$ | Score $y$ |
|---|---|
| 8 | 56 |
| 5 | 44 |
| 11 | 79 |
| 13 | 72 |
| 10 | 70 |
| 5 | 54 |
| 18 | 94 |
| 15 | 85 |
| 2 | 33 |
| 8 | 65 |

Calculate $r_S$.

### Solution

Ranking the $x$'s and the $y$'s, and proceeding as in the following table, we get

| Rank of x | Rank of y | $d$ | $d^2$ |
|---|---|---|---|
| 6.5 | 7 | −0.5 | 0.25 |
| 8.5 | 9 | −0.5 | 0.25 |
| 4 | 3 | 1.0 | 1.00 |
| 3 | 4 | 1.0 | 1.00 |
| 5 | 5 | 0.0 | 0.00 |
| 8.5 | 8 | 0.5 | 0.25 |
| 1 | 1 | 0.0 | 0.00 |
| 2 | 2 | 0.0 | 0.00 |
| 10 | 10 | 0.0 | 0.00 |
| 6.5 | 6 | 0.5 | 0.25 |
| | | | 3.00 |

Then, substitution into the formula for $r_S$ yields

$$r_S = 1 - \frac{6 \cdot 3}{10(10^2 - 1)} = 0.98 \quad \blacktriangle$$

As can be seen from this example, $r_S$ is easily determined; indeed, it is sometimes used instead of $r$ mainly because of its computational ease. Had we calculated $r$ for the data of the preceding example, we would have obtained $r = 0.96$, and this is very close to $r_S = 0.98$.

For small values of $n$ ($n \leqslant 10$), the test of the null hypothesis of no correlation, indeed, the test of the null hypothesis that the $x$'s and $y$'s are randomly matched, may be based on special tables determined from the exact sampling distributions of $R_S$ (see references on page 606). Most of the time, though, we use the fact that the sampling distribution of $R_S$ can be approximated closely with a normal distribution, and to this end we need the following results.

---

**THEOREM 16.4** Under the null hypothesis of no correlation, the mean and the variance of $R_S$ are

$$E(R_S) = 0 \quad \text{and} \quad \text{var}(R_S) = \frac{1}{n - 1}$$

---

A proof of this theorem may be found in the book by J. D. Gibbons listed among the references at the end of this chapter. Strictly speaking, the theorem applies when there are no ties, but it can be used unless the number of ties is large.

## EXAMPLE 16.13

With reference to Example 16.12, test at the 0.01 level of significance whether the value obtained for $r_S$, 0.98, is significant.

### Solution

1. $H_0$:  There is no correlation.
   $H_1$:  There is a correlation.
   $\alpha = 0.01$

2. Reject the null hypothesis if $z \leqslant -2.575$ or $z \geqslant 2.575$, where

$$z = r_S \sqrt{n - 1}$$

3.  Substituting $n = 10$ and $r_S = 0.98$, we get

$$z = 0.98\sqrt{10 - 1} = 2.94$$

4.  Since $z = 2.94$ exceeds 2.575, the null hypothesis must be rejected; we conclude that there is a real (positive) relationship between study time and scores.    ▲

## EXERCISE

**16.53**  Given a set of $k$-tuples $(x_{11}, x_{12}, \ldots , x_{1k})$, $(x_{21}, x_{22}, \ldots , x_{2k})$, $\ldots$ , and $(x_{n1}, x_{n2}, \ldots , x_{nk})$, the extent of their association, or agreement, may be measured by means of the **coefficient of concordance**

$$W = \frac{12}{k^2 n(n^2 - 1)} \cdot \sum_{i=1}^{n} \left[ R_i - \frac{k(n + 1)}{2} \right]^2$$

where $R_i$ is the sum of the ranks assigned to $x_{i1}, x_{i2}, \ldots$ , and $x_{ik}$ when the $x$'s with the second subscript 1 are ranked among themselves, and so are the $x$'s with the second subscript 2, $\ldots$ , and the $x$'s with the second subscript $k$. What are the maximum and minimum values of $W$, and what do they reflect with respect to the agreement, or lack of agreement, of the values of the $k$ random variables?

## APPLICATIONS

**16.54**  Calculate $r_S$ for the following data representing the statistics grades, $x$, and psychology grades, $y$, of 18 students:

| $x$ | $y$ | $x$ | $y$ |
|-----|-----|-----|-----|
| 78 | 80 | 97 | 90 |
| 86 | 74 | 74 | 85 |
| 49 | 63 | 53 | 71 |
| 94 | 85 | 58 | 67 |
| 53 | 55 | 62 | 64 |
| 89 | 86 | 74 | 69 |
| 94 | 90 | 74 | 71 |
| 71 | 84 | 70 | 67 |
| 70 | 71 | 74 | 71 |

**16.55**  With reference to the preceding exercise, test at the 0.05 level of significance whether the value obtained for $r_S$ is significant.

**16.56** The following shows how a panel of nutrition experts and a panel of housewives ranked 15 breakfast foods on their palatability:

| Breakfast food | Nutrition experts | Housewives |
|---|---|---|
| A | 3 | 5 |
| B | 7 | 4 |
| C | 11 | 8 |
| D | 9 | 14 |
| E | 1 | 2 |
| F | 4 | 6 |
| G | 10 | 12 |
| H | 8 | 7 |
| I | 5 | 1 |
| J | 13 | 15 |
| K | 12 | 9 |
| L | 2 | 3 |
| M | 15 | 10 |
| N | 6 | 11 |
| O | 14 | 13 |

Calculate $r_S$ as a measure of the consistency of the two rankings.

**16.57** Calculate $r_S$ for the data of Exercise 14.17 and test the null hypothesis of no correlation at the 0.05 level of significance.

**16.58** The following are the rankings given by three judges to the works of 10 artists:

| Judge A | Judge B | Judge C |
|---|---|---|
| 6 | 2 | 7 |
| 4 | 5 | 3 |
| 2 | 4 | 1 |
| 5 | 8 | 2 |
| 9 | 10 | 10 |
| 3 | 1 | 6 |
| 1 | 6 | 4 |
| 8 | 9 | 9 |
| 10 | 7 | 8 |
| 7 | 3 | 5 |

Calculate the value of $W$, the coefficient of concordance of Exercise 16.53, as a measure of the agreement of the three sets of rankings.

**16.59** With reference to Exercise 16.58, calculate the $k = 3$ pairwise rank correlation coefficients and verify that the relationship between their mean $\bar{r}_S$ and the coefficient of concordance (see Exercise 16.53) is given by

$$\bar{r}_S = \frac{kW - 1}{k - 1}$$

## REFERENCES

Detailed tables for the most widely used nonparametric tests, including the ones discussed in this chapter, may be found in

OWEN, D. B., *Handbook of Statistical Tables*. Reading, Mass.: Addison-Wesley Publishing Company, Inc., 1962.

In particular, tables for small-sample tests of the significance of the rank correlation coefficient are given in

KENDALL, M. G., *Rank Correlation Methods*. New York: Hafner Publishing Co., Inc., 1962.

A wealth of information about the various nonparametric tests may be found in

GIBBONS, J. D., *Nonparametric Methods for Quantitative Analysis*, 2nd ed. Syracuse, N.Y.: American Sciences Press, 1985.
LEHMANN, E. L., *Nonparametrics: Statistical Methods Based on Ranks*. San Francisco: Holden-Day, Inc., 1975,
MOSTELLER, F., and ROURKE, R. E. K., *Sturdy Statistics*. Reading, Mass.: Addison-Wesley Publishing Company, Inc., 1973,
NOETHER, G. E., *Introduction to Statistics: A Nonparametric Approach*, 2nd ed. Boston: Houghton Mifflin Company, 1976,
SIEGEL, S., *Nonparametric Statistics for the Behavioral Sciences*. New York: McGraw-Hill Book Company, 1956.

# *Appendix: Sums and Products*

## A.1  RULES FOR SUMS AND PRODUCTS

To simplify expressions involving sums and products, the $\Sigma$ and $\Pi$ notations are widely used in statistics. In the usual notation we write

$$\sum_{i=a}^{b} x_i = x_a + x_{a+1} + x_{a+2} + \cdots + x_b$$

and

$$\prod_{i=a}^{b} x_i = x_a \cdot x_{a+1} \cdot x_{a+2} \cdot \ldots \cdot x_b$$

for any nonnegative integers $a$ and $b$ with $a \le b$.

When working with sums or products, it is often helpful to apply the following rules, which can all be verified by writing the respective expressions in full, that is, without the $\Sigma$ or $\Pi$ notation:

**607**

**THEOREM A.1**

1. $\displaystyle\sum_{i=1}^{n} kx_i = k \cdot \sum_{i=1}^{n} x_i$

2. $\displaystyle\sum_{i=1}^{n} k = nk$

3. $\displaystyle\sum_{i=1}^{n} (x_i + y_i) = \sum_{i=1}^{n} x_i + \sum_{i=1}^{n} y_i$

4. $\displaystyle\prod_{i=1}^{n} kx_i = k^n \cdot \prod_{i=1}^{n} x_i$

5. $\displaystyle\prod_{i=1}^{n} k = k^n$

6. $\displaystyle\prod_{i=1}^{n} x_i y_i = \left(\prod_{i=1}^{n} x_i\right)\left(\prod_{i=1}^{n} y_i\right)$

7. $\displaystyle\ln \prod_{i=1}^{n} x_i = \sum_{i=1}^{n} \ln x_i$

Double sums, triple sums, ... , are also widely used in statistics, and if we repeatedly apply the definition of $\Sigma$ given above, we have, for example,

$$\sum_{i=1}^{m} \sum_{j=1}^{n} x_{ij} = \sum_{i=1}^{m} (x_{i1} + x_{i2} + \cdots + x_{in})$$

$$= (x_{11} + x_{12} + \cdots + x_{1n})$$

$$+ (x_{21} + x_{22} + \cdots + x_{2n})$$

$$\cdot \qquad \cdot \qquad \cdot \qquad \cdot \qquad \cdot$$

$$+ (x_{m1} + x_{m2} + \cdots + x_{mn})$$

Note that when the $x_{ij}$ are thus arranged in a rectangular array, the first subscript denotes the row to which a particular element belongs, and the second subscript denotes the column.

When we work with double sums, the following theorem is of special interest; it is an immediate consequence of the multinomial expansion of

$$(x_1 + x_2 + \cdots + x_n)^2$$

**THEOREM A.2**

$$\sum\sum_{i<j} x_i x_j = \frac{1}{2}\left[\left(\sum_{i=1}^{n} x_i\right)^2 - \sum_{i=1}^{n} x_i^2\right]$$

where

$$\sum\sum_{i<j} x_i x_j = \sum_{i=1}^{n-1}\sum_{j=i+1}^{n} x_i x_j$$

## A.2 SPECIAL SUMS

In the theory of nonparametric statistics, particularly when we deal with rank sums, we often need expressions for the sums of powers of the first $n$ positive integers; namely, expressions for

$$S(n, r) = 1^r + 2^r + 3^r + \cdots + n^r$$

for $r = 0, 1, 2, 3, \ldots$ . The following theorem, which the reader will be asked to prove in Exercise A.1, provides a convenient way of obtaining these sums.

**THEOREM A.3**

$$\sum_{r=0}^{k-1}\binom{k}{r}S(n, r) = (n + 1)^k - 1$$

for any positive integers $n$ and $k$.

A disadvantage of this theorem is that we have to find the sums $S(n, r)$ one at a time, first for $r = 0$, then for $r = 1$, then for $r = 2$, and so forth. For instance, for $k = 1$ we get

$$\binom{1}{0}S(n, 0) = (n + 1) - 1 = n$$

and, hence, $S(n, 0) = 1^0 + 2^0 + \cdots + n^0 = n$. Similarly, for $k = 2$ we get

$$\binom{2}{0} S(n, 0) + \binom{2}{1} S(n, 1) = (n + 1)^2 - 1$$

$$n + 2S(n, 1) = n^2 + 2n$$

and, hence, $S(n, 1) = 1^1 + 2^1 + \cdots + n^1 = \frac{1}{2} n(n + 1)$. Using the same technique, the reader will be asked to show in Exercise A.2 that

$$S(n, 2) = \frac{1}{6} n(n + 1)(2n + 1)$$

and

$$S(n, 3) = \frac{1}{4} n^2(n + 1)^2$$

### EXERCISES

**A.1** Prove Theorem A.3 by making use of the fact that

$$(m + 1)^k - m^k = \sum_{r=0}^{k-1} \binom{k}{r} m^r$$

which follows from the binomial expansion of $(m + 1)^k$.

**A.2** Verify the formulas for $S(n, 2)$ and $S(n, 3)$ given above, and find an expression for $S(n, 4)$.

**A.3** Given $x_1 = 1$, $x_2 = 3$, $x_3 = -2$, $x_4 = 4$, $x_5 = -1$, $x_6 = 2$, $x_7 = 1$, and $x_8 = 2$, find

(a) $\displaystyle\sum_{i=1}^{8} x_i$;     (b) $\displaystyle\sum_{i=1}^{8} x_i^2$.

**A.4** Given $x_1 = 3$, $x_2 = 4$, $x_3 = 5$, $x_4 = 6$, $x_5 = 7$, $f_1 = 3$, $f_2 = 7$, $f_3 = 10$, $f_4 = 5$, and $f_5 = 2$, find

(a) $\displaystyle\sum_{i=1}^{5} x_i$;     (b) $\displaystyle\sum_{i=1}^{5} f_i$;

(c) $\displaystyle\sum_{i=1}^{5} x_i f_i$;     (d) $\displaystyle\sum_{i=1}^{5} x_i^2 f_i$.

**A.5** Given $x_1 = 2$, $x_2 = -3$, $x_3 = 4$, $x_4 = -2$, $y_1 = 5$, $y_2 = -3$, $y_3 = 2$, and $y_4 = -1$, find

(a) $\displaystyle\sum_{i=1}^{4} x_i$;     (b) $\displaystyle\sum_{i=1}^{4} y_i$;

(c) $\displaystyle\sum_{i=1}^{4} x_i^2$;    (d) $\displaystyle\sum_{i=1}^{4} y_i^2$;

(e) $\displaystyle\sum_{i=1}^{4} x_i y_i$.

**A.6** Given $x_{11} = 3$, $x_{12} = 1$, $x_{13} = -2$, $x_{14} = 2$, $x_{21} = 1$, $x_{22} = 4$, $x_{23} = -2$, $x_{24} = 5$, $x_{31} = 3$, $x_{32} = -1$, $x_{33} = 2$, and $x_{34} = 3$, find

(a) $\displaystyle\sum_{i=1}^{3} x_{ij}$ separately for $j = 1, 2, 3,$ and 4;

(b) $\displaystyle\sum_{j=1}^{4} x_{ij}$ separately for $i = 1, 2,$ and 3.

**A.7** With reference to Exercise A.6, evaluate the double summation $\displaystyle\sum_{i=1}^{3}\sum_{j=1}^{4} x_{ij}$ using

(a) the results of part (a) of that exercise;
(b) the results of part (b) of that exercise.

# *Statistical Tables*

# TABLE I

## Binomial Probabilities[†]

| n | x | .05 | .10 | .15 | .20 | .25 | θ .30 | .35 | .40 | .45 | .50 |
|---|---|------|------|------|------|------|------|------|------|------|------|
| 1 | 0 | .9500 | .9000 | .8500 | .8000 | .7500 | .7000 | .6500 | .6000 | .5500 | .5000 |
|   | 1 | .0500 | .1000 | .1500 | .2000 | .2500 | .3000 | .3500 | .4000 | .4500 | .5000 |
| 2 | 0 | .9025 | .8100 | .7225 | .6400 | .5625 | .4900 | .4225 | .3600 | .3025 | .2500 |
|   | 1 | .0950 | .1800 | .2550 | .3200 | .3750 | .4200 | .4550 | .4800 | .4950 | .5000 |
|   | 2 | .0025 | .0100 | .0225 | .0400 | .0625 | .0900 | .1225 | .1600 | .2025 | .2500 |
| 3 | 0 | .8574 | .7290 | .6141 | .5120 | .4219 | .3430 | .2746 | .2160 | .1664 | .1250 |
|   | 1 | .1354 | .2430 | .3251 | .3840 | .4219 | .4410 | .4436 | .4320 | .4084 | .3750 |
|   | 2 | .0071 | .0270 | .0574 | .0960 | .1406 | .1890 | .2389 | .2880 | .3341 | .3750 |
|   | 3 | .0001 | .0010 | .0034 | .0080 | .0156 | .0270 | .0429 | .0640 | .0911 | .1250 |
| 4 | 0 | .8145 | .6561 | .5220 | .4096 | .3164 | .2401 | .1785 | .1296 | .0915 | .0625 |
|   | 1 | .1715 | .2916 | .3685 | .4096 | .4219 | .4116 | .3845 | .3456 | .2995 | .2500 |
|   | 2 | .0135 | .0486 | .0975 | .1536 | .2109 | .2646 | .3105 | .3456 | .3675 | .3750 |
|   | 3 | .0005 | .0036 | .0115 | .0256 | .0469 | .0756 | .1115 | .1536 | .2005 | .2500 |
|   | 4 | .0000 | .0001 | .0005 | .0016 | .0039 | .0081 | .0150 | .0256 | .0410 | .0625 |
| 5 | 0 | .7738 | .5905 | .4437 | .3277 | .2373 | .1681 | .1160 | .0778 | .0503 | .0312 |
|   | 1 | .2036 | .3280 | .3915 | .4096 | .3955 | .3602 | .3124 | .2592 | .2059 | .1562 |
|   | 2 | .0214 | .0729 | .1382 | .2048 | .2637 | .3087 | .3364 | .3456 | .3369 | .3125 |
|   | 3 | .0011 | .0081 | .0244 | .0512 | .0879 | .1323 | .1811 | .2304 | .2757 | .3125 |
|   | 4 | .0000 | .0004 | .0022 | .0064 | .0146 | .0284 | .0488 | .0768 | .1128 | .1562 |
|   | 5 | .0000 | .0000 | .0001 | .0003 | .0010 | .0024 | .0053 | .0102 | .0185 | .0312 |
| 6 | 0 | .7351 | .5314 | .3771 | .2621 | .1780 | .1176 | .0754 | .0467 | .0277 | .0156 |
|   | 1 | .2321 | .3543 | .3993 | .3932 | .3560 | .3025 | .2437 | .1866 | .1359 | .0938 |
|   | 2 | .0305 | .0984 | .1762 | .2458 | .2966 | .3241 | .3280 | .3110 | .2780 | .2344 |
|   | 3 | .0021 | .0146 | .0415 | .0819 | .1318 | .1852 | .2355 | .2765 | .3032 | .3125 |
|   | 4 | .0001 | .0012 | .0055 | .0154 | .0330 | .0595 | .0951 | .1382 | .1861 | .2344 |
|   | 5 | .0000 | .0001 | .0004 | .0015 | .0044 | .0102 | .0205 | .0369 | .0609 | .0938 |
|   | 6 | .0000 | .0000 | .0000 | .0001 | .0002 | .0007 | .0018 | .0041 | .0083 | .0156 |
| 7 | 0 | .6983 | .4783 | .3206 | .2097 | .1335 | .0824 | .0490 | .0280 | .0152 | .0078 |
|   | 1 | .2573 | .3720 | .3960 | .3670 | .3115 | .2471 | .1848 | .1306 | .0872 | .0547 |
|   | 2 | .0406 | .1240 | .2097 | .2753 | .3115 | .3177 | .2985 | .2613 | .2140 | .1641 |
|   | 3 | .0036 | .0230 | .0617 | .1147 | .1730 | .2269 | .2679 | .2903 | .2918 | .2734 |
|   | 4 | .0002 | .0026 | .0109 | .0287 | .0577 | .0972 | .1442 | .1935 | .2388 | .2734 |
|   | 5 | .0000 | .0002 | .0012 | .0043 | .0115 | .0250 | .0466 | .0774 | .1172 | .1641 |
|   | 6 | .0000 | .0000 | .0001 | .0004 | .0013 | .0036 | .0084 | .0172 | .0320 | .0547 |
|   | 7 | .0000 | .0000 | .0000 | .0000 | .0001 | .0002 | .0006 | .0016 | .0037 | .0078 |
| 8 | 0 | .6634 | .4305 | .2725 | .1678 | .1001 | .0576 | .0319 | .0168 | .0084 | .0039 |
|   | 1 | .2793 | .3826 | .3847 | .3355 | .2670 | .1977 | .1373 | .0896 | .0548 | .0312 |
|   | 2 | .0515 | .1488 | .2376 | .2936 | .3115 | .2965 | .2587 | .2090 | .1569 | .1094 |
|   | 3 | .0054 | .0331 | .0839 | .1468 | .2076 | .2541 | .2786 | .2787 | .2568 | .2188 |
|   | 4 | .0004 | .0046 | .0185 | .0459 | .0865 | .1361 | .1875 | .2322 | .2627 | .2734 |

[†]Based on *Tables of the Binomial Probability Distribution*, National Bureau of Standards Applied Mathematics Series No. 6. Washington, D.C.: U.S. Government Printing Office, 1950.

# TABLE I (continued)

| n | x | .05 | .10 | .15 | .20 | .25 | .30 | .35 | .40 | .45 | .50 |
|---|---|-----|-----|-----|-----|-----|-----|-----|-----|-----|-----|
| 8 | 5 | .0000 | .0004 | .0026 | .0092 | .0231 | .0467 | .0808 | .1239 | .1719 | .2188 |
|   | 6 | .0000 | .0000 | .0002 | .0011 | .0038 | .0100 | .0217 | .0413 | .0703 | .1094 |
|   | 7 | .0000 | .0000 | .0000 | .0001 | .0004 | .0012 | .0033 | .0079 | .0164 | .0312 |
|   | 8 | .0000 | .0000 | .0000 | .0000 | .0000 | .0001 | .0002 | .0007 | .0017 | .0039 |
| 9 | 0 | .6302 | .3874 | .2316 | .1342 | .0751 | .0404 | .0207 | .0101 | .0046 | .0020 |
|   | 1 | .2985 | .3874 | .3679 | .3020 | .2253 | .1556 | .1004 | .0605 | .0339 | .0176 |
|   | 2 | .0629 | .1722 | .2597 | .3020 | .3003 | .2668 | .2162 | .1612 | .1110 | .0703 |
|   | 3 | .0077 | .0446 | .1069 | .1762 | .2336 | .2668 | .2716 | .2508 | .2119 | .1641 |
|   | 4 | .0006 | .0074 | .0283 | .0661 | .1168 | .1715 | .2194 | .2508 | .2600 | .2461 |
|   | 5 | .0000 | .0008 | .0050 | .0165 | .0389 | .0735 | .1181 | .1672 | .2128 | .2461 |
|   | 6 | .0000 | .0001 | .0006 | .0028 | .0087 | .0210 | .0424 | .0743 | .1160 | .1641 |
|   | 7 | .0000 | .0000 | .0000 | .0003 | .0012 | .0039 | .0098 | .0212 | .0407 | .0703 |
|   | 8 | .0000 | .0000 | .0000 | .0000 | .0001 | .0004 | .0013 | .0035 | .0083 | .0176 |
|   | 9 | .0000 | .0000 | .0000 | .0000 | .0000 | .0000 | .0001 | .0003 | .0008 | .0020 |
| 10 | 0 | .5987 | .3487 | .1969 | .1074 | .0563 | .0282 | .0135 | .0060 | .0025 | .0010 |
|   | 1 | .3151 | .3874 | .3474 | .2684 | .1877 | .1211 | .0725 | .0403 | .0207 | .0098 |
|   | 2 | .0746 | .1937 | .2759 | .3020 | .2816 | .2335 | .1757 | .1209 | .0763 | .0439 |
|   | 3 | .0105 | .0574 | .1298 | .2013 | .2503 | .2668 | .2522 | .2150 | .1665 | .1172 |
|   | 4 | .0010 | .0112 | .0401 | .0881 | .1460 | .2001 | .2377 | .2508 | .2384 | .2051 |
|   | 5 | .0001 | .0015 | .0085 | .0264 | .0584 | .1029 | .1536 | .2007 | .2340 | .2461 |
|   | 6 | .0000 | .0001 | .0012 | .0055 | .0162 | .0368 | .0689 | .1115 | .1596 | .2051 |
|   | 7 | .0000 | .0000 | .0001 | .0008 | .0031 | .0090 | .0212 | .0425 | .0746 | .1172 |
|   | 8 | .0000 | .0000 | .0000 | .0001 | .0004 | .0014 | .0043 | .0106 | .0229 | .0439 |
|   | 9 | .0000 | .0000 | .0000 | .0000 | .0000 | .0001 | .0005 | .0016 | .0042 | .0098 |
|   | 10 | .0000 | .0000 | .0000 | .0000 | .0000 | .0000 | .0000 | .0001 | .0003 | .0010 |
| 11 | 0 | .5688 | .3138 | .1673 | .0859 | .0422 | .0198 | .0088 | .0036 | .0014 | .0005 |
|   | 1 | .3293 | .3835 | .3248 | .2362 | .1549 | .0932 | .0518 | .0266 | .0125 | .0054 |
|   | 2 | .0867 | .2131 | .2866 | .2953 | .2581 | .1998 | .1395 | .0887 | .0513 | .0269 |
|   | 3 | .0137 | .0710 | .1517 | .2215 | .2581 | .2568 | .2254 | .1774 | .1259 | .0806 |
|   | 4 | .0014 | .0158 | .0536 | .1107 | .1721 | .2201 | .2428 | .2365 | .2060 | .1611 |
|   | 5 | .0001 | .0025 | .0132 | .0388 | .0803 | .1321 | .1830 | .2207 | .2360 | .2256 |
|   | 6 | .0000 | .0003 | .0023 | .0097 | .0268 | .0566 | .0985 | .1471 | .1931 | .2256 |
|   | 7 | .0000 | .0000 | .0003 | .0017 | .0064 | .0173 | .0379 | .0701 | .1128 | .1611 |
|   | 8 | .0000 | .0000 | .0000 | .0002 | .0011 | .0037 | .0102 | .0234 | .0462 | .0806 |
|   | 9 | .0000 | .0000 | .0000 | .0000 | .0001 | .0005 | .0018 | .0052 | .0126 | .0269 |
|   | 10 | .0000 | .0000 | .0000 | .0000 | .0000 | .0000 | .0002 | .0007 | .0021 | .0054 |
|   | 11 | .0000 | .0000 | .0000 | .0000 | .0000 | .0000 | .0000 | .0000 | .0002 | .0005 |
| 12 | 0 | .5404 | .2824 | .1422 | .0687 | .0317 | .0138 | .0057 | .0022 | .0008 | .0002 |
|   | 1 | .3413 | .3766 | .3012 | .2062 | .1267 | .0712 | .0368 | .0174 | .0075 | .0029 |
|   | 2 | .0988 | .2301 | .2924 | .2835 | .2323 | .1678 | .1088 | .0639 | .0339 | .0161 |
|   | 3 | .0173 | .0852 | .1720 | .2362 | .2581 | .2397 | .1954 | .1419 | .0923 | .0537 |
|   | 4 | .0021 | .0213 | .0683 | .1329 | .1936 | .2311 | .2367 | .2128 | .1700 | .1208 |
|   | 5 | .0002 | .0038 | .0193 | .0532 | .1032 | .1585 | .2039 | .2270 | .2225 | .1934 |
|   | 6 | .0000 | .0005 | .0040 | .0155 | .0401 | .0792 | .1281 | .1766 | .2124 | .2256 |
|   | 7 | .0000 | .0000 | .0006 | .0033 | .0115 | .0291 | .0591 | .1009 | .1489 | .1934 |
|   | 8 | .0000 | .0000 | .0001 | .0005 | .0024 | .0078 | .0199 | .0420 | .0762 | .1208 |
|   | 9 | .0000 | .0000 | .0000 | .0001 | .0004 | .0015 | .0048 | .0125 | .0277 | .0537 |

TABLE I (continued)

| n | x | .05 | .10 | .15 | .20 | .25 | .30 | .35 | .40 | .45 | .50 |
|---|---|-----|-----|-----|-----|-----|-----|-----|-----|-----|-----|
| 12 | 10 | .0000 | .0000 | .0000 | .0000 | .0000 | .0002 | .0008 | .0025 | .0068 | .0161 |
|    | 11 | .0000 | .0000 | .0000 | .0000 | .0000 | .0000 | .0001 | .0003 | .0010 | .0029 |
|    | 12 | .0000 | .0000 | .0000 | .0000 | .0000 | .0000 | .0000 | .0000 | .0001 | .0002 |
| 13 | 0 | .5133 | .2542 | .1209 | .0550 | .0238 | .0097 | .0037 | .0013 | .0004 | .0001 |
|    | 1 | .3512 | .3672 | .2774 | .1787 | .1029 | .0540 | .0259 | .0113 | .0045 | .0016 |
|    | 2 | .1109 | .2448 | .2937 | .2680 | .2059 | .1388 | .0836 | .0453 | .0220 | .0095 |
|    | 3 | .0214 | .0997 | .1900 | .2457 | .2517 | .2181 | .1651 | .1107 | .0660 | .0349 |
|    | 4 | .0028 | .0277 | .0838 | .1535 | .2097 | .2337 | .2222 | .1845 | .1350 | .0873 |
|    | 5 | .0003 | .0055 | .0266 | .0691 | .1258 | .1803 | .2154 | .2214 | .1989 | .1571 |
|    | 6 | .0000 | .0008 | .0063 | .0230 | .0559 | .1030 | .1546 | .1968 | .2169 | .2095 |
|    | 7 | .0000 | .0001 | .0011 | .0058 | .0186 | .0442 | .0833 | .1312 | .1775 | .2095 |
|    | 8 | .0000 | .0000 | .0001 | .0011 | .0047 | .0142 | .0336 | .0656 | .1089 | .1571 |
|    | 9 | .0000 | .0000 | .0000 | .0001 | .0009 | .0034 | .0101 | .0243 | .0495 | .0873 |
|    | 10 | .0000 | .0000 | .0000 | .0000 | .0001 | .0006 | .0022 | .0065 | .0162 | .0349 |
|    | 11 | .0000 | .0000 | .0000 | .0000 | .0000 | .0001 | .0003 | .0012 | .0036 | .0095 |
|    | 12 | .0000 | .0000 | .0000 | .0000 | .0000 | .0000 | .0000 | .0001 | .0005 | .0016 |
|    | 13 | .0000 | .0000 | .0000 | .0000 | .0000 | .0000 | .0000 | .0000 | .0000 | .0001 |
| 14 | 0 | .4877 | .2288 | .1028 | .0440 | .0178 | .0068 | .0024 | .0008 | .0002 | .0001 |
|    | 1 | .3593 | .3559 | .2539 | .1539 | .0832 | .0407 | .0181 | .0073 | .0027 | .0009 |
|    | 2 | .1229 | .2570 | .2912 | .2501 | .1802 | .1134 | .0634 | .0317 | .0141 | .0056 |
|    | 3 | .0259 | .1142 | .2056 | .2501 | .2402 | .1943 | .1366 | .0845 | .0462 | .0222 |
|    | 4 | .0037 | .0349 | .0998 | .1720 | .2202 | .2290 | .2022 | .1549 | .1040 | .0611 |
|    | 5 | .0004 | .0078 | .0352 | .0860 | .1468 | .1963 | .2178 | .2066 | .1701 | .1222 |
|    | 6 | .0000 | .0013 | .0093 | .0322 | .0734 | .1262 | .1759 | .2066 | .2088 | .1833 |
|    | 7 | .0000 | .0002 | .0019 | .0092 | .0280 | .0618 | .1082 | .1574 | .1952 | .2095 |
|    | 8 | .0000 | .0000 | .0003 | .0020 | .0082 | .0232 | .0510 | .0918 | .1398 | .1833 |
|    | 9 | .0000 | .0000 | .0000 | .0003 | .0018 | .0066 | .0183 | .0408 | .0762 | .1222 |
|    | 10 | .0000 | .0000 | .0000 | .0000 | .0003 | .0014 | .0049 | .0136 | .0312 | .0611 |
|    | 11 | .0000 | .0000 | .0000 | .0000 | .0000 | .0002 | .0010 | .0033 | .0093 | .0222 |
|    | 12 | .0000 | .0000 | .0000 | .0000 | .0000 | .0000 | .0001 | .0005 | .0019 | .0056 |
|    | 13 | .0000 | .0000 | .0000 | .0000 | .0000 | .0000 | .0000 | .0001 | .0002 | .0009 |
|    | 14 | .0000 | .0000 | .0000 | .0000 | .0000 | .0000 | .0000 | .0000 | .0000 | .0001 |
| 15 | 0 | .4633 | .2059 | .0874 | .0352 | .0134 | .0047 | .0016 | .0005 | .0001 | .0000 |
|    | 1 | .3658 | .3432 | .2312 | .1319 | .0668 | .0305 | .0126 | .0047 | .0016 | .0005 |
|    | 2 | .1348 | .2669 | .2856 | .2309 | .1559 | .0916 | .0476 | .0219 | .0090 | .0032 |
|    | 3 | .0307 | .1285 | .2184 | .2501 | .2252 | .1700 | .1110 | .0634 | .0318 | .0139 |
|    | 4 | .0049 | .0428 | .1156 | .1876 | .2252 | .2186 | .1792 | .1268 | .0780 | .0417 |
|    | 5 | .0006 | .0105 | .0449 | .1032 | .1651 | .2061 | .2123 | .1859 | .1404 | .0916 |
|    | 6 | .0000 | .0019 | .0132 | .0430 | .0917 | .1472 | .1906 | .2066 | .1914 | .1527 |
|    | 7 | .0000 | .0003 | .0030 | .0138 | .0393 | .0811 | .1319 | .1771 | .2013 | .1964 |
|    | 8 | .0000 | .0000 | .0005 | .0035 | .0131 | .0348 | .0710 | .1181 | .1647 | .1964 |
|    | 9 | .0000 | .0000 | .0001 | .0007 | .0034 | .0116 | .0298 | .0612 | .1048 | .1527 |
|    | 10 | .0000 | .0000 | .0000 | .0001 | .0007 | .0030 | .0096 | .0245 | .0515 | .0916 |
|    | 11 | .0000 | .0000 | .0000 | .0000 | .0001 | .0006 | .0024 | .0074 | .0191 | .0417 |
|    | 12 | .0000 | .0000 | .0000 | .0000 | .0000 | .0001 | .0004 | .0016 | .0052 | .0139 |
|    | 13 | .0000 | .0000 | .0000 | .0000 | .0000 | .0000 | .0001 | .0003 | .0010 | .0032 |
|    | 14 | .0000 | .0000 | .0000 | .0000 | .0000 | .0000 | .0000 | .0000 | .0001 | .0005 |
|    | 15 | .0000 | .0000 | .0000 | .0000 | .0000 | .0000 | .0000 | .0000 | .0000 | .0000 |

TABLE I (continued)

| n | x | .05 | .10 | .15 | .20 | .25 | .30 | .35 | .40 | .45 | .50 |
|---|---|-----|-----|-----|-----|-----|-----|-----|-----|-----|-----|
| 16 | 0 | .4401 | .1853 | .0743 | .0281 | .0100 | .0033 | .0010 | .0003 | .0001 | .0000 |
| | 1 | .3706 | .3294 | .2097 | .1126 | .0535 | .0228 | .0087 | .0030 | .0009 | .0002 |
| | 2 | .1463 | .2745 | .2775 | .2111 | .1336 | .0732 | .0353 | .0150 | .0056 | .0018 |
| | 3 | .0359 | .1423 | .2285 | .2463 | .2079 | .1465 | .0888 | .0468 | .0215 | .0085 |
| | 4 | .0061 | .0514 | .1311 | .2001 | .2252 | .2040 | .1553 | .1014 | .0572 | .0278 |
| | 5 | .0008 | .0137 | .0555 | .1201 | .1802 | .2099 | .2008 | .1623 | .1123 | .0667 |
| | 6 | .0001 | .0028 | .0180 | .0550 | .1101 | .1649 | .1982 | .1983 | .1684 | .1222 |
| | 7 | .0000 | .0004 | .0045 | .0197 | .0524 | .1010 | .1524 | .1889 | .1969 | .1746 |
| | 8 | .0000 | .0001 | .0009 | .0055 | .0197 | .0487 | .0923 | .1417 | .1812 | .1964 |
| | 9 | .0000 | .0000 | .0001 | .0012 | .0058 | .0185 | .0442 | .0840 | .1318 | .1746 |
| | 10 | .0000 | .0000 | .0000 | .0002 | .0014 | .0056 | .0167 | .0392 | .0755 | .1222 |
| | 11 | .0000 | .0000 | .0000 | .0000 | .0002 | .0013 | .0049 | .0142 | .0337 | .0667 |
| | 12 | .0000 | .0000 | .0000 | .0000 | .0000 | .0002 | .0011 | .0040 | .0115 | .0278 |
| | 13 | .0000 | .0000 | .0000 | .0000 | .0000 | .0000 | .0002 | .0008 | .0029 | .0085 |
| | 14 | .0000 | .0000 | .0000 | .0000 | .0000 | .0000 | .0000 | .0001 | .0005 | .0018 |
| | 15 | .0000 | .0000 | .0000 | .0000 | .0000 | .0000 | .0000 | .0000 | .0001 | .0002 |
| | 16 | .0000 | .0000 | .0000 | .0000 | .0000 | .0000 | .0000 | .0000 | .0000 | .0000 |
| 17 | 0 | .4181 | .1668 | .0631 | .0225 | .0075 | .0023 | .0007 | .0002 | .0000 | .0000 |
| | 1 | .3741 | .3150 | .1893 | .0957 | .0426 | .0169 | .0060 | .0019 | .0005 | .0001 |
| | 2 | .1575 | .2800 | .2673 | .1914 | .1136 | .0581 | .0260 | .0102 | .0035 | .0010 |
| | 3 | .0415 | .1556 | .2359 | .2393 | .1893 | .1245 | .0701 | .0341 | .0144 | .0052 |
| | 4 | .0076 | .0605 | .1457 | .2093 | .2209 | .1868 | .1320 | .0796 | .0411 | .0182 |
| | 5 | .0010 | .0175 | .0668 | .1361 | .1914 | .2081 | .1849 | .1379 | .0875 | .0472 |
| | 6 | .0001 | .0039 | .0236 | .0680 | .1276 | .1784 | .1991 | .1839 | .1432 | .0944 |
| | 7 | .0000 | .0007 | .0065 | .0267 | .0668 | .1201 | .1685 | .1927 | .1841 | .1484 |
| | 8 | .0000 | .0001 | .0014 | .0084 | .0279 | .0644 | .1134 | .1606 | .1883 | .1855 |
| | 9 | .0000 | .0000 | .0003 | .0021 | .0093 | .0276 | .0611 | .1070 | .1540 | .1855 |
| | 10 | .0000 | .0000 | .0000 | .0004 | .0025 | .0095 | .0263 | .0571 | .1008 | .1484 |
| | 11 | .0000 | .0000 | .0000 | .0001 | .0005 | .0026 | .0090 | .0242 | .0525 | .0944 |
| | 12 | .0000 | .0000 | .0000 | .0000 | .0001 | .0006 | .0024 | .0081 | .0215 | .0472 |
| | 13 | .0000 | .0000 | .0000 | .0000 | .0000 | .0001 | .0005 | .0021 | .0068 | .0182 |
| | 14 | .0000 | .0000 | .0000 | .0000 | .0000 | .0000 | .0001 | .0004 | .0016 | .0052 |
| | 15 | .0000 | .0000 | .0000 | .0000 | .0000 | .0000 | .0000 | .0001 | .0003 | .0010 |
| | 16 | .0000 | .0000 | .0000 | .0000 | .0000 | .0000 | .0000 | .0000 | .0000 | .0001 |
| | 17 | .0000 | .0000 | .0000 | .0000 | .0000 | .0000 | .0000 | .0000 | .0000 | .0000 |
| 18 | 0 | .3972 | .1501 | .0536 | .0180 | .0056 | .0016 | .0004 | .0001 | .0000 | .0000 |
| | 1 | .3763 | .3002 | .1704 | .0811 | .0338 | .0126 | .0042 | .0012 | .0003 | .0001 |
| | 2 | .1683 | .2835 | .2556 | .1723 | .0958 | .0458 | .0190 | .0069 | .0022 | .0006 |
| | 3 | .0473 | .1680 | .2406 | .2297 | .1704 | .1046 | .0547 | .0246 | .0095 | .0031 |
| | 4 | .0093 | .0700 | .1592 | .2153 | .2130 | .1681 | .1104 | .0614 | .0291 | .0117 |
| | 5 | .0014 | .0218 | .0787 | .1507 | .1988 | .2017 | .1664 | .1146 | .0666 | .0327 |
| | 6 | .0002 | .0052 | .0301 | .0816 | .1436 | .1873 | .1941 | .1655 | .1181 | .0708 |
| | 7 | .0000 | .0010 | .0091 | .0350 | .0820 | .1376 | .1792 | .1892 | .1657 | .1214 |
| | 8 | .0000 | .0002 | .0022 | .0120 | .0376 | .0811 | .1327 | .1734 | .1864 | .1669 |
| | 9 | .0000 | .0000 | .0004 | .0033 | .0139 | .0386 | .0794 | .1284 | .1694 | .1855 |

# TABLE I (continued)

| n | x | θ .05 | .10 | .15 | .20 | .25 | .30 | .35 | .40 | .45 | .50 |
|---|---|---|---|---|---|---|---|---|---|---|---|
| 18 | 10 | .0000 | .0000 | .0001 | .0008 | .0042 | .0149 | .0385 | .0771 | .1248 | .1669 |
|  | 11 | .0000 | .0000 | .0000 | .0001 | .0010 | .0046 | .0151 | .0374 | .0742 | .1214 |
|  | 12 | .0000 | .0000 | .0000 | .0000 | .0002 | .0012 | .0047 | .0145 | .0354 | .0708 |
|  | 13 | .0000 | .0000 | .0000 | .0000 | .0000 | .0002 | .0012 | .0045 | .0134 | .0327 |
|  | 14 | .0000 | .0000 | .0000 | .0000 | .0000 | .0000 | .0002 | .0011 | .0039 | .0117 |
|  | 15 | .0000 | .0000 | .0000 | .0000 | .0000 | .0000 | .0000 | .0002 | .0009 | .0031 |
|  | 16 | .0000 | .0000 | .0000 | .0000 | .0000 | .0000 | .0000 | .0000 | .0001 | .0006 |
|  | 17 | .0000 | .0000 | .0000 | .0000 | .0000 | .0000 | .0000 | .0000 | .0000 | .0001 |
|  | 18 | .0000 | .0000 | .0000 | .0000 | .0000 | .0000 | .0000 | .0000 | .0000 | .0000 |
| 19 | 0 | .3774 | .1351 | .0456 | .0144 | .0042 | .0011 | .0003 | .0001 | .0000 | .0000 |
|  | 1 | .3774 | .2852 | .1529 | .0685 | .0268 | .0093 | .0029 | .0008 | .0002 | .0000 |
|  | 2 | .1787 | .2852 | .2428 | .1540 | .0803 | .0358 | .0138 | .0046 | .0013 | .0003 |
|  | 3 | .0533 | .1796 | .2428 | .2182 | .1517 | .0869 | .0422 | .0175 | .0062 | .0018 |
|  | 4 | .0112 | .0798 | .1714 | .2182 | .2023 | .1491 | .0909 | .0467 | .0203 | .0074 |
|  | 5 | .0018 | .0266 | .0907 | .1636 | .2023 | .1916 | .1468 | .0933 | .0497 | .0222 |
|  | 6 | .0002 | .0069 | .0374 | .0955 | .1574 | .1916 | .1844 | .1451 | .0949 | .0518 |
|  | 7 | .0000 | .0014 | .0122 | .0443 | .0974 | .1525 | .1844 | .1797 | .1443 | .0961 |
|  | 8 | .0000 | .0002 | .0032 | .0166 | .0487 | .0981 | .1489 | .1797 | .1771 | .1442 |
|  | 9 | .0000 | .0000 | .0007 | .0051 | .0198 | .0514 | .0980 | .1464 | .1771 | .1762 |
|  | 10 | .0000 | .0000 | .0001 | .0013 | .0066 | .0220 | .0528 | .0976 | .1449 | .1762 |
|  | 11 | .0000 | .0000 | .0000 | .0003 | .0018 | .0077 | .0233 | .0532 | .0970 | .1442 |
|  | 12 | .0000 | .0000 | .0000 | .0000 | .0004 | .0022 | .0083 | .0237 | .0529 | .0961 |
|  | 13 | .0000 | .0000 | .0000 | .0000 | .0001 | .0005 | .0024 | .0085 | .0233 | .0518 |
|  | 14 | .0000 | .0000 | .0000 | .0000 | .0000 | .0001 | .0006 | .0024 | .0082 | .0222 |
|  | 15 | .0000 | .0000 | .0000 | .0000 | .0000 | .0000 | .0001 | .0005 | .0022 | .0074 |
|  | 16 | .0000 | .0000 | .0000 | .0000 | .0000 | .0000 | .0000 | .0001 | .0005 | .0018 |
|  | 17 | .0000 | .0000 | .0000 | .0000 | .0000 | .0000 | .0000 | .0000 | .0001 | .0003 |
|  | 18 | .0000 | .0000 | .0000 | .0000 | .0000 | .0000 | .0000 | .0000 | .0000 | .0000 |
|  | 19 | .0000 | .0000 | .0000 | .0000 | .0000 | .0000 | .0000 | .0000 | .0000 | .0000 |
| 20 | 0 | .3585 | .1216 | .0388 | .0115 | .0032 | .0008 | .0002 | .0000 | .0000 | .0000 |
|  | 1 | .3774 | .2702 | .1368 | .0576 | .0211 | .0068 | .0020 | .0005 | .0001 | .0000 |
|  | 2 | .1887 | .2852 | .2293 | .1369 | .0669 | .0278 | .0100 | .0031 | .0008 | .0002 |
|  | 3 | .0596 | .1901 | .2428 | .2054 | .1339 | .0716 | .0323 | .0123 | .0040 | .0011 |
|  | 4 | .0133 | .0898 | .1821 | .2182 | .1897 | .1304 | .0738 | .0350 | .0139 | .0046 |
|  | 5 | .0022 | .0319 | .1028 | .1746 | .2023 | .1789 | .1272 | .0746 | .0365 | .0148 |
|  | 6 | .0003 | .0089 | .0454 | .1091 | .1686 | .1916 | .1712 | .1244 | .0746 | .0370 |
|  | 7 | .0000 | .0020 | .0160 | .0545 | .1124 | .1643 | .1844 | .1659 | .1221 | .0739 |
|  | 8 | .0000 | .0004 | .0046 | .0222 | .0609 | .1144 | .1614 | .1797 | .1623 | .1201 |
|  | 9 | .0000 | .0001 | .0011 | .0074 | .0271 | .0654 | .1158 | .1597 | .1771 | .1602 |
|  | 10 | .0000 | .0000 | .0002 | .0020 | .0099 | .0308 | .0686 | .1171 | .1593 | .1762 |
|  | 11 | .0000 | .0000 | .0000 | .0005 | .0030 | .0120 | .0336 | .0710 | .1185 | .1602 |
|  | 12 | .0000 | .0000 | .0000 | .0001 | .0008 | .0039 | .0136 | .0355 | .0727 | .1201 |
|  | 13 | .0000 | .0000 | .0000 | .0000 | .0002 | .0010 | .0045 | .0146 | .0366 | .0739 |
|  | 14 | .0000 | .0000 | .0000 | .0000 | .0000 | .0002 | .0012 | .0049 | .0150 | .0370 |
|  | 15 | .0000 | .0000 | .0000 | .0000 | .0000 | .0000 | .0003 | .0013 | .0049 | .0148 |
|  | 16 | .0000 | .0000 | .0000 | .0000 | .0000 | .0000 | .0000 | .0003 | .0013 | .0046 |
|  | 17 | .0000 | .0000 | .0000 | .0000 | .0000 | .0000 | .0000 | .0000 | .0002 | .0011 |
|  | 18 | .0000 | .0000 | .0000 | .0000 | .0000 | .0000 | .0000 | .0000 | .0000 | .0002 |
|  | 19 | .0000 | .0000 | .0000 | .0000 | .0000 | .0000 | .0000 | .0000 | .0000 | .0000 |
|  | 20 | .0000 | .0000 | .0000 | .0000 | .0000 | .0000 | .0000 | .0000 | .0000 | .0000 |

# TABLE II

## Poisson Probabilities[†]

### λ

| x | 0.1 | 0.2 | 0.3 | 0.4 | 0.5 | 0.6 | 0.7 | 0.8 | 0.9 | 1.0 |
|---|-----|-----|-----|-----|-----|-----|-----|-----|-----|-----|
| 0 | .9048 | .8187 | .7408 | .6703 | .6065 | .5488 | .4966 | .4493 | .4066 | .3679 |
| 1 | .0905 | .1637 | .2222 | .2681 | .3033 | .3293 | .3476 | .3595 | .3659 | .3679 |
| 2 | .0045 | .0164 | .0333 | .0536 | .0758 | .0988 | .1217 | .1438 | .1647 | .1839 |
| 3 | .0002 | .0011 | .0033 | .0072 | .0126 | .0198 | .0284 | .0383 | .0494 | .0613 |
| 4 | .0000 | .0001 | .0002 | .0007 | .0016 | .0030 | .0050 | .0077 | .0111 | .0153 |
| 5 | .0000 | .0000 | .0000 | .0001 | .0002 | .0004 | .0007 | .0012 | .0020 | .0031 |
| 6 | .0000 | .0000 | .0000 | .0000 | .0000 | .0000 | .0001 | .0002 | .0003 | .0005 |
| 7 | .0000 | .0000 | .0000 | .0000 | .0000 | .0000 | .0000 | .0000 | .0000 | .0001 |

### λ

| x | 1.1 | 1.2 | 1.3 | 1.4 | 1.5 | 1.6 | 1.7 | 1.8 | 1.9 | 2.0 |
|---|-----|-----|-----|-----|-----|-----|-----|-----|-----|-----|
| 0 | .3329 | .3012 | .2725 | .2466 | .2231 | .2019 | .1827 | .1653 | .1496 | .1353 |
| 1 | .3662 | .3614 | .3543 | .3452 | .3347 | .3230 | .3106 | .2975 | .2842 | .2707 |
| 2 | .2014 | .2169 | .2303 | .2417 | .2510 | .2584 | .2640 | .2678 | .2700 | .2707 |
| 3 | .0738 | .0867 | .0998 | .1128 | .1255 | .1378 | .1496 | .1607 | .1710 | .1804 |
| 4 | .0203 | .0260 | .0324 | .0395 | .0471 | .0551 | .0636 | .0723 | .0812 | .0902 |
| 5 | .0045 | .0062 | .0084 | .0111 | .0141 | .0176 | .0216 | .0260 | .0309 | .0361 |
| 6 | .0008 | .0012 | .0018 | .0026 | .0035 | .0047 | .0061 | .0078 | .0098 | .0120 |
| 7 | .0001 | .0002 | .0003 | .0005 | .0008 | .0011 | .0015 | .0020 | .0027 | .0034 |
| 8 | .0000 | .0000 | .0001 | .0001 | .0001 | .0002 | .0003 | .0005 | .0006 | .0009 |
| 9 | .0000 | .0000 | .0000 | .0000 | .0000 | .0000 | .0001 | .0001 | .0001 | .0002 |

### λ

| x | 2.1 | 2.2 | 2.3 | 2.4 | 2.5 | 2.6 | 2.7 | 2.8 | 2.9 | 3.0 |
|---|-----|-----|-----|-----|-----|-----|-----|-----|-----|-----|
| 0 | .1225 | .1108 | .1003 | .0907 | .0821 | .0743 | .0672 | .0608 | .0550 | .0498 |
| 1 | .2572 | .2438 | .2306 | .2177 | .2052 | .1931 | .1815 | .1703 | .1596 | .1494 |
| 2 | .2700 | .2681 | .2652 | .2613 | .2565 | .2510 | .2450 | .2384 | .2314 | .2240 |
| 3 | .1890 | .1966 | .2033 | .2090 | .2138 | .2176 | .2205 | .2225 | .2237 | .2240 |
| 4 | .0992 | .1082 | .1169 | .1254 | .1336 | .1414 | .1488 | .1557 | .1622 | .1680 |
| 5 | .0417 | .0476 | .0538 | .0602 | .0668 | .0735 | .0804 | .0872 | .0940 | .1008 |
| 6 | .0146 | .0174 | .0206 | .0241 | .0278 | .0319 | .0362 | .0407 | .0455 | .0504 |
| 7 | .0044 | .0055 | .0068 | .0083 | .0099 | .0118 | .0139 | .0163 | .0188 | .0216 |
| 8 | .0011 | .0015 | .0019 | .0025 | .0031 | .0038 | .0047 | .0057 | .0068 | .0081 |
| 9 | .0003 | .0004 | .0005 | .0007 | .0009 | .0011 | .0014 | .0018 | .0022 | .0027 |
| 10 | .0001 | .0001 | .0001 | .0002 | .0002 | .0003 | .0004 | .0005 | .0006 | .0008 |
| 11 | .0000 | .0000 | .0000 | .0000 | .0000 | .0001 | .0001 | .0001 | .0002 | .0002 |
| 12 | .0000 | .0000 | .0000 | .0000 | .0000 | .0000 | .0000 | .0000 | .0000 | .0001 |

[†]Based on E. C. Molina, *Poisson's Exponential Binomial Limit,* 1973 Reprint, Robert E. Krieger Publishing Company, Melbourne, Fla., by permission of the publisher.

# TABLE II (continued)

$\lambda$

| $x$ | 3.1 | 3.2 | 3.3 | 3.4 | 3.5 | 3.6 | 3.7 | 3.8 | 3.9 | 4.0 |
|---|---|---|---|---|---|---|---|---|---|---|
| 0 | .0450 | .0408 | .0369 | .0334 | .0302 | .0273 | .0247 | .0224 | .0202 | .0183 |
| 1 | .1397 | .1304 | .1217 | .1135 | .1057 | .0984 | .0915 | .0850 | .0789 | .0733 |
| 2 | .2165 | .2087 | .2008 | .1929 | .1850 | .1771 | .1692 | .1615 | .1539 | .1465 |
| 3 | .2237 | .2226 | .2209 | .2186 | .2158 | .2125 | .2087 | .2046 | .2001 | .1954 |
| 4 | .1734 | .1781 | .1823 | .1858 | .1888 | .1912 | .1931 | .1944 | .1951 | .1954 |
| 5 | .1075 | .1140 | .1203 | .1264 | .1322 | .1377 | .1429 | .1477 | .1522 | .1563 |
| 6 | .0555 | .0608 | .0662 | .0716 | .0771 | .0826 | .0881 | .0936 | .0989 | .1042 |
| 7 | .0246 | .0278 | .0312 | .0348 | .0385 | .0425 | .0466 | .0508 | .0551 | .0595 |
| 8 | .0095 | .0111 | .0129 | .0148 | .0169 | .0191 | .0215 | .0241 | .0269 | .0298 |
| 9 | .0033 | .0040 | .0047 | .0056 | .0066 | .0076 | .0089 | .0102 | .0116 | .0132 |
| 10 | .0010 | .0013 | .0016 | .0019 | .0023 | .0028 | .0033 | .0039 | .0045 | .0053 |
| 11 | .0003 | .0004 | .0005 | .0006 | .0007 | .0009 | .0011 | .0013 | .0016 | .0019 |
| 12 | .0001 | .0001 | .0001 | .0002 | .0002 | .0003 | .0003 | .0004 | .0005 | .0006 |
| 13 | .0000 | .0000 | .0000 | .0000 | .0001 | .0001 | .0001 | .0001 | .0002 | .0002 |
| 14 | .0000 | .0000 | .0000 | .0000 | .0000 | .0000 | .0000 | .0000 | .0000 | .0001 |

$\lambda$

| $x$ | 4.1 | 4.2 | 4.3 | 4.4 | 4.5 | 4.6 | 4.7 | 4.8 | 4.9 | 5.0 |
|---|---|---|---|---|---|---|---|---|---|---|
| 0 | .0166 | .0150 | .0136 | .0123 | .0111 | .0101 | .0091 | .0082 | .0074 | .0067 |
| 1 | .0679 | .0630 | .0583 | .0540 | .0500 | .0462 | .0427 | .0395 | .0365 | .0337 |
| 2 | .1393 | .1323 | .1254 | .1188 | .1125 | .1063 | .1005 | .0948 | .0894 | .0842 |
| 3 | .1904 | .1852 | .1798 | .1743 | .1687 | .1631 | .1574 | .1517 | .1460 | .1404 |
| 4 | .1951 | .1944 | .1933 | .1917 | .1898 | .1875 | .1849 | .1820 | .1789 | .1755 |
| 5 | .1600 | .1633 | .1662 | .1687 | .1708 | .1725 | .1738 | .1747 | .1753 | .1755 |
| 6 | .1093 | .1143 | .1191 | .1237 | .1281 | .1323 | .1362 | .1398 | .1432 | .1462 |
| 7 | .0640 | .0686 | .0732 | .0778 | .0824 | .0869 | .0914 | .0959 | .1002 | .1044 |
| 8 | .0328 | .0360 | .0393 | .0428 | .0463 | .0500 | .0537 | .0575 | .0614 | .0653 |
| 9 | .0150 | .0168 | .0188 | .0209 | .0232 | .0255 | .0280 | .0307 | .0334 | .0363 |
| 10 | .0061 | .0071 | .0081 | .0092 | .0104 | .0118 | .0132 | .0147 | .0164 | .0181 |
| 11 | .0023 | .0027 | .0032 | .0037 | .0043 | .0049 | .0056 | .0064 | .0073 | .0082 |
| 12 | .0008 | .0009 | .0011 | .0014 | .0016 | .0019 | .0022 | .0026 | .0030 | .0034 |
| 13 | .0002 | .0003 | .0004 | .0005 | .0006 | .0007 | .0008 | .0009 | .0011 | .0013 |
| 14 | .0001 | .0001 | .0001 | .0001 | .0002 | .0002 | .0003 | .0003 | .0004 | .0005 |
| 15 | .0000 | .0000 | .0000 | .0000 | .0001 | .0001 | .0001 | .0001 | .0001 | .0002 |

$\lambda$

| $x$ | 5.1 | 5.2 | 5.3 | 5.4 | 5.5 | 5.6 | 5.7 | 5.8 | 5.9 | 6.0 |
|---|---|---|---|---|---|---|---|---|---|---|
| 0 | .0061 | .0055 | .0050 | .0045 | .0041 | .0037 | .0033 | .0030 | .0027 | .0025 |
| 1 | .0311 | .0287 | .0265 | .0244 | .0225 | .0207 | .0191 | .0176 | .0162 | .0149 |
| 2 | .0793 | .0746 | .0701 | .0659 | .0618 | .0580 | .0544 | .0509 | .0477 | .0446 |
| 3 | .1348 | .1293 | .1239 | .1185 | .1133 | .1082 | .1033 | .0985 | .0938 | .0892 |
| 4 | .1719 | .1681 | .1641 | .1600 | .1558 | .1515 | .1472 | .1428 | .1383 | .1339 |

# TABLE II (continued)

| $x$ | 5.1 | 5.2 | 5.3 | 5.4 | 5.5 | 5.6 | 5.7 | 5.8 | 5.9 | 6.0 |
|---|---|---|---|---|---|---|---|---|---|---|
| 5 | .1753 | .1748 | .1740 | .1728 | .1714 | .1697 | .1678 | .1656 | .1632 | .1606 |
| 6 | .1490 | .1515 | .1537 | .1555 | .1571 | .1584 | .1594 | .1601 | .1605 | .1606 |
| 7 | .1086 | .1125 | .1163 | .1200 | .1234 | .1267 | .1298 | .1326 | .1353 | .1377 |
| 8 | .0692 | .0731 | .0771 | .0810 | .0849 | .0887 | .0925 | .0962 | .0998 | .1033 |
| 9 | .0392 | .0423 | .0454 | .0486 | .0519 | .0552 | .0586 | .0620 | .0654 | .0688 |
| 10 | .0200 | .0220 | .0241 | .0262 | .0285 | .0309 | .0334 | .0359 | .0386 | .0413 |
| 11 | .0093 | .0104 | .0116 | .0129 | .0143 | .0157 | .0173 | .0190 | .0207 | .0225 |
| 12 | .0039 | .0045 | .0051 | .0058 | .0065 | .0073 | .0082 | .0092 | .0102 | .0113 |
| 13 | .0015 | .0018 | .0021 | .0024 | .0028 | .0032 | .0036 | .0041 | .0046 | .0052 |
| 14 | .0006 | .0007 | .0008 | .0009 | .0011 | .0013 | .0015 | .0017 | .0019 | .0022 |
| 15 | .0002 | .0002 | .0003 | .0003 | .0004 | .0005 | .0006 | .0007 | .0008 | .0009 |
| 16 | .0001 | .0001 | .0001 | .0001 | .0001 | .0002 | .0002 | .0002 | .0003 | .0003 |
| 17 | .0000 | .0000 | .0000 | .0000 | .0000 | .0001 | .0001 | .0001 | .0001 | .0001 |

| $x$ | 6.1 | 6.2 | 6.3 | 6.4 | 6.5 | 6.6 | 6.7 | 6.8 | 6.9 | 7.0 |
|---|---|---|---|---|---|---|---|---|---|---|
| 0 | .0022 | .0020 | .0018 | .0017 | .0015 | .0014 | .0012 | .0011 | .0010 | .0009 |
| 1 | .0137 | .0126 | .0116 | .0106 | .0098 | .0090 | .0082 | .0076 | .0070 | .0064 |
| 2 | .0417 | .0390 | .0364 | .0340 | .0318 | .0296 | .0276 | .0258 | .0240 | .0223 |
| 3 | .0848 | .0806 | .0765 | .0726 | .0688 | .0652 | .0617 | .0584 | .0552 | .0521 |
| 4 | .1294 | .1249 | .1205 | .1162 | .1118 | .1076 | .1034 | .0992 | .0952 | .0912 |
| 5 | .1579 | .1549 | .1519 | .1487 | .1454 | .1420 | .1385 | .1349 | .1314 | .1277 |
| 6 | .1605 | .1601 | .1595 | .1586 | .1575 | .1562 | .1546 | .1529 | .1511 | .1490 |
| 7 | .1399 | .1418 | .1435 | .1450 | .1462 | .1472 | .1480 | .1486 | .1489 | .1490 |
| 8 | .1066 | .1099 | .1130 | .1160 | .1188 | .1215 | .1240 | .1263 | .1284 | .1304 |
| 9 | .0723 | .0757 | .0791 | .0825 | .0858 | .0891 | .0923 | .0954 | .0985 | .1014 |
| 10 | .0441 | .0469 | .0498 | .0528 | .0558 | .0588 | .0618 | .0649 | .0679 | .0710 |
| 11 | .0245 | .0265 | .0285 | .0307 | .0330 | .0353 | .0377 | .0401 | .0426 | .0452 |
| 12 | .0124 | .0137 | .0150 | .0164 | .0179 | .0194 | .0210 | .0227 | .0245 | .0264 |
| 13 | .0058 | .0065 | .0073 | .0081 | .0089 | .0098 | .0108 | .0119 | .0130 | .0142 |
| 14 | .0025 | .0029 | .0033 | .0037 | .0041 | .0046 | .0052 | .0058 | .0064 | .0071 |
| 15 | .0010 | .0012 | .0014 | .0016 | .0018 | .0020 | .0023 | .0026 | .0029 | .0033 |
| 16 | .0004 | .0005 | .0005 | .0006 | .0007 | .0008 | .0010 | .0011 | .0013 | .0014 |
| 17 | .0001 | .0002 | .0002 | .0002 | .0003 | .0003 | .0004 | .0004 | .0005 | .0006 |
| 18 | .0000 | .0001 | .0001 | .0001 | .0001 | .0001 | .0001 | .0002 | .0002 | .0002 |
| 19 | .0000 | .0000 | .0000 | .0000 | .0000 | .0000 | .0000 | .0001 | .0001 | .0001 |

| $x$ | 7.1 | 7.2 | 7.3 | 7.4 | 7.5 | 7.6 | 7.7 | 7.8 | 7.9 | 8.0 |
|---|---|---|---|---|---|---|---|---|---|---|
| 0 | .0008 | .0007 | .0007 | .0006 | .0006 | .0005 | .0005 | .0004 | .0004 | .0003 |
| 1 | .0059 | .0054 | .0049 | .0045 | .0041 | .0038 | .0035 | .0032 | .0029 | .0027 |
| 2 | .0208 | .0194 | .0180 | .0167 | .0156 | .0145 | .0134 | .0125 | .0116 | .0107 |
| 3 | .0492 | .0464 | .0438 | .0413 | .0389 | .0366 | .0345 | .0324 | .0305 | .0286 |
| 4 | .0874 | .0836 | .0799 | .0764 | .0729 | .0696 | .0663 | .0632 | .0602 | .0573 |
| 5 | .1241 | .1204 | .1167 | .1130 | .1094 | .1057 | .1021 | .0986 | .0951 | .0916 |
| 6 | .1468 | .1445 | .1420 | .1394 | .1367 | .1339 | .1311 | .1282 | .1252 | .1221 |
| 7 | .1489 | .1486 | .1481 | .1474 | .1465 | .1454 | .1442 | .1428 | .1413 | .1396 |
| 8 | .1321 | .1337 | .1351 | .1363 | .1373 | .1382 | .1388 | .1392 | .1395 | .1396 |
| 9 | .1042 | .1070 | .1096 | .1121 | .1144 | .1167 | .1187 | .1207 | .1224 | .1241 |

**TABLE II (continued)**

|  | λ | | | | | | | | | |
|---|---|---|---|---|---|---|---|---|---|---|
| x | 7.1 | 7.2 | 7.3 | 7.4 | 7.5 | 7.6 | 7.7 | 7.8 | 7.9 | 8.0 |
| 10 | .0740 | .0770 | .0800 | .0829 | .0858 | .0887 | .0914 | .0941 | .0967 | .0993 |
| 11 | .0478 | .0504 | .0531 | .0558 | .0585 | .0613 | .0640 | .0667 | .0695 | .0722 |
| 12 | .0283 | .0303 | .0323 | .0344 | .0366 | .0388 | .0411 | .0434 | .0457 | .0481 |
| 13 | .0154 | .0168 | .0181 | .0196 | .0211 | .0227 | .0243 | .0260 | .0278 | .0296 |
| 14 | .0078 | .0086 | .0095 | .0104 | .0113 | .0123 | .0134 | .0145 | .0157 | .0169 |
| 15 | .0037 | .0041 | .0046 | .0051 | .0057 | .0062 | .0069 | .0075 | .0083 | .0090 |
| 16 | .0016 | .0019 | .0021 | .0024 | .0026 | .0030 | .0033 | .0037 | .0041 | .0045 |
| 17 | .0007 | .0008 | .0009 | .0010 | .0012 | .0013 | .0015 | .0017 | .0019 | .0021 |
| 18 | .0003 | .0003 | .0004 | .0004 | .0005 | .0006 | .0006 | .0007 | .0008 | .0009 |
| 19 | .0001 | .0001 | .0001 | .0002 | .0002 | .0002 | .0003 | .0003 | .0003 | .0004 |
| 20 | .0000 | .0000 | .0001 | .0001 | .0001 | .0001 | .0001 | .0001 | .0001 | .0002 |
| 21 | .0000 | .0000 | .0000 | .0000 | .0000 | .0000 | .0000 | .0000 | .0001 | .0001 |

|  | λ | | | | | | | | | |
|---|---|---|---|---|---|---|---|---|---|---|
| x | 8.1 | 8.2 | 8.3 | 8.4 | 8.5 | 8.6 | 8.7 | 8.8 | 8.9 | 9.0 |
| 0 | .0003 | .0003 | .0002 | .0002 | .0002 | .0002 | .0002 | .0002 | .0001 | .0001 |
| 1 | .0025 | .0023 | .0021 | .0019 | .0017 | .0016 | .0014 | .0013 | .0012 | .0011 |
| 2 | .0100 | .0092 | .0086 | .0079 | .0074 | .0068 | .0063 | .0058 | .0054 | .0050 |
| 3 | .0269 | .0252 | .0237 | .0222 | .0208 | .0195 | .0183 | .0171 | .0160 | .0150 |
| 4 | .0544 | .0517 | .0491 | .0466 | .0443 | .0420 | .0398 | .0377 | .0357 | .0337 |
| 5 | .0882 | .0849 | .0816 | .0784 | .0752 | .0722 | .0692 | .0663 | .0635 | .0607 |
| 6 | .1191 | .1160 | .1128 | .1097 | .1066 | .1034 | .1003 | .0972 | .0941 | .0911 |
| 7 | .1378 | .1358 | .1338 | .1317 | .1294 | .1271 | .1247 | .1222 | .1197 | .1171 |
| 8 | .1395 | .1392 | .1388 | .1382 | .1375 | .1366 | .1356 | .1344 | .1332 | .1318 |
| 9 | .1256 | .1269 | .1280 | .1290 | .1299 | .1306 | .1311 | .1315 | .1317 | .1318 |
| 10 | .1017 | .1040 | .1063 | .1084 | .1104 | .1123 | .1140 | .1157 | .1172 | .1186 |
| 11 | .0749 | .0776 | .0802 | .0828 | .0853 | .0878 | .0902 | .0925 | .0948 | .0970 |
| 12 | .0505 | .0530 | .0555 | .0579 | .0604 | .0629 | .0654 | .0679 | .0703 | .0728 |
| 13 | .0315 | .0334 | .0354 | .0374 | .0395 | .0416 | .0438 | .0459 | .0481 | .0504 |
| 14 | .0182 | .0196 | .0210 | .0225 | .0240 | .0256 | .0272 | .0289 | .0306 | .0324 |
| 15 | .0098 | .0107 | .0116 | .0126 | .0136 | .0147 | .0158 | .0169 | .0182 | .0194 |
| 16 | .0050 | .0055 | .0060 | .0066 | .0072 | .0079 | .0086 | .0093 | .0101 | .0109 |
| 17 | .0024 | .0026 | .0029 | .0033 | .0036 | .0040 | .0044 | .0048 | .0053 | .0058 |
| 18 | .0011 | .0012 | .0014 | .0015 | .0017 | .0019 | .0021 | .0024 | .0026 | .0029 |
| 19 | .0005 | .0005 | .0006 | .0007 | .0008 | .0009 | .0010 | .0011 | .0012 | .0014 |
| 20 | .0002 | .0002 | .0002 | .0003 | .0003 | .0004 | .0004 | .0005 | .0005 | .0006 |
| 21 | .0001 | .0001 | .0001 | .0001 | .0001 | .0002 | .0002 | .0002 | .0002 | .0003 |
| 22 | .0000 | .0000 | .0000 | .0000 | .0001 | .0001 | .0001 | .0001 | .0001 | .0001 |

|  | λ | | | | | | | | | |
|---|---|---|---|---|---|---|---|---|---|---|
| x | 9.1 | 9.2 | 9.3 | 9.4 | 9.5 | 9.6 | 9.7 | 9.8 | 9.9 | 10 |
| 0 | .0001 | .0001 | .0001 | .0001 | .0001 | .0001 | .0001 | .0001 | .0001 | .0000 |
| 1 | .0010 | .0009 | .0009 | .0008 | .0007 | .0007 | .0006 | .0005 | .0005 | .0005 |
| 2 | .0046 | .0043 | .0040 | .0037 | .0034 | .0031 | .0029 | .0027 | .0025 | .0023 |
| 3 | .0140 | .0131 | .0123 | .0115 | .0107 | .0100 | .0093 | .0087 | .0081 | .0076 |
| 4 | .0319 | .0302 | .0285 | .0269 | .0254 | .0240 | .0226 | .0213 | .0201 | .0189 |

TABLE II (continued)

| | | | | | $\lambda$ | | | | | |
|---|---|---|---|---|---|---|---|---|---|---|
| $x$ | 9.1 | 9.2 | 9.3 | 9.4 | 9.5 | 9.6 | 9.7 | 9.8 | 9.9 | 10 |
| 5 | .0581 | .0555 | .0530 | .0506 | .0483 | .0460 | .0439 | .0418 | .0398 | .0378 |
| 6 | .0881 | .0851 | .0822 | .0793 | .0764 | .0736 | .0709 | .0682 | .0656 | .0631 |
| 7 | .1145 | .1118 | .1091 | .1064 | .1037 | .1010 | .0982 | .0955 | .0928 | .0901 |
| 8 | .1302 | .1286 | .1269 | .1251 | .1232 | .1212 | .1191 | .1170 | .1148 | .1126 |
| 9 | .1317 | .1315 | .1311 | .1306 | .1300 | .1293 | .1284 | .1274 | .1263 | .1251 |
| 10 | .1198 | .1210 | .1219 | .1228 | .1235 | .1241 | .1245 | .1249 | .1250 | .1251 |
| 11 | .0991 | .1012 | .1031 | .1049 | .1067 | .1083 | .1098 | .1112 | .1125 | .1137 |
| 12 | .0752 | .0776 | .0799 | .0822 | .0844 | .0866 | .0888 | .0908 | .0928 | .0948 |
| 13 | .0526 | .0549 | .0572 | .0594 | .0617 | .0640 | .0662 | .0685 | .0707 | .0729 |
| 14 | .0342 | .0361 | .0380 | .0399 | .0419 | .0439 | .0459 | .0479 | .0500 | .0521 |
| 15 | .0208 | .0221 | .0235 | .0250 | .0265 | .0281 | .0297 | .0313 | .0330 | .0347 |
| 16 | .0118 | .0127 | .0137 | .0147 | .0157 | .0168 | .0180 | .0192 | .0204 | .0217 |
| 17 | .0063 | .0069 | .0075 | .0081 | .0088 | .0095 | .0103 | .0111 | .0119 | .0128 |
| 18 | .0032 | .0035 | .0039 | .0042 | .0046 | .0051 | .0055 | .0060 | .0065 | .0071 |
| 19 | .0015 | .0017 | .0019 | .0021 | .0023 | .0026 | .0028 | .0031 | .0034 | .0037 |
| 20 | .0007 | .0008 | .0009 | .0010 | .0011 | .0012 | .0014 | .0015 | .0017 | .0019 |
| 21 | .0003 | .0003 | .0004 | .0004 | .0005 | .0006 | .0006 | .0007 | .0008 | .0009 |
| 22 | .0001 | .0001 | .0002 | .0002 | .0002 | .0002 | .0003 | .0003 | .0004 | .0004 |
| 23 | .0000 | .0001 | .0001 | .0001 | .0001 | .0001 | .0001 | .0001 | .0002 | .0002 |
| 24 | .0000 | .0000 | .0000 | .0000 | .0000 | .0000 | .0000 | .0001 | .0001 | .0001 |

| | | | | | $\lambda$ | | | | | |
|---|---|---|---|---|---|---|---|---|---|---|
| $x$ | 11 | 12 | 13 | 14 | 15 | 16 | 17 | 18 | 19 | 20 |
| 0 | .0000 | .0000 | .0000 | .0000 | .0000 | .0000 | .0000 | .0000 | .0000 | .0000 |
| 1 | .0002 | .0001 | .0000 | .0000 | .0000 | .0000 | .0000 | .0000 | .0000 | .0000 |
| 2 | .0010 | .0004 | .0002 | .0001 | .0000 | .0000 | .0000 | .0000 | .0000 | .0000 |
| 3 | .0037 | .0018 | .0008 | .0004 | .0002 | .0001 | .0000 | .0000 | .0000 | .0000 |
| 4 | .0102 | .0053 | .0027 | .0013 | .0006 | .0003 | .0001 | .0001 | .0000 | .0000 |
| 5 | .0224 | .0127 | .0070 | .0037 | .0019 | .0010 | .0005 | .0002 | .0001 | .0001 |
| 6 | .0411 | .0255 | .0152 | .0087 | .0048 | .0026 | .0014 | .0007 | .0004 | .0002 |
| 7 | .0646 | .0437 | .0281 | .0174 | .0104 | .0060 | .0034 | .0018 | .0010 | .0005 |
| 8 | .0888 | .0655 | .0457 | .0304 | .0194 | .0120 | .0072 | .0042 | .0024 | .0013 |
| 9 | .1085 | .0874 | .0661 | .0473 | .0324 | .0213 | .0135 | .0083 | .0050 | .0029 |
| 10 | .1194 | .1048 | .0859 | .0663 | .0486 | .0341 | .0230 | .0150 | .0095 | .0058 |
| 11 | .1194 | .1144 | .1015 | .0844 | .0663 | .0496 | .0355 | .0245 | .0164 | .0106 |
| 12 | .1094 | .1144 | .1099 | .0984 | .0829 | .0661 | .0504 | .0368 | .0259 | .0176 |
| 13 | .0926 | .1056 | .1099 | .1060 | .0956 | .0814 | .0658 | .0509 | .0378 | .0271 |
| 14 | .0728 | .0905 | .1021 | .1060 | .1024 | .0930 | .0800 | .0655 | .0514 | .0387 |
| 15 | .0534 | .0724 | .0885 | .0989 | .1024 | .0992 | .0906 | .0786 | .0650 | .0516 |
| 16 | .0367 | .0543 | .0719 | .0866 | .0960 | .0992 | .0963 | .0884 | .0772 | .0646 |
| 17 | .0237 | .0383 | .0550 | .0713 | .0847 | .0934 | .0963 | .0936 | .0863 | .0760 |
| 18 | .0145 | .0256 | .0397 | .0554 | .0706 | .0830 | .0909 | .0936 | .0911 | .0844 |
| 19 | .0084 | .0161 | .0272 | .0409 | .0557 | .0699 | .0814 | .0887 | .0911 | .0888 |
| 20 | .0046 | .0097 | .0177 | .0286 | .0418 | .0559 | .0692 | .0798 | .0866 | .0888 |
| 21 | .0024 | .0055 | .0109 | .0191 | .0299 | .0426 | .0560 | .0684 | .0783 | .0846 |
| 22 | .0012 | .0030 | .0065 | .0121 | .0204 | .0310 | .0433 | .0560 | .0676 | .0769 |
| 23 | .0006 | .0016 | .0037 | .0074 | .0133 | .0216 | .0320 | .0438 | .0559 | .0669 |
| 24 | .0003 | .0008 | .0020 | .0043 | .0083 | .0144 | .0226 | .0328 | .0442 | .0557 |

## TABLE II (continued)

| $x$ | 11 | 12 | 13 | 14 | 15 | 16 | 17 | 18 | 19 | 20 |
|---|---|---|---|---|---|---|---|---|---|---|
| 25 | .0001 | .0004 | .0010 | .0024 | .0050 | .0092 | .0154 | .0237 | .0336 | .0446 |
| 26 | .0000 | .0002 | .0005 | .0013 | .0029 | .0057 | .0101 | .0164 | .0246 | .0343 |
| 27 | .0000 | .0001 | .0002 | .0007 | .0016 | .0034 | .0063 | .0109 | .0173 | .0254 |
| 28 | .0000 | .0000 | .0001 | .0003 | .0009 | .0019 | .0038 | .0070 | .0117 | .0181 |
| 29 | .0000 | .0000 | .0001 | .0002 | .0004 | .0011 | .0023 | .0044 | .0077 | .0125 |
| 30 | .0000 | .0000 | .0000 | .0001 | .0002 | .0006 | .0013 | .0026 | .0049 | .0083 |
| 31 | .0000 | .0000 | .0000 | .0000 | .0001 | .0003 | .0007 | .0015 | .0030 | .0054 |
| 32 | .0000 | .0000 | .0000 | .0000 | .0001 | .0001 | .0004 | .0009 | .0018 | .0034 |
| 33 | .0000 | .0000 | .0000 | .0000 | .0000 | .0001 | .0002 | .0005 | .0010 | .0020 |
| 34 | .0000 | .0000 | .0000 | .0000 | .0000 | .0000 | .0001 | .0002 | .0006 | .0012 |
| 35 | .0000 | .0000 | .0000 | .0000 | .0000 | .0000 | .0000 | .0001 | .0003 | .0007 |
| 36 | .0000 | .0000 | .0000 | .0000 | .0000 | .0000 | .0000 | .0001 | .0002 | .0004 |
| 37 | .0000 | .0000 | .0000 | .0000 | .0000 | .0000 | .0000 | .0000 | .0001 | .0002 |
| 38 | .0000 | .0000 | .0000 | .0000 | .0000 | .0000 | .0000 | .0000 | .0000 | .0001 |
| 39 | .0000 | .0000 | .0000 | .0000 | .0000 | .0000 | .0000 | .0000 | .0000 | .0001 |

**TABLE III**

**Standard Normal Distribution**

| z | .00 | .01 | .02 | .03 | .04 | .05 | .06 | .07 | .08 | .09 |
|---|-----|-----|-----|-----|-----|-----|-----|-----|-----|-----|
| 0.0 | .0000 | .0040 | .0080 | .0120 | .0160 | .0199 | .0239 | .0279 | .0319 | .0359 |
| 0.1 | .0398 | .0438 | .0478 | .0517 | .0557 | .0596 | .0636 | .0675 | .0714 | .0753 |
| 0.2 | .0793 | .0832 | .0871 | .0910 | .0948 | .0987 | .1026 | .1064 | .1103 | .1141 |
| 0.3 | .1179 | .1217 | .1255 | .1293 | .1331 | .1368 | .1406 | .1443 | .1480 | .1517 |
| 0.4 | .1554 | .1591 | .1628 | .1664 | .1700 | .1736 | .1772 | .1808 | .1844 | .1879 |
| 0.5 | .1915 | .1950 | .1985 | .2019 | .2054 | .2088 | .2123 | .2157 | .2190 | .2224 |
| 0.6 | .2257 | .2291 | .2324 | .2357 | .2389 | .2422 | .2454 | .2486 | .2517 | .2549 |
| 0.7 | .2580 | .2611 | .2642 | .2673 | .2704 | .2734 | .2764 | .2794 | .2823 | .2852 |
| 0.8 | .2881 | .2910 | .2939 | .2967 | .2995 | .3023 | .3051 | .3078 | .3106 | .3133 |
| 0.9 | .3159 | .3186 | .3212 | .3238 | .3264 | .3289 | .3315 | .3340 | .3365 | .3389 |
| 1.0 | .3413 | .3438 | .3461 | .3485 | .3508 | .3531 | .3554 | .3577 | .3599 | .3621 |
| 1.1 | .3643 | .3665 | .3686 | .3708 | .3729 | .3749 | .3770 | .3790 | .3810 | .3830 |
| 1.2 | .3849 | .3869 | .3888 | .3907 | .3925 | .3944 | .3962 | .3980 | .3997 | .4015 |
| 1.3 | .4032 | .4049 | .4066 | .4082 | .4099 | .4115 | .4131 | .4147 | .4162 | .4177 |
| 1.4 | .4192 | .4207 | .4222 | .4236 | .4251 | .4265 | .4279 | .4292 | .4306 | .4319 |
| 1.5 | .4332 | .4345 | .4357 | .4370 | .4382 | .4394 | .4406 | .4418 | .4429 | .4441 |
| 1.6 | .4452 | .4463 | .4474 | .4484 | .4495 | .4505 | .4515 | .4525 | .4535 | .4545 |
| 1.7 | .4554 | .4564 | .4573 | .4582 | .4591 | .4599 | .4608 | .4616 | .4625 | .4633 |
| 1.8 | .4641 | .4649 | .4656 | .4664 | .4671 | .4678 | .4686 | .4693 | .4699 | .4706 |
| 1.9 | .4713 | .4719 | .4726 | .4732 | .4738 | .4744 | .4750 | .4756 | .4761 | .4767 |
| 2.0 | .4772 | .4778 | .4783 | .4788 | .4793 | .4798 | .4803 | .4808 | .4812 | .4817 |
| 2.1 | .4821 | .4826 | .4830 | .4834 | .4838 | .4842 | .4846 | .4850 | .4854 | .4857 |
| 2.2 | .4861 | .4864 | .4868 | .4871 | .4875 | .4878 | .4881 | .4884 | .4887 | .4890 |
| 2.3 | .4893 | .4896 | .4898 | .4901 | .4904 | .4906 | .4909 | .4911 | .4913 | .4916 |
| 2.4 | .4918 | .4920 | .4922 | .4925 | .4927 | .4929 | .4931 | .4932 | .4934 | .4936 |
| 2.5 | .4938 | .4940 | .4941 | .4943 | .4945 | .4946 | .4948 | .4949 | .4951 | .4952 |
| 2.6 | .4953 | .4955 | .4956 | .4957 | .4959 | .4960 | .4961 | .4962 | .4963 | .4964 |
| 2.7 | .4965 | .4966 | .4967 | .4968 | .4969 | .4970 | .4971 | .4972 | .4973 | .4974 |
| 2.8 | .4974 | .4975 | .4976 | .4977 | .4977 | .4978 | .4979 | .4979 | .4980 | .4981 |
| 2.9 | .4981 | .4982 | .4982 | .4983 | .4984 | .4984 | .4985 | .4985 | .4986 | .4986 |
| 3.0 | .4987 | .4987 | .4987 | .4988 | .4988 | .4989 | .4989 | .4989 | .4990 | .4990 |

Also, for $z = 4.0$, $5.0$, and $6.0$, the probabilities are $0.49997$, $0.4999997$, and $0.499999999$.

**TABLE IV**

Values of $t_{\alpha,\nu}$ [†]

| $\nu$ | $\alpha = .10$ | $\alpha = .05$ | $\alpha = .025$ | $\alpha = .01$ | $\alpha = .005$ | $\nu$ |
|---|---|---|---|---|---|---|
| 1 | 3.078 | 6.314 | 12.706 | 31.821 | 63.657 | 1 |
| 2 | 1.886 | 2.920 | 4.303 | 6.965 | 9.925 | 2 |
| 3 | 1.638 | 2.353 | 3.182 | 4.541 | 5.841 | 3 |
| 4 | 1.533 | 2.132 | 2.776 | 3.747 | 4.604 | 4 |
| 5 | 1.476 | 2.015 | 2.571 | 3.365 | 4.032 | 5 |
| 6 | 1.440 | 1.943 | 2.447 | 3.143 | 3.707 | 6 |
| 7 | 1.415 | 1.895 | 2.365 | 2.998 | 3.499 | 7 |
| 8 | 1.397 | 1.860 | 2.306 | 2.896 | 3.355 | 8 |
| 9 | 1.383 | 1.833 | 2.262 | 2.821 | 3.250 | 9 |
| 10 | 1.372 | 1.812 | 2.228 | 2.764 | 3.169 | 10 |
| 11 | 1.363 | 1.796 | 2.201 | 2.718 | 3.106 | 11 |
| 12 | 1.356 | 1.782 | 2.179 | 2.681 | 3.055 | 12 |
| 13 | 1.350 | 1.771 | 2.160 | 2.650 | 3.012 | 13 |
| 14 | 1.345 | 1.761 | 2.145 | 2.624 | 2.977 | 14 |
| 15 | 1.341 | 1.753 | 2.131 | 2.602 | 2.947 | 15 |
| 16 | 1.337 | 1.746 | 2.120 | 2.583 | 2.921 | 16 |
| 17 | 1.333 | 1.740 | 2.110 | 2.567 | 2.898 | 17 |
| 18 | 1.330 | 1.734 | 2.101 | 2.552 | 2.878 | 18 |
| 19 | 1.328 | 1.729 | 2.093 | 2.539 | 2.861 | 19 |
| 20 | 1.325 | 1.725 | 2.086 | 2.528 | 2.845 | 20 |
| 21 | 1.323 | 1.721 | 2.080 | 2.518 | 2.831 | 21 |
| 22 | 1.321 | 1.717 | 2.074 | 2.508 | 2.819 | 22 |
| 23 | 1.319 | 1.714 | 2.069 | 2.500 | 2.807 | 23 |
| 24 | 1.318 | 1.711 | 2.064 | 2.492 | 2.797 | 24 |
| 25 | 1.316 | 1.708 | 2.060 | 2.485 | 2.787 | 25 |
| 26 | 1.315 | 1.706 | 2.056 | 2.479 | 2.779 | 26 |
| 27 | 1.314 | 1.703 | 2.052 | 2.473 | 2.771 | 27 |
| 28 | 1.313 | 1.701 | 2.048 | 2.467 | 2.763 | 28 |
| 29 | 1.311 | 1.699 | 2.045 | 2.462 | 2.756 | 29 |
| inf. | 1.282 | 1.645 | 1.960 | 2.326 | 2.576 | inf. |

[†]Based on Richard A. Johnson, Dean W. Wichern, *Applied Multivariate Statistical Analysis,* 2nd ed., © 1988, Table 2, p. 592. By permission of Prentice-Hall, Inc., Englewood Cliffs, N.J.

# TABLE V

## Values of $\chi^2_{\alpha,\nu}$ [†]

| $\nu$ | $\alpha=.995$ | $\alpha=.99$ | $\alpha=.975$ | $\alpha=.95$ | $\alpha=.05$ | $\alpha=.025$ | $\alpha=.01$ | $\alpha=.005$ | $\nu$ |
|---|---|---|---|---|---|---|---|---|---|
| 1 | .0000393 | .000157 | .000982 | .00393 | 3.841 | 5.024 | 6.635 | 7.879 | 1 |
| 2 | .0100 | .0201 | .0506 | .103 | 5.991 | 7.378 | 9.210 | 10.597 | 2 |
| 3 | .0717 | .115 | .216 | .352 | 7.815 | 9.348 | 11.345 | 12.838 | 3 |
| 4 | .207 | .297 | .484 | .711 | 9.488 | 11.143 | 13.277 | 14.860 | 4 |
| 5 | .412 | .554 | .831 | 1.145 | 11.070 | 12.832 | 15.086 | 16.750 | 5 |
| 6 | .676 | .872 | 1.237 | 1.635 | 12.592 | 14.449 | 16.812 | 18.548 | 6 |
| 7 | .989 | 1.239 | 1.690 | 2.167 | 14.067 | 16.013 | 18.475 | 20.278 | 7 |
| 8 | 1.344 | 1.646 | 2.180 | 2.733 | 15.507 | 17.535 | 20.090 | 21.955 | 8 |
| 9 | 1.735 | 2.088 | 2.700 | 3.325 | 16.919 | 19.023 | 21.666 | 23.589 | 9 |
| 10 | 2.156 | 2.558 | 3.247 | 3.940 | 18.307 | 20.483 | 23.209 | 25.188 | 10 |
| 11 | 2.603 | 3.053 | 3.816 | 4.575 | 19.675 | 21.920 | 24.725 | 26.757 | 11 |
| 12 | 3.074 | 3.571 | 4.404 | 5.226 | 21.026 | 23.337 | 26.217 | 28.300 | 12 |
| 13 | 3.565 | 4.107 | 5.009 | 5.892 | 22.362 | 24.736 | 27.688 | 29.819 | 13 |
| 14 | 4.075 | 4.660 | 5.629 | 6.571 | 23.685 | 26.119 | 29.141 | 31.319 | 14 |
| 15 | 4.601 | 5.229 | 6.262 | 7.261 | 24.996 | 27.488 | 30.578 | 32.801 | 15 |
| 16 | 5.142 | 5.812 | 6.908 | 7.962 | 26.296 | 28.845 | 32.000 | 34.267 | 16 |
| 17 | 5.697 | 6.408 | 7.564 | 8.672 | 27.587 | 30.191 | 33.409 | 35.718 | 17 |
| 18 | 6.265 | 7.015 | 8.231 | 9.390 | 28.869 | 31.526 | 34.805 | 37.156 | 18 |
| 19 | 6.844 | 7.633 | 8.907 | 10.117 | 30.144 | 32.852 | 36.191 | 38.582 | 19 |
| 20 | 7.434 | 8.260 | 9.591 | 10.851 | 31.410 | 34.170 | 37.566 | 39.997 | 20 |
| 21 | 8.034 | 8.897 | 10.283 | 11.591 | 32.671 | 35.479 | 38.932 | 41.401 | 21 |
| 22 | 8.643 | 9.542 | 10.982 | 12.338 | 33.924 | 36.781 | 40.289 | 42.796 | 22 |
| 23 | 9.260 | 10.196 | 11.689 | 13.091 | 35.172 | 38.076 | 41.638 | 44.181 | 23 |
| 24 | 9.886 | 10.856 | 12.401 | 13.848 | 36.415 | 39.364 | 42.980 | 45.558 | 24 |
| 25 | 10.520 | 11.524 | 13.120 | 14.611 | 37.652 | 40.646 | 44.314 | 46.928 | 25 |
| 26 | 11.160 | 12.198 | 13.844 | 15.379 | 38.885 | 41.923 | 45.642 | 48.290 | 26 |
| 27 | 11.808 | 12.879 | 14.573 | 16.151 | 40.113 | 43.194 | 46.963 | 49.645 | 27 |
| 28 | 12.461 | 13.565 | 15.308 | 16.928 | 41.337 | 44.461 | 48.278 | 50.993 | 28 |
| 29 | 13.121 | 14.256 | 16.047 | 17.708 | 42.557 | 45.722 | 49.588 | 52.336 | 29 |
| 30 | 13.787 | 14.953 | 16.791 | 18.493 | 43.773 | 46.979 | 50.892 | 53.672 | 30 |

[†]Based on Table 8 of *Biometrika Tables for Statisticians*, Vol. 1, Cambridge University Press, 1954, by permission of the *Biometrika* trustees.

**TABLE VI**

Values of $f_{.05, \nu_1, \nu_2}$ [†]

$\nu_1$ = Degrees of freedom for numerator

| $\nu_2$ | 1 | 2 | 3 | 4 | 5 | 6 | 7 | 8 | 9 | 10 | 12 | 15 | 20 | 24 | 30 | 40 | 60 | 120 | ∞ |
|---|---|---|---|---|---|---|---|---|---|---|---|---|---|---|---|---|---|---|---|
| 1 | 161 | 200 | 216 | 225 | 230 | 234 | 237 | 239 | 241 | 242 | 244 | 246 | 248 | 249 | 250 | 251 | 252 | 253 | 254 |
| 2 | 18.5 | 19.0 | 19.2 | 19.2 | 19.3 | 19.3 | 19.4 | 19.4 | 19.4 | 19.4 | 19.4 | 19.4 | 19.4 | 19.5 | 19.5 | 19.5 | 19.5 | 19.5 | 19.5 |
| 3 | 10.1 | 9.55 | 9.28 | 9.12 | 9.01 | 8.94 | 8.89 | 8.85 | 8.81 | 8.79 | 8.74 | 8.70 | 8.66 | 8.64 | 8.62 | 8.59 | 8.57 | 8.55 | 8.53 |
| 4 | 7.71 | 6.94 | 6.59 | 6.39 | 6.26 | 6.16 | 6.09 | 6.04 | 6.00 | 5.96 | 5.91 | 5.86 | 5.80 | 5.77 | 5.75 | 5.72 | 5.69 | 5.66 | 5.63 |
| 5 | 6.61 | 5.79 | 5.41 | 5.19 | 5.05 | 4.95 | 4.88 | 4.82 | 4.77 | 4.74 | 4.68 | 4.62 | 4.56 | 4.53 | 4.50 | 4.46 | 4.43 | 4.40 | 4.37 |
| 6 | 5.99 | 5.14 | 4.76 | 4.53 | 4.39 | 4.28 | 4.21 | 4.15 | 4.10 | 4.06 | 4.00 | 3.94 | 3.87 | 3.84 | 3.81 | 3.77 | 3.74 | 3.70 | 3.67 |
| 7 | 5.59 | 4.74 | 4.35 | 4.12 | 3.97 | 3.87 | 3.79 | 3.73 | 3.68 | 3.64 | 3.57 | 3.51 | 3.44 | 3.41 | 3.38 | 3.34 | 3.30 | 3.27 | 3.23 |
| 8 | 5.32 | 4.46 | 4.07 | 3.84 | 3.69 | 3.58 | 3.50 | 3.44 | 3.39 | 3.35 | 3.28 | 3.22 | 3.15 | 3.12 | 3.08 | 3.04 | 3.01 | 2.97 | 2.93 |
| 9 | 5.12 | 4.26 | 3.86 | 3.63 | 3.48 | 3.37 | 3.29 | 3.23 | 3.18 | 3.14 | 3.07 | 3.01 | 2.94 | 2.90 | 2.86 | 2.83 | 2.79 | 2.75 | 2.71 |
| 10 | 4.96 | 4.10 | 3.71 | 3.48 | 3.33 | 3.22 | 3.14 | 3.07 | 3.02 | 2.98 | 2.91 | 2.85 | 2.77 | 2.74 | 2.70 | 2.66 | 2.62 | 2.58 | 2.54 |
| 11 | 4.84 | 3.98 | 3.59 | 3.36 | 3.20 | 3.09 | 3.01 | 2.95 | 2.90 | 2.85 | 2.79 | 2.72 | 2.65 | 2.61 | 2.57 | 2.53 | 2.49 | 2.45 | 2.40 |
| 12 | 4.75 | 3.89 | 3.49 | 3.26 | 3.11 | 3.00 | 2.91 | 2.85 | 2.80 | 2.75 | 2.69 | 2.62 | 2.54 | 2.51 | 2.47 | 2.43 | 2.38 | 2.34 | 2.30 |
| 13 | 4.67 | 3.81 | 3.41 | 3.18 | 3.03 | 2.92 | 2.83 | 2.77 | 2.71 | 2.67 | 2.60 | 2.53 | 2.46 | 2.42 | 2.38 | 2.34 | 2.30 | 2.25 | 2.21 |
| 14 | 4.60 | 3.74 | 3.34 | 3.11 | 2.96 | 2.85 | 2.76 | 2.70 | 2.65 | 2.60 | 2.53 | 2.46 | 2.39 | 2.35 | 2.31 | 2.27 | 2.22 | 2.18 | 2.13 |
| 15 | 4.54 | 3.68 | 3.29 | 3.06 | 2.90 | 2.79 | 2.71 | 2.64 | 2.59 | 2.54 | 2.48 | 2.40 | 2.33 | 2.29 | 2.25 | 2.20 | 2.16 | 2.11 | 2.07 |
| 16 | 4.49 | 3.63 | 3.24 | 3.01 | 2.85 | 2.74 | 2.66 | 2.59 | 2.54 | 2.49 | 2.42 | 2.35 | 2.28 | 2.24 | 2.19 | 2.15 | 2.11 | 2.06 | 2.01 |
| 17 | 4.45 | 3.59 | 3.20 | 2.96 | 2.81 | 2.70 | 2.61 | 2.55 | 2.49 | 2.45 | 2.38 | 2.31 | 2.23 | 2.19 | 2.15 | 2.10 | 2.06 | 2.01 | 1.96 |
| 18 | 4.41 | 3.55 | 3.16 | 2.93 | 2.77 | 2.66 | 2.58 | 2.51 | 2.46 | 2.41 | 2.34 | 2.27 | 2.19 | 2.15 | 2.11 | 2.06 | 2.02 | 1.97 | 1.92 |
| 19 | 4.38 | 3.52 | 3.13 | 2.90 | 2.74 | 2.63 | 2.54 | 2.48 | 2.42 | 2.38 | 2.31 | 2.23 | 2.16 | 2.11 | 2.07 | 2.03 | 1.98 | 1.93 | 1.88 |
| 20 | 4.35 | 3.49 | 3.10 | 2.87 | 2.71 | 2.60 | 2.51 | 2.45 | 2.39 | 2.35 | 2.28 | 2.20 | 2.12 | 2.08 | 2.04 | 1.99 | 1.95 | 1.90 | 1.84 |
| 21 | 4.32 | 3.47 | 3.07 | 2.84 | 2.68 | 2.57 | 2.49 | 2.42 | 2.37 | 2.32 | 2.25 | 2.18 | 2.10 | 2.05 | 2.01 | 1.96 | 1.92 | 1.87 | 1.81 |
| 22 | 4.30 | 3.44 | 3.05 | 2.82 | 2.66 | 2.55 | 2.46 | 2.40 | 2.34 | 2.30 | 2.23 | 2.15 | 2.07 | 2.03 | 1.98 | 1.94 | 1.89 | 1.84 | 1.78 |
| 23 | 4.28 | 3.42 | 3.03 | 2.80 | 2.64 | 2.53 | 2.44 | 2.37 | 2.32 | 2.27 | 2.20 | 2.13 | 2.05 | 2.01 | 1.96 | 1.91 | 1.86 | 1.81 | 1.76 |
| 24 | 4.26 | 3.40 | 3.01 | 2.78 | 2.62 | 2.51 | 2.42 | 2.36 | 2.30 | 2.25 | 2.18 | 2.11 | 2.03 | 1.98 | 1.94 | 1.89 | 1.84 | 1.79 | 1.73 |
| 25 | 4.24 | 3.39 | 2.99 | 2.76 | 2.60 | 2.49 | 2.40 | 2.34 | 2.28 | 2.24 | 2.16 | 2.09 | 2.01 | 1.96 | 1.92 | 1.87 | 1.82 | 1.77 | 1.71 |
| 30 | 4.17 | 3.32 | 2.92 | 2.69 | 2.53 | 2.42 | 2.33 | 2.27 | 2.21 | 2.16 | 2.09 | 2.01 | 1.93 | 1.89 | 1.84 | 1.79 | 1.74 | 1.68 | 1.62 |
| 40 | 4.08 | 3.23 | 2.84 | 2.61 | 2.45 | 2.34 | 2.25 | 2.18 | 2.12 | 2.08 | 2.00 | 1.92 | 1.84 | 1.79 | 1.74 | 1.69 | 1.64 | 1.58 | 1.51 |
| 60 | 4.00 | 3.15 | 2.76 | 2.53 | 2.37 | 2.25 | 2.17 | 2.10 | 2.04 | 1.99 | 1.92 | 1.84 | 1.75 | 1.70 | 1.65 | 1.59 | 1.53 | 1.47 | 1.39 |
| 120 | 3.92 | 3.07 | 2.68 | 2.45 | 2.29 | 2.18 | 2.09 | 2.02 | 1.96 | 1.91 | 1.83 | 1.75 | 1.66 | 1.61 | 1.55 | 1.50 | 1.43 | 1.35 | 1.25 |
| ∞ | 3.84 | 3.00 | 2.60 | 2.37 | 2.21 | 2.10 | 2.01 | 1.94 | 1.88 | 1.83 | 1.75 | 1.67 | 1.57 | 1.52 | 1.46 | 1.39 | 1.32 | 1.22 | 1.00 |

$\nu_2$ = Degrees of freedom for denominator

[†]Reproduced from M. Merrington and C. M. Thompson, "Tables of percentage points of the inverted beta ($F$) distribution," *Biometrika*, Vol. 33 (1943), by permission of the *Biometrika* trustees.

627

**TABLE VI (continued)**

Values of $f_{.01,\nu_1,\nu_2}$

$\nu_1$ = Degrees of freedom for numerator

| $\nu_2$ | 1 | 2 | 3 | 4 | 5 | 6 | 7 | 8 | 9 | 10 | 12 | 15 | 20 | 24 | 30 | 40 | 60 | 120 | ∞ |
|---|---|---|---|---|---|---|---|---|---|---|---|---|---|---|---|---|---|---|---|
| 1 | 4,052 | 5,000 | 5,403 | 5,625 | 5,764 | 5,859 | 5,928 | 5,982 | 6,023 | 6,056 | 6,106 | 6,157 | 6,209 | 6,235 | 6,261 | 6,287 | 6,313 | 6,339 | 6,366 |
| 2 | 98.5 | 99.0 | 99.2 | 99.2 | 99.3 | 99.3 | 99.4 | 99.4 | 99.4 | 99.4 | 99.4 | 99.4 | 99.4 | 99.5 | 99.5 | 99.5 | 99.5 | 99.5 | 99.5 |
| 3 | 34.1 | 30.8 | 29.5 | 28.7 | 28.2 | 27.9 | 27.7 | 27.5 | 27.3 | 27.2 | 27.1 | 26.9 | 26.7 | 26.6 | 26.5 | 26.4 | 26.3 | 26.2 | 26.1 |
| 4 | 21.2 | 18.0 | 16.7 | 16.0 | 15.5 | 15.2 | 15.0 | 14.8 | 14.7 | 14.5 | 14.4 | 14.2 | 14.0 | 13.9 | 13.8 | 13.7 | 13.7 | 13.6 | 13.5 |
| 5 | 16.3 | 13.3 | 12.1 | 11.4 | 11.0 | 10.7 | 10.5 | 10.3 | 10.2 | 10.1 | 9.89 | 9.72 | 9.55 | 9.47 | 9.38 | 9.29 | 9.20 | 9.11 | 9.02 |
| 6 | 13.7 | 10.9 | 9.78 | 9.15 | 8.75 | 8.47 | 8.26 | 8.10 | 7.98 | 7.87 | 7.72 | 7.56 | 7.40 | 7.31 | 7.23 | 7.14 | 7.06 | 6.97 | 6.88 |
| 7 | 12.2 | 9.55 | 8.45 | 7.85 | 7.46 | 7.19 | 6.99 | 6.84 | 6.72 | 6.62 | 6.47 | 6.31 | 6.16 | 6.07 | 5.99 | 5.91 | 5.82 | 5.74 | 5.65 |
| 8 | 11.3 | 8.65 | 7.59 | 7.01 | 6.63 | 6.37 | 6.18 | 6.03 | 5.91 | 5.81 | 5.67 | 5.52 | 5.36 | 5.28 | 5.20 | 5.12 | 5.03 | 4.95 | 4.86 |
| 9 | 10.6 | 8.02 | 6.99 | 6.42 | 6.06 | 5.80 | 5.61 | 5.47 | 5.35 | 5.26 | 5.11 | 4.96 | 4.81 | 4.73 | 4.65 | 4.57 | 4.48 | 4.40 | 4.31 |
| 10 | 10.0 | 7.56 | 6.55 | 5.99 | 5.64 | 5.39 | 5.20 | 5.06 | 4.94 | 4.85 | 4.71 | 4.56 | 4.41 | 4.33 | 4.25 | 4.17 | 4.08 | 4.00 | 3.91 |
| 11 | 9.65 | 7.21 | 6.22 | 5.67 | 5.32 | 5.07 | 4.89 | 4.74 | 4.63 | 4.54 | 4.40 | 4.25 | 4.10 | 4.02 | 3.94 | 3.86 | 3.78 | 3.69 | 3.60 |
| 12 | 9.33 | 6.93 | 5.95 | 5.41 | 5.06 | 4.82 | 4.64 | 4.50 | 4.39 | 4.30 | 4.16 | 4.01 | 3.86 | 3.78 | 3.70 | 3.62 | 3.54 | 3.45 | 3.36 |
| 13 | 9.07 | 6.70 | 5.74 | 5.21 | 4.86 | 4.62 | 4.44 | 4.30 | 4.19 | 4.10 | 3.96 | 3.82 | 3.66 | 3.59 | 3.51 | 3.43 | 3.34 | 3.25 | 3.17 |
| 14 | 8.86 | 6.51 | 5.56 | 5.04 | 4.70 | 4.46 | 4.28 | 4.14 | 4.03 | 3.94 | 3.80 | 3.66 | 3.51 | 3.43 | 3.35 | 3.27 | 3.18 | 3.09 | 3.00 |
| 15 | 8.68 | 6.36 | 5.42 | 4.89 | 4.56 | 4.32 | 4.14 | 4.00 | 3.89 | 3.80 | 3.67 | 3.52 | 3.37 | 3.29 | 3.21 | 3.13 | 3.05 | 2.96 | 2.87 |
| 16 | 8.53 | 6.23 | 5.29 | 4.77 | 4.44 | 4.20 | 4.03 | 3.89 | 3.78 | 3.69 | 3.55 | 3.41 | 3.26 | 3.18 | 3.10 | 3.02 | 2.93 | 2.84 | 2.75 |
| 17 | 8.40 | 6.11 | 5.19 | 4.67 | 4.34 | 4.10 | 3.93 | 3.79 | 3.68 | 3.59 | 3.46 | 3.31 | 3.16 | 3.08 | 3.00 | 2.92 | 2.83 | 2.75 | 2.65 |
| 18 | 8.29 | 6.01 | 5.09 | 4.58 | 4.25 | 4.01 | 3.84 | 3.71 | 3.60 | 3.51 | 3.37 | 3.23 | 3.08 | 3.00 | 2.92 | 2.84 | 2.75 | 2.66 | 2.57 |
| 19 | 8.19 | 5.93 | 5.01 | 4.50 | 4.17 | 3.94 | 3.77 | 3.63 | 3.52 | 3.43 | 3.30 | 3.15 | 3.00 | 2.92 | 2.84 | 2.76 | 2.67 | 2.58 | 2.49 |
| 20 | 8.10 | 5.85 | 4.94 | 4.43 | 4.10 | 3.87 | 3.70 | 3.56 | 3.46 | 3.37 | 3.23 | 3.09 | 2.94 | 2.86 | 2.78 | 2.69 | 2.61 | 2.52 | 2.42 |
| 21 | 8.02 | 5.78 | 4.87 | 4.37 | 4.04 | 3.81 | 3.64 | 3.51 | 3.40 | 3.31 | 3.17 | 3.03 | 2.88 | 2.80 | 2.72 | 2.64 | 2.55 | 2.46 | 2.36 |
| 22 | 7.95 | 5.72 | 4.82 | 4.31 | 3.99 | 3.76 | 3.59 | 3.45 | 3.35 | 3.26 | 3.12 | 2.98 | 2.83 | 2.75 | 2.67 | 2.58 | 2.50 | 2.40 | 2.31 |
| 23 | 7.88 | 5.66 | 4.76 | 4.26 | 3.94 | 3.71 | 3.54 | 3.41 | 3.30 | 3.21 | 3.07 | 2.93 | 2.78 | 2.70 | 2.62 | 2.54 | 2.45 | 2.35 | 2.26 |
| 24 | 7.82 | 5.61 | 4.72 | 4.22 | 3.90 | 3.67 | 3.50 | 3.36 | 3.26 | 3.17 | 3.03 | 2.89 | 2.74 | 2.66 | 2.58 | 2.49 | 2.40 | 2.31 | 2.21 |
| 25 | 7.77 | 5.57 | 4.68 | 4.18 | 3.86 | 3.63 | 3.46 | 3.32 | 3.22 | 3.13 | 2.99 | 2.85 | 2.70 | 2.62 | 2.53 | 2.45 | 2.36 | 2.27 | 2.17 |
| 30 | 7.56 | 5.39 | 4.51 | 4.02 | 3.70 | 3.47 | 3.30 | 3.17 | 3.07 | 2.98 | 2.84 | 2.70 | 2.55 | 2.47 | 2.39 | 2.30 | 2.21 | 2.11 | 2.01 |
| 40 | 7.31 | 5.18 | 4.31 | 3.83 | 3.51 | 3.29 | 3.12 | 2.99 | 2.89 | 2.80 | 2.66 | 2.52 | 2.37 | 2.29 | 2.20 | 2.11 | 2.02 | 1.92 | 1.80 |
| 60 | 7.08 | 4.98 | 4.13 | 3.65 | 3.34 | 3.12 | 2.95 | 2.82 | 2.72 | 2.63 | 2.50 | 2.35 | 2.20 | 2.12 | 2.03 | 1.94 | 1.84 | 1.73 | 1.60 |
| 120 | 6.85 | 4.79 | 3.95 | 3.48 | 3.17 | 2.96 | 2.79 | 2.66 | 2.56 | 2.47 | 2.34 | 2.19 | 2.03 | 1.95 | 1.86 | 1.76 | 1.66 | 1.53 | 1.38 |
| ∞ | 6.63 | 4.61 | 3.78 | 3.32 | 3.02 | 2.80 | 2.64 | 2.51 | 2.41 | 2.32 | 2.18 | 2.04 | 1.88 | 1.79 | 1.70 | 1.59 | 1.47 | 1.32 | 1.00 |

$\nu_2$ = Degrees of freedom for denominator

## TABLE VII

**Factorials**

| $n$ | $n!$ | $\log n!$ |
|---|---|---|
| 0 | 1 | 0.0000 |
| 1 | 1 | 0.0000 |
| 2 | 2 | 0.3010 |
| 3 | 6 | 0.7782 |
| 4 | 24 | 1.3802 |
| 5 | 120 | 2.0792 |
| 6 | 720 | 2.8573 |
| 7 | 5,040 | 3.7024 |
| 8 | 40,320 | 4.6055 |
| 9 | 362,880 | 5.5598 |
| 10 | 3,628,800 | 6.5598 |
| 11 | 39,916,800 | 7.6012 |
| 12 | 479,001,600 | 8.6803 |
| 13 | 6,227,020,800 | 9.7943 |
| 14 | 87,178,291,200 | 10.9404 |
| 15 | 1,307,674,368,000 | 12.1165 |

**Binomial Coefficients**

| $n$ | $\binom{n}{0}$ | $\binom{n}{1}$ | $\binom{n}{2}$ | $\binom{n}{3}$ | $\binom{n}{4}$ | $\binom{n}{5}$ | $\binom{n}{6}$ | $\binom{n}{7}$ | $\binom{n}{8}$ | $\binom{n}{9}$ | $\binom{n}{10}$ |
|---|---|---|---|---|---|---|---|---|---|---|---|
| 0 | 1 | | | | | | | | | | |
| 1 | 1 | 1 | | | | | | | | | |
| 2 | 1 | 2 | 1 | | | | | | | | |
| 3 | 1 | 3 | 3 | 1 | | | | | | | |
| 4 | 1 | 4 | 6 | 4 | 1 | | | | | | |
| 5 | 1 | 5 | 10 | 10 | 5 | 1 | | | | | |
| 6 | 1 | 6 | 15 | 20 | 15 | 6 | 1 | | | | |
| 7 | 1 | 7 | 21 | 35 | 35 | 21 | 7 | 1 | | | |
| 8 | 1 | 8 | 28 | 56 | 70 | 56 | 28 | 8 | 1 | | |
| 9 | 1 | 9 | 36 | 84 | 126 | 126 | 84 | 36 | 9 | 1 | |
| 10 | 1 | 10 | 45 | 120 | 210 | 252 | 210 | 120 | 45 | 10 | 1 |
| 11 | 1 | 11 | 55 | 165 | 330 | 462 | 462 | 330 | 165 | 55 | 11 |
| 12 | 1 | 12 | 66 | 220 | 495 | 792 | 924 | 792 | 495 | 220 | 66 |
| 13 | 1 | 13 | 78 | 286 | 715 | 1287 | 1716 | 1716 | 1287 | 715 | 286 |
| 14 | 1 | 14 | 91 | 364 | 1001 | 2002 | 3003 | 3432 | 3003 | 2002 | 1001 |
| 15 | 1 | 15 | 105 | 455 | 1365 | 3003 | 5005 | 6435 | 6435 | 5005 | 3003 |
| 16 | 1 | 16 | 120 | 560 | 1820 | 4368 | 8008 | 11440 | 12870 | 11440 | 8008 |
| 17 | 1 | 17 | 136 | 680 | 2380 | 6188 | 12376 | 19448 | 24310 | 24310 | 19448 |
| 18 | 1 | 18 | 153 | 816 | 3060 | 8568 | 18564 | 31824 | 43758 | 48620 | 43758 |
| 19 | 1 | 19 | 171 | 969 | 3876 | 11628 | 27132 | 50388 | 75582 | 92378 | 92378 |
| 20 | 1 | 20 | 190 | 1140 | 4845 | 15504 | 38760 | 77520 | 125970 | 167960 | 184756 |

## TABLE VIII

**Values of $e^x$ and $e^{-x}$**

| $x$ | $e^x$ | $e^{-x}$ | $x$ | $e^x$ | $e^{-x}$ |
|-----|-------|----------|-----|-------|----------|
| 0.0 | 1.000 | 1.000 | 2.5 | 12.18 | 0.082 |
| 0.1 | 1.105 | 0.905 | 2.6 | 13.46 | 0.074 |
| 0.2 | 1.221 | 0.819 | 2.7 | 14.88 | 0.067 |
| 0.3 | 1.350 | 0.741 | 2.8 | 16.44 | 0.061 |
| 0.4 | 1.492 | 0.670 | 2.9 | 18.17 | 0.055 |
| 0.5 | 1.649 | 0.607 | 3.0 | 20.09 | 0.050 |
| 0.6 | 1.822 | 0.549 | 3.1 | 22.20 | 0.045 |
| 0.7 | 2.014 | 0.497 | 3.2 | 24.53 | 0.041 |
| 0.8 | 2.226 | 0.449 | 3.3 | 27.11 | 0.037 |
| 0.9 | 2.460 | 0.407 | 3.4 | 29.96 | 0.033 |
| 1.0 | 2.718 | 0.368 | 3.5 | 33.12 | 0.030 |
| 1.1 | 3.004 | 0.333 | 3.6 | 36.60 | 0.027 |
| 1.2 | 3.320 | 0.301 | 3.7 | 40.45 | 0.025 |
| 1.3 | 3.669 | 0.273 | 3.8 | 44.70 | 0.022 |
| 1.4 | 4.055 | 0.247 | 3.9 | 49.40 | 0.020 |
| 1.5 | 4.482 | 0.223 | 4.0 | 54.60 | 0.018 |
| 1.6 | 4.953 | 0.202 | 4.1 | 60.34 | 0.017 |
| 1.7 | 5.474 | 0.183 | 4.2 | 66.69 | 0.015 |
| 1.8 | 6.050 | 0.165 | 4.3 | 73.70 | 0.014 |
| 1.9 | 6.686 | 0.150 | 4.4 | 81.45 | 0.012 |
| 2.0 | 7.389 | 0.135 | 4.5 | 90.02 | 0.011 |
| 2.1 | 8.166 | 0.122 | 4.6 | 99.48 | 0.010 |
| 2.2 | 9.025 | 0.111 | 4.7 | 109.95 | 0.009 |
| 2.3 | 9.974 | 0.100 | 4.8 | 121.51 | 0.008 |
| 2.4 | 11.023 | 0.091 | 4.9 | 134.29 | 0.007 |

**TABLE VIII (continued)**

| $x$ | $e^x$ | $e^{-x}$ | $x$ | $e^x$ | $e^{-x}$ |
|-----|-------|----------|-----|-------|----------|
| 5.0 | 148.4 | 0.0067 | 7.5 | 1,808.0 | 0.00055 |
| 5.1 | 164.0 | 0.0061 | 7.6 | 1,998.2 | 0.00050 |
| 5.2 | 181.3 | 0.0055 | 7.7 | 2,208.3 | 0.00045 |
| 5.3 | 200.3 | 0.0050 | 7.8 | 2,440.6 | 0.00041 |
| 5.4 | 221.4 | 0.0045 | 7.9 | 2,697.3 | 0.00037 |
| 5.5 | 244.7 | 0.0041 | 8.0 | 2,981.0 | 0.00034 |
| 5.6 | 270.4 | 0.0037 | 8.1 | 3,294.5 | 0.00030 |
| 5.7 | 298.9 | 0.0033 | 8.2 | 3,641.0 | 0.00027 |
| 5.8 | 330.3 | 0.0030 | 8.3 | 4,023.9 | 0.00025 |
| 5.9 | 365.0 | 0.0027 | 8.4 | 4,447.1 | 0.00022 |
| 6.0 | 403.4 | 0.0025 | 8.5 | 4,914.8 | 0.00020 |
| 6.1 | 445.9 | 0.0022 | 8.6 | 5,431.7 | 0.00018 |
| 6.2 | 492.8 | 0.0020 | 8.7 | 6,002.9 | 0.00017 |
| 6.3 | 544.6 | 0.0018 | 8.8 | 6,634.2 | 0.00015 |
| 6.4 | 601.8 | 0.0017 | 8.9 | 7,332.0 | 0.00014 |
| 6.5 | 665.1 | 0.0015 | 9.0 | 8,103.1 | 0.00012 |
| 6.6 | 735.1 | 0.0014 | 9.1 | 8,955.3 | 0.00011 |
| 6.7 | 812.4 | 0.0012 | 9.2 | 9,897.1 | 0.00010 |
| 6.8 | 897.8 | 0.0011 | 9.3 | 10,938 | 0.00009 |
| 6.9 | 992.3 | 0.0010 | 9.4 | 12,088 | 0.00008 |
| 7.0 | 1,096.6 | 0.0009 | 9.5 | 13,360 | 0.00007 |
| 7.1 | 1,212.0 | 0.0008 | 9.6 | 14,765 | 0.00007 |
| 7.2 | 1,339.4 | 0.0007 | 9.7 | 16,318 | 0.00006 |
| 7.3 | 1,480.3 | 0.0007 | 9.8 | 18,034 | 0.00006 |
| 7.4 | 1,636.0 | 0.0006 | 9.9 | 19,930 | 0.00005 |

## TABLE IX

### Critical Values for the Signed-Rank Test[†]

| $n$ | $T_{.10}$ | $T_{.05}$ | $T_{.02}$ | $T_{.01}$ |
|---|---|---|---|---|
| 4 | | | | |
| 5 | 1 | | | |
| 6 | 2 | 1 | | |
| 7 | 4 | 2 | 0 | |
| 8 | 6 | 4 | 2 | 0 |
| 9 | 8 | 6 | 3 | 2 |
| 10 | 11 | 8 | 5 | 3 |
| 11 | 14 | 11 | 7 | 5 |
| 12 | 17 | 14 | 10 | 7 |
| 13 | 21 | 17 | 13 | 10 |
| 14 | 26 | 21 | 16 | 13 |
| 15 | 30 | 25 | 20 | 16 |
| 16 | 36 | 30 | 24 | 19 |
| 17 | 41 | 35 | 28 | 23 |
| 18 | 47 | 40 | 33 | 28 |
| 19 | 54 | 46 | 38 | 32 |
| 20 | 60 | 52 | 43 | 37 |
| 21 | 68 | 59 | 49 | 43 |
| 22 | 75 | 66 | 56 | 49 |
| 23 | 83 | 73 | 62 | 55 |
| 24 | 92 | 81 | 69 | 61 |
| 25 | 101 | 90 | 77 | 68 |

[†]From F. Wilcoxon and R. A. Wilcox, *Some Rapid Approximate Statistical Procedures,* American Cyanamid Company, Pearl River, N.Y., 1964. Reproduced with permission of American Cyanamid Company.

## TABLE X

**Critical Values for the $U$ Test**[†]

### Values of $U_{.10}$

| $n_1$ \ $n_2$ | 2 | 3 | 4 | 5 | 6 | 7 | 8 | 9 | 10 | 11 | 12 | 13 | 14 | 15 |
|---|---|---|---|---|---|---|---|---|---|---|---|---|---|---|
| 2 | | | | 0 | 0 | 0 | 1 | 1 | 1 | 1 | 2 | 2 | 3 | 3 |
| 3 | | 0 | 0 | 1 | 2 | 2 | 3 | 4 | 4 | 5 | 5 | 6 | 7 | 7 |
| 4 | | 0 | 1 | 2 | 3 | 4 | 5 | 6 | 7 | 8 | 9 | 10 | 11 | 12 |
| 5 | 0 | 1 | 2 | 4 | 5 | 6 | 8 | 9 | 11 | 12 | 13 | 15 | 16 | 18 |
| 6 | 0 | 2 | 3 | 5 | 7 | 8 | 10 | 12 | 14 | 16 | 17 | 19 | 21 | 23 |
| 7 | 0 | 2 | 4 | 6 | 8 | 11 | 13 | 15 | 17 | 19 | 21 | 24 | 26 | 28 |
| 8 | 1 | 3 | 5 | 8 | 10 | 13 | 15 | 18 | 20 | 23 | 26 | 28 | 31 | 33 |
| 9 | 1 | 4 | 6 | 9 | 12 | 15 | 18 | 21 | 24 | 27 | 30 | 33 | 36 | 39 |
| 10 | 1 | 4 | 7 | 11 | 14 | 17 | 20 | 24 | 27 | 31 | 34 | 37 | 41 | 44 |
| 11 | 1 | 5 | 8 | 12 | 16 | 19 | 23 | 27 | 31 | 34 | 38 | 42 | 46 | 50 |
| 12 | 2 | 5 | 9 | 13 | 17 | 21 | 26 | 30 | 34 | 38 | 42 | 47 | 51 | 55 |
| 13 | 2 | 6 | 10 | 15 | 19 | 24 | 28 | 33 | 37 | 42 | 47 | 51 | 56 | 61 |
| 14 | 3 | 7 | 11 | 16 | 21 | 26 | 31 | 36 | 41 | 46 | 51 | 56 | 61 | 66 |
| 15 | 3 | 7 | 12 | 18 | 23 | 28 | 33 | 39 | 44 | 50 | 55 | 61 | 66 | 72 |

### Values of $U_{.05}$

| $n_1$ \ $n_2$ | 2 | 3 | 4 | 5 | 6 | 7 | 8 | 9 | 10 | 11 | 12 | 13 | 14 | 15 |
|---|---|---|---|---|---|---|---|---|---|---|---|---|---|---|
| 2 | | | | | | | 0 | 0 | 0 | 0 | 1 | 1 | 1 | 1 |
| 3 | | | | 0 | 1 | 1 | 2 | 2 | 3 | 3 | 4 | 4 | 5 | 5 |
| 4 | | | 0 | 1 | 2 | 3 | 4 | 4 | 5 | 6 | 7 | 8 | 9 | 10 |
| 5 | | 0 | 1 | 2 | 3 | 5 | 6 | 7 | 8 | 9 | 11 | 12 | 13 | 14 |
| 6 | | 1 | 2 | 3 | 5 | 6 | 8 | 10 | 11 | 13 | 14 | 16 | 17 | 19 |
| 7 | | 1 | 3 | 5 | 6 | 8 | 10 | 12 | 14 | 16 | 18 | 20 | 22 | 24 |
| 8 | 0 | 2 | 4 | 6 | 8 | 10 | 13 | 15 | 17 | 19 | 22 | 24 | 26 | 29 |
| 9 | 0 | 2 | 4 | 7 | 10 | 12 | 15 | 17 | 20 | 23 | 26 | 28 | 31 | 34 |
| 10 | 0 | 3 | 5 | 8 | 11 | 14 | 17 | 20 | 23 | 26 | 29 | 30 | 36 | 39 |
| 11 | 0 | 3 | 6 | 9 | 13 | 16 | 19 | 23 | 26 | 30 | 33 | 37 | 40 | 44 |
| 12 | 1 | 4 | 7 | 11 | 14 | 18 | 22 | 26 | 29 | 33 | 37 | 41 | 45 | 49 |
| 13 | 1 | 4 | 8 | 12 | 16 | 20 | 24 | 28 | 30 | 37 | 41 | 45 | 50 | 54 |
| 14 | 1 | 5 | 9 | 13 | 17 | 22 | 26 | 31 | 36 | 40 | 45 | 50 | 55 | 59 |
| 15 | 1 | 5 | 10 | 14 | 19 | 24 | 29 | 34 | 39 | 44 | 49 | 54 | 59 | 64 |

[†]This table is based on D. Auble, "Extended tables for the Mann-Whitney Statistics," *Bulletin of the Institute of Educational Research at Indiana University*, Vol. 1, 1953. By permission of the author.

**TABLE X (continued)**

Values of $U_{.02}$

| $n_1$ \ $n_2$ | 2 | 3 | 4 | 5 | 6 | 7 | 8 | 9 | 10 | 11 | 12 | 13 | 14 | 15 |
|---|---|---|---|---|---|---|---|---|---|---|---|---|---|---|
| 2 | | | | | | | | | | | | 0 | 0 | 0 |
| 3 | | | | | | 0 | 0 | 1 | 1 | 1 | 2 | 2 | 2 | 3 |
| 4 | | | 0 | 1 | 1 | 2 | 3 | 3 | 4 | 5 | 5 | 6 | 7 | |
| 5 | | | 0 | 1 | 2 | 3 | 4 | 5 | 6 | 7 | 8 | 9 | 10 | 11 |
| 6 | | | 1 | 2 | 3 | 4 | 6 | 7 | 8 | 9 | 11 | 12 | 13 | 15 |
| 7 | | 0 | 1 | 3 | 4 | 6 | 7 | 9 | 11 | 12 | 14 | 16 | 17 | 19 |
| 8 | | 0 | 2 | 4 | 6 | 7 | 9 | 11 | 13 | 15 | 17 | 20 | 22 | 24 |
| 9 | | 1 | 3 | 5 | 7 | 9 | 11 | 14 | 16 | 18 | 21 | 23 | 26 | 28 |
| 10 | | 1 | 3 | 6 | 8 | 11 | 13 | 16 | 19 | 22 | 24 | 27 | 30 | 33 |
| 11 | | 1 | 4 | 7 | 9 | 12 | 15 | 18 | 22 | 25 | 28 | 31 | 34 | 37 |
| 12 | | 2 | 5 | 8 | 11 | 14 | 17 | 21 | 24 | 28 | 31 | 35 | 38 | 42 |
| 13 | 0 | 2 | 5 | 9 | 12 | 16 | 20 | 23 | 27 | 31 | 35 | 39 | 43 | 47 |
| 14 | 0 | 2 | 6 | 10 | 13 | 17 | 22 | 26 | 30 | 34 | 38 | 43 | 47 | 51 |
| 15 | 0 | 3 | 7 | 11 | 15 | 19 | 24 | 28 | 33 | 37 | 42 | 47 | 51 | 56 |

Values of $U_{.01}$

| $n_1$ \ $n_2$ | 3 | 4 | 5 | 6 | 7 | 8 | 9 | 10 | 11 | 12 | 13 | 14 | 15 |
|---|---|---|---|---|---|---|---|---|---|---|---|---|---|
| 3 | | | | | | | 0 | 0 | 0 | 1 | 1 | 1 | 2 |
| 4 | | | | 0 | 0 | 1 | 1 | 2 | 2 | 3 | 3 | 4 | 5 |
| 5 | | | 0 | 1 | 1 | 2 | 3 | 4 | 5 | 6 | 7 | 7 | 8 |
| 6 | | 0 | 1 | 2 | 3 | 4 | 5 | 6 | 7 | 9 | 10 | 11 | 12 |
| 7 | | 0 | 1 | 3 | 4 | 6 | 7 | 9 | 10 | 12 | 13 | 15 | 16 |
| 8 | | 1 | 2 | 4 | 6 | 7 | 9 | 11 | 13 | 15 | 17 | 18 | 20 |
| 9 | 0 | 1 | 3 | 5 | 7 | 9 | 11 | 13 | 16 | 18 | 20 | 22 | 24 |
| 10 | 0 | 2 | 4 | 6 | 9 | 11 | 13 | 16 | 18 | 21 | 24 | 26 | 29 |
| 11 | 0 | 2 | 5 | 7 | 10 | 13 | 16 | 18 | 21 | 24 | 27 | 30 | 33 |
| 12 | 1 | 3 | 6 | 9 | 12 | 15 | 18 | 21 | 24 | 27 | 31 | 34 | 37 |
| 13 | 1 | 3 | 7 | 10 | 13 | 17 | 20 | 24 | 27 | 31 | 34 | 38 | 42 |
| 14 | 1 | 4 | 7 | 11 | 15 | 18 | 22 | 26 | 30 | 34 | 38 | 42 | 46 |
| 15 | 2 | 5 | 8 | 12 | 16 | 20 | 24 | 29 | 33 | 37 | 42 | 46 | 51 |

## TABLE XI

**Critical Values for the Runs Test[†]**

### Values of $u'_{.025}$

| $n_1$ \ $n_2$ | 2 | 3 | 4 | 5 | 6 | 7 | 8 | 9 | 10 | 11 | 12 | 13 | 14 | 15 |
|---|---|---|---|---|---|---|---|---|---|---|---|---|---|---|
| 2 | | | | | | | | | | | 2 | 2 | 2 | 2 |
| 3 | | | | | 2 | 2 | 2 | 2 | 2 | 2 | 2 | 2 | 2 | 3 |
| 4 | | | | 2 | 2 | 2 | 3 | 3 | 3 | 3 | 3 | 3 | 3 | 3 |
| 5 | | | 2 | 2 | 3 | 3 | 3 | 3 | 3 | 4 | 4 | 4 | 4 | 4 |
| 6 | | 2 | 2 | 3 | 3 | 3 | 3 | 4 | 4 | 4 | 4 | 5 | 5 | 5 |
| 7 | | 2 | 2 | 3 | 3 | 3 | 4 | 4 | 5 | 5 | 5 | 5 | 5 | 6 |
| 8 | | 2 | 3 | 3 | 3 | 4 | 4 | 5 | 5 | 5 | 6 | 6 | 6 | 6 |
| 9 | | 2 | 3 | 3 | 4 | 4 | 5 | 5 | 5 | 6 | 6 | 6 | 7 | 7 |
| 10 | | 2 | 3 | 3 | 4 | 5 | 5 | 5 | 6 | 6 | 7 | 7 | 7 | 7 |
| 11 | | 2 | 3 | 4 | 4 | 5 | 5 | 6 | 6 | 7 | 7 | 7 | 8 | 8 |
| 12 | 2 | 2 | 3 | 4 | 4 | 5 | 6 | 6 | 7 | 7 | 7 | 8 | 8 | 8 |
| 13 | 2 | 2 | 3 | 4 | 5 | 5 | 6 | 6 | 7 | 7 | 8 | 8 | 9 | 9 |
| 14 | 2 | 2 | 3 | 4 | 5 | 5 | 6 | 7 | 7 | 8 | 8 | 9 | 9 | 9 |
| 15 | 2 | 3 | 3 | 4 | 5 | 6 | 6 | 7 | 7 | 8 | 8 | 9 | 9 | 10 |

### Values of $u_{.025}$

| $n_1$ \ $n_2$ | 4 | 5 | 6 | 7 | 8 | 9 | 10 | 11 | 12 | 13 | 14 | 15 |
|---|---|---|---|---|---|---|---|---|---|---|---|---|
| 4 | | 9 | 9 | | | | | | | | | |
| 5 | 9 | 10 | 10 | 11 | 11 | | | | | | | |
| 6 | 9 | 10 | 11 | 12 | 12 | 13 | 13 | 13 | 13 | | | |
| 7 | | 11 | 12 | 13 | 13 | 14 | 14 | 14 | 14 | 15 | 15 | 15 |
| 8 | | 11 | 12 | 13 | 14 | 14 | 15 | 15 | 16 | 16 | 16 | 16 |
| 9 | | | 13 | 14 | 14 | 15 | 16 | 16 | 16 | 17 | 17 | 18 |
| 10 | | | 13 | 14 | 15 | 16 | 16 | 17 | 17 | 18 | 18 | 18 |
| 11 | | | 13 | 14 | 15 | 16 | 17 | 17 | 18 | 19 | 19 | 19 |
| 12 | | | 13 | 14 | 16 | 16 | 17 | 18 | 19 | 19 | 20 | 20 |
| 13 | | | | 15 | 16 | 17 | 18 | 19 | 19 | 20 | 20 | 21 |
| 14 | | | | 15 | 16 | 17 | 18 | 19 | 20 | 20 | 21 | 22 |
| 15 | | | | 15 | 16 | 18 | 18 | 19 | 20 | 21 | 22 | 22 |

[†]This table is adapted, by permission, from F. S. Swed and C. Eisenhart, "Tables for testing randomness of grouping in a sequence of alternatives," *Annals of Mathematical Statistics,* Vol. 14.

**TABLE XI (continued)**

Values of $u'_{.005}$

| $n_1$ \ $n_2$ | 3 | 4 | 5 | 6 | 7 | 8 | 9 | 10 | 11 | 12 | 13 | 14 | 15 |
|---|---|---|---|---|---|---|---|---|---|---|---|---|---|
| 3 | | | | | | | | | | 2 | 2 | 2 | 2 |
| 4 | | | | | | 2 | 2 | 2 | 2 | 2 | 2 | 2 | 3 |
| 5 | | | 2 | 2 | 2 | 2 | 3 | 3 | 3 | 3 | 3 | 3 | 3 |
| 6 | | | 2 | 2 | 2 | 3 | 3 | 3 | 3 | 3 | 3 | 4 | 4 |
| 7 | | | 2 | 2 | 3 | 3 | 3 | 3 | 4 | 4 | 4 | 4 | 4 |
| 8 | | 2 | 2 | 3 | 3 | 3 | 3 | 4 | 4 | 4 | 5 | 5 | 5 |
| 9 | | 2 | 2 | 3 | 3 | 3 | 4 | 4 | 5 | 5 | 5 | 5 | 6 |
| 10 | | 2 | 3 | 3 | 3 | 4 | 4 | 5 | 5 | 5 | 5 | 6 | 6 |
| 11 | | 2 | 3 | 3 | 4 | 4 | 5 | 5 | 5 | 6 | 6 | 6 | 7 |
| 12 | 2 | 2 | 3 | 3 | 4 | 4 | 5 | 5 | 6 | 6 | 6 | 7 | 7 |
| 13 | 2 | 2 | 3 | 3 | 4 | 5 | 5 | 5 | 6 | 6 | 7 | 7 | 7 |
| 14 | 2 | 2 | 3 | 4 | 4 | 5 | 5 | 6 | 6 | 7 | 7 | 7 | 8 |
| 15 | 2 | 3 | 3 | 4 | 4 | 5 | 6 | 6 | 7 | 7 | 7 | 8 | 8 |

Values of $u_{.005}$

| $n_1$ \ $n_2$ | 5 | 6 | 7 | 8 | 9 | 10 | 11 | 12 | 13 | 14 | 15 |
|---|---|---|---|---|---|---|---|---|---|---|---|
| 5 | | 11 | | | | | | | | | |
| 6 | 11 | 12 | 13 | 13 | | | | | | | |
| 7 | | 13 | 13 | 14 | 15 | 15 | 15 | | | | |
| 8 | | 13 | 14 | 15 | 15 | 16 | 16 | 17 | 17 | 17 | |
| 9 | | | 15 | 15 | 16 | 17 | 17 | 18 | 18 | 18 | 19 |
| 10 | | | 15 | 16 | 17 | 17 | 18 | 19 | 19 | 19 | 20 |
| 11 | | | 15 | 16 | 17 | 18 | 19 | 19 | 20 | 20 | 21 |
| 12 | | | | 17 | 18 | 19 | 19 | 20 | 21 | 21 | 22 |
| 13 | | | | 17 | 18 | 19 | 20 | 21 | 21 | 22 | 22 |
| 14 | | | | 17 | 18 | 19 | 20 | 21 | 22 | 23 | 23 |
| 15 | | | | | 19 | 20 | 21 | 22 | 22 | 23 | 24 |

# Answers to Odd-Numbered Exercises

## CHAPTER 1

**1.1** (a) $\displaystyle\sum_{i=1}^{n_1} n_{2i}$.

**1.5** (b) 6, 20, and 70.

**1.9** $\dbinom{r + n - 1}{r}$ and 21.

**1.11** (b) The seventh row is 1, 6, 15, 20, 15, 6, and 1, and the eighth row is 1, 7, 21, 35, 35, 21, 7, and 1. Also, $(x + y)^6 = x^6 + 6x^5y + 15x^4y^2 + 20x^3y^3 + 15x^2y^4 + 6xy^5 + y^5$ and $(x + y)^7 = x^7 + 7x^6y + 21x^5y^2 + 35x^4y^3 + 35x^3y^4 + 21x^2y^5 + 7xy^6 + y^7$.

**1.19** (a) $-\frac{15}{384}$ and $-10$; (b) 2.230.

**1.21** 560.

**1.25** (a) 5; (b) 4.

**1.29** (a) 30; (b) 36.

**1.31** (a) 40; (b) 90.

**1.33** (a) 5,040; (b) 210.

**1.35** 90.

**1.37** 120 and 72.

**1.39** (a) 120; (b) 60.

**1.41** 1,260.

**1.43** (a) 77,520; (b) 184,756; (c) 1,351.

**1.45** 70.

**1.47**  8,211,173,256.
**1.49**  59,049.
**1.51**  6,188.
**1.53**  120.

# CHAPTER 2

**2.5**   (a) {6, 8, 9}; (b) {8}; (c) {1, 2, 3, 4, 5, 8}; (d) {1, 5}; (e) {2, 4, 8}; (f) $\varnothing$.
**2.7**   (a) {Car 5, Car 6, Car 7, Car 8}; (b) {Car 2, Car 4, Car 5, Car 7}; (c) {Car 1, Car 8}; (d) {Car 3, Car 4, Car 7, Car 8}.
**2.9**   (a) The house has fewer than three baths. (b) The house does not have a fireplace. (c) The house does not cost more than $100,000. (d) The house is not new. (e) The house has three or more baths and a fireplace. (f) The house has three or more baths and costs more than $100,000. (g) The house costs more than $100,000 but has no fireplace. (h) The house is new or costs more than $100,000. (i) The house is new or costs at most $100,000. (j) The house has three or more baths and/or a fireplace. (k) The house has three or more baths and/or costs more than $100,000. (l) The house is new and costs more than $100,000.
**2.11**  (a) (H, 1), (H, 2), (H, 3), (H, 4), (H, 5), (H, 6) (T, H, H), (T, H, T), (T, T, H), and (T, T, T); (b) (H, 1), (H, 2), (H, 3), (H, 4), (H, 5), (H, 6), (T, H, T), and (T, T, H); (c) (H, 5), (H, 6), (T, H, T), (T, T, H), and (T, T, T).
**2.13**  (a) $5^{k-1}$; (b) $\dfrac{5^k - 1}{4}$.
**2.15**  $\{(x, y)|(x - 2)^2 + (y + 3)^2 \leqslant 9\}$.
**2.17**  (a) The event that a driver has liability insurance. (b) The event that a driver does not have collision insurance. (c) The event that a driver has liability insurance or collision insurance, but not both. (d) The event that a driver does not have both kinds of insurance.
**2.19**  (a) Region 5; (b) regions 1 and 2 together; (c) regions 3, 5, and 6 together; (d) regions 1, 3, 4, and 6 together.
**2.21**  The figures are inconsistent and the results of the study should be questioned.
**2.35**  (a) Permissible; (b) not permissible because the sum of the probabilities exceeds 1; (c) permissible; (d) not permissible because $P(E)$ is negative; (e) not permissible because the sum of the probabilities is less than 1.
**2.37**  (a) The probability that she will not pass cannot be negative. (b) $0.77 + 0.08 = 0.85 \neq 0.95$; (c) $0.12 + 0.25 + 0.36 + 0.14 + 0.09 + 0.07 = 1.03 > 1$; (d) $0.08 + 0.21 + 0.29 + 0.40 = 0.98 < 1$.
**2.39**  (a) 0.29; (b) 0.80; (c) 0.63; (d) 0.71.
**2.41**  (a) $\frac{3}{8}$; (b) $\frac{1}{4}$; (c) $\frac{1}{10}$; (d) $\frac{1}{10}$; (e) $\frac{11}{40}$.
**2.43**  $\frac{20}{221}$.
**2.45**  (a) $\frac{25}{108}$; (b) $\frac{25}{162}$; (c) $\frac{25}{648}$; (d) $\frac{25}{1296}$.
**2.47**  (a) $P(A \cup B)$ is less than $P(A)$. (b) $P(A \cap B)$ exceeds $P(A)$. (c) $P(A \cup B)$ exceeds 1.
**2.49**  0.34.
**2.51**  $\frac{13}{36}$.

**2.53**   0.94.
**2.55**   (a) 3 to 2;   (b) 11 to 5;   (c) 7 to 2 against it.
**2.57**   5 to 3.
**2.59**   (a) For instance, let $P(A \cap B) = P(A \cap B') = P(A' \cap B) = P(A' \cap B') = \frac{1}{4}$.   (b) For
        instance, let $P(A \cap B) = P(A \cap B') = P(A' \cap B) = \frac{1}{5}$ and $P(A' \cap B') = \frac{2}{5}$.
**2.75**   $\frac{15}{28}$.
**2.77**   $\frac{1}{3}$.
**2.79**   0.44.
**2.81**   (a) 0.096;   (b) 0.048;   (c) 0.0512;   (d) 0.76.
**2.83**   (a) The events are pairwise independent.   (b) the events are not independent.
**2.85**   (a) 0.1406;   (b) 0.1198.
**2.87**   $\frac{1}{12}$.
**2.89**   0.729.
**2.91**   0.735.
**2.93**   0.475.
**2.95**   (a) 0.0616;   (b) 0.5325.
**2.97**   0.3818.
**2.99**   (a) Most likely cause is sabotage;   (b) least likely cause is static electricity.
**2.101**  (a) $P(M) = 5P(Y) + 2$;   (b) 0.12.

# CHAPTER 3

**3.1**   (a) No, because $f(4)$ is negative;   (b) yes;   (c) no, because the sum of the probabilities is less than 1.
**3.5**   $0 < k < 1$.
**3.9**   (a) No, because $F(4)$ exceeds 1;   (b) no, because $F(2)$ is less than $F(1)$;   (c) yes.

**3.11**
$$F(x) = \begin{cases} 0 & \text{for } x < 1 \\ \frac{1}{15} & \text{for } 1 \leq x < 2 \\ \frac{3}{15} & \text{for } 2 \leq x < 3 \\ \frac{6}{15} & \text{for } 3 \leq x < 4 \\ \frac{10}{15} & \text{for } 4 \leq x < 5 \\ 1 & \text{for } x \geq 5 \end{cases}$$

**3.13**   (a) $\frac{3}{4}$;   (b) $\frac{1}{4}$;   (c) $\frac{1}{2}$;   (d) $\frac{3}{4}$;   (e) $\frac{1}{2}$;   (f) $\frac{1}{4}$.
**3.17**   (a) $f(3) = \frac{1}{6}$, $f(4) = \frac{1}{6}$, $f(5) = \frac{1}{3}$, $f(6) = \frac{1}{6}$, and $f(7) = \frac{1}{6}$;

$$\text{(b) } F(z) = \begin{cases} 0 & \text{for } z < 3 \\ \frac{1}{6} & \text{for } 3 \leq z < 4 \\ \frac{1}{3} & \text{for } 4 \leq z < 5 \\ \frac{2}{3} & \text{for } 5 \leq z < 6 \\ \frac{5}{6} & \text{for } 6 \leq z < 7 \\ 1 & \text{for } z \geq 7 \end{cases}$$

**3.19**  $F(x) = \begin{cases} 0 & \text{for } x < 0 \\ \frac{1}{27} & \text{for } 0 \leqslant x < 1 \\ \frac{7}{27} & \text{for } 1 \leqslant x < 2 \\ \frac{19}{27} & \text{for } 2 \leqslant x < 3 \\ 1 & \text{for } x \geqslant 3 \end{cases}$

(a) $\frac{20}{27}$;  (b) $\frac{8}{27}$.

**3.21**  (a) 0.30;  (b) 0.30.

**3.23**  $F(x) = \begin{cases} 0 & \text{for } x \leqslant 2 \\ \frac{1}{5}(x - 2) & \text{for } 2 < x < 7; \quad \frac{2}{5}. \\ 1 & \text{for } x \geqslant 7 \end{cases}$

**3.25**  $F(y) = \begin{cases} 0 & \text{for } y \leqslant 2 \\ \frac{1}{16}(y^2 + 2y - 8) & \text{for } 2 < y < 4 \\ 1 & \text{for } y \geqslant 4 \end{cases}$

(a) 0.54;  (b) 0.1519.

**3.27**  $F(x) = \begin{cases} 0 & \text{for } x \leqslant 0 \\ \frac{1}{2}\sqrt{x} & \text{for } 0 < x < 4; \quad \text{(b) } \frac{1}{4} \text{ and } \frac{1}{2}. \\ 1 & \text{for } x \geqslant 4 \end{cases}$

**3.29**  $F(z) = \begin{cases} 0 & \text{for } z \leqslant 0 \\ 1 - e^{-z^2} & \text{for } z > 0 \end{cases}$

**3.31**  $G(x) = \begin{cases} 0 & \text{for } x \leqslant 0 \\ 3x^2 - 2x^3 & \text{for } 0 < x < 1 \\ 1 & \text{for } x \geqslant 1 \end{cases}$

The probabilities are $\frac{5}{32}$ and $\frac{1}{2}$.

**3.33**  $F(x) = \begin{cases} 0 & \text{for } x \leqslant 0 \\ \dfrac{x^2}{2} & \text{for } 0 < x < 1 \\ 2x - \dfrac{x^2}{2} - 1 & \text{for } 1 \leqslant x < 2 \\ 1 & \text{for } x \geqslant 2 \end{cases}$

**3.35**  $F(x) = \begin{cases} 0 & \text{for } x \leqslant 0 \\ \dfrac{x^2}{4} & \text{for } 0 < x \leqslant 1 \\ \frac{1}{4}(2x - 1) & \text{for } 1 < x \leqslant 2 \\ \frac{1}{4}(6x - x^2 - 5) & \text{for } 2 < x < 3 \\ 1 & \text{for } x \geqslant 3 \end{cases}$

**3.37**   $f(x) = \frac{1}{2}$ for $-1 < x < 1$ and $f(x) = 0$ elsewhere.

**3.39**   $f(y) = 18/y^3$ for $y > 0$ and $f(y) = 0$ elsewhere; the two probabilities are $\frac{16}{25}$ and $\frac{9}{64}$.

**3.41**   The three probabilities are $1 - 3e^{-2}$, $2e^{-1} - 4e^{-3}$, and $5e^{-5}$.

**3.43**   (a) $F(x) = 0$;   (b) $F(x) = \frac{1}{2}x$;   (c) $F(x) = \frac{1}{2}(x + 1)$;   (d) $F(x) = 0$.

**3.45**   The probabilities are $\frac{1}{4}$, $\frac{1}{4}$, $\frac{3}{8}$, and $\frac{1}{2}$.

**3.47**   (a) $\frac{7}{27}$;   (b) $\frac{325}{864}$;   (c) $\frac{95}{432}$;   (d) 0.

**3.49**   (a) 0.4512;   (b) 0.1054;   (c) 0.2019.

**3.51**   (a) $\frac{1}{4}$;   (b) $\frac{39}{64}$;   (c) $\frac{1}{16}$.

**3.53**   (a) $\frac{1}{4}$;   (b) 0;   (c) $\frac{7}{24}$;   (d) $\frac{119}{120}$.

**3.55**   (a) $\frac{29}{89}$;   (b) $\frac{5}{89}$;   (c) $\frac{55}{89}$.

**3.57**

|   |   | $x$ | | | |
|---|---|---|---|---|---|
|   |   | 0 | 1 | 2 | 3 |
|   | 0 | 0 | $\frac{1}{30}$ | $\frac{1}{10}$ | $\frac{1}{5}$ |
| $y$ | 1 | $\frac{1}{30}$ | $\frac{2}{15}$ | $\frac{3}{10}$ | $\frac{8}{15}$ |
|   | 2 | $\frac{1}{10}$ | $\frac{3}{10}$ | $\frac{3}{5}$ | 1 |

**3.59**   $k = 2$.

**3.61**   (a) $\frac{1}{2}$;   (b) $\frac{5}{9}$;   (c) $\frac{1}{3}$.

**3.63**   $1 - \frac{1}{2} \cdot \ln 2 = 0.6534$.

**3.65**   $(e^{-1} - e^{-4})^2$.

**3.67**   $(e^{-2} - e^{-3})^2$.

**3.73**   (a) $\frac{1}{18}$;   (b) $\frac{7}{27}$.

**3.75**   $k = 144$.

**3.79**   (b)

|   |   | $x$ | | |
|---|---|---|---|---|
|   |   | 0 | 1 | 2 |
|   | 0 | $\frac{3}{28}$ | $\frac{9}{28}$ | $\frac{3}{28}$ |
| $y$ | 1 | $\frac{6}{28}$ | $\frac{6}{28}$ | |
|   | 2 | $\frac{1}{28}$ | | |

**3.81**

|   |   | $x$ | | | |
|---|---|---|---|---|---|
|   |   | 0 | 1 | 2 | 3 |
|   | $-3$ | $\frac{1}{8}$ | | | |
|   | $-1$ | | $\frac{3}{8}$ | | |
| $y$ | 1 | | | $\frac{3}{8}$ | |
|   | 3 | | | | $\frac{1}{8}$ |

**3.83**   (a) 0.064;   (b) 0.102.

**3.85**   (a) $g(-1) = \frac{1}{4}$ and $g(1) = \frac{3}{4}$;   (b) $h(-1) = \frac{5}{8}$, $h(0) = \frac{1}{4}$, and $h(1) = \frac{1}{8}$;   (c) $f(-1|-1) = \frac{1}{5}$ and $f(1|-1) = \frac{4}{5}$.

**3.87**   (a) $m(x, y) = \dfrac{xy}{36}$ for $x = 1, 2, 3$, and $y = 1, 2, 3$;   (b) $n(x, z) = \dfrac{xz}{18}$ for $x = 1$, 2, 3, and $z = 1, 2$;   (c) $g(x) = \dfrac{x}{6}$ for $x = 1, 2, 3$;   (d) $\phi(z|1, 2) = \dfrac{z}{3}$ for $z = 1, 2$;   (e) $\psi(y, z|3) = \dfrac{yz}{18}$ for $y = 1, 2, 3$, and $z = 1, 2$.

**3.89**   (a) Independent;   (b) not independent.

**3.91**   (a) $h(y) = \frac{1}{4}(1 + y)$ for $0 < y < 2$ and $h(y) = 0$ elsewhere;   (b) $f(x|1) = \frac{1}{2}(2x + 1)$ for $0 < x < 1$ and $f(x|1) = 0$ elsewhere.

**3.93**   (a) $g(x) = -\ln x$ for $0 < x < 1$ and $g(x) = 0$ elsewhere;   (b) $h(y) = 1$ for $0 < y < 1$ and $h(y) = 0$ elsewhere. The two random variables are not independent.

**3.95**   $G(x) = 1 - e^{-x^2}$ for $x > 0$ and $G(x) = 0$ elsewhere.

**3.99**   (a) $g(0) = \frac{5}{14}$, $g(1) = \frac{15}{28}$, and $g(2) = \frac{3}{28}$;   (b) $\phi(0|0) = \frac{3}{10}$, $\phi(1|0) = \frac{6}{10}$, and $\phi(2|0) = \frac{1}{10}$.

**3.101**   (a) 0.742;   (b) 0.273.

**3.103**   (a) $g(x) = \dfrac{20 - x}{50}$ for $10 < x < 20$ and $g(x) = 0$ elsewhere;   (b) $\phi(y|12) = \frac{1}{6}$ for $6 < x < 12$ and $\phi(y|12) = 0$ elsewhere;   (c) $\frac{1}{3}$.

**3.105**   (a) $f(x_1, x_2, x_3) = \dfrac{(20,000)^3}{(x_1 + 100)^3(x_2 + 100)^3(x_3 + 100)^3}$ for $x_1 > 0$, $x_2 > 0$, $x_3 > 0$, and $f(x_1, x_2, x_3) = 0$ elsewhere;   (b) $\frac{1}{16}$.

## CHAPTER 4

**4.1**   (a) $g_1 = 0$, $g_2 = 1$, $g_3 = 4$, and $g_4 = 9$;   (b) $f(0)$, $f(-1) + f(1)$, $f(-2) + f(2)$, and $f(3)$;   (c) $0 \cdot f(0) + 1 \cdot \{f(-1) + f(1)\} + 4 \cdot \{f(-2) + f(2)\} + 9 \cdot f(3) = (-2)^2 \cdot f(-2) + (-1)^2 \cdot f(-1) + 0^2 \cdot f(0) + 1^2 \cdot f(1) + 2^2 \cdot f(2) + 3^2 \cdot f(3) = \sum_x g(x) \cdot f(x)$.

**4.3**   Replace $\int$ by $\sum$ in the proof of Theorem 4.3.

**4.5**   (a) $E(X) = \displaystyle\int_{-\infty}^{\infty} \int_{-\infty}^{\infty} x f(x, y)\, dy\, dx$;   (b) $E(x) = \displaystyle\int_{-\infty}^{\infty} x g(x)\, dx$.

**4.7**   $E(Y) = \frac{37}{12}$.

**4.9**   (a) 2.4 and 6.24;   (b) 88.96.

**4.11**   $-\frac{11}{6}$.

**4.13**   $\frac{1}{2}$.

**4.15**   $\frac{1}{12}$.

**4.17**   $750.

**4.19**   (a) $0.77;   (b) $1.30;   (c) $1.60;   (d) $150;   (e) $150. The expected profit is greatest if he bakes four cakes.

**4.21**   30,000 kilometers.

**4.23**  $10,000.

**4.27**  $\mu = \frac{4}{3}$, $\mu_2' = 2$, and $\sigma^2 = \frac{2}{9}$.

**4.33**  $\mu_3 = \mu_3' - 3\mu\mu_2' + 2\mu^3$ and $\mu_4 = \mu_4' - 4\mu\mu_3' + 6\mu^2\mu_2' - 3\mu^4$.

**4.35**  (a) 3.2;   (b) 2.6.

**4.39**  (a) $k = \sqrt{20}$;   (b) $k = 10$.

**4.41**  $M_X(t) = \dfrac{2e^t}{3 - e^t}$, $\mu_1' = \frac{3}{2}$, and $\mu_2' = 3$.

**4.43**  $\mu = 4$ and $\sigma^2 = 4$.

**4.49**  $\frac{15}{44}$.

**4.51**  $\mu = 0$ and $\sigma^2 = 7.2$.

**4.53**  (a) 0.631;   (b) 0.553.

**4.55**  (a) Between 0.230 and 0.290;   (b) between 0.200 and 0.320.

**4.57**  8.

**4.59**  0.

**4.61**  For instance, $f(0, 0) = \frac{1}{6}$ and $g(0)h(0) = \frac{1}{24}$, so that $f(0, 0) \neq g(0)h(0)$.

**4.63**  (c) $E(e^{t_1 X + t_2 Y}) = \dfrac{1}{(1 - t_1)(1 - t_2)}$, $E(XY) = 1$, $E(X) = 1$, $E(Y) = 1$, and $\text{cov}(X, Y)$
= 0.

**4.65**  (a) 143;   (b) 54.

**4.69**  75.

**4.71**  $\mu_{X|-1} = \frac{3}{5}$ and $\sigma^2_{X|-1} = \frac{16}{25}$.

**4.73**  $\mu_{Y|1/4} = \frac{11}{9}$ and $\sigma^2_{Y|1/4} = \frac{23}{81}$.

**4.77**  0.0224.

**4.79**  (a) $\mu = 0.74$ and $\sigma = 0.68$;   (b) $\mu = 1.91$ and $\sigma = 1.05$.

**4.81**  0.8.

**4.83**  2.95 minutes.

# CHAPTER 5

**5.11**  $\mu_2' = \mu_{(2)}' + \mu_{(1)}'$, $\mu_3' = \mu_{(3)}' + 3\mu_{(2)}' + \mu_{(1)}'$, and $\mu_4' = \mu_{(4)}' + 6\mu_{(3)}' + 7\mu_{(2)}' + \mu_{(1)}'$.

**5.13**  (a) $F_X(t) = 1 - \theta + \theta t$;   (b) $F_X(t) = [1 + \theta(t - 1)]^n$.

**5.15**  (a) $\alpha_3 = 0$ when $\theta = \frac{1}{2}$;   (b) $\alpha_3 \to 0$ when $n \to \infty$.

**5.17**  0.0086.

**5.19**  (a) 0.3025;   (b) 0.3025.

**5.21**  (a) 0.2205;   (b) 0.2206.

**5.23**  0.2041.

**5.25**  0.9222.

**5.27**  $f(y) = \dbinom{y + k - 1}{k - 1}\theta^k(1 - \theta)^y$ for $y = 0, 1, 2, \ldots$.

**5.37**  $h(0; 4, 9, 5) = \frac{1}{126}$, $h(1; 4, 9, 5) = \frac{20}{126}$, $h(2; 4, 9, 5) = \frac{60}{126}$, $h(3; 4, 9, 5) = \frac{40}{126}$, and
$h(4; 4, 9, 5) = \frac{5}{126}$.

**5.41**  (a) 0.0060;   (b) 0.0076.

**5.47**  $\mu_2 = \lambda$, $\mu_3 = \lambda$, and $\mu_4 = \lambda + 3\lambda^2$.

**5.49**  (a) 0.1298;   (b) 0.1101.

**5.51**  (a) 0.0180;   (b) 0.0180.

**5.53** 0.4529.
**5.55** (a) 0.5948;   (b) 0.2941;   (c) 0.0504.
**5.57** (a) $\mu = \frac{15}{8}$ and $\sigma^2 = \frac{39}{64}$;   (b) $\mu = \frac{15}{8}$ and $\sigma^2 = \frac{39}{64}$.
**5.59** (a) The condition is not satisfied.   (b) The condition is satisfied.   (c) The condition is satisfied.
**5.61** (a) 0.2478;   (b) 0.2458.
**5.63** (a) Neither rule of thumb is satisfied.   (b) The rule of thumb for good approximation is satisfied.   (c) The rule of thumb for excellent approximation is satisfied.   (d) Neither rule of thumb is satisfied.
**5.65** 0.2700.
**5.67** (a) 0.1606;   (b) 0.1512.
**5.69** 0.2015.
**5.71** (a) 0.1653;   (b) 0.2975.
**5.73** (a) 0.9098;   (b) 0.9105.
**5.77** 0.0103.
**5.79** 0.0970.

# CHAPTER 6

**6.3**  $F(x) = \begin{cases} 0 & \text{for } x \leqslant \alpha \\ \dfrac{x - \alpha}{\beta - \alpha} & \text{for } \alpha < x < \beta \\ 1 & \text{for } x \geqslant \beta \end{cases}$

**6.5** $\alpha_3 = 0$ and $\alpha_4 = \frac{9}{5}$.
**6.11** For $0 < \alpha < 1$ the function $\to \infty$ when $x \to 0$; for $\alpha = 1$ the function has an absolute maximum at $x = 0$.
**6.13** $\mu_1' = \alpha\beta$, $\mu_2' = \alpha(\alpha + 1)\beta^2$, $\mu_3' = \alpha(\alpha + 1)(\alpha + 2)\beta^3$, and $\mu_4' = \alpha(\alpha + 1)(\alpha + 2)(\alpha + 3)\beta^4$.
**6.17** $M_Y(t) = \dfrac{e^{-\theta t}}{1 - \theta t}$.
**6.19** For $0 < \nu < 2$ the function $\to \infty$ when $x \to 0$; for $\nu = 2$ the function has an absolute maximum at $x = 0$.
**6.23** (a) $k = \alpha\beta$.
**6.31** $\frac{1}{2}$.
**6.33** $n = 100$.
**6.35** (a) 0.6065;   (b) 0.0620.
**6.37** 0.1827.
**6.39** 0.2231.
**6.41** (a) $\mu = 0.2$;   (b) 0.3164.
**6.45** $\mu_3 = 0$ and $\mu_4 = 3\sigma^4$.
**6.55** (a) 0.1271;   (b) 0.6406;   (c) 0.1413;   (d) 0.5876.
**6.57** (a) $z = 1.92$;   (b) $z = 2.22$;   (c) $z = 1.12$;   (d) $z = \pm 1.44$.
**6.59** (a) 0.6826;   (b) 0.9544;   (c) 0.9984;   (d) 0.99994.
**6.61** (a) 0.0668;   (b) 0.2860;   (c) 0.6490.

**6.63**  6.094 ounces.
**6.65**  (a) No;   (b) yes;   (c) no.
**6.69**  0.1446.
**6.71**  (a) 0.24;   (b) 0.49;   (c) 0.96.
**6.75**  $\sigma_1 = 6$, $\sigma_2 = 3$, and $\rho = -\frac{1}{2}$.
**6.77**  $\rho_{UV} = \dfrac{\sigma_1^2 - \sigma_2^2}{\sqrt{(\sigma_1^2 + \sigma_2^2)^2 - 4\rho^2\sigma_1^2\sigma_2^2}}$.
**6.79**  (a) 0.2990;   (b) 0.18.
**6.81**  (a) 14.5 pounds;   (b) 23.625 inches.

# CHAPTER 7

**7.1**  (a) $G(y) = 1 - e^{-y}$ for $y > 0$ and $G(y) = 0$ elsewhere;   (b) $g(y) = e^{-y}$ for $y > 0$ and $g(y) = 0$ elsewhere.

**7.3**  $g(y) = 2y$ for $0 < y < 1$ and $g(y) = 0$ elsewhere.

**7.5**  (a) $f(y) = \dfrac{1}{\theta_1 - \theta_2} \cdot (e^{-y/\theta_1} - e^{-y/\theta_2})$ for $y > 0$ and $f(y) = 0$ elsewhere;   (b) $f(y) = \dfrac{1}{\theta^2} \cdot ye^{-y/\theta}$ for $y > 0$ and $f(y) = 0$ elsewhere.

**7.7**  (a) $F(y) = 0$;   (b) $F(y) = \frac{1}{2}y^2$;   (c) $F(y) = 1 - \frac{1}{2}(2 - y)^2$;   (d) $F(y) = 1$. Also, $f(y) = 0$ for $y \leq 0$, $f(y) = y$ for $0 < y \leq 1$, $f(y) = 2 - y$ for $1 < y < 2$, and $f(y) = 0$ for $y \geq 2$.

**7.9**  $g(v) = e^{-v}$ for $v > 0$ and $g(v) = 0$ elsewhere.

**7.11**  $g(z) = \frac{1}{25} \cdot (20 \cdot \ln 2 - 10)$ for $0 < z \leq 5$,
$g(z) = \frac{1}{25} \cdot \left(2z - 20 - 20 \cdot \ln \dfrac{z}{10}\right)$ for $5 < z < 10$, and $g(z) = 0$ elsewhere.

**7.13**  $g(y) = \frac{9}{11} \cdot y^2$ for $0 < y \leq 1$, $g(y) = \dfrac{3(2 - y)(7y - 4)}{22}$ for $1 < y < 2$, and $g(y) = 0$ elsewhere.

**7.15**  $h(0) = \frac{3}{5}$ and $h(1) = \frac{2}{5}$.

**7.17**  $g(y) = \theta(1 - \theta)^{-(1+y)/5}$ for $y = -1, -6, -11, \ldots$.

**7.21**  $g(y) = \dfrac{1}{\sigma\sqrt{2\pi}} \cdot \dfrac{1}{y} \cdot e^{-\frac{1}{2}\left(\frac{\ln y - \mu}{\sigma}\right)^2}$ for $y > 0$ and $g(y) = 0$ elsewhere.

**7.25**  $g(y) = \dfrac{k}{16} y^3(1 - y)$ for $0 < y < 1$ and $g(y) = 0$ elsewhere;   this is a beta distribution with $\alpha = 4$ and $\beta = 2$;   $k = 320$.

**7.29**  (a) $g(y) = 1/2$ for $0 < y < 1$ and $g(y) = 1/4$ for $1 < y < 3$;   (b) $h(z) = \frac{1}{16} \cdot z^{-3/4}$ for $1 < z < 81$ and $h(z) = 0$ elsewhere.

**7.31**  (a) $f(2, 0) = \frac{1}{36}$, $f(3, -1) = \frac{2}{36}$, $f(3, 1) = \frac{2}{36}$, $f(4, -2) = \frac{3}{36}$, $f(4, 0) = \frac{4}{36}$, $f(4, 2) = \frac{3}{36}$, $f(5, -1) = \frac{6}{36}$, $f(5, 1) = \frac{6}{36}$, and $f(6, 0) = \frac{9}{36}$;
(b) $g(2) = \frac{1}{36}$, $g(3) = \frac{4}{36}$, $g(4) = \frac{10}{36}$, $g(5) = \frac{12}{36}$, and $g(6) = \frac{9}{36}$.

**7.33**  $g(0, 0, 2) = \frac{25}{144}$, $g(1, -1, 1) = \frac{5}{18}$, $g(1, 1, 1) = \frac{5}{24}$, $g(2, -2, 0) = \frac{1}{9}$, $g(2, 0, 0) = \frac{1}{6}$, and $g(2, 2, 0) = \frac{1}{16}$.

**7.37**  $\mu = 0$ and $\sigma^2 = 2$.

**7.39** $g(z, u) = 12z(u^{-3} - u^{-2})$ over the region bounded by $z = 0$, $u = 1$, and $z = u^2$, and $g(z, u) = 0$ elsewhere; $h(z) = 6z + 6 - 12\sqrt{z}$ for $0 < z < 1$ and $h(z) = 0$ elsewhere.

**7.41** The marginal distribution is the Cauchy distribution $g(y) = \dfrac{1}{\pi} \cdot \dfrac{2}{4 + y^2}$ for $-\infty < y < \infty$.

**7.43** $f(u, v) = \frac{1}{2}$ over the region bounded by $v = 0$, $u = -v$, and $2v + u = 2$, and $f(u, v) = 0$ elsewhere; $g(u) = \frac{1}{4}(2 + u)$ for $-2 < u \le 0$, $g(u) = \frac{1}{4}(2 - u)$ for $0 < y < 2$, and $g(u) = 0$ elsewhere.

**7.45** $g(w, z) = 24w(z - w)$ over the region bounded by $w = 0$, $z = 1$, and $z = w$; $g(w, z) = 0$ elsewhere.

**7.51** $h(z) = 8z^2$ for $0 < z \le \frac{1}{2}$, $h(z) = 8z(1 - z)$ for $\frac{1}{2} < z < 1$, and $h(z) = 0$ elsewhere; $h(\frac{1}{2}) = 2$ to make function continuous.

**7.55** It is a gamma distribution with the parameters $\alpha n$ and $\beta$.

**7.59** (a) 0.1021; (b) 0.0259.

**7.61** 0.0222.

**7.63** (a) 0.1781; (b) 0.9523; (c) 0.1189.

**7.65** (a) 0.1086; (b) 0.3492; (c) 0.6168.

## CHAPTER 8

**8.11** When we sample with replacement from a finite population, we satisfy the conditions for random sampling from an infinite population; that is, the random variables are independent and identically distributed.

**8.17** $s = \dfrac{\sum\limits_{i=1}^{n} x_i^2}{n - 1} - \dfrac{n}{n - 1} \cdot \bar{x}^2$.

**8.19** (a) $\frac{1}{495}$; (b) $\frac{1}{77}$.

**8.21** (a) It is divided by 2. (b) It is divided by 1.5. (c) It is multiplied by 3. (d) It is multiplied by 2.5.

**8.23** The probability is at least 0.96.

**8.25** (a) The probability is at most 0.25; (b) 0.0456.

**8.27** (a) 0.0388; (b) 0.7108; (c) 0.1736.

**8.29** The value will fall between $-18.03$ and 18.03.

**8.31** 0.2743.

**8.33** The value will fall between 0.0268 and 0.2732.

**8.43** 21.9% and 5.53%.

**8.59** 0.216.

**8.61** 0.05.

**8.63** $t = -1.347$; since $t = -1.347$ is fairly small (close to $-t_{.10,11} = -1.363$), the data tend to support the claim.

**8.65** 0.99.

**8.67** $g_1(y_1) = n(1 - y_1)^{n-1}$ for $0 < y_1 < 1$ and $g_1(y_1) = 0$ elsewhere; $g_n(y_n) = ny_n^{n-1}$ for $0 < y_n < 1$ and $g_n(y_n) = 0$ elsewhere.

**8.69** $E(Y_1) = \dfrac{1}{n + 1}$ and $\text{var}(Y_1) = \dfrac{n}{(n + 1)^2(n + 2)}$.

**8.71**  $h(\tilde{x}) = \dfrac{12(2m+1)!}{m!m!} \cdot \tilde{x}^{3m+2}(1-\tilde{x})(4-3\tilde{x})^m(1-4\tilde{x}^3+3\tilde{x}^4)^m$ for $0 < \tilde{x} < 1$ and

$h(\tilde{x}) = 0$ elsewhere.

**8.73**  (a) $g(y_1, y_n) = \dfrac{n(n-1)}{\theta^2} \cdot e^{-\frac{1}{\theta}(y_1+y_n)} \left[ e^{-\frac{y_1}{\theta}} - e^{-\frac{y_n}{\theta}} \right]^{n-2}$ for $0 < y_1 < y_n < \infty$ and

$g(y_1, y_n) = 0$ elsewhere.

(b) $g(y_1, y_n) = n(n-1)(y_n - y_1)^{n-2}$ for $0 < y_1 < y_n < 1$ and $g(y_1, y_n) = 0$

elsewhere.

**8.75**  $h(y_1, R) = n(n-1)f(y_1)f(y_1 + R) \left[ \int_{y_1}^{y_1+R} f(x)\,dx \right]^{n-1}$ for $0 < y_1 < 1 - R < 1$

and $h(y_1, R) = 0$ elsewhere.

**8.77**  $g(R) = n(n-1)(1-R)R^{n-2}$ for $0 < R < 1$ and $g(R) = 0$ elsewhere.

**8.81**  0.4096.

**8.83**  0.3672.

**8.85**  $n = 46$.

## CHAPTER 9

**9.3**  (a) The decision would be reversed.    (b) The decision would be the same.

**9.5**  (a) He should go to the construction site which is 33 miles from the lumber-yard.    (b) He should go to the construction site which is 27 miles from the lumberyard.    (c) It does not matter.

**9.7**  (a) He should expand his plant capacity now.    (b) She should choose Hotel $Y$.    (c) He should go to the construction site which is 27 miles from the lumberyard.

**9.9**  (a) She should choose Hotel $Y$.    (b) He should go to the construction site which is 27 miles from the lumberyard.

**9.11**  (a) The optimum strategies are I and 2 and the value of the game is 5.    (b) The optimum strategies are II and 1 and the value is 11.    (c) The optimum strategies are I and 1 and the value is −5.    (d) The optimum strategies are I and 2 and the value is 8.

**9.13**  (a) The payoffs are 0 and −6 for the first row of the table and 8 and 3 for the second row of the table.    (b) The optimum strategies are for Station $A$ to give away the glasses and for Station $B$ to give away the knives.

**9.15**  (a) $\frac{5}{11}$ and $\frac{6}{11}$;    (b) $\frac{4}{11}$ and $\frac{7}{11}$;    (c) $-\frac{9}{11}$.

**9.17**  The defending country should randomize its strategies with probabilities $\frac{1}{6}$ and $\frac{5}{6}$, and the enemy should randomize its strategies with probabilities $\frac{5}{6}$ and $\frac{1}{6}$;    the value is \$10,333,333.

**9.19**  (a) He should lower the prices.    (b) They could accomplish this by lowering their prices on alternate days.

**9.23**  $\theta = \dfrac{2ab - n}{2(b^2 - n)}$, $a = \frac{1}{2}\sqrt{n}$, and $b = \sqrt{n}$.

**9.27**  (a) The values in the first row of the table are 0, 50, and 100, those in the second row are 50, 0, and 50, and those in the third row are 100, 50, and 0.    (b) $d_1(0)$ $= 0$ and $d_1(1) = 0$, $d_2(0) = 0$ and $d_2(1) = \frac{1}{2}$, $d_3(0) = 0$ and $d_3(1) = 1$, $d_4(0) = \frac{1}{2}$ and $d_4(1) = 0$, $d_5(0) = \frac{1}{2}$ and $d_5(1) = \frac{1}{2}$, $d_6(0) = \frac{1}{2}$ and $d_6(1) = 1$, $d_7(0) = 1$ and

$d_7(1) = 0$, $d_8(0) = 1$ and $d_8(1) = \frac{1}{2}$, and $d_9(0) = 1$ and $d_9(1) = 1$.    (d) decision function $d_3$.

**9.29**  (a) The values in the first row are 0, $\beta$, and $2\beta$, those in the second row are $\alpha + 2\beta + \phi$, $\frac{1}{2}\alpha + 2\beta + \phi$, and $2\beta + \phi$, and those in the third row are $\alpha + 2\beta + 2\phi$, $2\beta + 2\phi$, and $2\beta + 2\phi$.    (b) She should inspect both.    (c) She should ship the items without inspection.

# CHAPTER 10

**10.1**  $\displaystyle\sum_{i=1}^{n} a_i = 1$.

**10.9**  $(n + 1)Y_1$.

**10.25**  $\frac{8}{9}$.

**10.29**  (a) $\frac{3}{4}$;    (b) $\frac{3}{5}$.

**10.33**  28.

**10.35**  279.

**10.47**  It is not a sufficient estimator of $\theta$.

**10.53**  $\hat{\mu} = m_1'$ and $\hat{\sigma}^2 = m_2' - (m_1')^2$.

**10.55**  $\hat{\beta} = 2m_1'$.

**10.57**  $\hat{\alpha} = \dfrac{m_1'}{1 - m_1'}$.

**10.59**  $\hat{\theta} = \sqrt{m_2' - (m_1')^2}$ and $\hat{\delta} = m_1' - \sqrt{m_2' - (m_1')^2}$.

**10.61**  $\hat{\theta} = \dfrac{n_1 + 2n_2 + 3n_3}{3N}$.

**10.63**  $\hat{\alpha} = \dfrac{-n}{\displaystyle\sum_{i=1}^{n} \ln x_i}$.

**10.65**  $\hat{\sigma} = \sqrt{\dfrac{\Sigma(x - \mu)^2}{n}}$.

**10.67**  $\hat{\alpha} = \dfrac{n}{\Sigma x^2}$.

**10.69**  $\hat{\delta} = y_1$ (smallest sample value) and $\hat{\theta} = \bar{x} - y_1$.

**10.71**  $\hat{\theta} = \dfrac{n_1 + 2n_2 + 3n_3}{3N}$.

**10.73**  $\hat{\alpha} = \dfrac{\bar{v} + \bar{w}}{2}$ and $\hat{\beta} = \dfrac{\bar{v} - \bar{w}}{2}$.

**10.77**  $\hat{\alpha} = 4.627$ and $\hat{\beta} = 1.556$.

**10.79**  $\hat{\theta} = 40{,}200$ miles.

**10.81**  $\hat{\theta} = 47.69$ and $\hat{\delta} = 412.64$.

**10.83**  $\hat{\alpha} = 3.83$ and $\hat{\beta} = 11.95$.

**10.85**  $\hat{\theta} = \frac{11}{60}$.

**10.87**  $\hat{\theta} = 0.30$.

**10.93**  $E(\Theta|38) = 0.29$.

**10.95** 0.4786.

**10.97** (a) $\hat{\mu} = 100$; (b) $\hat{\mu} = 112$; (c) $\hat{\mu} = 108$.

## CHAPTER 11

**11.1** $k = \dfrac{-1}{\ln(1 - \alpha)}$.

**11.3** $c = \dfrac{1 \pm \sqrt{1 - \alpha}}{\alpha}$.

**11.7** If $\overline{X}_1$ and $\overline{X}_2$ are the means of independent random samples of size $n_1$ and $n_2$ from normal populations with $\mu_1$, $\mu_2$, $\sigma_1$, and $\sigma_2$, and $\overline{X}_1 - \overline{X}_2$ is to be used as an estimator of $\mu_1 - \mu_2$, the probability is $1 - \alpha$ that the error will be less than

$$z_{\alpha/2} \cdot \sqrt{\frac{\sigma_1^2}{n_1} + \frac{\sigma_2^2}{n_2}}.$$

**11.11** $59.82 < \mu < 63.78$.

**11.13** $139.57 < \mu < 144.03$.

**11.15** With 95% confidence, maximum error is 0.83 minute.

**11.17** $59.99 < \mu < 63.61$.

**11.19** $n = 355$.

**11.21** $61.96 < \mu < 65.72$ gallons.

**11.23** $-7.485 < \mu_1 - \mu_2 < -2.915$.

**11.25** $-0.198 < \mu_1 - \mu_2 < 1.998$ feet.

**11.27** With 98% confidence, maximum error is 0.0023 ohm.

**11.35** With 99% confidence, maximum error is 0.069.

**11.37** With 95% confidence, maximum error is 0.053.

**11.39** With 90% confidence, maximum error is 0.075.

**11.41** $n = 2,401$.

**11.43** $n = 1,037$.

**11.45** $-0.372 < \theta_1 - \theta_2 < -0.204$.

**11.47** With 98% confidence, maximum error is 0.053.

**11.51** $0.04 < \sigma^2 < 0.28$.

**11.53** $3.67 < \sigma < 5.83$.

**11.55** $0.58 < \dfrac{\sigma_1^2}{\sigma_2^2} < 1.96$.

**11.57** $0.233 < \dfrac{\sigma_1^2}{\sigma_2^2} < 9.506$.

## CHAPTER 12

**12.1** (a) Simple; (b) composite; (c) composite; (d) composite.

**12.3** $\alpha = \frac{1}{21}$ and $\beta = \frac{5}{7}$.

**12.5** $\alpha = (1 - \theta_0)^{k-1}$ and $\beta = 1 - (1 - \theta_1)^{k-1}$.

**12.7** $\alpha = 0.08$.

**12.9** $1 - \beta = 0.114$.

**12.11**  $\sum_{i=1}^{n} x_i \geqslant K$, where $K$ can be determined by making use of the fact that $\sum_{i=1}^{n} X_i$ has the gamma distribution with $\alpha = n$ and $\beta = \theta_0$.

**12.13**  $\beta = 0.37$.

**12.15**  $\sum_{i=1}^{n} x_i^2 \geqslant K$, where $K$ can be determined by making use of the formula for the sum of $n$ terms of a geometric progression.

**12.17**  They would be committing a type I error if they erroneously reject the null hypothesis that 60 percent of their passengers object to smoking inside the plane. They would be committing a type II error if they erroneously accept the null hypothesis that 60 percent of their passengers object to smoking in the plane.

**12.19**  (a) The manufacturer should use the alternative hypothesis $\mu < 20$ and make the modification only if the null hypothesis can be rejected.
(b) The manufacturer should use the alternative hypothesis $\mu > 20$ and make the modification unless the null hypothesis can be rejected.

**12.21**  (a) He will erroneously accept the null hypothesis.
(b) He will correctly accept the null hypothesis.

**12.23**  (a) Yes;   (b) yes.

**12.25**  The corresponding values of the power function are $0.0025, 0.0433, 0.1702, 0.3704$, $0.5852, 0.7626, 0.8817, 0.9491, 0.9812$, and $0.9941$.

**12.27**  (a)  $\lambda = \dfrac{\left(\dfrac{n}{2}\right)^{n}}{x^{x}(n-x)^{n-x}}$.

**12.31**  $\lambda = \left[\dfrac{\sum_{i=1}^{n}(x_i - \bar{x})^2}{n\sigma_0^2}\right]^{\frac{n}{2}} \cdot e^{-\frac{1}{2}\left[\frac{\sum_{i=1}^{n}(x_i-\bar{x})^2}{\sigma_0^2} - n\right]}$.

**12.35**  (a) $0.852$;   (b) $0.016, 0.086, 0.129, 0.145, 0.144, 0.134$, and $0.122$.

**12.37**  (a) $0.0375, 0.0203, 0.0107, 0.0055$, and $0.0027$;
(b) $0.9329, 0.7585, 0.3840$, and $0.0420$.

**12.39**  $-2 \cdot \ln \lambda = 44.37$;   the null hypothesis must be rejected.

## CHAPTER 13

**13.1**  Use the critical region $\dfrac{n(\bar{x} - \mu_0)^2}{\sigma^2} \geqslant \chi_{\alpha,1}^2$.

**13.3**  $n = 52$.

**13.5**  $n = 151$.

**13.7**  (a) No;   (b) yes;   (c) yes.

**13.11**  $z = 2.73$;   the null hypothesis must be rejected.

**13.13**  $z = 3.02$;   the null hypothesis must be rejected.

**13.15**  $z = -2.11$;   the null hypothesis must be rejected.

**13.19**  Reject the null hypothesis if $\bar{x}_1 - \bar{x}_2 \leqslant 0.145$ or $\bar{x}_1 - \bar{x}_2 \geqslant 0.255$. (a) $\beta = 0.18$;   (b) $\beta = 0.71$;   (c) $\beta = 0.71$;   (d) $\beta = 0.18$.

**13.21**  $P$-value $= 0.0094$;   the null hypothesis must be rejected.

**13.23**  $P$-value $= 0.1112;$   the null hypothesis cannot be rejected.

**13.27**  $P$-value $= 0.6144;$   the null hypothesis cannot be rejected.

**13.29**  $t = 4.03;$   the null hypothesis must be rejected.

**13.31**  For $H_1$:   $\sigma^2 < \sigma_0^2$, the critical region is $s^2 \leq \sigma_0^2 \left[ 1 - z_\alpha \sqrt{\dfrac{2}{n-1}} \right];$ for $H_1$:

$\sigma^2 \neq \sigma_0^2$ the critical region is $s^2 \leq \sigma_0^2 \left[ 1 - z_{\alpha/2} \sqrt{\dfrac{2}{n-1}} \right]$ or

$s^2 \geq \sigma_0^2 \left[ 1 + z_{\alpha/2} \sqrt{\dfrac{2}{n-1}} \right].$

**13.33**  $\chi^2 = 5.92;$   the null hypothesis cannot be rejected.

**15.35**  $\chi^2 = 22.85;$   the null hypothesis cannot be rejected.

**13.37**  $z = 1.93;$   the null hypothesis must be rejected.

**13.39**  $\dfrac{s_1^2}{s_2^2} = 1.42;$   the null hypothesis cannot be rejected.

**13.41**  $\dfrac{s_1^2}{s_2^2} = 1.80;$   the null hypothesis cannot be rejected.

**13.43**  For $H_1$:   $\lambda > \lambda_0$, reject the null hypothesis if $\sum\limits_{i=1}^{n} x_i \geq k_\alpha$, where $k_\alpha$ is the smallest

integer for which $\sum\limits_{y=k_\alpha}^{\infty} p(y; n\lambda_0) \leq \alpha$. For $H_1$:   $\lambda < \lambda_0$, reject the null hypothesis

if $\sum\limits_{i=1}^{n} x_i \leq k_\alpha'$, where $k_\alpha'$ is the largest integer for which $\sum\limits_{y=0}^{k_\alpha'} p(y; n\lambda_0) \leq \alpha$. For

$H_1$:   $\lambda \neq \lambda_0$, reject the null hypothesis if $\sum\limits_{i=1}^{n} x_i \leq k_{\alpha/2}'$ or $\sum\limits_{i=1}^{n} x_i \geq k_{\alpha/2}$.

**13.49**  $P$-value $= 0.1348;$   the null hypothesis cannot be rejected.

**13.51**  $P$-value $= 0.0104;$   the null hypothesis must be rejected.

**13.53**  $P$-value $= 0.0012;$   the null hypothesis must be rejected.

**13.55**  $P$-value $= 0.0194;$   the null hypothesis cannot be rejected.

**13.57**  $z = -2.05;$   the null hypothesis must be rejected.

**13.59**  $\chi^2 = 2.92;$   the null hypothesis cannot be rejected.

**13.61**  $\chi^2 = 0.86;$   the null hypothesis cannot be rejected.

**13.63**  $z = 2.78;$   the null hypothesis must be rejected.

**13.65**  $\chi^2 = 0.75;$   the null hypothesis cannot be rejected.

**13.71**  $\chi^2 = 4.0;$   the null hypothesis cannot be rejected.

**13.73**  $\chi^2 = 1.3;$   the null hypothesis cannot be rejected.

**13.75**  $\chi^2 = 21.9;$   the null hypothesis must be rejected.

**13.77**  $\chi^2 = 13.6;$   the null hypothesis must be rejected.

## CHAPTER 14

**14.3**  $\mu_{Y|x} = \dfrac{1+x}{2}$ and $\mu_{X|y} = \dfrac{2y}{3}.$

**14.5**  $\mu_{X|1} = \frac{4}{7}$ and $\mu_{Y|0} = \frac{9}{8}.$

**14.13**  $\hat{\beta} = \dfrac{\displaystyle\sum_{i=1}^{n} x_i y_i}{\displaystyle\sum_{i=1}^{n} x_i^2}$.

**14.17**  (a) $\hat{y} = 31.609 + 0.5816x$;   (b) $\hat{y} = 80.46$.

**14.19**  (a) $\hat{y} = 1.8999 - 0.0857x$;   (b) $\hat{y} = 1.4714$.

**14.21**  (a) $\hat{y} = 1.3 - 0.0857u$ (coded);   (b) $\hat{y} = 1.4714$.

**14.23**  $\hat{y} = 1.371(1.383)^x$.

**14.27**  (a) $t = \dfrac{\hat{\beta} - \beta}{s_e/\sqrt{S_{xx}}}$;   (b) $\hat{\beta} - t_{\alpha/2,n-2} \cdot \dfrac{s_e}{\sqrt{S_{xx}}} < \beta < \hat{\beta} + t_{\alpha/2,n-2} \cdot \dfrac{s_e}{\sqrt{S_{xx}}}$.

**14.35**  $t = 3.72$;   the null hypothesis must be rejected.

**14.37**  (a) $\hat{y} = 12.2471 + 1.4927x$;   (b) $t = 3.41$;   the null hypothesis must be rejected.

**14.39**  $-0.1217 < \beta < -0.0497$.

**14.41**  (a) $\hat{y} = 1.2594 + 1.4826x$;   (b) $t = 3.10$;   the null hypothesis cannot be rejected.

**14.43**  $-2.2846 < \alpha < 65.5026$.

**14.45**  $6.452 < \mu_{Y|9} < 9.7634$;   (b) $3.4777$ and $12.7009$.

**14.49**  $\dfrac{1 + r - (1 - r)e^{-2z_{\alpha/2}/\sqrt{n-3}}}{1 + r + (1 - r)e^{-2z_{\alpha/2}/\sqrt{n-3}}} < \rho < \dfrac{1 + r - (1 - r)e^{2z_{\alpha/2}/\sqrt{n-3}}}{1 + r + (1 - r)e^{2z_{\alpha/2}/\sqrt{n-3}}}$.

**14.51**  $r = 0.55$;   $z = 2.565$ and the value of $r$ is significant.

**14.53**  $r = 0.727$;   $z = 5.05$ and the value of $r$ is significant.

**14.55**  $2.84 < \beta < 4.10$.

**14.57**  $r = 0.772$;   $z = 4.81$ and the value of $r$ is significant.

**14.59**  $r = 0.285$;   $z = 5.55$ and the value of $r$ is significant.

**14.67**  (b) $B'X_0 \pm t_{\alpha/2,n-k-1}\hat{\sigma} \cdot \sqrt{\dfrac{n[X_0'(X'X)^{-1}X_0]}{n - k - 1}}$.

**14.69**  (a) $\hat{\beta}_0 = 14.56$, $\hat{\beta}_1 = 30.109$, and $\hat{\beta}_2 = 12.16$;   (b) $\hat{y} = \$101.41$.

**14.71**  (a) $\hat{\beta}_0 = -124.57$, $\hat{\beta}_1 = 1.659$, and $\hat{\beta}_2 = 1.439$;   (b) $\hat{y} = 63.24$.

**14.73**  $\hat{y} = 69.73 + 2.975z_1 - 11.97z_2$ (coded);   $\hat{y} = 71.2$.

**14.75**  $\hat{y} = 10.5 - 2.0x + 0.2x^2$;   $\hat{y} = 5.95$.

**14.77**  $t = 2.94$;   the null hypothethis cannot be rejected and there is no real evidence that it is worthwhile to fit a parabola rather than a straight line.

**14.79**  $t = 0.16$;   the null hypothesis cannot be rejected.

**14.81**  $t = -4.18$;   the null hypothesis must be rejected.

**14.83**  $\$78,568 < \mu_{Y|3,2} < \$79,649$.

**14.85**  $\$74.5 < \mu_{Y|2.4,1.2} < \$128.3$.


# CHAPTER 15

**15.7**  $f = 17.0$    the differences in effectiveness are significant.

**15.9**  $f = 39.3$ and the differences in dosage do have an effect;   $\hat{\mu} = 18.5$, $\hat{\alpha}_1 = 4.0$, $\hat{\alpha}_2 = 1.5$, and $\hat{\alpha}_3 = -5.5$.

**15.11**  $f = 1.02$;   the differences among the sample means can be attributed to chance.

**15.13**  $f = 3.00$;   the null hypothesis cannot be rejected (as in Exercise 13.74).

**15.19**  $f_{\mathrm{Tr}} = 4.43$ and the null hypothesis for launchers cannot be rejected;
$f_B = 17.05$ and the null hypothesis for fuels must be rejected.

**15.21**  $f_{\mathrm{Tr}} = 7.99$ and the null hypothesis for threads must be rejected;
$f_B = 0.81$ and the null hypothesis for measuring instruments cannot be rejected.

**15.23**  The doctor who is a Westerner is a Republican.

## CHAPTER 16

**16.3**  The mean is 0 and the variance is $\dfrac{n(n + 1)(2n + 1)}{6}$.

**16.5**  (a) The $P$-value is 0.0059;   the null hypothesis must be rejected;
(b) $z = -2.68$;   the null hypothesis must be rejected.

**16.7**  The $P$-value is 0.0381;   the null hypothesis must be rejected.

**16.9**  $z = 2.16$ with continuity correction;   the null hypothesis cannot be rejected.

**16.11**  The $P$-value is 0.1937;   the null hypothesis cannot be rejected.

**16.13**  $z = -1.84$ with continuity correction;   the null hypothesis must be rejected.

**16.15**  (a) Reject the null hypothesis if $T \leq 3$;
(b) reject the null hypothesis if $T^- \leq 5$;
(c) reject the null hypothesis if $T^+ \leq 5$.

**16.17**  $T = 15$;   the null hypothesis cannot be rejected.

**16.19**  (a) $T = 98.5$;   the null hypothesis cannot be rejected;
(b) $z = 0.26$;   the null hypothesis cannot be rejected.

**16.27**  $U_1 = 5.5$;   the null hypothesis must be rejected.

**16.29**  $z = -0.10$;   the null hypothesis cannot be rejected.

**16.31**  $z = 1.62$;   the null hypothesis cannot be rejected.

**16.33**  $U = 3$.

**16.35**  $H = 0.86$;   the null hypothesis cannot be rejected.

**16.39**  $\frac{11}{42}$.

**16.41**  $u = 5$;   the null hypothesis of randomness must be rejected.

**16.43**  $u = 4$;   the null hypothesis of randomness cannot be rejected.

**16.45**  $z = 1.31$;   the null hypothesis of randomness cannot be rejected.

**16.47**  $u = 13$;   the null hypothesis of randomness cannot be rejected.

**16.49**  $u = 5$;   the null hypothesis of randomness must be rejected.

**16.51**  $z = 2.50$;   the null hypothesis must be rejected and we conclude that there is a cyclical pattern.

**16.53**  The minimum value is $W = 0$ and it reflects a complete lack of association;   the maximum value is $W = 1$ and it reflects a perfect agreement.

**16.55**  $z = 3.55$;   $r_S = 0.86$ is significant.

**16.57**  $z = 1.79$;   $r_S = 0.54$ is not significant.

## APPENDIX

**A.3**  (a) 10;   (b) 40.

**A.5**  (a) 1;   (b) 3;   (c) 33;   (d) 39;   (e) 29.

**A.7**  (a) 19;   (b) 19.

# Index

# LIST OF NUMBERED THEOREMS (*continued*)